Revision of the
COCCOID
MYXOPHYCEAE

Revision of the
COCCOID
MYXOPHYCEAE

FRANCIS DROUET
WILLIAM A. DAILY

(FACSIMILE OF THE 1956 EDITION)

HAFNER PRESS
NEW YORK
1973

Originally published in 1956
by Butler University

Reprinted by arrangement

Published by
Hafner Press
A Division of Macmillan Publishing Co., Inc.
866 Third Avenue
New York, N.Y. 10022

Library of Congress Catalog Card Number: 73-76177
ISBN: 02-844070-6

Printed in the U.S.A. by
Edwards Brothers
Ann Arbor, Michigan

1

Butler University Botanical Studies

Volume XII

June, 1956

REVISION OF THE COCCOID MYXOPHYCEAE

by

FRANCIS DROUET

and

WILLIAM A. DAILY

BUTLER UNIVERSITY BOTANICAL STUDIES

Edited by

J. E. POTZGER

VOLUME XII

1956

TO

DONALD RICHARDS

AND

FAY KENOYER DAILY

WITHOUT WHOSE ASSISTANCE AND CONSTANT ENCOURAGEMENT
THIS STUDY COULD NOT HAVE BEEN THUS FAR ADVANCED
AND THIS VOLUME COULD NOT HAVE BEEN COMPLETED

Revision of the
COCCOID
MYXOPHYCEAE

REVISION OF THE COCCOID MYXOPHYCEAE

Francis Drouet and William A. Daily

This study originated as an inquiry, some twenty years ago, into what names should be employed for species of coccoid Myxophyceae found in general collections of algae from various parts of the world. Gradually we accumulated and examined many thousands of specimens. Equally gradually it became apparent to us that only a carefully executed revision of the group, with sufficient attention paid to the morphological variation and life history of each species and with strict adherence to the stern discipline of the science of historical taxonomy, would produce a classification and a nomenclature which would satisfy our desire.

Our method consisted essentially of studying and re-studying every specimen which we could collect, borrow, purchase, or otherwise secure. Among these specimens were included the original material of many species described by various authors. During the latter years we made an ambitious attempt to find the original specimens of as many previously described species as possible. With funds provided by Mr. E. J. Richards of Chicago, a search was made in the larger European herbaria for such historical specimens. The nomenclature, synonymy, descriptions, notes, *nomina excludenda,* and photographs in the treatment of genera and species below record the results of this work.

It early became apparent that the various species are of very broad distribution in appropriate habitats over most of the earth. Our interpretations of their morphology and life histories were (and perhaps still are to a certain extent) often qualified by prejudices learned from previous classifications, and it has required considerable audacity to recognize such now readily apparent phenomena as the properties of the sheath material to be-

1

come hydrolyzed and to develop pigments which change color easily. It was a difficult admission finally to note hyelloid filaments at the bases of the cushions of Entophysalis, although it had been wholly natural to see entophysaloid cushions at the surface of plants of Hyella. We came to realize the fact that big plants grow from little plants and that (in our time) all plants come from other plants. In this group where the cells differentiate no hard parts, it became reasonable to expect that no two plants, each with a separate history of growth in separate environments, could look exactly alike; the same had to be admitted for each cell of each plant. In every microenvironment there occur at times catastrophic events (freezing, thawing, inundation, drying, heating, intense insolation, shading, parasitization, depredation by animals, changes in chemical nature of the substratum or medium, etc.) peculiar to its small area; it finally became apparent that these events often cause radical changes in the appearance and mode of growth of both plants and cells. The classification outlined here attempts to take these matters into consideration.

Over a thousand species, varieties and forms have been described within, or at one time or another transferred into, this group of plants. Authors of these taxa published before 1935, when the type method was incorporated into the International Rules of Nomenclature, did not, except in rare instances, indicate type specimens; and almost as seldom since 1935 have authors of novelties in microscopic algae designated types. It has therefore been a major task of this project for us to select type specimens for all specific and subspecific taxa previously untypified. In making such selections, we have wherever possible chosen the material originally studied by the authors; if such material were not found, then a specimen named later or reasonably assumed to have been seen by the author has been designated as the type.

There are perhaps numerous species for which no original specimens were preserved. The number of these is probably far less than this paper indicates, since we have been somewhat restricted in time, in patience, in assistance, and in financial, political, and psychological means to search out and examine the historical bases for all such species. For the purpose of completeness, we are designating the original descriptions of these species as temporary types (to serve until original specimens may be found); and these descriptions have been interpreted by us with some sense of responsibility in special lists of synonymy under the appropriate taxa. Descriptions (even with illustrations) are of course only ink and paper, not algae; the application of iodine to them will not indicate what to the taxonomist will be of primary concern: whether or not the cells of the new taxon contain starch. Especially of the algae are descriptions and illustrations matters of personal interpretation. It would be eminently unscientific of us and unfair

2

to the author if we presumed to comprehend fully a description without seeing the original specimen also. We have concluded that names of species with only descriptions and illustrations for types cannot be employed as names for taxa of algae in nature.

HISTORY OF CLASSIFICATION

If botanists prior to 1777 encountered species of coccoid Myxophyceae, they probably referred them to the Linnaean genera *Byssus, Tremella,* or *Ulva.* J. Lightfoot in his *Flora Scotica* (1777) named a conspicuous gelatinous alga from wet places on the Isle of Skye *Ulva montana {Anacystis montana* of this paper]. K. Sprengel in *Flora Halensis, Mantissa* (1807) described green globules floating in a lake near Halle as *Coccochloris stagnina.* During the period 1790—1850, numerous genera and species were published by J. B. Bory de St. Vincent, C. A. Agardh, H. C. Lyngbye, J. P. Vaucher, C. Sommerfelt, S. F. Gray, H. F. Link, C. G. Ehrenberg, A. J. C. Corda, R. K. Greville, E. Fries, P. J. Turpin, B. Biasoletto, F. G. F. Meyen, B. Gaillon, M. J. Berkeley, A. de Brébisson, A. Braun, V. Trevisan, A. H. Hassall, J. J. Roemer, F. C. Mertens, A. G. Roth, and others to accomodate species of coccoid algae. In the 1830's, G. Meneghini and G. Zanardini in Italy, F. T. Kützing and L. Rabenhorst in Germany, W. H. Harvey in England, J. G. Agardh in Sweden, and C. Montagne in France began the careful description and classification of all algae, revising at the same time the work done by colleagues and by past authors. Without separating the Myxophycean forms as a group, they created classifications which would account for a large proportion of the unornamented coccoid algae which we know today.

Carl Nägeli, studying collections of microscopic algae principally from Switzerland, wrote a short paper in 1849 entitled *Gattungen Einzelliger Algen.* He separated the family *Chroococcaceae* from the more obviously green and red algae and divided the species in it among genera characterized by planes of cell division and the resultant distribution of the cells within the gelatinous matrix. The limitations which he set to the variability of the gelatinous matrix have remained with us as an incontrovertible orthodoxy. His classification, an expression of philosophical commitment to his theory of the cell as the unit of structure and function in plants, his descriptions, and his illustrations were so logically and mechanically executed that their material bases have remained almost completely unquestioned until now.

In his *Flora Europaea Algarum* (1864—68), L. Rabenhorst attempted to resolve the classifications of coccoid algae by Kützing, Meneghini, Nägeli, and others, chiefly along Nägelian lines. With this publication as

3

a basis, and encouraged by the indefatiguable Rabenhorst personally and through the media of his journals and published sets of exsiccatae, numerous workers throughout Europe and on other continents devoted themselves to the study of the microscopic algae. Many new genera and species of coccoid Myxophyceae were described, and considerable research in morphology, physiology and geographic distribution was done by J. E. Areschoug, E. Askenasy, J. W. Bailey, G. de Beck, E. Bornet, A. Borzi, A. Braun, C. G. Brügger, V. Cesati, F. Cohn, F. S. Collins, C. Cramer, P. L. and H. M. Crouan, A. de Bary, G. B. de Toni, E. de Wildeman, G. Dickie, W. G. Farlow, C. Flahault, J. Flotow, M. Foslie, C. Gobi, M. Gomont, A. Grunow, R. Gutwinski, A. Hansgirg, C. A. Hantzsch, P. Hariot, F. Hauck, G. Hieronymus, L. Hilse, L. Heufler, F. Hy, H. Itzigsohn, O. Kirchner, L. K. Rosenvinge, G. Lagerheim, A. Le Jolis, E. Lemmermann, P. Magnus, G. Martens, W. Migula, M. Möbius, A. Mougeot, O. Nordstedt, G. de Notaris, M. Perty, M. Raciborski, J. Reinke, P. Richter, J. Rostafinski, J. Schröter, W. A. Setchell, E. Stizenberger, W. F. R. Suringar, G. Thuret, J. E. Tilden, F. Unger, E. Wartmann, A. A. Weber-van Bosse, W. West, N. Wille, V. B. Wittrock, F. Wolle, and others. Thuret and Bornet, as their herbarium indicates, made an abortive attempt to revise the coccoid Myxophyceae; they contributed materially to our understanding of the life histories of numerous species. Itzigsohn and Hansgirg developed theories that among microscopic algae the species are polymorphic and thus not comparable with those of larger plants. This is an inversion of the more common assumption that all morphological variations and transformations through which an organism passes must necessarily be proper to a single species.

Since the early 1890's, many phycologists have occupied themselves with the ecology, cytology, and physiology of the microscopic algae. Historical taxonomy as a field of scientific investigation passed generally out of vogue in the universities toward 1900. At the same time, however, the describing of new taxa was very much accelerated. Until this period, the practice of preserving specimens permanently as indispensable records had been scrupulously observed by students of algae. Now, freed from a tradition of re-examining the material bases for descriptions of these microscopic plants, phycologists evolved an authoritarian taxonomic "science" about algae represented by idealized proxies in manuscript and published descriptions and illustrations, not about real algae. If original specimens were preserved at all, they were stored in liquid, usually in inadequately labeled containers. Some authors described taxa from living material, which was forthwith discarded after camera-lucida drawings had been made. Since 1900 the amount of literature in descriptive taxonomy has become immense, and the tradition has established itself that historical taxonomy in microscopic algae is inconceivable. According to this tradition, the validity of a

4

species and the application of its name can be judged only from the literary excellence of its description and the artistic excellence of its illustration, not from the plants which the author studied.

A list of the active researchers on the coccoid Myxophyceae since 1890, in addition to those mentioned above, would be too long for inclusion here. Some of the more notable have been G. S. West, A. Forti, N. L. Gardner, O. Jaag, L. Geitler, H. Skuja, A. A. Elenkin, P. Frémy, M. M. Hollerbach, A. Ercegovic, Giuseppe de Toni, and Y. Bharadwaja.

General treatments of the classification of the coccoid Myxophyceae along Nägelian lines, with minor innovations, were made by O. Kirchner in Engler & Prantl, *Natürl. Pflanzenfam.*, vol. 1(1a) (1900), by A. Forti in de Toni, *Sylloge Algarum,* vol. 5 (1907), by J. Tilden in *Minnesota Algae,* vol. 1 (1910), and by others. In 1923, A. A. Elenkin in Not. Syst. Inst. Crypt. Hort. Bot. Petropol. 2 (5) introduced a classification of the Chroococcaceae which proceeded much farther than Nägeli's did to enhance the generic and suprageneric value of the gelatinous matrix. This system was elaborated in his *Monographia Algarum Cyanophycearum, Pars Specialis,* Fasc. 1 (1938).

Geitler in Beih. z. Bot. Centralbl., II., 41: 163—294 (1925) summarized the Nägelian classification of the Chroococcaceae with its attritions during seventy-five years and created a new series of families and genera in the groups treated in the present paper as Chamaesiphonaceae and Clastidiaceae. This classification has been further refined in his larger works in Rabenh. Kryptogamen-Fl., ed. 2, vol. 14 (1932) and in Engler & Prantl, *Natürl. Pflanzenfam.,* ed. 2, vol. 1b (1942).

In 1939 we (in Field Mus. Bot. Ser. 20: 67—83) published the results of a tentative inquiry into the life histories and nomenclature of species of Microcystis. Daily, using a similar procedure, treated the Chroococcaceae of Ohio, Kentucky, and Indiana in Amer. Midl. Nat. 27: 636—661 (1942). In *Manual of Phycology* (G. M. Smith, ed.), p. 159 ff. (1951), Drouet briefly summarized the morphology of the three families included in the present paper. We later published a synopsis of the present classification, with keys, and formally introduced the new family *Clastidiaceae* in Butler Univ. Bot. Stud. 10: 220—223 (1952).

MORPHOLOGY

THE PLANT.—This term is employed here to indicate a single free cell or a group of cells joined together, usually in a gelatinous matrix. A specimen or single collection may contain many plants of one or few cells mixed with other algae; it may be a single entire globule, cushion, or stratum, or a part of one of these. In the Chroococcaceae and some species of Entophysalis, plants of the same species may be found in the form of strata, cushions, or globules (both microscopic and macroscopic), or as free unattached cells.

THE CELL.—In this paper, the term is restricted to the protoplast. Essentially the cell is a mass of protoplasm, containing no differentiated nucleus, and surrounded by a membrane. Gelatinous material is secreted through this membrane to form a sheath or gelatinous matrix about the cell. The protoplasm contains various chlorophyll, xanthophyll, and phycoerythrin pigments, and always the blue pigment phycocyanin. Granules of diverse natures and sizes may be distributed through the protoplasm. Glycoproteins are the end-products of photosynthesis. In some species, especially those of the plankton and those persisting under anaerobic conditions, pseudovacuoles are found. These are irregular in outline, black in transmitted light and red in reflected light. Cell division proceeds by fission, a process in which the cell constricts gradually until the two daughter cells separate and are pushed apart by the gelatinous material secreted; in other cells, membranes grow centripetally through the protoplasm as the constriction proceeds, until the daughter cells are separated by membranes, not by gelatinous material. After division has taken place, the daughter cells enlarge and change in shape so that they resemble the former mother cell. The rate of such regeneration is of course dependent upon environmental conditions, just as the rate of division must be. It is therefore reasonable to admit considerable variation in size of cells in a species, even in various parts of the same plant, according to the nature of the contemporary environment during growth. In the Chamaesiphonaceae, certain cells enlarge and divide internally into many small endospores. The single cells in the Clastidiaceae divide internally into usually uniseriate filaments of numerous cells. Geitler in Engler & Prantl, *Natürl. Pflanzenfam.,* ed. 2., 1b (1942) and F. E. Fritsch in *Structure and Reproduction of the Algae,* vol. 2 (1945) have reviewed the literature on the cytology of these plants.

THE GELATINOUS MATRIX.—This term is here applied to the sheaths about the cells. In a several—many-celled plant, the matrix consists of the sheaths of not only the cells present but also those of the ancestral mother cells for as many previous generations as the plant has existed. The persistence of the ancestral cell-sheaths in the matrix thus often governs the size of the plant itself. The matrix consists of pectins, which, according to their chemical natures and the environmental conditions under which they were produced and have survived, exhibit various types of lamellations about the cells. Like all pectic substances, they hydrolyze readily and may

6

become dissipated into the surrounding medium; or they remain firm, in parts lamellose, in parts homogeneous. Red and blue coloring matter often develops in the matrices of plants subject to direct insolation and to drying, especially in *Anacystis montana* and several species of Entophysalis; these are "indicator" pigments, turning one or the other color in direct response to changes in acidity and alkalinity of the medium. Brown and yellow pigments develop in these and other species (especially in *Coccochloris stagnina* and *Entophysalis deusta*) under similar natural conditions; these pigments also seem to accompany the infestation of the matrices and the parasitization of the cells by fungi. In some species, as in *Coccochloris aeruginosa,* the gelatinous material appears always to hydrolyze as it is extruded. However, this condition of complete hydrolyzation of the matrix and dissociation of the cells occurs frequently in all the species of Chroococcaceae and Entophysalidaceae. In the Clastidiaceae, the thin gelatinous sheath eventually hydrolyzes in part or *in toto.*

PARASITIZATION BY FUNGI. — Except in purified cultures, fungi are present in most habitats occupied by these algae. Where algal growth is rapid, conditions favoring equally rapid growth of the fungi do not often obtain. But in habitats subject to periodic drying, the sheathed Myxophyceae, usually perennial, eventually become infested with fungi. Hyphae of saprophytic fungi, as well as bacteria and other algae, are to be found in the gelatinous matrices of most mature plants of the coccoid blue-green algae. Such fungi appear to be responsible for the production of brown pigments in the sheaths, as well as for changes in the consistency and in the acidity and alkalinity of the sheath material. Other fungi parasitize the cells, causing the latter to enlarge, change color, and cease the production of gelatinous material. In *Anacystis montana,* the hyphae often completely surround the cell and encase it in a wall. In most instances, parasitization by fungi results in the death of the Myxophycean cells. However, with certain species of fungi, the algae continue to live and grow; and together they develop as lichens. Lichens containing blue-green algae appear more often to result from an association of heterocystous, rather than coccoid, Myxophyceae with fungi.

ENDOPHYTISM. —Although various cases of intracellular endophytism of these plants within other organisms have been described and discussed in the literature [see Geitler in Rabenh. Kryptogamen-Fl., ed. 2., vol. 14 (1932) and in Engler & Prantl, *Natürl. Pflanzenfam.,* ed. 2., vol. 1b (1942)], we find no real evidence that the blue-green structures involved, if autonomous organisms at all, are coccoid Myxophyceae. If they are not chromataphores containing phycocyanin or a similarly colored pigment, then the behavior of the Nostocaceae and other heterocystous blue-green algae in culture and in conditions where they are parasitized by fungi would

7

suggest that the filamentous Myxophyceae are concerned here. Perhaps when they are cultured in a free state outside their host organisms, these alleged endophytes can be properly disposed in a system of classification.

PRESERVATION AND MICROSCOPICAL TECHNIQUES

Permanent preservation of these algae is most adequately assured by drying. The dried specimens can be placed in paper packets, labeled, and stored in the herbarium. Since most of the species thrive best in subaerial or temporarily inundated habitats, they exist for much of the year in the dried condition in nature. Wet algae can be laid top-side up on paper (newspaper is excellent) and allowed to dry in the open air. Drying in a plant press excludes the air, so that autolyzation of the protoplasm takes place; drying with heat often destroys the cells by cooking. If possible, part of the substratum should be preserved as part of the specimen. If the plants are microscopic and mixed with other algae, a good specimen should contain a sufficient number of the plants so that at least several may be found in every field of every mount made for microscopic study. Such small plants, as well as those in plankton collections, may be dried directly on sheets of mica or clear plastic or on glass slides. Plankton organisms are sometimes destroyed during the process of drying unless they are first killed by the addition of formalin. Old collections preserved in formalin etc. can likewise be dried on mica; the material should be spread in a film thin enough for light to pass through it under the microscope.

Dried specimens can be examined microscopically by soaking a fragment in water on the slide, then by crushing it down under a cover glass. If present, sand grains can be teased out with the forceps and removed. Dilute solutions of the common household detergents are excellent mounting media; they soften most of the sheathed Myxophyceae quickly and remove all traces of air imprisoned in and among the algae. By this method even many poorly preserved specimens can be restored to an appearance similar to that of living material. On the other hand, the detergents will remove permanently the pseudovacuoles from the cells of plankton algae. Carbonates can be removed by mounting in dilute solutions of nitric, hydrochloric, citric, or acetic acids or in vinegar or lemon juice. A dilute solution of iodine or a solution of methyl green in slightly acidulated water is useful for staining structures in the protoplasm and sheaths. If a specimen has been dried on mica, plastic, or glass, a drop of water and a cover glass can be placed on the alga; after the examination, the cover glass can be removed and the alga can be allowed to dry again in the open air.

If material preserved in formalin has dried in the container, a solution of a household detergent will usually soften it sufficiently for the making of microscopic mounts. This or a dilute solution of sodium or potassium hydroxide will dissolve the white polymerized formaldehyde deposited on and inside the algae. If microscopic mounts have been made in liquids beneath a cover glass sealed to the slide with a cement, the liquid eventually dries out. The mount can be made useful again by the removal of bits of the cement at opposite corners of the cover glass; water or a dilute solution of a detergent can then be allowed to seep under the cover glass. After examination, the mount can be allowed to dry out before being stored away.

LOCATION OF SPECIMENS AND ACKNOWLEDGMENTS

Herbaria in which specimens cited are to be found are indicated in the lists below by means of the following abbreviations.[1] We are greatly indebted to the persons and to the institutions and their staffs who have made their collections available for this study.

B, Botanisches Museum, Berlin-Dahlem; BIRM, University of Birmingham, Birmingham, England; BKL, Brooklyn Botanic Garden; BM, British Museum (Natural History), London; BR, Jardin Botanique de l'État, Brussels; BUT, Butler University, Indianapolis; C, Botaniske Museum, Copenhagen; CAS, California Academy of Sciences, San Francisco; D, herbarium of Francis Drouet, Chicago; DA, herbarium of William A. Daily, Indianapolis; DT, Algarium de Toni, Brescia, Italy; E, Royal Botanic Garden, Edinburgh; EAR, Earlham College, Richmond, Indiana; FC, Cryptogamic Herbarium, Chicago Natural History Museum; FH, Farlow Herbarium, Cambridge, Massachusetts; FI, Istituto Botanico, Universitá di Firenze; G, Conservatoire Botanique, Geneva; HA, herbarium of Herbert Habeeb, Grand Falls, New Brunswick; K, Royal Botanic Gardens, Kew, England; KI, herbarium of Walter Kiener, Lincoln, Nebraska; L, Rijksherbarium, Leiden; LD, Botaniska Museum, Lund, Sweden; M, Botanische Staatssammlung, Munich; MICH, University Museums, Ann Arbor, Michigan; MIN, University of Minnesota, Minneapolis; MO, Missouri Botanical Garden, St. Louis; NEB, University of Nebraska, Lincoln; NY, New York Botanical Garden; O, Botaniske Museum, Oslo; PC, Laboratoire de Cryptogamie, Muséum National d'Histoire Naturelle, Paris; PENN, University of Pennsylvania, Philadelphia; PH, Academy of Natural Sciences, Philadelphia; PHI, herbarium of Harry K. Phinney, Corvallis, Oregon; PRC, Botanický Ústav, University Karlovy, Prague; PUH, University of the Philippines, Quezon City; S, Botaniska Avdelningen,

[1]These are essentially the abbreviations proposed by Lanjouw & Stafleu in *Index Herbariorum*, Part 1, ed. 2 (1954), adapted to the special requirements of this paper.

Naturhistoriska Riksmuseet, Stockholm; ST, herbarium of J. C. Strickland, Richmond, Virginia; T, Department of Botany, Florida State University, Tallahassee; TA, herbarium of Wm. Randolph Taylor, Ann Arbor, Michigan; TEX, University of Texas, Austin; UC, Herbarium, University of California, Berkeley; UPS, Botaniska Museet, Upsala, Sweden; US, Smithsonian Institution, Washington; W, Naturhistorisches Museum, Vienna; WU, Botanisches Institut, Universität Wien; YU, Osborn Botanical Laboratory, Yale University, New Haven; Z, Botanisches Museum, Universität Zürich; ZT, Institut für Spezielle Botanik, Eidgenossische Technische Hochschule, Zürich.

We are indebted to numerous friends and colleagues who have given us assistance in various ways: Mrs. Fay K. Daily, Mr. Donald Richards, Dr. Harry K. Phinney, Dr. E. C. Kleiderer, Mr. E. J. Richards, Dr. Wm. Randolph Taylor, Dr. José Cuatrecasas, Dr. Ruth Patrick, Dr. Josephine T. Koster, Dr. Karl Rechinger, Mr. P. Bourrelly, the late Dr. T. Arwidsson, Mr. G. Tandy, Mr. Kung-Chu Fan, the late Dr. Ray C. Friesner, Dr. John E. Potzger, Dr. Paul Cundiff, the late Dr. William A. Setchell, Miss Rosalie Weikert, the late Dr. Giuseppe de Toni, Dr. Anna de Toni, Mr. Horst Smolenski, Mr. Harold B. Louderback, Mrs. Grace S. Phinney, Dr. B. E. Dahlgren, Dr. J. H. Hoskins, Mr. Robert Ross, Mrs. Linda Newton Irvine, Dr. C. L. Hickman, Dr. Chester S. Nielsen, Dr. and Mrs. Joseph Rubinstein, Dr. Gregorio T. Velasquez, Dr. Grace C. Madsen, the late Dr. F. E. Fritsch, Mr. Clare F. Massey, Dr. Maxwell S. Doty, Dr. Otto Jaag, Mr. Robert Lami, Dr. Josephine E. Tilden, Mr. Jan Newhouse, the late Dr. David H. Linder, Dr. Lothar Geitler, Dr. Ellsworth P. Killip, the late Dr. C. B. Lipman, Miss Susy Weckering, Dr. Georges Cufodontis, Mr. Peter A. Green, Mr. Bo Peterson, Dr. Lois Lillick, the late Dr. J. M. Greenman, Mr. Kurt Fitz, Mme. A. Hamel-Jukov, Dr. Paul C. Standley, Dr. Charles Baehni, Miss C. J. Dickinson, Dr. Helen Foot Buell, Mrs. Effie M. Schugman, Mr. Spencer Savage, Mr. Philip W. Wolle, Dr. O. Hagerup, Dr. Joventino D. Soriano, Miss Vera Nováckova, Dr. E. Yale Dawson, Dr. L. H. Tiffany, Dr. J. W. G. Lund, Dr. Heinrichs Skuja, Mr. J. Francis Macbride, Dr. Tycho Norlindh, Dr. T. Hasselrot, Dr. J. Ramsbottom, Dr. H. B. S. Womersley, the late Dr. Melvin A. Brannon, Mr. John H. Wallace, Dr. A. H. G. Alston, Dr. Bohuslav Fott, Mr. E. T. Prange, Dr. Gunnar Nygaard, Dr. Gerald W. Prescott, Mr. Roger Heim, Dr. H. K. Svenson, Dr. C. E. B. Bonner, Dr. J. C. Strickland, the late Dr. Karl Suessenguth, Dr. Herman S. Forest, the late Dr. O. C. Schmidt, the late Dr. L. P. Khanna, Dr. Lewis H. Flint, Dr. W. Robyns, Dr. Rodolpho Pichi-Sermolli, Dr. Ramón Margalef, Dr. Max E. Britton, Dr. George J. Hollenberg, Dr. Carl Keissler, Dr. C. E. B. Bremekamp, Dr. Richard D. Wood, Mr. E. Manguin, Dr. Arne Hässler, Mr. Jean Mabille, Dr. Eric Hultén, Miss Edith M. Vincent, Dr. P. Gonzales Guerrero, the late Dr.

Angel Maldonado, Dr. G. Evelyn Hutchinson, the late Dr. William R. Maxon, Dr. Fred A. Barkley, Dr. A. U. Däniker, Dr. F. Børgesen, Dr. Ove Sundene, Miss Hilda Harris, Dr. V. J. Chapman, Dr. Herbert Habeeb, Dr. Hao-Jan Chu, Dr. Rimo Bacigalupi, Dr. Rex N. Webster, Dr. J. W. Moore, Dr. C. Mervin Palmer, Dr. Harold C. Bold, Dr. Sidney F. Glassman, Dr. Cesare Trebeschi, Mr. Gianni Trebeschi, Dr. John Thomas Howell, Dr. Achille de Toni, Dr. Nando de Toni, Mr. Norman C. Lahti, Dr. Ralph A. Lewin, Dr. Annetta Carter, Miss Alice Middleton, Dr. G. Beauverd, Dr. Julian A. Steyermark, Mr. Alois Patzak, Dr. Paul C. Silva, Dr. G. M. Smith, the late Dr. N. L. Gardner, Dr. C. V. Morton, the late Dr. Marshall A. Howe, Miss Cécile Lanouette, Miss Nell Horner, Miss Edith Mepham, Miss Filomena Fortich, Dr. George F. Papenfuss, Dr. Walter Kiener, Dr. Lawrence J. King, Dr. Richard I. Evans, Dr. Jacques Rousseau, Dr. Kuno Thomasson, Dr. Stillman Wright, Dr. Ira La Rivers, Dr. Arland T. Hotchkiss, Dr. B. B. McInteer, Dr. Clarence E. Taft, Dr. B. H. Smith, Dr. C. C. Palmiter, Mr. Robert Runyon, Dr. Harold J. Humm, Dr. R. N. Ginsburg, Dr. Lillian Arnold, Dr. John D. Dodd, Dr. Valerie May, Mrs. R. Catala, Dr. Mary A. Pocock, Dr. V. W. Lindauer, Dr. Delzie Demaree, Dr. Albert J. Bernatowicz, Dr. H. L. Blomquist, Dr. Alan J. Brook, Dr. M. S. Markle, Dr. R. L. Caylor, Dr. John L. Blum, Dr. Louis G. Williams, Dr. John Pelton, Dr. G. Haglund, Dr. A. B. Cribb, Dr. Gerald B. Ownbey, Mr. Emil Sella, Dr. M. J. Groesbeck and Dr. Herbert L. Mason.

The cost of publication of this manuscript was met with funds provided by the Richards Foundation of Chicago. A contribution from Eli Lilly and Company, Indianapolis, made possible the engraving of the copper plates for the illustrations.

CRYPTOGAMIC HERBARIUM, CHICAGO NATURAL HISTORY MUSEUM,
AND
HERBARIUM, BUTLER UNIVERSITY

FAMILY I. CHROOCOCCACEAE

Nägeli, Gatt. Einzell. Alg., p. 44. 1849. *Chroococcaceae* Subfam. *Euchroococcaceae* Hansgirg, Notarisia 3: 589. 1888. *Chroococcaceae* Subfam. *Euchroococcaceae* Tribus *Coccineae* Hansgirg, loc. cit. 1888. *Chroococcaceae* III. *Gloeoc000ceae-stereometreae* B. *Heterogloeae* Elenkin, Not. Syst. Crypt. Horti Bot. Petropol. 2: 67. 1923. *Chroococcaceae* III. *Gloeococceae-stereometreae* B. *Heterogloeae* 1. *Tegumentotenuiores* Elenkin, ibid., p. 68. 1923. —Type genus: *Chroococcus* Näg.

Chroococcaceae Subfam. *Euchroococcaceae* Tribus *Phyllothecineae* Hansgirg, Notarisia 3: 589. 1888. *Chroococcaceae* II. *Gloeococceae-planimetreae* Elenkin, Not. Syst. Inst. Crypt. Horti Bot. Petropol. 2: 66. 1923. *Merismopoediaceae* Elenkin, Acta Inst. Bot. Acad. Sci. U. R. S. S., II, 1: 19. 1933. —Type genus: *Merismopoedia* Kütz.

Chroococcaceae I. *Coccobactreae* Elenkin, Not. Syst. Crypt. Hort. Bot. Petropol. 2: 65. 1923. —Type genus: *Synechococcus* Näg.

Chroococcaceae III. *Gloeococceae-stereometreae* A. *Homoeogloeae* 2. *Excavatae* Elenkin, Not. Syst.

11

Inst. Crypt. Horti Bot. Petropol. 2: 67. 1923. *Chroococcaceae* III. *Gloeococceae-stereometreae* A. *Homoeogloeae* 2. *Excavatae* a) *Distantes* Elenkin, loc. cit. 1923. *Gomphosphaeriaceae* Elenkin, Acta Inst. Bot. Acad. Sci. U. R. S. S., II, 1: 18. 1933. —Type genus: *Gomphosphaeria* Kütz. *Chroococcaceae* III. *Gloeococceae-stereometreae* B. *Heterogloeae* 2. *Tegumentocrassiores* Elenkin, Not. Syst. Inst. Crypt. Horti Bot. Petropol. 2: 68. 1923. *Gloeocapsaceae* Elenkin, Acta Inst. Bot. Acad. Sci. U. R. S. S., II, 1: 19. 1938. —Type genus: *Gloeocapsa* Kütz. *Holopediaceae* Elenkin, Acta Inst. Bot. Acad. Sci. U. R. S. S., II, 1: 19. 1933. *Beckiaceae* Elenkin, loc. cit. 1933. —Type genus: *Holopedium* Lagerh.

The following family names have type species of their type genera for which no original specimens have been available to us for study:

Woronichiniaceae Elenkin, Acta Inst. Bot. Acad. Sci. U. R. S. S., II, 1: 18. 1933. —Type genus: *Woronichinia* Elenk. *Coelosphaeriaceae* Elenkin, Acta Inst. Bot. Acad. Sci. U. R. S. S., II, 1: 18. 1933. —Type genus: *Coelosphaerium* Näg. (= *Coelocystis* Näg.).

Plantae uni—multi-cellulares, microscopicae vel macroscopicae, forma et magnitudine et ambitu diversae, libere in aqua natantes vel in stratis crescentes, cellulis sphaericis, discoideis, ovoideis, ellipticis, cylindraceis, vel pyriformibus, unaquidque in duas cellulas-filias aequales dividente, unaquidque mox ab aliis cum gelatino vaginale se separante; reproductione a fragmentatione.

Most of the species of Chroococcaceae are capable of developing in various ways: as microscopic or macroscopic globular aquatic plants with homogeneous or lamellose matrices; as aquatic, subaerial, or aerial strata or cushions; or as free single cells where the sheath material has completely hydrolyzed. The cells, their methods of division and regeneration, and the arrangement of the cells in the plant are the chief means of distinguishing genera and species. Division of a cell into two equal daughter-cells is characteristic of the family.

Many of the unornamented coccoid and palmelloid Chlorophyceae have been described or transferred into this family at one time or another, even in recent years. *Palmogloea protuberans* (Sm. & Sow.) Kütz. is perhaps the most often placed here by mistake. Palmelloid flagellates and the primordia of Ulotrichaceae, Chaetophoraceae, Ulvaceae, and even Trentepohliaceae, especially where preserved in liquids, have often been interpreted as Chroococcaceae. The presence of chloroplasts, pyrenoids, starch, and cellulose walls should make the Chlorophycean affinities obvious. Likewise, *Porphyridium, Asterocytis,* and *Porphyra* spp. are sometimes mistakenly supposed to be Chroococcaceae. Spores of Nostocaceae and few-celled hormogonia of the smaller species of Phormidium, Plectonema, and Scytonema have been confused with members of this family. Bacteria and certain spores and cells of fungi also simulate these plants.

The measurements for cells noted in the keys and descriptions here are those generally encountered. Where successive divisions procede at a rapid rate and enlargement at a slow rate, cells may be found which are smaller than those recorded here. With a reversal of these extremes, the largest measurements could be exceeded.

Key to genera:

1. Cells before division ovoid to cylindrical, longer than broad, each dividing in a plane perpendicular to the long axis ....1. COCCOCHLORIS

1. Cells before division spherical, ovoid, discoid, cylindrical, or pyriform, never dividing in planes perpendicular to the long axis................................2.

2. Cells before division spherical, irregularly distributed within the gelatinous matrix or arranged more or less regularly in series of rows in three planes perpendicular to each other; cell division proceeding successively in three planes perpendicular to each other ....2. ANACYSTIS

2. Cells before division discoid, distributed in a single linear series within the gelatinous matrix; division proceeding in a single plane through the diameter of the cell ....3. JOHANNESBAPTISTIA

2. Cells before division spherical, ovoid, cylindrical, or pyriform, distributed through a flat or curved surface; cell division proceeding successively in two planes perpendicular to each other .... 3.

3. Plant a flat or curved plate; cells spherical, ovoid, or cylindrical, arranged regularly in series of rows perpendicular to each other. ....4. AGMENELLUM

3. Plant a flat or curved plate; cells ovoid or cylindrical, irregularly arranged .... 5. MICROCROCIS

3. Plant spherical or ovoid; cells spherical to ovoid, cylindrical, or pyriform, regularly or irregularly arranged ....6. GOMPHOSPHAERIA

GENUS 1. COCCOCHLORIS

Sprengel, Linn. Syst. Vegetabil., ed. 16, 4(1): 314. 1827. *Aphanothece* Sectio *Coccochloris* Sprengel ex Kirchner in Engler & Prantl, Natürl. Pflanzenfam., ed. 1, 1(1a): 55. 1898. *Aphanothece* Subgenus *Coccochloris* Kirchner ex Forti, Syll. Myxophyc., p. 76. 1907. *Microcystis* ∝ *Macroscopicae* Elenkin, Not. Syst. Inst. Crypt. Horti Bot. Petropol. 2: 67. 1923. *Microcystis* ∝ *Macroscopicae* Subgenus *Eucoccochloris* Elenkin, loc. cit. 1923. —Type species: *Coccochloris stagnina* Spreng.

Synechococcus Nägeli, Gatt. Einzell. Alg., p. 56. 1849. *Chroothece* Sectio *Synechococcus* Nägeli ex Hansgirg, Oesterr. Bot. Zeitschr. 34: 352. 1884. *Synechococcus* Subgenus *Eusynechococcus* Elenkin, Not. Syst. Inst. Crypt. Horti Bot. Petropol. 2: 65. 1923. —Type species: *Synechococcus elongatus* Näg. (= *Protococcus elongatus* Näg.).

Gloeothece Nägeli, Gatt. Einzell. Alg., p. 57. 1849. *Gloeocapsa* Subgenus *Gloeothece* Elenkin, Not. Syst. Inst. Crypt. Horti Bot. Petropol. 2: 69. 1923. —Type species: *Gloeothece linearis* Näg. *Gloeocapsa* Sectio *Rhodocapsa* Hansgirg, Prodr. Algenfl. Böhmen 2: 147. 1892. —Type species: *Gloeocapsa purpurea* Kütz.

Gloeocapsa Sectio *Xanthocapsa* Nägeli ex Kirchner in Engler & Prantl, Natürl. Pflanzenfam., ed. 1, 1(1a): 54. 1898. —Type species: *Gloeocapsa fuscolutea* Näg.

Original specimens of the type species of the following generic and subgeneric names have not been available to us for study:

Aphanothece Nägeli, Gatt. Einzell. Alg., p. 59. 1849. *Aphanothece* Sectio *Aphanothece* Nägeli ex Kirchner in Engler & Prantl, Natürl Pflanzenfam., ed. 1, 1(1a): 55. 1898. *Aphanothece* Subgenus *Aphanothece* Kirchner ex Forti, Syll. Myxophyc., p. 79. 1907. —Type species: *Aphanothece microscopica* Näg.

Rhabdoderma Schmidle & Lauterborn, Ber. Deutsch. Bot. Ges. 18: 148. 1900. *Gloeocapsa* Subgenus *Rhabdoderma* Elenkin, Not. Syst. Inst. Crypt. Hort. Bot. Petropol. 2: 69. 1923. —Type species: *Rhabdoderma lineare* Schmidle & Lauterb.

Bacularia Borzi, Nuova Notarisia 1905: 21. 1905. — Type species: *B. coerulescens* Borzi.

Rhabdogloea Schröder, Ber. Deutsch. Bot. Ges. 35: 549. 1917. —Type species: *R. ellipsoidea* Schröder.

Spirillopsis Naumann, K. Sv. Vet.-Akad. Handl. 62(4): 18. 1921. —Type species: *S. irregularis* Naum.

Cyanocloster Kufferath, Ann. Crypt. Exot. 2: 49. 1929. —Type species: *C. muscicola* Kuff.

13

Krkia Pevalek, Acta Bot. Inst. Bot. Univ. Zagreb. 4: 55. 1929. —Type species: *K. croatica* Peval.

Dzensia Woronichin, Bull. Jard. Bot. Princip. U. R. S. S. 28: 155. 1929. —Type species: *D. salina* Woronich.

Tetrarcus Skuja, Acta Horti Bot. Univ. Latv. 7: 46. 1932. —Type species: *T. Ilsteri* Skuja.

Plantae microscopicae vel macroscopicae, 1—pluri-cellulares, cellulis in divisione subsphaericis usque ad cylindraceas, aetate provecta ovoideis usque ad longicylindraceas, divisione semper in uno plano ad axem cellulae perpendiculare; gelatino vaginale hyalino vel deinde (in *C. stagnina*) lutescente, homogeneo vel lamelloso.

The mature cells of all the species of Coccochloris are characteristically elongate; they divide only at right angles to their long axes. Considerable variation in lengths and diameters of cells of each species should be expected according to the histories of the microenvironments in which the plants were produced.

Subaerial and aerial strata of *Palmogloea protuberans* (Sm. & Sow.) Kütz. are often confused with species here. Likewise, certain bacteria and the one—few-celled hormogonia of Scytonema and the smaller species of Phormidium and Plectonema can be mistaken for species of Coccochloris.

Key to species of Coccochloris:

Cells before division ovoid to cylindrico-elliptic, 7—45μ broad, up to 3 times as long as broad .... 1. C. AERUGINOSA

Cells before division ovoid to cylindrico-elliptic, (3—)4—8μ broad, up to 3 times as long as broad .... 2. C. STAGNINA

Cells before division cylindrical (usually quasi-truncate at the ends), straight, 2—6μ broad, up to 8 times as long as broad .... 3. C. ELABENS

Cells before division cylindrical (rotund or tapering at the ends), often curved, 1—3μ broad, up to 12 times as long as broad .... 4. C. PENIOCYSTIS

1. COCCOCHLORIS AERUGINOSA Drouet & Daily, Butler Univ. Bot. Stud. 10: 222. 1952. *Synechococcus aeruginosus* Nägeli, Gatt. Einzell. Alg., p. 56. 1849. — Type from Lucerne, Switzerland (ZT).

Synechococcus major f. *crassior* Lagerheim, Bot. Not. 1886: 50. 1886. *S. major* var. *crassior* Lagerheim ex Hansgirg, Prodr. Algenfl. Böhmen 2: 139. 1892. *S. crassus* f. *crassior* Lagerheim ex Forti, Syll. Myxophyc., p. 27. 1907. *S. aeruginosus* f. *crassior* Lagerheim ex Wille, Nyt Mag. Naturvidenskab. 62: 203. 1925. —Type from Upsala, Sweden (S). FIG. 143.

Synechococcus aeruginosus f. *angustior* West, Proc. Roy. Irish Acad. 31: 40. 1912. —Type from Clare island, Ireland (BIRM).

Original specimens have not been available to us for the following names; their original descriptions are here designated as the Types until the specimens can be found:

Synechococcus crassus Archer, Quart. Journ. Microsc. Sci., N. S. 7: 87. 1867. *S. major* f. *crassus* Elenkin, Monogr. Algar. Cyanophyc., Pars Spec. 1: 35. 1938.

Synechococcus major Schröter, Jahresber. Schles. Ges. Vaterl. Cultur 1883: 188. 1884.

Synechococcus major var. *maximus* Lemmermann, Forschungsber. Biol. Sta. Plön 4: 130. 1896. *S. aeruginosus* var. *maximus* Lemmermann, Krypt.-Fl. Mark Brandenb. 3: 46. 1907; Lemmermann ex Forti, Syll. Myxophyc., p. 27. 1907. *S. major* f. *maximus* Lemmermann ex Elenkin, Monogr. Algar. Cyanophyc., Pars Spec. 1: 35. 1938.

Synechococcus roseo-purpureus G. S. West, Journ. of Bot. 37: 265. 1899. —The original sketch and notes from the West notebooks in the British Museum (Natural History) are reproduced here: FIG. 144.

Synechococcus grandis Playfair, Proc. Linn. Soc. New South Wales 43: 499. 1918.
Synechococcus aeruginosus var. *salinus* Meyer, Izvest. Gosudarstv. Gidrolog. Inst. (Bull. de l'Inst. Hydrol. Russie) 15: 37. 1925.
Synechococcus diachloros Skuja, Acta Horti Bot. Univ. Latv. 11—12: 43. 1939.

Plantae laete aerugineae vel olivaceae, microscopicae, 1—2 cellulares, cellulis in divisione binis truncato-sphaericis usque ad truncato-ovoideas vel truncato-sub-cylindricas, aetate provecta ovoideo-cylindraceis vel cylindrico-ellipticis, diametro 7—45μ crassis, usque ad 3-plo longiores; gelatino vaginale hyalino, tenue vel omnino diffluente; protoplasmate aerugineo vel olivaceo, saepe sparsim grosse-granuloso. FIGS. 143, 144.

In shallow water and seepage in bogs, ponds, and lakes, seldom found in the plankton, seldom in brackish water. Cells of *Anacystis dimidiata* with the gelatinous matrix almost completely hydrolyzed are often confused with this species. Isolated spores of *Anabaena* and *Cylindrospermum* spp. superficially resemble the cells of *Coccochloris aeruginosa*.

Specimens examined:

FINLAND: in scrobiculis rupium ad Mariehamn, *G. Lagerheim,* Aug. 1898 (S). NORWAY: prope Molde, *N. Wille,* Jul. 1919 (S). SWEDEN: in Lassby backar prope Upsaliam, *Lagerheim,* 18 May 1884 (Type of *Synechococcus major* f. *crassior* Lagerh. in Wittr. & Nordst., Alg. Exs. no. 792a, S; isotypes, FC [Fig. 143], L, MIN); in Norby skog prope Upsaliam, *Lagerheim,* May 1884 (as *S. major* in Wittr. & Nordst., Alg. Exs. no. 792b, FC, L, MIN, S); in stagno parvo ad Malmagen Herjedaliae, *Lagerheim,* 1897 (S); Djurö, Uplandia, *O. Borge,* Aug. 1908 (S); Marstrand, Bohuslan, *O. Nordstedt,* Jun. 1876 (S). GERMANY: Saxon Switzerland: *C. A. Hantzsch* (FC); aus der Schlucht, die vom Wildenstein (Kuhstall) abwärts nach dem Habichtsgrund führt, auch in grossen Zschand an einer Felswand, *C. Biene* (as *S. aeruginosus* in Rabenh. Alg. no. 1335, L, NY, UC). CZECHOSLOVAKIA: Bohemia: Dittersbach—Hinter-Dittersbach, *A. Hansgirg,* Jul. 1889 (FC, W); an einem Felsen bie Dittersbach, *Rostock,* Jun. 1861 (as *S. aeruginosus* with *Sirosiphon saxicola* in Rabenh. Alg. no. 1120, FC); mit *Urococcus insignis,* Bielathal, *W. Krieger,* Aug. 1881 (TA). SWITZERLAND: Luzern, an nassen Felsen, *C. von Nägeli,* Dec. 1847 (Type of *Synechococcus aeruginosus* Näg., ZT). IRELAND: Clare island (Type of *S. aeruginosus* f. *angustior* West, BIRM). MAINE: sphagnum pool, tableland on south spur of Mt. Katahdin, *B. M. Davis 13,* Aug. 1902 (FC, MICH). NEW HAMPSHIRE: the Flume, *W. G. Farlow* (FH); pool near Lake of the Clouds, summit of Mt. Washington, *C. Bullard,* 13 Sept. 1912 (as *S. crassus* in Coll., Hold., & Setch., Phyc. Bor.-Amer. no. 1951, FC, L, TA). VERMONT: on rocks in cave, Lake Willoughby, *W. G. Farlow,* Aug. 1880 (FC).

QUEBEC: falaise humide, pic à Rousseau dans les monts Otish, *J. Rousseau 389,* 10 Aug. 1949 (FC); roche moutonée, ou suinte une source, Rivière Payne, vers 71° 25′ long. O., *J. Rousseau 1134,* 11 Aug. 1948 (FC); mousses d'un étang, embouchure de la Kogaluk, baie d'Hudson, *Rousseau 208,* 16 Jul. 1948 (FC); sur les pierres, Ile Marie Victorin, Lac Mistassini, *J. & E. Rousseau 321a,* 14 Jul. 1944 (FC). NORTH CAROLINA: in rock pools, Mt. Whitesides and summit of Yellow Mountain, Jackson county, *H. C. Bold H95b, H140,* Jun. 1939 (FC). FLORIDA: on roots in ditch near Kissimmee, Osceola county, *P. O. Schallert 1853,* 1 Mar. 1948 (FC). KENTUCKY: on sandstone overhang, Tight Hollow, Pine Ridge, Wolfe county, *W. A. & F. K. Daily 849, 854,* 25 May 1941 (DA, FC.) TENNESSEE: wet sandstone in Falls Creek Canyon, Van Buren county, *H. Silva 675,* 4 May 1947 (FC, TENN). WYOMING: drying pool, bottom of small lake near Island lake, Shoshone national forest, Park county, *F. H. Rose & F. A. Barkley 4050,* 7 Sept. 1940 (FC). MONTANA: shallow pool, Rattlesnake creek, Missoula county, *Rose 4063,* 10 Oct. 1940 (FC). COLORADO: over mosses on a siliceous rock ledge, South Colony creek, Custer county, *W. Kiener 10393a,* 9 Jul. 1941 (FC, KI). BRITISH COLUMBIA: in wet moss near Glacier, Kootenay county, *W. R. Taylor 10,* Aug. 1921 (FH, NY, TA). PUERTO RICO: on shaded rocks at the "Campo", Maricao, *N. Wille 1235,* Feb. 1915 (NY).

2. COCCOCHLORIS STAGNINA Sprengel, Fl. Halens., Mantissa 1: 14. 1807.
Aphanothece stagnina A. Braun in Rabenhorst, Alg. Eur. 157 & 158: 1572. 1863.
—Type from Dieskau near Halle, Germany (FC). FIG. 154.

Palmella rupestris Lyngbye, Tent. Hydrophyt. Dan., p. 207. 1819. *Coccochloris rupestris*

15

Sprengel, Linn. Syst. Vegetab., ed. 16, 4(1): 373. 1827. *Microcystis rupestris* Kützing, Linnaea 8: 374. 1833. *Protococcus rupestris* Corda in Sturm, Deutschl. Fl., Abt. 2, 6: 45. 1833. *Microhaloa rupestris* Kützing, Phyc. Gener., p. 169. 1843; Linnaea 17: 84. 1843. *Haematococcus rupestris* Hassall, Hist. Brit. Freshw. Alg. 1: 326. 1845. *Micraloa rupestris* Kützing ex Trevisan, Sagg. Monogr. Alg. Coccot., p. 38. 1848. *Bichatia rupestris* Trevisan, ibid. p. 62. 1848. *Gloeocapsa rupicola* Kützing, Sp. Algar., p. 221. 1849. *Gloeocystis rupestris* Rabenhorst, Fl. Eur. Algar. 3: 30. 1868. *Gloeothece rupestris* Bornet in Wittrock & Nordstedt, Alg. Exs. 8: 399. 1880. *Bichatia rupicola* Kuntze, Rev. Gen. Pl. 2: 886. 1891. *Gloeocapsa rupestris* Dalla Torre & Sarnthein, Fl. Tirol 2: 157. 1901. *Anacystis rupestris* Drouet & Daily in Daily, Amer. Midl. Nat. 27: 650. 1942. *Coccochloris stagnina* f. *rupestris* Drouet & Daily, Lloydia 11: 78. 1948. —Type from Nass, Norway (C). FIG. 148.

Palmella obscura Sommerfelt, Suppl. Fl. Lappon., p. 203. 1826. —Type from Saltdal, Norway (C).

Palmella bullosa Kützing, Alg. Aq. Dulc. Dec. 16: 154. 1836. *Microcystis bullosa* Meneghini, Consp. Algol. Eugan., p. 324. 1837. *Bichatia bullosa* Trevisan, Nomencl. Algar. 1: 60. 1845. *Cagniardia bullosa* Trevisan, Sagg. Monogr. Alg. Coccot., p. 49. 1848. *Aphanothece bullosa* Rabenhorst, Fl. Eur. Algar. 2: 65. 1865. *Coccochloris bullosa* Kuntze, Rev. Gen. Pl. 3(2): 546. 1898. *Glaucocystis bullosa* Wille, Nyt Mag. Naturvidensk. 56(1): 38. 1918. —Type from Battaglia, Italy (L).

Microcystis gelatinosa Meneghini, Consp. Algol. Eugan., p. 324. 1837. —Type from the Euganean springs, Italy (FI).

Palmella Mooreana Harvey, Man. Brit. Alg., p. 178. 1841. *Coccochloris Mooreana* Hassall, Hist. Brit. Freshw. Alg. 1: 316. 1845. *Palmogloea Mooreana* Crouan fr., Fl. du Finistère, p. 110. 1867. *Aphanothece Mooreana* Lagerheim, Öfvers. K. Sv. Vet.-Akad. Forh. 30(2): 44. 1883. —Type from near Lough Neagh, Ireland (K). FIG. 158.

Microcystis paroliniana Meneghini, Mem. R. Accad. Sci. Torino, ser. 2, 5(Sci. Fis. & Mat.): 78. 1843. *Bichatia paroliniana* Trevisan, Nomencl. Algar. 1: 61. 1845. *Gloeocapsa paroliniana* Brébisson in Kützing, Tab. Phyc. 1: 25. 1846. *Gloeocystis paroliniana* Nägeli, Gatt. Einzell. Alg., p. 65. 1849. —Type from Oliero, Italy (FI). FIG. 152.

Microcystis microspora Meneghini, Mem. R. Accad. Torino, ser. 2, 5(Sci. Fis. & Mat.): 80. 1843. *Bichatia microspora* Trevisan, Nomencl. Algar. 1: 61. 1845. *Palmella microspora* Kützing, Tab. Phyc. 1: 11. 1846. *Aphanothece microspora* Rabenhorst, Krypt.-Fl. Sachsen 1: 76. 1863. *Palmogloea microspora* Crouan fr., Fl. du Finistère, p. 110. 1867. —Type from the Euganean springs (FI).

Protococcus minutus Kützing, Phyc. Gener., p. 168. 1843. *Pleurococcus minutus* Trevisan, Sagg. Monogr. Alg. Coccot., p. 33. 1848. *Chroococcus minutus* Nägeli, Gatt. Einzell. Alg., p. 46. 1849. *Protococcus minutus* var. *caerulescens* Brébisson in Rabenhorst, Alg. Eur. 201–204: 2035. 1867. *Gloeocapsa minuta* Hollerbach in Elenkin, Monogr. Algar. Cyanophyc., Pars Spec. 1: 233. 1938. —Type from Nordhausen, Germany (L). FIG. 145.

Gloeocapsa thermalis Kützing, Phyc. Gener., p. 173. 1843. *Bichatia thermalis* Trevisan, Nomencl. Algar. 1: 62. 1845. —Type from Battaglia, Italy (L).

Gloeocapsa gelatinosa Kützing, Phyc. Gener., p. 174. 1843. *Palmella gelatinosa* Meneghini *pro synon.* in Kützing, loc. cit. 1843. *Bichatia gelatinosa* Trevisan, Sagg. Monogr. Alg. Coccot., p. 63. 1848. *Gloeocapsa gelatinosa* ∝ *olivacea* Kützing, Sp. Algar., p. 219. 1849. —Type from Battaglia, Italy (L). FIG. 163.

Palmella pallida Kützing, Phyc. Germ., p. 149. 1845. *Cagniardia pallida* Trevisan, Sagg. Monogr. Alg. Coccot., p. 49. 1848. *Aphanothece pallida* Rabenhorst, Krypt.-Fl. Sachsen 1: 76. 1863. *Polycystis pallida* Thuret ex Farlow in Farlow, Anderson & Eaton, Alg. Exs. Amer. Bor. 4: 179. 1881. *Microcystis pallida* Lemmermann, Krypt.-Fl. Mark Brandenb. 3: 77. 1907. —Type from the Harz mountains, Germany (L).

Coccochloris cystifera Hassall, Hist. Brit. Freshw. Alg. 1: 441. 1845. *Bichatia cystiphora* Trevisan, Sagg. Monogr. Alg. Coccot., p. 66. 1848. *Gloeothece cystifera* Rabenhorst, Fl. Eur. Algar. 2: 61. 1865. —Type from Bristol, England (BM).

Protococcus julianus Kützing, Tab. Phyc. 1: 5. 1846. *P. Kuetzingianus* Trevisan, Sagg. Monogr. Alg. Coccot., p. 26. 1848. *Gloeocapsa fulva* Kützing, Sp. Algar., p. 224. 1849. *Bichatia fulva* Kuntze, Rev. Gen. Pl. 2: 886. 1891. —Type from Abano, Italy (L).

Coccochloris stagnina β *cuprina* Kützing, Sp. Algar., p. 216. 1849. *Palmella cuprina* Agardh *pro synon.* ex Kützing, loc. cit. 1849. *Aphanothece stagnina* β *cuprina* Kützing ex Forti, Syll. Myxophyc., p. 77. 1907. —Type from Scania, Sweden (FI).

Gloeocapsa gelatinosa β *aeruginea* Kützing, Sp. Algar., p. 219. 1849. *Gloeothece palea* var. *aeruginea* Hansgirg, Physiol. & Algol. Stud., p. 140. 1887. *G. palea aeruginosa* Hansgirg ex Gardner, New York Acad. Sci., Sci. Surv. Porto Rico 8: 259. 1932. *Gloeocapsa gelatinosa aeruginosa* Kützing ex Gardner, loc. cit. 1932. —Type from Abano, Italy (L).

Gloeocapsa fuscolutea Nägeli in Kützing, Sp. Algar., p. 224. 1849. *G. fuscolutea* ∝ *achroa*

16

Kützing, loc. cit. 1849. *Gloeothece fuscolutea* Nägeli, Gatt. Einzell. Alg., p. 57. 1849. —Type from Küsnacht, Zürich, Switzerland (L).

Gloeocapsa fuscolutea β *holochroa* Kützing, Sp. Algar., p. 224. 1849. *Gloeothece fuscolutea* var. *holochroa* Kützing ex Wartmann & Winter, Schweiz. Krypt. no. 847. —Type from Küsnacht, Zürich, Switzerland (L).

Gloeocapsa fuscolutea γ *hemichroa* Kützing, Sp. Algar., p. 224. 1849. —Type from Küsnacht, Zürich, Switzerland (L).

Gloeocapsa Kuetzingiana Nägeli in Kützing, Sp. Algar., p. 224. 1849; Nägeli, Gatt. Einzell. Alg., p. 51. 1849. *Bichatia Kuetzingiana* Kuntze, Rev. Gen. Pl. 2: 886. 1891. —Type from Küsnacht, Zürich, Switzerland (L). FIG. 159.

Gloeothece devia Nägeli, Gatt. Einzell. Alg., p. 58. 1849. —Type from Lucerne, Switzerland (ZT).

Gloeocapsa tepidariorum A. Braun in Rabenhorst, Alg. Sachs. 23 & 24: 221. 1852. *Gloeothece tepidariorum* Lagerheim, Öfvers. K. Sv. Vet.-Akad. Forh. 40(2): 44. 1883. *G rupestris* var. *tepidariorum* Hansgirg, Sitzungsber. K. Böhm. Ges. Wiss., Math.-Nat. Cl., 1891(1): 354. 1891. *Bichatia tepidariorum* Kuntze, Rev. Gen. Pl. 2: 886. 1891. *Gloeothece rupestris* b. *tepidariorum* Hansgirg, Prodr. Algenfl. Böhmen 2: 136. 1892. —Type from Berlin, Germany (L). FIG. 157.

Gloeocapsa aurata Stizenberger in Rabenhorst, Alg. Sachs. 61 & 62: 607. 1857. *Bichatia aurata* Kuntze, Rev. Gen. Pl. 2: 886. 1891. —Isotypes from Constanz, Germany (FC, L, MIN, PC, TA, UC).

Aphanothece Naegelii Wartmann in Rabenhorst, Alg. Sachs. 109-110: 1093. 1861. —Type from St. Gallen, Switzerland (ZT). FIG. 147.

Palmella heterococca Kützing, Osterprogram d. Realschule zu Nordhausen, p. 6. 1863; Hedwigia 2: 86. 1863. —Type from Vera Cruz, Mexico (L). FIG. 153.

Gloeocystis rupestris Rabenhorst, Krypt.-Fl. Sachsen 1: 128. 1863. —Type from Planitz, Germany (FC).

Aphanothece (stagnina var.) prasina A. Braun in Rabenhorst, Alg. Sachs. 57 & 58: 1572. 1863. *Coccochloris stagnina* var. *prasina* A. Braun ex Richter, Hedwigia 25: 254. 1886. *C. stagnina* b. *prasina* Richter ex Hansgirg, Prodr. Algenfl. Böhmen 2: 141. 1892. *Aphanothece stagnina* f. *prasina* Elenkin, Monogr. Algar. Cyanophyc., Pars Spec. 1: 144. 1938. *Anacystis rupestris* var. *prasina* Drouet & Daily in Daily, Amer. Midl. Nat. 27: 651. 1942. —Type from Berlin, Germany (FC). FIG. 161.

Gloeothece carpathica Grunow in Rabenhorst, Fl. Eur. Algar. 2: 63. 1865. —Type from Wallendorf, Slovakia (W).

Gloeothece Heufleri Grunow in Rabenhorst, Fl. Eur. Algar. 2: 63. 1865. —Type from Kindberg, Styria (W).

Aphanothece (stagnina var.) coerulescens A. Braun in Rabenhorst, Alg. Sachs. 157 & 158: 1572. 1863; A. Braun in Rabenhorst, Fl. Eur. Algar. 2: 66. 1865. —Type from Pillnitz, Germany (L).

Palmella Jesenii Wood, Proc. Amer. Philos. Soc. 11: 134. 1869. —Type from Philadelphia, Pennsylvania (in herb. Princeton University in NY).

Gloeothece decipiens A. Braun in Wittrock & Nordstedt, Alg. Exs. 12: 594. 1883. —Type from Berlin, Germany (L). FIG. 149.

Chroococcus montanus Hansgirg, Notarisia 3: 400. 1888. —Type from Libsic, Bohemia (W). FIG. 150.

Polycystis litoralis Hansgirg in Foslie, Contrib. Knowl. Mar. Alg. Norway 1: 169. 1890. *Microcystis litoralis* Forti, Syll. Myxophyc., p. 89. 1907. —Isotype from Bugönaes, Norway (L).

Gloeothece samoensis f. *major* Wille in Rechinger, Bot. & Zool. Ergebn. Samoa & Salomoninseln, Süsswasseralg., p. 7. 1914. —Type from Upolu, Samoa (O).

Gloeocapsa luteo-fusca Martens ex Wille, Nyt Mag. Naturvid. 62: 191. 1925. —Type from Pegu, Burma (LD).

Aphanothece gelatinosa Gardner, Univ. Calif. Publ. Bot. 14: 2. 1927. *A. Gardneri* J. de Toni, Noter. di Nomencl. Algol. VIII. 1936. —Type from Fukien province, China (FH).

Aphanocapsa Richteriana var. *major* Gardner, Mem. New York Bot. Bard. 7: 4. 1927. *A. Richteriana major* Gardner, New York Acad. Sci., Sci. Surv. Porto Rico 8: 253. 1932. —Type from Palmer, Puerto Rico (NY).

Aphanothece conferta var. *brevis* Gardner, Mem. New York Bot. Gard. 7: 5. 1927. *A. conferta brevis* Gardner, New York Acad. Sci., Sci. Surv. Porto Rico 8: 254. 1932. —Type from Maricao, Puerto Rico (NY). FIG. 160.

Aphanothece microscopica var. *granulosa* Gardner, Mem. New York Bot. Gard. 7: 5. 1927. *A. microscopia granulosa* Gardner, New York Acad. Sci., Sci. Surv. Porto Rico 8: 254. 1932. —Type from Penuelas, Puerto Rico (NY). FIG. 151.

Gloeocapsa quaternata var. *major* Gardner, Mem. New York Bot. Gard. 7: 12. 1927. *G. quaternata major* Gardner, New York Acad. Sci., Sci. Surv. Porto Rico 8: 258. 1932. —Type from San Juan, Puerto Rico (NY). FIG. 146.

17

Gloeothece interspersa Gardner, Mem. New York Bot. Gard. 7: 13. 1927. —Type from Santurce, Puerto Rico. (NY). FIG. 155.

Anacystis cylindracea Gardner, Mem. New York Bot. Gard. 7: 19. 1927. —Type from near Utuado, Puerto Rico (NY). FIG. 156.

Aphanothece bullosa var. *minor* Geitler, Arch. f. Hydrobiol., Suppl. XII, Trop. Binnengew. 4: 622. 1933. *A. bullosa* var. *major* Geitler, ibid. Suppl. XIV, 6: 374. 1935. —Type from Bukit Kili, Sumatra (in the collection of L. Geitler).

Gloeothece rupestris var. *major* Geitler, Arch. f. Hydrobiol., Suppl. XII, 4: 622. 1933. —Type from Panjingahan, Sumatra (in the collection of L. Geitler).

Cyanostylon banyolensis Margalef, Publ. Inst. Biol. Aplic. Barcelona 1: 60. 1946. —Type from Lago de Bañolas, Gerona, Spain (D).

Original specimens have not been available to us for the following names; their descriptions are here designated as the Types until the specimens can be found:

Gloeocapsa polyzonia Perty, Bewegung Durch Schwingende Mikroskop. Organe, p. 39. 1848. *G. polydermatica* b. *polyzonia* Perty ex Rabenhorst, Fl. Eur. Algar. 2: 37. 1865. *G. polydermatica* f. *polyzonia* Perty ex Forti, Syll. Myxophyc., p. 52. 1907.

Aphanothece microscopica Nägeli, Gatt. Einzell. Alg., p. 59. 1849. —No specimens labeled thus appear to be present in the Nägeli collections of either the Staatssammlung at Munich or the Institut für Spezielle Botanik, Eidgenossische Technische Hochschule at Zürich. The original notes and sketches of Nägeli in the Pflanzenphysiologisches Institut of the latter institution are reproduced here: FIG. 162.

Gloeothece nigrescens Rabenhorst, Krypt.-Fl. v. Sachsen 1: 76. 1863.

Aphanothece pallida var. *micrococca* Brügger, Jahresber. Naturf. Ges. Graubündens, N. F. 8: 245. 1863. *Aphanocapsa pallida* var. *micrococca* Brügger ex Rabenhorst, Hedwigia 1863: 181. 1863. *Aphanothece pallida* b. *micrococca* Brügger ex Rabenhorst, Fl. Eur. Algar. 2: 65. 1865.

Gloeocapsa aurata var. *alpicola* Brügger, Jahresber. Naturf. Ges. Graubündens, N. F. 8: 247. 1863; Hedwigia 1863: 181. 1863. *G. aurata* b. *alpicola* Brügger ex Rabenhorst, Fl. Eur. Algar. 2: 47. 1865.

Polycystis piscinalis Brügger, Jahresber. Naturf. Ges. Graubündens, N. F. 8: 249. 1863; Hedwigia 1863: 181. 1863. *Microcystis piscinalis* Forti, Syll. Myxophyc., p. 90. 1907.

Polycystis piscinalis var. *Microcystis* Brügger, Jahresber. Naturf. Ges. Graubündens, N. F. 8: 249. 1863; Hedwigia 1863: 181. 1863.

Gloeocapsa sparsa Wood, Proc. Amer. Philos. Soc. 11: 123. 1869.

Gloeothece ambigua A. Braun, Sitzungsber. Ges. Naturf. Fr. Berlin 1875: 99. 1875.

Gomphosphaeria anomala Bennett, Journ. Roy. Microsc. Soc. London, ser. 2, 8: 3. 1888. *Coelosphaerium anomalum* de Toni & Levi, Notarisia 3: 528. 1888.

Chroococcus consociatus Hariot, Bull. Soc. Bot. France 38: 416. 1891.

Microcystis maxima Bernard, Protococc. & Desmid. d'Eau Douce, p. 49. 1908.

Chroococcus Rochei Virieux, Soc. d'Hist. Nat. Doubs 21: 51. 1911.

Aphanothece prasina f. *minor* Wille, Deutsche Südpolar-Exped. 1901—03, 8: 414. 1928.

Microcystis kerguelensis Wille, Deutsche Südpolar-Exped. 1901—03, 8: 416. 1928.

Gloeocapsa rupestris var. *achroa* Beck-Mannagetta, Lotos 77: 99. 1929.

Aphanothece stagnina var. *nemathece* Frémy, Arch. de Bot. Caen 3(Mém. 2): 26. 1930.

Aphanocapsa rivularis f. *major* Petkoff, Bull. Soc. Bot. Bulgarie 4: 104. 1931.

Gloeothece rupestris var. *chalybea* Krieger, Hedwigia 70: 151. 1931.

Aphanothece pallida f. *minor* Dixit, Proc. Indian Acad. Sci. 3: 95. 1936.

Chroococcus minimus var. *crassus* B. Rao, Proc. Indian Acad. Sci. 3: 166. 1936.

Gloeocapsa stegophila var. *crassa* B. Rao, Proc. Indian Acad. Sci. 6: 344. 1937.

Chroococcus montanus var. *hyalinus* B. Rao, Proc. Indian Acad. Sci. 6: 345. 1937.

Gloeothece rupestris var. *minor* Jao, Sinensia 10: 178. 1939.

Gloeothece fuscolutea var. *unilamellaris* Chu, Ohio Journ. Sci. 52: 96. 1952.

Aphanothece Lemnae Chu, Ohio Journ. Sci. 52: 96. 1952.

Plantae aerugineae, olivaceae, luteolae, brunneae, roseae, violaceae, vel nigrae, microscopicae vel macroscopicae, (1—)pluri—multi-cellulares, cellulis in divisione binis truncato-hemisphaericis usque ad truncato-ovoideas, aetate provecta ovoideis, ellipticis, vel raro elliptico-cylindraceis, diametro (3—)4—8µ crassis, usque ad 3-plo longiores; gelatino vaginale primum hyalino deinde lutescente vel brunnescente, homogeneo vel saepe conspicue lamelloso, nonnumquam omnino diffluente;

protoplasmate aerugineo, olivaceo, luteolo, roseo, vel violaceo, plerumque homogeneo. FIGS. 145—163.

Commonly found on soil, rocks, and wood in aerial and subaerial situations and in shallow fresh (occasionally in brackish and marine) waters, or free-floating in lakes and ponds. These plants grow equally well and rapidly in and out of the water through a wide range of environmental conditions. Dissociated spores and vegetative cells of *Nostoc* spp. in old natural and laboratory cultures often resemble cells of this species superficially. Subaerial and aerial growth-forms of *Palmogloea protuberans* (Sm. & Sow.) Kütz. are often mistaken for plants of *Coccochloris stagnina*.

Specimens examined:

NORWAY: (in a lagoon at high water mark) Bugönaes in Sydvaranger, *M. Foslie* (isotype of *Polycystis litoralis* Hansg., L); Gudbrandsdalen, *N. Wille*, Sept. 1916 (S); Valders, in rupibus ad Skogstad, *Wille*, 31 Jul. 1879 (D, S); in rupibus udis earundenique fissuris ad Nass, *H. C. Lyngbye*, 4 Jul. 1816 (Type of *Palmella rupestris* Lyngb., C; isotypes, FC[Fig. 148], FI); Saltdal, Nordland, *C. Sommerfelt* (Type of *P. obscura* Sommerf., C; isotype "e Norvegia", PC). SWEDEN: in turfosis Scaniae, *J. Agardh* (Type of *Coccochloris stagnina β cuprina* Kütz., FI; isotype, PC); in stagno ad Qvikjock in Lapponia Lulensi, *G. Lagerheim*, Jul. 1883 (as *C. stagnina* in Wittr., Nordst., & Lagerh., Alg. Exs. no. 1595a, L); in lacu Langen in Nericia, *O. Borge*, Jul. 1898 (as *C. stagnina* in Wittr., Nordst., & Lagerh., Alg. Exs. no. 1595b, L); in amne Fyrisan in Ultuna, Uplandia, *C. Johansson & G. Lagerheim*, Nov. 1882 (as *Aphanothece Mooreana* in Wittr. & Nordst., Alg. Exs. no. 695, FC, L, NY, S); in stagno Marestrandi, *O. Nordstedt*, Aug. 1876 (as *Coccochloris prasina* in Aresch., Alg. Scand. Exs., Ser. Nov., no. 427, FC, L, S); Lurbo, *per. Näs*, Uppland, *V. Wittrock*, Jul. 1876 (D, S); Säbra, *H. W. Arnell*, Aug. 1877 (S); Stockholmen, *S. Areschoug*, May 1868 (S); Haparanda, *T. Krok*, Jul. 1868 (S); in stagno ad Bara Scaniae, *H. G. Simmons*, Sept. 1897 (S); in palude ad Kungsmarken prope Lund, *Simmons*, Jun. 1897 (S); Varberg, Halland, *D. E. Hylmö*, Apr. 1928, 1929 (S); Scania, *J. Agardh*, 1843 (DT); in insula Vermdö Uplandiae, *E. Warming & N. Wille*, Jul. 1883 (as *C. tuberculosa* in Wittr., Nordst., & Lagerh., Alg. Exs. no. 1599, NY); in spelunca "Silfvergrotten" montis Skullen Scaniae, *Simmons*, Jul. 1897 (NY); Upsala, in aquario, *Lagerheim*, Mar. 1882 (S); Transtrand, Dalarne, *Borge*, Jun. 1916 (S); ad marginem aquarii in caldario horti botanici Upsaliensis, *Lagerheim*, Mar. 1892 (as *Gloeothece tepidariorum* in Wittr., Nordst., & Lagerh., Alg. Exs. no. 1544a, L). POLAND: Silesia: in fossa ad Tscheichwitz prope Vratislaviam, *B. Schröder*, Sept. 1895 (as *Coccochloris stagnina* in Wittr., Nordst., & Lagerh., Alg. Exs. no. 1596, L); in einem Teiche an der Steinau bei Tillowitz, *O. Kirchner*, 4 Jul. 1875 (as *Aphanothece prasina* in Rabenh. Alg. no. 2442, FC, L, TA, UC); in einem Erdausstich bei Bischwitz bei Breslau, *Hilse*, Sept. 1864 (as *A. pallida* in Rabenh. Alg. no. 1831, FC, FH, NY, UC); an einem feuchten, grasigen Wege im Steinbruche auf dem Galgenberge bei Strehlen, *Hilse*, 1862 (as *Palmella testacea* in Rabenh. Alg. no. 1524, FC, L). ROMANIA: Chitila, in palude riv. Colintina, dist. Ilfov, *E. C. Teodorescu 1319*, 9 May 1903 (FC, W).

DENMARK: ad parietes calceos internos caldarii Orchidearum horti botanici hauniensis, *O. Nordstedt*, 21 Oct. 1878 (as *Gloeothece rupestris* in Wittr. & Nordst., Alg. Exs. no. 399, FC, L, NY, PC, S); Köpenhaven, *Nordstedt*, Oct. 1878 (S); Lyngby sø Selandiae, *L. K. Rosenvinge*, Sept. 1882 (S), Jun. 1881 (as *Coccochloris stagnina* in Wittr., Nordst., & Lagerh., Alg. Exs. no. 1597a, L). GERMANY: Baden, an den Speichen eines Wasserrades bei Constanz, *E. Stizenberger*, Aug. 1857 (isotypes of *Gloeocapsa aurata* Stizenb. in Rabenh. Alg. no. 607, FC, L, MIN, PC; and in Jack, Lein., & Stizenb., Krypt. Badens no. 1, TA, UC). Bavaria: an der Wand des feuchten Gewölbes der über den Rödelheim führenden Brücke des Donau-Mainkanales bei Erlangen, *P. Reinsch*, 26 Mar. 1864 (as *G. atrata* in Rabenh. Alg. no. 1914, FC, L, S); Kersbach bei Erlangen, *Glück*, Sept. 1890 (L, NY, S), *Reinsch* (D, L). Brandenburg: an einer feuchten Mauer in Warmhaus des Oberhofbuchdruckers Decker zu Berlin, *A. Braun*, Jul., Oct. 1852 (Type of *G. tepidariorum* A. Br. in Rabenh. Alg. no. 221, L; isotypes, FC[Fig. 157], PC, TA), Aug. 1852 (PC); in Warmhaus des Universitätsgartens zu Berlin, *Braun*, Oct. 1875 (Type of *Gloeothece decipiens* A. Br. in Rabenh. Alg. no. 2456, L; isotypes, FC, PC, S, TA); im Berliner botanischen Garten, *Braun*, Aug. 1852 (L), Jan. 1855 (L), Jan. 1875 (L), Apr. 1875 (as *G. decipiens* in Rabenh. Alg. no. 2456b, FC [Fig. 149], L, PC, S, TA), May 1875 (S), May 1879 (S), *P. Hennings*, Apr., May 1892 (as *G. decipiens* in Henn., Phyk. March. no 45, L); Berliner Treibhäuser, *Anon.*, Aug. 1863 (FH, S); im Tümpeln bei Weissensee unweit Berlin, *Braun*, Aug. 1863 (Type of *Aphanothece prasina* A. Br. in Rabenh. Alg. no. 1572, FC [Fig. 161]), *A. de Bary*, Aug. 1851 (as *Coccochloris stagnina* in Rabenh. Alg. Suppl. no. 3, L), *Bauer*, Aug. 1854 (FH, L); Berlin, *A. Braun*, Aug. 1851 (L, NY), Aug. 1863 (L, S); Berliner bot. Garten, *P. Magnus*, Jun. 1896 (S); Tümpel bei Schönborn, *A. Jenks*, Aug. 1892 (TA); Friedrichsfelde,

19

Jahn, Jul. 1856 (TA); Halensee und Hundekohlensee im Grunewald, *P. Hennings,* May, Jun. 1892 (as *C. stagnina* var. *prasina* in Henn., Phyk. March. no. 42, FH, L). HESSE: Bessungen bei Darmstadt, *G. Martens* (L); fontaine du jardin grandducal à Darmstadt, *Martens* (L); im Schlossgarten zu Darmstadt, *A. Braun 23,* Sept. 1853 (L); Darmstadt, *Braun,* 1851 (L). Mecklenburg (Pomerania): Wolgast, *M. Marsson* (NY). Saxony: Leipzig, *Reuter* (NY, UC); bei Auerbach in Voigtlande, *P. Richter* (UC); in Tümpeln bei Riesigk, Wörlitz (Anhalt), *R. Haritz,* Jun. 1890 (S); in piscinis ad Lipsiam et Dresdam, *Welwitsch* (S); in einem Teiche um Lübschütz bei Wurzen unweit Leipzig, *H. Reichelt,* Jun. 1886 (as *C. stagnina* var. *prasina* in Hauck & Richt., Phyk. Univ. no. 91, L, MICH, MIN, S); Nordhausen, *F. T. Kützing* (Type of *Protococcus minutus* Kütz., L[Fig. 145]; isotype, UC); an nassen Felsen des Innerstethales im Oberharze, *Römer 154* (Type of *Palmella pallida* Kütz., L; isotype, UC); Halle, *Kützing* (D, L); in stehenden Wässern bei Halle, *Kützing* (as *Coccochloris stagnina* in Kütz., Alg. Aq. Dulc. Dec., no. 29, FC, L, NY, UC); (in lacubus ad Dieskau) K. *Sprengel* (Type of *C. stagnina* Spreng., FC[Fig. 154]); Dresden, an der Wand eines Gewächshauses in Prinz Georg's Garten, *C. A. Hantzsch* (as *Gloeocapsa muralis* in Rabenh. Alg. no. 1216, FC, L, PC); Pillnitz, *Welwitsch* (BUT); im Schlossgarten bei Pillnitz, *L. Rabenhorst* (Type of *Aphanothece coerulescens* A. Br., as *Coccochloris stagnina* in Rabenh. Alg. no. 3, L; isotypes, FC, TA, UC); Mauerwände in einem Warmhause, Connewitz, Leipzig, *P. Richter,* Apr. 1879 (FC); ad parietes caldarii in Connewitz prope Lipsiam, *Richter,* 21 Apr. 1880 (as *Gloeothece decipiens* in Wittr. & Nordst., Alg. Exs. no. 594, D, FC, L, S); an feuchten Wänden eines Gewächshauses, Anger bei Leipzig, *Richter* (FC, S), Feb. 1884 (as *Gloeocapsa atrovirens* in Hauck & Richt., Phyk. Univ. no. 42, L, PC), 12 Apr. 1881 (as *G. decipiens* with *Chroococcus varius* in Wittr. & Nordst., Alg. Exs. no. 600, FC); an dem Mauerwerk der Treibgärtnerei, Planitz, *G. Kreischer* (designated as Type of *Gloeocystis rupestris* Rabenh. in Rabenh. Alg. no. 1790, FC; isotype, PC). Schleswig-Holstein: ex lacu prope Slesviciam, *J. N. Suhr,* 26 May 1846 (FC, S); ad Slesviciam, *C. Jessen* (FC, L, S); Megger Koog in Schleswig, *Jessen* (FC, FH, L, S); in Torfgruben bei Waldhausen unfern Lübeck, *R. Häcker* (as *Coccochloris stagnina* in Rabenh. Alg. Suppl. no. 3, L, UC). Thuringia: an feuchten Felsen unterhalb des Breitengescheid, Eisenach, *W. Migula,* Sept. 1905 (as *Gloeothece rupestris* in Mig., Crypt. Germ., Austr., & Helv. Exs., Alg. no. 99, MICH, NY, TA); an feuchten Felsen an der Strasse nach der hohen Sonne, Eisenach, *Migula,* 26 Sept. 1924 (as *Aphanothece Naegelii* in Mig., Crypt. Germ. Austr., & Helv. Exs., Alg. no. 179, FC, TA); an feuchten Felsen in der Ludwigsklamm, Eisenach, *Migula,* July, 1932 (as *A. Castagnei* in Mig., Krypt. Germ., Austr., & Helvet. Exs., Alg. no. 278, FC, TA).

CZECHOSLOVAKIA: Moravia: kultura na agaru, Trebic, *F. Novácek,* 7 Oct. 1928 (FC); serpentin, rokle, Mohelno, *Novácek,* 10 Mar. 1926 (FC). Slovakia: in einem feuchten Sandsteinfelsen in Wallendorfer Wald, *C. Kalkbrenner* (Type of *Gloeothece carpathica* Grun., W); murs humides, Wallendorf, *Kalchbrenner* (as *Gloeocapsa nigra* in Roumeguère, Alg. de Fr. no. 1041, L). Bohemia: Arnau, *A. Hansgirg,* Jul. 1885 (FC, W); Bystric nächst Beneschau, *Hansgirg,* Aug. 1884 (FC, W); Chotoviny u Tábora, *Hansgirg,* Aug. 1888 (FC, W); Labská Tynice, *Hansgirg,* 1891 (FC, W); in saxis madidis ad Libsic, *Hansgirg,* Oct. 1886 (Type of *Chroococcus montanus* Hansg., W; isotype, FC[Fig. 150]); in lacu silvatico prope Neu-Strezmir ad Stupcic, *Hansgirg,* 10 Sept. 1884 (as *Aphanothece stagnina* in Wittr. & Nordst., Alg. Exs. no. 794, FC, L, S, UC; and in Kerner, Fl. Exs. Austro-Hung. no. 2000, L, MO, S); Praha, Spolec. Zahrada, *Hansgirg,* Dec. 1883 (FC, W); ad Slichov prope Pragam, *Hansgirg,* May 1885 (FC, W); Hortus Botanicus, Smichow, *Hansgirg,* Dec. 1883, Jan. 1884, Feb. 1884, Jul. 1886 (FC, W); skaly groti, Srbsku u Berouna, *Hansgirg,* Aug. 1884 (FC, W); Stechovic nächst Prag, *Hansgirg,* Apr. 1887 (FC, UC); in rupibus subhumidis ad St. Kilian prope Stechovic, *Hansgirg* (as *A. pallida* in Kerner, Fl. Exs. Austro-Hung. no. 1999, L, MO, S); Stresmir u Stupsice, *Hansgirg,* Aug. 1884 (FC, W); Steinkirchen ad Budweis, *Hansgirg,* Aug. 1888 (FC, W). AUSTRIA: Lower Austria: in rupibus humidis prope Frankenfels, *S. Stockmayer* (as *Gloeothece fusco-lutea* in Mus. Vindob. Krypt. Exs. no. 148, FC, L, S); Tümpel in den Auer bei Marchegg, *G. de Beck,* Aug. 1887 (FC, W); Wien, in tepidariis horti botanici universitatis, *K. Rechinger* (as *G. tepidariorum* in Mus. Vindob. Krypt. Exs. no. 2138, S). Styria: an Moosen in einem Mühlgange der Mürz in Kindberg, *C. Heufler,* 11 Jul. 1856 (Type of *Gloeothece Heufleri* Grun., W); Judendorf-Gratwein prope Graz, *A. Hansgirg,* Sept. 1890 (FC, W). Carinthia: Kalkfelsen, Vorranach, Kreuzeckgebiet, *H. Simmer,* May 1898 (S); Klagenfurt, *Hansgirg,* Sept. 1889 (FC, W); Feldkirchen prope Villach, *Hansgirg,* Sept. 1889 (FC, W). Tyrol: Brixlegg, *Hansgirg,* 1891 (FC, W). HUNGARY: Budapest, in vitrinis caldariorum horti botanici, *F. Filárszky* (as *Gloeocapsa fenestralis* in Mus. Vindob. Krypt. Exs. no. 433, FC, L, S, Z). JUGOSLAVIA: Slovenia: St. Georgen—Pöltschach, Podnart, and Krainburg, *Hansgirg,* Aug. 1889 (FC, W); Marburg, Cilli, and Steinbrück, *Hansgirg,* Aug. 1890 (FC, W). TRIESTE: Zaule—Capo d'Istria, *Hansgirg,* Aug. 1889 (FC, W).

SWITZERLAND: an nassen Felsen, Luzern, *C. von Nägeli,* Dec. 1847 (Type of *Gloeothece devia* Näg,. with *Scytonema* sp., ZT); Dübendorf, auf dem Grunde in wenig tiefen Grüben, *Nägeli,* Feb. 1849 (ZT); Thalwerk, *C. Cramer,* Apr. 1863 (ZT); St. Gallen (MICH); auf Moosen, Zweibrücker-Tobel bei St. Gallen, *B. Wartmann* (as *G. fuscolutea* var. *holochroa* in Wartm. & Wint., Schweiz.

20

Krypt, no. 847, ZT), 15 Jul. 1860 (Type of *Aphanothece Naegelii* Wartm. in Wartm. & Schenk, Schweiz. Krypt. no. 36, ZT; isotypes in Rabenh. Alg. no. 1093, FC[Fig. 147], FH); an Nagelfluhfelsen der Bernegg bei St. Gallen, *Wartmann* (as *Gloeothece rupestris* in Wartm. & Wint., Schweiz. Krypt. no. 846, ZT); feuchte Felsen, Küsnacht. Zürich, *Nägeli 368*, Nov. 1847 (Type of *Gloeocapsa fuscolutea* ∝ *achroa* Näg., β *holochroa* Näg., and γ *hemichroa* Näg., L; isotypes, M, ZT); ad terram inter muscos, Zürich, *G. Winter*, May 1878 (TA); Küsnachter Wasserfall um Zürich, *Nägeli* (PC); feuchte Felsen, Küsnacht, Zürich, *Nägeli 258* (Type of *G. Kuetzingiana* Näg., L; isotypes, M, O, UC), Nov. 1847 (FC [Fig. 159], M); Küsnacht bei Zürich, *B. Wartmann*, 28 Jul. 1856 (as *G. Kuetzingiana* in Rabenh. Alg. no. 630, FC, L, M, PC; and in Wartm. & Schenk, Schweiz. Krypt. no. 133, M, Z). ITALY: Sagrado prope Görz, *A. Hansgirg*, 1891 (FC, W); sui muri delle serre dell'Orto Botanico di Pisa, *A. Mori* (as *G. muralis* in Erb. Critt. Ital., ser. II, no. 1245, FC, PC); Oliero, ex herb. Zanardini (Type of *Microcystis paroliniana* Menegh., FI [Fig. 152]; isotype, PC); margine della fontana di Viali in Euganeis, *G. Meneghini*, Aug. 1841 (Type of *Microcystis microspora* Menegh., FI; Isotypes, L, S, UC); St. Elena prope Battaglia, *Meneghini* (Type of *Gloeocapsa thermalis* Kütz., L); with *Scytonema thermale*, Abano, in herb. Kützing (Type of *Protococcus julianus* Kütz., L); in canalibus ubi effluit aqua calida thermarum ad Battaglia, *F. T. Kützing*, May 1835 (Type of *Palmella bullosa* Kütz., L; isotypes in Kütz., Alg. Aq. Dulc. Dec. no. 154, L, UC); Abano, *G. Meneghini* (Type of *Gloeocapsa gelatinosa* β *aeruginea* Kütz., L.; isotypes, S, UC); in aqua thermali, Battaglia, *Meneghini* (Type of *G. gelatinosa* Kütz., L [Fig. 163]; isotypes, S, UC); in aquis thermalibus Aponi, *Meneghini* (Type of *Microcystis gelatinosa* Menegh., FI; isotypes, PC, UC).

NETHERLANDS: Goes, *van den Bosch 365* (UC); den Haag, in Slotte, *Vrijdagsijnen* (L); Delft, warme kas van het Laboratorium voor Technische Botanie, *A. J. Meeuse*, Apr. 1943 (L); Delft, culturtuin, *G. van Rossem*, Oct. 1942 (L); Botshol bij Abcoude, *V. Westhoff*, Jul., Aug. 1944 (L); Heemstede, Groenendaal, *P. Leeutnaar*, May 1937 (L); Groeneveld, *A. Weber-van Bosse 1216*, Apr. 1893 (L); Nieuwkoop, verbinding van Oosthout van de Wijdte met Kaarwetering, *Westhoff 96*, Sept. 1943 (L); plassen van Brenkleveen en Tienhoven, *W. Vervoort*, May 1942 (L); Westbroek, in Sloot, *R. A. Maas Geesteranus & Westhoff 2762*, Aug. 1943 (L), *Maas Geesteranus 2764* (L); Kaloot near Borsele, Zeeland, *W. G. Beeftink*, 2 Aug. 1950 (FC); Hortus Botanicus, Leiden, *J. T. Koster 19*, 24 Oct. 1935 (D, L). BELGIUM: dans les fossés le long de la route de Gand à Tronchiennes, *Scheidweiler* (as *Coccochloris stagnina* in Westend. & Wall., Herb. Crypt. Belg. no. 1099, NY, UC); Jardin Botanique de Bruxelles, *E. de Wildeman*, Jun. 1888 (as *Gloeocapsa polydermatica* in Roumeguère, Alg. de Fr. no. 1015, L). FAEROES ISLANDS: *H. C. Lyngbye* (PC). BRITISH ISLES: Scotland: in the bottom of Dunmore loch near Pitlochry, Perthshire, *A. J. Brook*, 5 Jan. 1954 (D). England: near Bristol, 2 Sept. 1847, *herb. Thwaites* (Type of *Coccochloris cystifera* Hass., BM); Sutton Park, Warwickshire, *G. S. West*, 6 May 1908 (FC, UC); in small pond near River Coln at Fairford, Gloucestershire, *W. L. Tolstead 8487*, 1 Oct. 1944 (FC). Ireland: boggy hole by Shane's Castle near Lough Neagh, county Antrim, *D. Moore* (Type of *Palmella Mooreana* Harv., K [Fig. 158]; isotype, BM); in brackish pool, Clare island, *A. D. Cotton*, May 1911 (NY). FRANCE: Aisne: sur un mur de casemate, Berthenicourt-par-Moy, *J. Mabille 4*, May 1952 (FC); forêt de St. Gobain, *Mabille*, Jul., Aug. 1952 (FC). Alpes Maritimes: Antibes, *E. Bornet*, Mar. 1874 (FH), Jan. 1875 (NY); fond d'un ruisseau à Menton, *G. Thuret* (as *Aphanocapsa Roeseana* with *A. membranacea* in Roumeguère, Alg. de Fr. no. 827, L). Alsace: jardin botanique de Strasbourg, *A. Braun* (as *Gloeothece decipiens* in Roumeguère, Alg. de Fr. no. 520, L); fossés, St. Marie aux Mines, *Caspary* (as *Aphanothece stagnina* in Roumeguère, Alg. de Fr. no. 728, L); fossés aux environs de Colmar, *herb. Buchinger* (as *Aphanocapsa Roeseana* in Roumeguère, Alg. de Fr. No. 516, L). Calvados: Vire, *herb. Lenormand* (FC, D, L, S), *Chauvin* (L); anciènnes carrières d'Allemagne, Caen, *P. Frémy 737*, Apr. 1924 (DT). Hérault: St. Pons, *C. Flahault 396*, May 1882 (L, S). Manche: Carentan (S); ad muros, Saint Lô, *Frémy 319*, Feb. 1921 (NY); ad rupes, Agneaux, *Frémy*, Jun. 1927 (NY); mare de Vauville, *Frémy*, 21 Aug. 1931 (as *Aphanothece stagnina* in Hamel, Alg. de Fr. no. 152, FC). Maine et Loire: rochers de Murs, *F. Hy*, 10 Nov. 1890, Nov. 1892 (PC), Mar. 1894 (as *Gloeothece Rabenhorstii* in Wittr., Nordst., & Lagerh., Alg. Exs. no. 1543, L, NY), Jul. 1894 (as *G. tepidariorum* in Wittr., Nordst., & Lagerh., Alg. Exs. no. 1544b, L); in stagno prope St. Georges sur Loire, *Hy*, Jul. 1891 (as *Coccochloris stagnina* in Wittr., Nordst., & Lagerh., Alg. Exs. no. 1596b, L). Nord: in canalibus Flandriae litoralis prope Dunkerque, *C. Flahault*, 14 Aug. 1885 (as *Polycystis aeruginosa* in Wittr. & Nordst., Alg. Exs. no. 795, FC, L, S); in fossis prope Bourbourg, *Flahault*, Aug. 1886 (as *Coccochloris stagnina* in Wittr., Nordst., & Lagerh., Alg. Exs. no. 1597, L). Seine et Marne: ad parietes aquaeductus, Meloduno, *Roussel*, 21 Jul. 1851 (FI). SPAIN: sobre piedras y tallos, 30 cm. de profundidad, Lago de Bañolas, prov. Gerona, *R. Margalef*, 7 Jul. 1944 (isotype of *Cyanostylon banyolensis* Margal., D). CANARY ISLANDS: Funchal, *C. Lindman*, 7 Jan. 1885 (D, S). ICELAND: in stagnant water in a branch to Grimsá, Vallanes, East Iceland, *J. Boye Petersen*, 26 Jun. 1914 (FC). BERMUDA: on ground near the seashore, Hamilton, *W. G. Farlow* (FC, UC); on ground, Spanish Point, *Farlow*, Jan. 1900 (S; as *Gloeothece rupestris* in Coll., Hold., & Setch. Phyc. Bor.-Amer. no. 703, FC, L, MICH, TA); on rocks near shore, *F. S. Collins 7002*, Apr. 1912 (FH); in a freshwater pool near Old Ferry

Road, *Collins,* 9 Aug. 1913 (as *G. rupestris* in Coll., Hold., & Setch., Phyc. Bor.-Amer. no. 2154, FC, D, MICH, L). NEW BRUNSWICK: Island Lake, 40—50 miles due south of Dalhousie, *M. Le Mesurier 3,* Aug. 1953 (FC); wet ground, ledges, etc., Grand Falls, *H. Habeeb 10063, 10206, 10378, 10844,* Jul., Aug. 1947, Jul., Sept. 1948 (FC, HA). MAINE: Fox island, Hunnewell's Point, mouth of the Kennebec river, *F. S. Collins,* Aug. 1880 (FC, FH, UC); lagoon, Eagle island, *Collins 2954,* Jul. 1894 (NY); on cliffs of Piscataquis river, Dover, *Collins,* 8 Jul. 1897 (FC, UC). NEW HAMPSHIRE: alpine cascade, Gorham, *W. G. Farlow,* Sept. 1887 (FH). MASSACHUSETTS: in greenhouse, Biological Institute, Harvard University, Cambridge, *R. M. Whelden & G. D. Darker 6457,* 23 Sept. 1937 (D, FH); on flower pots, Botanic Garden, Cambridge, *H. M. Richards,* 18 Feb. 1895 (as *Aphanothece microscopica* in Coll., Hold., & Setch., Phyc. Bor.-Amer. no. 552, FC, L, TA); in salt water kept in the laboratory more than a year, *Farlow,* Feb. 1911 (FC, FH); Nahant, *B. M. Davis,* May 1893 (FH); wet rocks, Pine Banks park, Melrose, *Collins 3591,* Apr. 1899 (FH, UC); cliffs, Eastern Head, Gloucester, *Farlow,* Jul. 1902 (FH); on the ground beyond the guano factory, Woods Hole, *Farlow,* Jul. 1889 (FH); shore of Herring pond, Eastham, *Collins 5894, 5896, 5897,* Aug. 1908 (NY); pond, Eastham, *Collins 5900,* Sept. 1907 (NY); Jemima pond, Eastham, *Collins,* Aug. 1908 (NY); Fay's ditch, Woods Hole, *F. Drouet 1876,* 12 Jul. 1936 (D, NY); salt marsh pools, Pasque island, Gosnold, *W. R. Taylor,* 5 Jul. 1932 (D, TA); rocks, Essex street, Saugus, *Collins 1696,* Jun. 1890 (UC); Tashmoo pond, Marthas Vineyard, *G. T. Moore,* Sept. 1907 (FH); Naushon island, *Taylor,* Jul. 1917 (TA); Salt pond, Falmouth, *H. T. Croasdale,* Jul. 1935 (TA); pond southeast of Oyster pond, Falmouth, *Taylor,* Jul. 1925 (TA); Fresh pond, Cambridge, *W. Trelease,* Oct. 1881 (MO); Stony Brook pond, *G. T. Moore,* Oct. 1895 (MO). RHODE ISLAND: *W. J. V. Osterhout,* 1894 (FH).

CONNECTICUT: shore of Fresh pond, Stratford, *I. Holden 729,* Sept. 1892 (UC); Norwich, *W. A. Setchell,* 2 Sept. 1892 (as *Aphanothece prasina* in Coll., Hold., & Setch., Phyc. Bor.-Amer. no. 251, FC, L, TA, UC); in a small pond, Norwich, *Setchell 31,* Jul. 1885 (UC); in a pool on Derby avenue, New Haven, *Setchell 574,* 1 Oct. 1892 (UC); Tyler lake, Litchfield county, *H. K. Phinney 1122,* 4 Sept. 1946 (FC, PHI); in a temporary pool at the head of the lake in the Yale bird preserve, New Haven, *Phinney 1074,* 28 Jul. 1946 (FC, PHI); in pond, New Haven, *D. C. Eaton,* Oct. 1880 (D, YU); in a jar in the laboratory, Yale University, New Haven, *Setchell 737,* 10 Nov. 1893 (FC, UC). QUEBEC: Grenville, Argenteuil county, *J. Brunel 304,* 25 Jun. 1931 (FC); Rawdon, Montcalm county, *S. M. J. Eudes 253-7* (FC); on wet calcareous cliff on Mount Royal, Montreal, *Brunel 457,* 18 Sept. 1937 (FC); Ile du Large dans le fleuve St. Laurent entre Longueuil et Boucherville, *J. Brunel 64,* 11 Sept. 1930 (FC, D); in an aquarium, McGill University, Montreal, *R. D. Gibbs,* Oct. 1950 (FC); sur rocher humide, embouchure de la Kogaluk, baie d'Hudson, *J. Rousseau 209,* 16 Jul. 1948 (FC). NEW YORK: in the greenhouse, Barnard College, New York City, *H. C. Bold B23,* Feb. 1941 (FC); on rocks and walls, Felter's Glen and gorge below Rensselaerville, *M. S. Markle 16, 30,* 28 Jul. 1946, Jul. 1947 (EAR, FC); Queechy lake, Canaan, *E. S. Deevey,* 7 Jun. 1937 (D, DT, NY, S); in Hidden lake, Indian Falls, Genesee county, *J. Blum 127,* 4 Oct. 1948 (FC); in Hundred Acre pond, Mendon Ponds, Monroe county, *Blum 407,* 15 Oct. 1949 (FC); on dripping rocks along Genesee river, Glen Iris, north of Portageville, Wyoming county, *Blum 296,* 1 Jun. 1949 (FC). PENNSYLVANIA: [prope Philadelphia] "coll. Dr. Wood misit Billings" (Type of *Palmella Jesenii* Wood in herb. Princeton University, NY); greenhouses [Bethlehem?], *F. Wolle* (NY); aquarium, greenhouse, University of Pennsylvania, Philadelphia, *F. R. Fosberg & Drouet 2199,* Mar. 1938 (D); on wet rocks, Silverthread falls, Dingmans Ferry, *Drouet & H. Herpers 3670,* 16 Jul. 1941 (FC). MARYLAND: in rock pool, Plummers island in the Potomac river near Cabin John, Montgomery county, *E. C. Leonard 3071,* 1 Nov. 1953 (FC, US); pool, Cove Point, Calvert county, *H. C. Bold 8,* 1936 (FC).

VIRGINIA: on dripping rocks by spring northeast of Warm Springs, Bath County, *E. S. Luttrell & J. C. Strickland 942,* 26 Jul. 1941 (FC, ST); on wet soil near railroad at Ivey road west of Charlottesville, *B. F. D. Runk & Strickland 560,* 13 Nov. 1939 (FC, ST); on rocks in stream below cave at Goodman's Ferry, Giles county, *Strickland 527,* 16 Aug. 1939 (FC, ST); concrete spring mask, Old Mill near Newport, Giles county, *H. Silva 2776,* 26 Jun. 1953 (FC, TENN); dripping sandstone in woods west of Fredericksburg, *H. H. Iltis V2,* 26 Dec. 1947 (FC, TENN); in pool and on wet soil, Back Bay national wild-life refuge, Princess Anne county, *H. A. Bailey, Luttrell, & Strickland 782, 784,* 1 Jul. 1941 (FC, ST). WEST VIRGINIA: on cliff east of Stollings, Logan county, *A. T. Cross,* 14 Nov. 1943 (FC); seepage from a limestone quarry near Athens, Mercer county, *E. M. McNeill 216,* 9 Nov. 1944 (FC). NORTH CAROLINA: on wood flume of waterwheel, Cherokee, Swain county, *H. Silva 1713,* 13 Aug. 1949 (FC, TENN); freshwater pond on Shackleford Banks, Beaufort, *H. J. Humm 5,* 11 Sept. 1947 (FC); outlet draining a salt marsh, Hatteras, Dare county, *L. G. Williams 2a,* 12 Jul. 1949 (FC); on moist soil at 924 Urban avenue, Durham, *F. A. Wolf,* 1 Oct. 1946 (FC); on wet cliffs, Jackson county, *H. C. Bold H29,* 7 May 1939 (FC); on rocks, White Oak and Otter creeks, Macon county, *Bold H204a, H209,* 27 Jun. 1939 (FC); on overhanging rocks below Dry Falls, Highlands, Macon county, *Bold H219,* 29 Jun. 1939 (FC). SOUTH CAROLINA: Eastitoe river, Pickens county, *Bold H261, H271,* Jul.

1939 (FC); Walhalla tunnel, Waihalla, Oconee county, *Bold H126, H126b*, 12 Jun. 1939 (FC). GEORGIA: on wet rocks, Tallulah Gorge, Rabun county, *Bold H296*, 5 Jul. 1939 (FC). FLORIDA: Gainesville, *M. A. Brannon T, 369, 384, 399*, Jul.—Oct. 1946, 1947 (FC, PC); mangrove swamp south of South lake, Hollywood, Broward county, *F. Drouet 10302*, 29 Dec. 1948 (FC); in a depression in sand dunes west of Eastpoint, Franklin county, *Drouet & C. S. Nielsen 10960*, 16 Jan. 1949 (FC, T); wall of canyon, Aspalaga, Gadsden county, *J. E. Harmon 7*, 4 Nov. 1950 (FC, T); Florida Caverns state park, *Florida State University class*, 28 Jul. 1952 (FC, T); Deep Lake, Lee county, *R. Patrick 10*, 31 Dec. 1937 (D, FH, PENN, UC); region of Hendry Creek about 10 miles south of Fort Myers, *P. C. Standley 73187, 73448, 73467, 73481*, Mar. 1940 (FC); Bonita Beach, Lee county, *Standley 92791*, 10 Mar. 1946 (FC); limestone, Jackson Bluff, Leon county, *C. Jackson*, 9 Nov. 1950 (FC, T); Natural Well, northeast of Woodville, Leon county, *Nielsen, G. C. Madsen, & D. Crowson 590*, 30 Oct. 1948 (FC, T); pool, Teatable Key, *L. B. Isham 19*, 1 Jul. 1952 (FC); Everglades national park, *Isham 12*, 1952 (FC); hammock 1 mile northeast of Bear Lake trail, Everglades national park, *D Blake & R. Hauke*, 3 Feb. 1953 (FC, T); buttonwood swamps northwest and northeast of Inn, Big Pine key, Monroe county, *E. P. Killip 41684, 41931, 41944*, Jan.-Feb. 1952 (FC, US); depression in sand dunes east of Pensacola Beach, *Drouet, Nielsen, Madsen, Crowson, & A. Pates 10594*, 8 Jan. 1949 (FC, T); in the lane leading to the wharf in St. Marks river, Port Leon, Wakulla county, *Drouet & E. M. Atwood 11452*, 26 Jan. 1949 (FC, T).

ONTARIO: on wet stonework in greenhouse, University of Toronto, *W. R. Watson* (FH), in a culture jar, Oct. 1937 (FH). OHIO: Adams county: on wet cliff, Beaver pond, *L. Lillick & I. Lee 88*, 29 Apr. 1933 (FC, DA). Athens county: state hospital ponds, Athens, *A. H. Blickle*, 3 Oct. 1940 (FC, DA); near Canaanville, *Blickle*, Jul. 1941 (FC, DA); sandstone on Wilson's Farm near Athens, *W. A. Daily 130*, 7 Oct. 1939 (FC, DA). Coshocton county: rocks at Butternut falls, North Appalachian Experimental Watershed, *L. J. King 718, 1027, 1037*, 26 Aug. 1942, 30 May 1943 (EAR, FC). Cuyahoga county: sandstone cliff, Olmsted falls, Olmsted, *W. A. & F. K. Daily 564—566, 568—572*, 26 Jul. 1940 (FC, DA). Erie county: quarry pool, Kelley's Island, *L. H. Tiffany*, 18 Jun. 1931 (D); on soil, Kelley's Island, *C. E. Taft* (FC, DA). Fairfield county: on sandstone at "B. I. S.", *J. Wolfe*, 20 Oct. 1935 (FC, DA). Greene county: on moist limestone cliff, Clifton Gorge, *M. Britton*, 3 Apr. 1937 (FC, DA). Hamilton county: soil near pond, Addyston, *W. A. & F. K. Daily 167, W. A. Daily 171, 345, 346*, 19 Oct. 1939, 20 Jun. 1940 (DA, FC); wet stone wall under bridge near Addyston, *J. H. Hoskins 705*, 23 Sept. 1933 (DA, FC); greenhouse, University of Cincinnati, *L. Lillick 605*, Mar. 1934 (DA, FC); *Daily 735—739*, 9 Sept. 1940 (DA, FC); on flower pot in Lakeview greenhouse, Cincinnati, *Daily 435*, 8 Jul. 1940 (DA, FC); cement wall, Peterson's greenhouse, Cincinnati, *Daily 285, 439, 443, 444, 847*, 23 Apr., 9 Jul. 1940, 11 Jan. 1941 (DA, FC); Burnet Woods, Cincinnati, *Daily 105, 306, Daily & R. Kosanke 342*, 1 Oct. 1939, 7 & 19 Jun. 1940 (DA, FC); rocks in Eden Park conservatory, Cincinnati, *Daily 180, 183*, 24 Oct. 1939 (DA, FC); on ground and cement walls, reservoir, Eden Park, Cincinnati, *Daily 411, 413, 414*, 3 Jul. 1940 (DA, FC). Highland county: on rock, Cave canyon, Seven Caves, Paint, *W. A. & F. K. Daily 315*, 16 Jun. 1940 (DA, FC). Hocking county: sandstones, Old Man's Cave state park, *W. A. & F. K. Daily 515, 531, 542, 551*, 27 Jul. 1940 (DA, FC); sandstone overhang, Cantwell Cliffs, *W. A. & F. K. Daily 689*, 31 Aug. 1940 (DA, FC). Ottawa county: quarry near Marblehead, Catawba island, *W. A. & F. K. Daily 577, 591*, 25 Jul. 1940 (DA, FC). Vinton county: cliffs 5 miles east of McArthur, *Daily 121*, 7 Oct. 1939 (DA, FC), *Blickle & Kosanke*, 10 Jun. 1940 (DA, FC).

KENTUCKY: jar of water and flowerpot in greenhouse, Lexington, *B. B. McInteer 1043, 1136*, 19 Sept. 1939, 26 Sept. 1940 (DA, FC); on rock ledge, Natural Bridge, Powell county, *Daily 272*, 14 Apr. 1940 (DA, FC); limestone, Slade, Powell county, *Daily, A. T. Cross, & Tucker 744—746, 748—753, 755, 757, 760*, 5 Sept. 1940 (DA, FC); at and near Sky Bridge, Powell county, *Daily, Cross & Tucker 825, 827, 829, 830, 834, 836*, 6 Sept. 1940 (DA, FC); sandstone, Torrent, Wolfe county, *W. A. & F. K. Daily & R. Kosanke 477, 487*, 14 Jul. 1940 (DA, FC); sandstone, Tight Hollow near Pine Ridge, Wolfe county, *Daily, Cross, & Tucker 782*, 6 Sept. 1940 (DA, FC), *Daily 839a*, 5 Oct. 1940 (DA, FC). TENNESSEE: in microcosm, aquarium, and greenhouse, University of Tennessee, Knoxville, *H. Silva 1708, 1709, 2346b*, 11 Aug. 1949, 9 Sept. 1950 (FC, TENN); rocks around spring, Summertown, Lawrence county, *Silva 1066*, 29 Jun. 1949 (FC, TENN); Dosters Cave no. 2 and Bellamy Cave, Montgomery county, *Silva 1401a, 1868*, 15 Jul. 1949 (FC, TENN); wet dam wall, Cumberland Springs, Moore county, *Silva 1034*, 29 Jun. 1949 (FC, TENN); mud by roadside slough west of Union City, Obion county, *Silva 2248*, 14 Jun. 1950 (FC, TENN); wet rocks, Greenbrier Cove and Ramsey Cascade, Sevier county, *Silva 358, 413*, 24 Nov., 29 Dec. 1946 (FC, TENN); aquarium, State Teachers College, Johnson City, Washington county, *Silva 458*, 14 Jan. 1947 (FC. TENN); Little River Gorge, Sevier county, *A. J. Sharp 4115, 4116*, 8 Sept. 1941 (FC, TENN); on dripping rocks near Gatlinburg, *R. F. Smart*, 21 Aug. 1938 (FC, ST); greenhouse, Vanderbilt University, Nashville, *H. C. Bold 9*, 5 Oct. 1936 (FC). MICHIGAN: on *Stereodon* sp., Mackinac island, *G. E. Nichols*, 9 Aug. 1930 (D, NY, YU); on ground, Warren Dunes, Berrien county, *G. E. Scharf*, 12 Oct. 1946 (FC); Sodon lake west of Bloomfield

Hills, Oakland county, *S. A. Cain,* 9 Dec. 1947 (FC); Lake Ann Louise on highway M77, Schoolcraft county, *H. K. Phinney 9m41/16-1,* 19 Jul. 1941 (FC, PHI); roadside ditch 1.5 miles north of Mackinaw City, Cheboygan county, *W. L. Culberson 830,* 9 Jul. 1951 (FC); Mud lake, Yankee Springs, Hastings, *G. T. Velasquez 14,* 4 Aug. 1936 (D); Walnut lake, Oakland county, *T. L. Hankinson,* 6 May 1906 (as *Aphanothece stagnina* in Coll., Hold., & Setch., Phyc. Bor.-Amer. no. 1302, FC, L, TA).

INDIANA: limestone cliff, Clifty Falls state park, Jefferson county, *W. A. & F. K. Daily 669,* 4 Aug. 1940 (DA, FC); on rock below dam in Muscatatuck river at North Vernon, Jennings county, *F. K. & W. A. Daily 911,* 23 May 1942 (DA, FC); Nicoson Quarry pond southwest of Alexandria, Madison county, *F. K. & W. A. Daily 2655,* 23 Aug. 1953 (DA, FC); in jar, Butler University, Indianapolis, *C. M. Palmer 98,* 8 Dec. 1926 (DA, FC), *Daily 235,* 24 Nov. 1939 (DA, FC); Riverside fish hatchery ponds, Indianapolis, *Palmer 72,* 4 Aug. 1931 (D, DA, FC), *F. K. & W. A. Daily 1570,* 18 Aug. 1946 (DA, FC); in Lost lake, Culver, Marshall county, *F. K. & W. A. Daily 1514, 1522, 1523,* 3 July 1946 (DA, FC); stone tank in greenhouse, Bloomington, *Palmer 150,* Jan. 1934 (D); Cree lake, 4 miles north of Hendersonville, Noble county, *F. K. & W. A. Daily 1913,* 11 Sept. 1947 (DA, FC); at falls, McCormicks Creek state park, Owen county, *F. K. & W. A. Daily 1831, 1833—1835,* 10 Aug. 1947 (DA, FC); Turkey Run state park, Parke county, *Palmer 12,* 25 Jul. 1931 (DA, FC), *F. Drouet & D. Richards 2469,* 16 Jun. 1939 (FC), *F. K. & W. A. Daily & G. Plummer 1868, 1875,* 16 Aug. 1947 (DA, FC); wet limestone in quarry, St. Paul, Shelby county, *F. K. & W. A. Daily 1171,* 3 Oct. 1943 (DA, FC); Lime lake, adjacent to Lake Gage, 8.5 miles northwest of Angola, Steuben county, *F. K. & W. A. Daily 2468,* 18 Jun. 1951 (DA, FC); in aquaria and laboratory jars, Terre Haute, *B. H. Smith 4, 25, 26, 93, 96, 228, 229,* Apr. 1928—Dec. 1929 (DA, FC); laboratory cultures, Earlham College, Richmond, *M. S. Markle 2, 3, 5a* (DA, EAR, FC), 1947 (FC), *L. J. King 2326, 2327,* Aug. 1944 (EAR, FC); in School street gravel pit west of Richmond, *King 12,* 9 Jul. 1940 (EAR, FC); moist rock under Elk falls near Richmond, *King 209,* 8 Sept. 1940 (EAR, FC).

WISCONSIN: slough, Arbor Vitae lake, Vilas county, *G. W. Prescott 2W254,* 9 Jul. 1937 (D), *Prescott 3W25,* 8 Aug. 1938 (FC); Mountain lake, *C. Juday 3W1,* 31 Jul. 1938 (FC); wet soil in swamp, Two Rivers, Manitowoc county, *H. C. Benke 6295,* 15 Sept. 1944 (FC). ILLINOIS: greenhouse, Northwestern University, Evanston, *K. Damann,* 12 Mar. 1939 (FC), *H. K. Phinney,* Nov. 1943 (FC, PHI); tank in greenhouse, University of Chicago, *L. J. King,* 20 Jan. 1941 (FC), *K. C. Fan 10600,* 27 Aug. 1954 (D); tanks in greenhouse, Garfield park, Chicago, *J. A. Steyermark, C. Shoop, & Drouet 2444,* 30 Oct. 1938 (FC); culture in Chicago Natural History Museum, *F. A. Barkley 151115,* 24 Aug. 1945 (FC); on clam shells in a field west of Fox river in the north part of Geneva, Kane county, *Drouet & H. B. Louderback 5405,* 24 Jun. 1944 (FC); in a depression in the sand dunes, Dunes state park, Lake county, *Drouet, P. C. Standley, & Steyermark 3515,* 14 Jun. 1940 (FC, FH); pond, Charleston, Coles county, *E. N. Transeau,* Nov. 1910 (NY). MISSISSIPPI: wet ground beside the freshwater reservoir in the western part of Gulfport, Harrison county, *Drouet & R. L. Caylor 9943,* 12 Dec. 1948 (FC). MINNESOTA: on cement wall, Minnehaha creek, Minneapolis, *K. Damann,* 28 Aug. 1936 (FC); Itasca state park, *R. Staub,* 5 July 1954 (D), *K. C. Fan 10255, Drouet & Fan 11878, Drouet, T. Morley & Fan 11896, Drouet 12045, 12074,* Jun. 1954 (D, MIN); in tank, University of Minnesota, Minneapolis, *J. E. Tilden,* Mar. 1909 (as *Gloeocapsa arenaria* in Tild., Amer. Alg. no. 650, FC). MISSOURI: in Chinn's lake, 17 miles west of Columbia, in Howard county, *J. R. Hurt C1,* 1939 (FC); on mud in small pool by stream, Thayer, Oregon county, *N. L. Gardner 1410,* Jul. 1904 (FC, UC). LOUISIANA: in ditch, Louisiana State University campus, Baton Rouge, *G. W. Prescott La13,* 18 Jun. 1938 (FC); barren ground behind Broussard's Beach on the Gulf of Mexico east of Cameron, *Drouet 8823, 8881,* Nov. 1948 (FC); in a drying pool on a road-embankment in the marsh between the arms of the Calcasieu river west of Cameron, *Drouet 8866,* 3 Nov. 1948 (FC); Jefferson Island, Iberia parish, *R. P. Ehrhardt & Drouet 28,* 12 Nov. 1948 (FC); on a shell ridge between Leeville and Chenier Caminada, Lafourche parish, *Drouet & P. Viosca Jr. 9447,* 26 Nov. 1948 (FC); pool at "Singing Waters", Holden, Livingston parish, *L. H. Flint,* 4 Oct. 1946 (FC); on an embankment, U. S. highway no. 90 east of the road to Chalmette, Orleans parish, *Drouet & Viosca 9350,* 24 Nov. 1948 (FC); in a pond, Little Lake Club, Plaquemines parish, *J. N. Gowanloch,* Sept. 1944 (FC); in a ditch entering Bayou du Large just north of Theriot, Terrebonne parish, *Drouet 9287,* 22 Nov. 1948 (FC).

NEBRASKA: low ground, Beeken lake south of Arthur, Arthur county, *W. Kiener 21030,* 5 Jul. 1946 (FC, KI); culture from floodplain pond of Platte river south of Kearney, Buffalo county, *Kiener 16496d,* 6 Nov. 1945 (FC, KI); sandy soil, Platte river floodplain north of David City, Butler county, *Kiener 21575,* 8 Nov. 1946 (FC, KI); soil in grassland west of Champion, Chase county, *Kiener 10620,* 31 Jul. 1941 (FC, KI); in Dewey lake, Cherry county, *E. Palmatier & T. R. Porter 13774a,* 28 Jun. 1936 (FC, KI); soil at edge of gravel pit, Fremont, Dodge county, *Kiener 23919,* 1 Jul. 1948 (FC, KI); wet sand, sand pit lakes, Fremont, *Kiener 13930a, 14648, 14650,* 23 Apr., 23 Jul. 1943 (FC, KI), and laboratory culture, *Kiener 21475b,* 15 Feb. 1947 (FC, KI); Twin lakes, northwest of Benkelman, Dundy county, *Kiener 21224*, 30 Jul. 1946 (FC, KI); grassland 5 miles north of Haigler, Dundy county, *Kiener 10498*, 25 Jul. 1941 (FC, KI); wet ground near Grand Island, *Kiener 15123*, 28 Aug. 1943 (FC, KI); wet meadow, Palisade, Hayes county, *Kiener 10828*, 5 Aug. 1941 (FC, KI); bank of Elkhorn river west of O'Neill, Holt county, *Kiener 11210*, 21 Aug. 1941 (FC, KI); culture from spring, Lonergan creek, Lemoyne, Keith county, *Kiener 23075a*, 20 Aug. 1948 (FC, KI); low ground, Lincoln, *Kiener 9734*, 11 Dec. 1940 (FC, KI); on limestone, University greenhouse, Lincoln, *Kiener 13627*, 16 Jan. 1943 (FC, KI); swampy soil 10 miles north of North Platte, *Kiener 17525*, 1 Sept. 1944 (FC, KI); cultures from creek south of Gretna, Sarpy county, *Kiener 13686, 13686c*, 16 Jul., 11 Aug. 1944 (FC, KI). TEXAS: Clear lake, Harris county, *H. K. Phinney 4T41/3b*, 3 Sept. 1941 (FC, PHI); on sand at Blind bayou, Galveston island, *Phinney 2T41/2*, 30 Aug. 1941 (FC, PHI); moist sinkhole, Granite mountain, 2 miles north of Marble Falls, Burnet county, *F. A. Barkley & H. V. Copeland 8*, 25 Jul. 1946 (FC, TEX); permanent pool, Austin, *Barkley 44alg23*, 25 Jul. 1944 (FC, TEX). WYOMING: in swamp near Towner lake, Albany county, *W. G. Solheim 24*, 13 Jul. 1940 (FC, KI). COLORADO: in Lily lake near Estes Park, Larimer county, *W. Kiener 13175*, 22 Sept. 1942 (FC, KI); Hoosier Pass, Park county, *Kiener 6705a*, 1 Sept. 1938 (FC, KI). UTAH: dripping cliffs, Mount Timpanogos, Utah county, *A. O. Garrett 46*, 25 Jul. 1925 (FC, UC). ARIZONA: on rocks in spray from Havasu falls, Coconino county, *E. U. Clover*, 28 Sept. 1940 (FC, MICH). ALASKA: Pt. Barrow (FH); lakes, Mileposts 70 and 127 (from Anchorage), Glenallen highway, *D. K. Hilliard 34, 57*, Sept. 1953 (FC); Mileposts 220 and 248 (from Anchorage), Richardson highway, *Hilliard 24, 26*, 6 Sept. 1953 (FC). WASHINGTON: Friday Harbor, *J. H. Hoskins*, Summer 1921 (DA, FC). OREGON: in aquarium, Lewis and Clark College, Portland, *S. E. Flint 3*, 21 Nov. 1947 (FC). CALIFORNIA: *W. J. V. Osterhout 1777*, 7 Jul. 1906 (FC, UC). Alameda county: on dripping boards, Lake Chabot, San Leandro, *Osterhout & N. L. Gardner*, 14 Jun. 1902 (as *Gloeothece cystifera* in Coll., Hold., & Setch., Phyc. Bor.-Amer. no. 1204, FC, L, TA); spring on hillside, Arroyo Mocho, Mt. Hamilton range, *Gardner & A. Carter 8006*, 22 Nov. 1936 (FC, UC); cultures in the botanical laboratory, University of California, Berkeley, *Gardner 1462* (UC), *2251* (UC), *6572, 6726, 6953, 7143, 7197, 7390, 7808, 1905—1933* (FC, UC); on soil and in rice culture, Agricultural greenhouse, University of California, Berkeley, *Gardner 7194, 7209*, Mar. 1933 (FC, UC); on flowerpots in the greenhouse, Department of Botany, University of California, Berkeley, *Gardner 7171, 7566a*, 15 Jan., 3 Sept. 1933 (FC, UC); in a flowerpot and cultivated in a hothouse in the courtyard of the Life Sciences Building, Berkeley, *Gardner 6951, 7023*, Jan. 1932 (FC, UC). Contra Costa county: culture from shaded ditch by mud-bath house, Byron Hot Springs, *Gardner 7715*, 11 Apr. 1933 (FC, UC). Humboldt county: culture from Eureka collected by H. E. Parks, *Gardner 6728a, 7152*, 17 Jan. 1932, 8 Jan. 1933 (FC, UC). Inyo county: Nevares spring near Cow creek, Death Valley, *G. J. Hollenberg 2279*, 7 Apr. 1937 (D). Los Angeles county: Pasadena, *A. J. McClatchie 1090* (UC); Puddingstone creek and dam, La Verne, *Hollenberg 1614a*, 19 Dec. 1933 (FC, UC), *M. Kneeland 1664*, 9 Dec. 1936 (FC, UC); culture, University of Southern California, Los Angeles, *G. R. Johnstone*, 27 Jan. 1940 (FC); on wet concrete, dam in San Dimas canyon near La Verne, *Hollenberg 2077A*, 20 Apr. 1934 (D, NY). Marin county: *V. Duran 6670, 6946*, Jan. 1931, 1932 (FC, UC); culture from Almonte and Alto collected by H. E. Parks, *Gardner 6566, 6570*, 13 Jan. 1931 (FC, UC); on moist soil, Lake Lagunitas, *Osterhout*, Jul. 1906 (as *Aphanothece microscopica* in Coll., Hold., & Setch., Phyc. Bor.-Amer. no. 1702c, FC, L, TA), *W. A. Setchell & Osterhout*, 23 Sept. 1908 (UC), *Setchell 6429a*, 1908 (FC, UC); on dripping rocks, Mill Valley works, *Gardner*, 20 Jul. 1906 (as *Gloeothece rupestris* in Coll., Hold., & Setch., Phyc. Bor.-Amer. no. 1703b, FC, MICH, TA), *Gardner 1791* (TA, UC); Mill Valley, *Gardner 4185*, Mar. 1918 (UC), *Osterhout*, Nov. 1906 (UC); on dripping rocks, Mt. Tamalpais, *Osterhout & H. D. Densmore*, 4 Jul. 1906 (as *G. rupestris* in Coll., Hold., & Setch., Phyc. Bor.-Amer. no. 1703a, FC, L, MICH, TA), *Gardner 1775*, 4 Jul. 1906 (FC, UC), *Gardner 3113*, May 1916 (UC). Riverside county: dripping rocks, Palm Canyon, *Hollenberg 1632*, 29 Mar. 1935 (FC, UC). San Bernardino county: Arrowhead Hot Springs, *Setchell 1539, 1545*, 19 Dec. 1896 (FC, UC), *Gardner 17, 18*, 27 May 1930 (FC, UC). San Diego county: along a stream west of Dulzura, *F. Drouet & J. F. Macbride 4827*, 18 Oct. 1941 (FC). San Francisco: Fort Point, *Gardner 1778* (UC), on rocks, *Gardner* (as *Gloeocapsa polydermatica* in Coll., Hold., & Setch., Phyc. Bor.-Amer. no. 1751b, FC, L, TA); on moist sand, Mountain lake, *Gardner 1462*, 17 May 1905 (UC; as *Aphanothece microscopica* in Coll., Hold., & Setch., Phyc. Bor.-Amer. no. 1702a, FC, L, TA); in the conservatory, Golden Gate park, *Gardner 7263, 7592*, 3 May, 19 Sept. 1933 (FC, UC); in Plath's greenhouse, *Duran 6957*, 8 Jan. 1932 (FC, UC), *Gardner 7246*, 20 Apr. 1933 (FC, UC); floating in Lake Merced, *Gardner 1583a*, 13 Nov. 1905 (UC; as *A. microscopica* in Coll., Hold., & Setch., Phyc. Bor.-Amer. no. 1702b, FC, L, TA); on a bank of soft sandstone, Lake Merced, *Gardner 844*, Nov. 1902 (UC; as *Gloeothece rupestris* in Coll., Hold., & Setch., Phyc. Bor.-Amer. no. 1703c, FC, L, MICH, TA); culture from Lake Merced, *Gardner 7427*, 24 Jul. 1933 (FC, UC). Santa Clara county: culture from Milpitas collected by H. E. Parks, *Gardner 6659*, 1931 (UC). Tulare county: on damp soil, Exeter, *H. E. Parks*, Dec. 1930 (UC); in shallow water of Tule river, Porterville, *Drouet & M. J. Groesbeck 4448, 4450*, 4 Oct. 1941 (FC).

25

PUERTO RICO: on rocks, Arroyo de los Corchos, Adjuntas to Jayuya, N. *Wille 1697, 1703,*
Mar. 1915 (NY); on limestone, Arecibo to Utuado, *Wille 1469b, 1470,* Mar. 1915 (NY); Cabo
Rojo, *A. A. Heller 4435,* Jan. 1890 (FC); rocks 7 km. east of Coamo, *Wille 1873,* Mar. 1915
(NY); Coamo Springs, *Wille 372, II,* Jan. 1915 (NY), *1861a,* Mar. 1915 (FC, UC); on stones
near Maricao, *Wille 1129,* Feb. 1915 (NY); on a water pipe near a stream, Maricao, *Wille*
1147 (Type of *Aphanothece conferta* var. *brevis* Gardn., NY [Fig. 160]; isotype, D), *1147a,* Feb.
1915 (NY, D); on brick wall, Experiment Station, Mayaguez, *Wille 967,* Feb. 1915 (NY);
Hacienda Catalina near Palmer, *Wille 754a,* Feb. 1915 (Type of *Aphanocapsa Richteriana* var. *major*
Gardn., NY); on rocks near Penuelas, *Wille 1848a,* Mar. 1915 (Type of *Aphanothece microscopica*
var. *granulosa* Gardn., NY [Fig. 151]); on wall of Fort Cristobal, San Juan, *Wille 2003c* (NY),
A. Lutz 2018 (FC, UC), Mar. 1915; on wall of a cemetery, San Juan, *Wille 131a,* Dec. 1914 (Type
of *Gloeocapsa quaternata* var. *major* Gardn., NY [Fig. 146]); on a wall by Hotel Nava, Santurce,
Wille 54a, Dec. 1914 (Type of *Gloeothece interspersa* Gardn., NY [Fig. 155]); on tree trunk on
hills near Santurce, *Heller 4333,* Jan. 1890 (FC); on limestone at the hacienda, Laguna Tortuguero,
Wille 866a, Feb. 1915 (NY); on limestone, Utuado to Adjuntas, *Wille 1640a,* Mar. 1915 (Type of
Anacystis cylindracea Gardn., NY [Fig. 156]); on rocks between Utuado and Adjuntas, *Wille 1635a*
(D, NY), *1637a* (FH, NY), *1650a,b* (NY), Mar. 1915; on rocks about 10 km. north of
Utuado, *Wille 1525* (FC, UC), *1528a, 1566, 1568* (NY), Mar. 1915. JAMAICA: Middle Morant
cay, *V. J. Chapman 11, 14,* 1939 (FC); on concrete wall, Arntully, *C. R. Orcutt,* 19 May
1928 (FC, US). MEXICO: Eisenbahnsümpfe bei Vera Cruz, *F. Müller 167,* Feb. 1853 (Type of
Palmella heterococca Kütz., L [Fig. 153]); paredes húmedas, Cueva de Sambula, Motul, Yucatan, *Exp.*
Rec. Hid. 97-143, 18 Mar. 1947 (FC); island in Rio Santa Catarina near Santa Catarina, Nuevo Leon,
F. A. Barkley 146-09, 16 Aug. 1944 (FC, TEX); in crevices of rocks at Aurora, south of Baviácora,
Sonora, *F. Drouet, D. Richards, & W. A. Lockhart 2949,* 16 Nov. 1939 (FC); in seepage near
highway between Morélia and Ciudad México, *F. B. Plummer,* 5 Sept. 1943 (FC, TEX); on sides
of cave and on bank of Rio Tuxpan, San José de Purúa, Michoacan, *R. Patrick 431, 443,* 4 Aug.
1947 (FC, PH); Lake Patzcuaro, Michoacan, *Patrick 510, 517,* 10 Aug. 1947 (FC, PH).
    GUATEMALA: on limestone bluff, Cerro de Agua Tortuga near Cubilguitz, dept. Alta Verapaz,
J. A. Steyermark 44588, 4 Mar. 1942 (FC); damp bank north of Chiantla, dept. Huehuetenango,
P. C. Standley 82585, 6 Jan. 1941 (FC); on wet banks along stream, Jalapa, *Steyermark 32149,* 28
Nov. 1939 (FC); in cold trickle by road between San Francisco El Alto and Momostenango, dept.
Totonicapán, *Standley 83987,* 19 Jan. 1941 (FC); on mud near the electric plant of Rio Hondo,
base of Sierra de las Minas, dept. Zacapa, *Standley 74057,* 11 Oct. 1940 (FC). HONDURAS: in
shallow running water near Yuscarán, dept. El Paraíso, *Standley 1218,* 11 Dec. 1946 (FC); on wet
masonry and banks, El Zamorano, dept. Morazán, *Standley 4413, 4416, 4422, 4423a, 4426,* Feb.
Mar. 1947 (FC). NICARAGUA: on wet bark, Chichigalpa, dept. Chinandega, *Standley 11461,*
Jul. 1947 (FC); on tree trunk, La Libertad, dept. Chontales, *Standley 8845,* May—June 1947 (FC);
on tree trunks, Casa Colorada near El Crucero, summit of Sierra de Managua, *Standley 8191, 8213,*
May 1947 (FC). PANAMA: in north arm of Gigante bay, Canal Zone, *C. W. Dodge,* 9 Aug. 1925
(D, MO); in cave along San Carlos river, *G. W. Prescott & R. L. Caylor CZ125,* 10 Sept. 1939
(FC). BRAZIL: Pará: on moist wall of mill-race, Santa Izabel, east of Belém, *F. Drouet 1534,* 10 Jul.
1935 (D); on rocks in seepage from artificial spring, Bosque Rodriguez Alves, Belém, *Drouet 1530,*
1 Jul. 1935 (D). Ceará: in pools by road 3 km. south of Porangaba, Fortaleza, *Drouet 1378,*
23 Aug. 1935 (D, FH, MICH, NY); in a temporary reservoir in Rio Maranguapinho, Barro
Vermelho, Fortaleza, *Drouet 1499,* 22 Nov. 1935 (D, FH, NY). ECUADOR: in rupibus humidis
prope Baños (Tungurahua), *G. Lagerheim,* Dec. 1891 (S; as *Gloeothece tepidariorum* in Wittr.,
Nordst., & Lagerh., Alg. Exs. No. 1545, L). PERU: agua alcalina, Chan-Chan y Buenos Aires,
Trujillo, *N. Ibañez H. "P", "W",* 19 & 27 Oct. 1952 (FC). ARGENTINA: Cap. Federal, *S. A.*
Guarrera 2201, 20 Sept. 1938 (FC); tapizando una pared. húmeda, Buenos Aires, *Guarrera*
3343, 25 Aug. 1939 (FC); culture in Instituto Bacteriológico, Dept. Nacional de Higiene, Buenos
Aires, *D. Rabinovich CP20,* Dec. 1943 (FC). CHILE: in pool, Punta Arenas, *R. Thaxter,* 1905—
1906 (D, FH).
    HAWAIIAN ISLANDS: rocks on cliff, Nuuanu point, Honolulu, *G. T. Shigeura,* 4 Mar. 1939
(FC, PH); in a rice field, Aiea, Oahu, *J. E. Tilden,* 2 June 1900 (as *Aphanothece prasina* in Tild.,
Amer. Alg. no. 498, FC; as *Gloeothece fuscolutea* in Tild., Amer. Alg. no. 500, FC); in brackish
stagnant water, Meheiwi, Makao, Koolauloa, Oahu, *Tilden,* 20 Jun. 1900 (as *Aphanothece prasina*
in Tild., Amer. Alg. no. 498B, FC); on wet cliffs south of Laupahoehoe, Hawaii, *Tilden,* 10 Jul.
1900 (as *Gloeocapsa quaternata* in Tild., Amer. Alg. no. 499, FC). MARSHALL ISLANDS: on
soil of open coconut grove, Taka islet, Taka atoll, *F. R. Fosberg 33733,* 5 Dec. 1951 (FC, US); on flat
on south end of Bikar islet, Bikar atoll, *Fosberg 34576,* 6 Aug. 1952 (FC, US); ocean side of
Enen-Edrik island, Arno atoll, *L. Horwitz 9024,* 7 Jul. 1951 (FC). SOCIETY ISLANDS: Tahiti,
J. E. Tilden, Oct. 1909 (as *Aphanothece microspora* in Tild., So. Pacific Alg. no. 3, L, MIN, UC);
on dripping rocks, Tahara mountain, Tahiti, *W. A. Setchell & H. E. Parks 5067a,* 24 May 1922
(UC); moist cliffs near mouth of Papenu river, Tahiti, *Setchell & Parks 5351,* 5 Jun. 1922 (UC).

26

SAMOA: on shaded tree-trunks, roadside beyond Aua, Tutuila island, *Setchell 1178*, 4 Jul. 1920 (UC); with *Hassallia Rechingeri* Wille, Upolu, *K. H. Rechinger 5137*, 1915 (Type of *Gloeothece samoensis* f. *major* Wille in the slide collection of N. Wille, O). GILBERT ISLANDS: surface of a Babai pit, center of North island, Onotoa, *E. T. Moul 8324*, 8 Aug. 1951 (FC); bottom of abandoned Babai pit, Tabuarorae island, Onotoa, *Moul 8242*, 26 Jul. 1951 (FC). NEW ZEALAND: Rangitoto island (Waitemata), *V. J. Chapman*, Jul. 1946 (FC); Orongo Bay (Bay of Islands), *Chapman*, Jan. 1948 (FC); Brown's island, Auckland harbor, *Chapman*, 18 Mar. 1950 (FC); in cave, fresh water, Kororareka Point (Bay of Islands), *I. B. & V. R. Warnock*, 16 Jan. 1935 (as *Coccochloris stagnina* in Tild., So. Pacific Pl., 2nd Ser., no. 358, FC).

PHILIPPINES: Bataan: on damp stone wall at Balanga, *G. T. Velasquez 2404*, 12 Jul. 1950 (FC, PUH); soil by the river, Panilao, Pilar, *Velasquez 2412*, 13 Jul. 1950 (FC, PUH). Batangas: floating in rice fields at Lipa, *Velasquez 595a*, 12 Jan. 1941 (FC). Iloilo: soil in ruined church, Dueñas, *J. D. Soriano 1082*, 20 Jul. 1952 (FC, PUH); in Delicana's fishpond, Dumangas, *Soriano 1595, 1596*, 25 Jan. 1953 (FC, PUH); on adobe wall of churchyard, Guimbal, *Soriano 1570*, 10 Jan. 1953 (FC, PUH); on aquarium, Iloilo College, Iloilo City, *Soriano 1543*, 10 Dec. 1952 (FC, PUH); in Griños fishponds, Leganes, *Soriano 1591*, 11 Jan. 1953 (FC, PUH); on damp walk at Catholic Church, Janiuay, *Soriano 1034*, 23 Jun. 1952 (FC, PUH); on interior walls of ruined Catholic Church, Leon, *Soriano 1820*, 15 Feb. 1953 (FC, PUH); on sides of dam, Oganon river irrigation system, San Miguel, *Soriano 1050*, 29 Jun. 1952 (FC, PUH); on interior walls of ruined church, Santa Barbara, *Soriano 1038*, 23 Jun. 1952 (FC, PUH). Leyte: in pond in coral pit near airstrip, Tacloban, *M. E. Britton 153*, 10 Oct. 1945 (FC). Manila: *Soriano 890*, 5 May 1951 (FC, PUH); near Manila, *R. C. McGregor 86*, 23 Oct. 1904 (FC, UC); on sandy soil at Dewey boulevard, *Velasquez 90*, 12 Nov. 1939 (FC); on stone wall along Pennsylvania street, *Velasquez 108a*, 18 Nov. 1939 (FC); on walls and in an aquarium and a tank, *Velasquez 468, 574, 591*, 18 Mar., 26 Nov. 1940, 11 Jan. 1941 (FC); in concrete tank at Pennsylvania and California streets, Camina, *Soriano 415*, 16 Oct. 1949 (FC, PUH); in standing water in front of old Sto. Domingo church, Intramuros, *Soriano 409*, 11 Oct. 1949 (FC, PUH); on soil of garden, San Andres elementary school, Malate, *Soriano 422*, 17 Oct. 1949 (FC, PUH); on high wall of Malate church, *Soriano 529*, 3 Dec. 1949 (FC, PUH); on bark of acacia tree, 281 Cristobal street, Paco, *Soriano 644*, 26 Mar. 1950 (FC, PUH); in a rice field at Pandacan, *Velasquez 572*, 10 Aug. 1940 (FC); on adobe walls and soil and in standing water, Pandacan, *Soriano 348, 350—352*, 17 Aug. 1949 (FC, PUH); on wall under roof garden, Quiapo, *Soriano 482*, 22 Dec. 1949 (FC, PUH); on soil and adobe walls, Sampaloc, *Soriano 70, 391, 393—695, 680, 751, 758*, Jun. 1948—Nov. 1950 (FC, PUH); moist soil in school grounds, San Nicolas, *Soriano 437*, 21 Oct. 1949 (FC, PUH); in standing water, 2523 Herran street, Santa Ana, *Soriano 445, 446*, 24 Oct. 1940 (FC, PUH); on adobe fence, La Concordia College, Santa Ana, *Soriano 551*, 11 Dec. 1949 (FC, PUH); on inundated lot, Santa Cruz, *Soriano 429*, 19 Oct. 1949 (FC, PUH); in canal, Singalong, *Velasquez 260*, 3 Jan. 1940 (FC); in standing water and on an adobe fence, Singalong, *Soriano 367, 588*, 12 Sept., 26 Dec. 1949 (FC, PUH); on soil and walls, Tondo, *Soriano 330, 517*, 3 Aug., 27 Nov. 1949 (FC, PUH). Laguna: hot spring, Talakay, College, *Velasquez 2464, 2465*, 8 Aug. 1948 (FC, PUH). Mindoro: artesian well and pond, Tabinay Maliit, Puerto Galera, *Velasquez 728, 729, 926*, Apr. 1941 (FC); on a drying pond of brackish water at Balatero Malaki, Puerto Galera, *Velasquez 944*, 22 Apr. 1941 (FC); near the waterfall at Bisayaan, Puerto Galera, *Velasquez 1954*, 25 May 1949 (FC, PUH); Big Balatero, Puerto Galera, *Velasquez 7*, 25 Apr. 1953 (FC, PUH). Palawan: Cuyo, *Velasquez 2848*, 9 Jun. 1951 (FC, PUH); Araceli, *Velasquez 2933*, 15 Jun. 1951 (FC, PUH). Rizal: on drainage canal, University of the Philippines, Diliman, *Soriano 236-3*, 20 Jan. 1949 (FC, PUH); rice paddy near Km. 17, Marikina valley, *Soriano 935*, 18 May 1952 (FC, PUH); rice paddy, Barrio Mayanbayanan, Marikina, *Soriano 921*, 16 May 1952 (FC, PUH); rice field at Barrio Mangahan, Pasig, *Soriano 937*, 18 May 1952 (FC, PUH); rice field, Broadway street, Quezon City, *Velasquez 178*, 13 Dec. 1939 (FC); moist ground, University of the Philippines campus, Diliman, Quezon City, *Velasquez 2425, 2454*, 22 & 28 Jul. 1950 (FC, PUH); in creek behind Christ the King Seminary, Kamuning, Quezon City, *Velasquez 2538, 2540*, 19 Aug. 1950 (FC, PUH); rice paddies along road to Taytay, Pasig, *Soriano 930*, 18 May 1952 (FC, PUH). Sulu Archipelago: on adobe, Sanga-Sanga, Tawi-Tawi, *Velasquez 3218*, 12 May 1952 (FC, PUH).

CHINA: Fukien: Fulung hot springs, *H. H. Chung A439* (Type of *Aphanothece gelatinosa* Gardn., FH; isotype, D), *A454* (MICH, NY), Sept. 1926; Huangchun near Foochow, *Chung A432*, Sept. 1926 (D, FH). Kiangsu: on rock and in ditch, Nanking, *C. C. Wang 317, 373*, 16 & 20 Apr. 1930 (FC, UC). Szechwan: on wet rock, Lung-chi bridges, Nanzechwan, *H. Chu 716*, 7 Aug. 1945 (FC); on wet rock, Thousand Gods Cliff, Nanzechwan, *H. Chu 774*, 9 Aug. 1945 (FC); under dripping rock, Ko-Fu-Tung, Mt. King-Fu, Nan-chwan, *Chu 849*, 15 Apr. 1945 (FC); on rock in rapid stream, Shan-Shan-Kow, Omei, *Chu 1329*, 21 Aug. 1942 (FC); on rock under dripping cliff, Hu-Lung-Kiang, Omei, *Chu 1336*, 22 Aug. 1942 (FC); on dripping cliff, Elephant pond, Omei, *Chu 1356*, 23 Aug. 1942 (FC); on wet rock, Lung-Men-Tung, Omei, *Chu 1389*, 25 Aug. 1942 (FC). NEW GUINEA: open creek, *H. J. Rogers NG23*, 2 Oct. 1944

27

(FC). INDONESIA: Sumatra: an warmem Stein 10 cm. ober Wasser, am Boden des unteren Beckens, Bukit Kili, *Ruttner*, 1928-29 (Type of *Aphanothece bullosa* var. *minor* Geitl., slide no. *SKW1g* in the collection of L. Geitler); Bukit Kili Ketgil, 5 cm. ober Wasser an Steinen, *Thienemann*, 7 Mar. 1939 (in slide collection of L. Geitler); Kalkstalaktiten, Grosser Wasserfall, Kalktuffe von Panjingahan am Westufer des Sees, *Ruttner*, 6 Mar. 1929 (Type of *Gloeothece rupestris* var. *major* Geitl., slide no. SkB3aδ in the collection of L. Geitler). Flores: Bari, *A. Weber-van Bosse 1192*, Nov. 1888 (D, L). Celebes: Tempe, *Weber-van Bosse 856*, Oct. 1888 (L). AUSTRALIA: on dripping rocks, Paradise cave, Noosa, Queensland, *A. B. Cribb 99*, 24 May 1949 (FC); Point Lonsdale, Victoria, *A. Nash & T. T. Earle*, Nov. 1934 (as *Aphanothece stagnina* in Tild., So. Pac. Pl., 2nd Ser., no. 169, FC); Muston, American River inlet, Kangaroo island, South Australia, *H. B. S. Womersley A6718b (Kl2305b)*, 10 Jan. 1948 (FC). BURMA: Pegu, Northern Yomah, Mayzelee Choung?, *S. Kurz {1854*, 20 Jan. 1868] (Type of *Gloeocapsa luteo-fusca* Mart. in herb. Agardh, LD; isotypes, K, L); Arracan, Kolodyne river, *S. Kurz 1960* (L); Royal lake, Rangoon, *L. P. Khanna 572*, 2 Dec. 1935 (FC); Maymyo, Northern Shan States, *Khanna 639*, 12 May 1936 (FC); Monywa lake, Monywa, *Khanna 857—859, 864, 867, 880, 883, 889—891*, 15 Mar. 1938 (FC). INDIA: on paths in the Royal Botanic Garden, Calcutta, *S. Kurz 7882*, Jul. 1868 (L).

3. Coccochloris elabens Drouet & Daily, Lloydia 11: 77. 1948. *Micraloa elabens* Brébisson in Meneghini, Mem. R. Accad. Torino, 2nd ser., 5 (Sci. Fis. & Mat.): 104. 1843. *Microcystis elabens* Kützing, Tab. Phyc. 1: 6. 1846. *Diplocystis elabens* Trevisan, Sagg. Monogr. Alg. Coccot., p. 40. 1848. *Polycystis elabens* Kützing, Sp. Algar., p. 210. 1849. *Microhaloa elabens* Brébisson ex Ainé, Pl. Crypt.-Cellul. du Dept. Saone-et-Loire, p. 259. 1863. *Anacystis elabens* Setch. & Gardn. in Gardn., Univ. Calif. Publ. Bot. 6: 455. 1918. *Aphanothece elabens* Elenkin, Monogr. Algar. Cyanophyc., Pars Spec. 1: 146. 1938. —Type from Falaise, France (L). FIG. 168.

Palmella Castagnei Kützing, Tab. Phyc. 1: 9. 1846. *Oncobyrsa Castagnei* Brébisson *pro synon.* in Kützing, loc. cit. 1846. *Cagniardia Castagnei* Trevisan, Sagg. Monogr. Alg. Coccot., p. 51. 1848. *Aphanocapsa Castagnei* Rabenhorst, Fl. Eur. Algar. 2: 50. 1865. *Aphanothece Castagnei* Rabenhorst, Fl. Eur. Algar. 2: 64. 1865. *Coccochloris Castagnei* Drouet & Daily, Lloydia 11: 77. 1948. —Type from Aix, France (L).

Palmella pulchra Kützing, Sp. Algar., p. 214. 1849. *Aphanocapsa pulchra* Rabenhorst, Fl. Eur. Algar. 2: 49. 1865. *Microcystis Grevillei* f. *pulchra* Elenkin, Monogr. Algar. Cyanophyc., Pars Spec. 1: 129. 1938. —Type from Hanau, Germany (L).

Protococcus elongatus Nägeli in Kützing, Sp. Algar., p. 197. 1849. *Synechococcus elongatus* Nägeli, Gatt. Einz. Alg., p. 56. 1849. —Type from Zürich, Switzerland (ZT).

Synechococcus violaceus Grunow in Rabenhorst, Fl. Eur. Algar. 3: 419. 1868. —Type from Wagrain, Austria (W).

Polycystis Packardii Farlow in Packard, Amer. Nat. 13: 702. 1879. *Microcystis Packardii* Tilden, Minn. Alg. 1: 36. 1910. *Aphanothece Packardii* Setchell in Daniels, Amer. Nat. 51: 502. 1917. —Type from Great Salt lake, Utah (FH). FIG. 164.

Aphanothece utahensis Tilden, Amer. Alg. 3: 297. 1898. *A. salina* f. *utahensis* Elenkin, Monogr. Algar. Cyanophyc., Pars Spec. 1: 151. 1938 . —Type from Great Salt lake, Utah (MIN). FIG. 169.

Synechococcus parvus Migula, Crypt. Germ., Austr. & Helvet. Exs. 26-27 (Alg.): 123. 1906. —Isotypes from Lainz near Vienna, Austria (MICH, NY, TA).

Synechococcus curtus Setchell in Collins, Holden & Setchell, Phyc. Bor.-Amer. 28: 1351. 1907. —Type from Oakland, California (UC). FIG. 167.

Microcystis elabens var. *minor* Nygaard, Vidensk. Medd. Dansk Naturh. Foren. Kjøbenh. 82: 204. 1926. —Type from Kai islands, Indonesia (in the collection of G. Nygaard).

Aphanothece microscopica var. *congesta* W. R. Taylor, Proc. Acad. Nat. Sci. Phila. 80: 85. 1928. —Type from Kootenay county, British Columbia (TA).

Aphanothece conglomerata Rich, Trans. Roy. Soc. S. Africa 20: 185. 1932. —Type from Cape Town, South Africa (D). FIG. 166.

Aphanothece Lebrunii Duvigneaud & Symoens, Inst. Parcs Nat. Congo Belge, Explor. Parc Nat. Albert, Mission J. Lebrun (1937—1938) 10: 9. 1948. —Type from Parc National Albert, Belgian Congo (BR). FIG. 165.

Aphanothece minor Frémy ex J. de Toni, Diagn. Alg. Nov. I. Myxophyc. 10: 959. 1955. — Type from Dutch West Indies (L).

Original specimens have not been available to us for the following names; their original descriptions are here designated as the Types until the specimens can be found:

Aphanothece curvata Lagerheim, Öfvers. K. Sv. Vet.-Akad. Förh. 40(2): 44. 1888.

28

Synechococcus Cedrorum Sauvageau, Bull. Soc. Bot. France, ser. 2, 14: cxv. 1892.
Aphanothece Stuhlmannii Hieronymus in Engler, Pflanzenwelt Ostafr. C: 8. 1895. *A. Kuhlmannii* Hieronymus ex Geitler, Rabenh. Krypt.-Fl. 14: 171. 1932.
Bacularia coerulescens Borzi, Nuova Notarisia 1905: 21. 1905. *Aphanothece coerulescens* Geitler, Rabenh. Krypt.-Fl. 14: 173. 1932. *A. Geitleri* J. de Toni, Noter. Nomencl. Algol. 3. 1936.
Aphanothece salina Elenkin & Danilov, Bull. Jard. Imp. Bot. Pierre le Grand 15: 180. 1915.
Gloeothece linearis var. *composita* G. M. Smith, Bull. Wisconsin Geol. & Nat. Hist. Surv. 57: 46. 1920. *Rhabdoderma lineare* f. *compositum* Hollerbach in Elenkin, Monogr. Algar. Cyanophyc., Pars Spec. 1: 44. 1938.
Microcystis elabentoides Zalessky, Rév. Gén. de Bot. 38: 33. 1926.
Microcystis aphanothecoides Zalessky, Rév. Gén. de Bot. 38: 33. 1926.
Microcystis angulata Zalessky, Rév. Gén. de Bot. 38: 34. 1926.
Microcystis floccosa Zalessky, Rév. Gén. de Bot. 38: 34. 1926.
Microcystis globosa Zalessky, Rév. Gén. de Bot. 38: 34. 1926.
Aphanothece glebulenta Zalessky, Rév. Gén. de Bot. 38: 34. 1926.
Aphanothece cancellata Zalessky, Rév. Gén. de Bot. 38: 35. 1926.
Chroococcus kerguelensis Wille, Deutsche Südpolar-Exped. 1901—03, 8: 414. 1928.
Synechococcus elongatus f. *minor* Wille, Deutsche Südpolar-Exped. 1901—03, 8: 417. 1928.
Dzensia salina Woronichin, Bull. Jard. Bot. Princip. U. R. S. S. 28: 155. 1929.
Aphanothece nostocopsis Skuja, Acta Horti Bot. Univ. Latv. 7: 45. 1932.
Aphanothece halophytica Frémy in Hof & Frémy, Rec. Trav. Bot. Néerl. 30: 152. 1933.
Aphanocapsa benaresensis Bharadwaja, Proc. Indian Acad. Sci. 2: 96. 1935.
Aphanothece karukerae Lami, Rév. Algol. 11: 221. 1938.
Synechococcus subsalsus Skuja, Acta Horti Bot. Univ. Latv. 11—12: 44. 1939.
Synechococcus Koidzumii Yoneda, Acta Phytotax. & Geobot. 13: 97. 1943.

Plantae aeruginosae, olivaceae, luteolae, violaceae, vel roseae, microscopicae vel macroscopicae, 1—pluri—multi-cellulares, cellulis in divisione binis quadrato-sphaericis usque ad elliptico-cylindraceas, aetate provecta cylindraceis, ad apices truncato-rotundis, diametro 2—6μ crassis, usque ad 8-plo longiores, rectis, in matrice gelatinosa saepe conferte completis; gelatino vaginale hyalino, homogeneo, non-numquam omnino diffluente; protoplasmate aerugineo, olivaceo, luteolo, violaceo, vel roseo, homogeneo vel granuloso. FIGS. 164—169.

On wet rocks, wood, and soil, in seepage, in shallow fresh, brackish, and marine waters, and in the plankton. The smaller growth-forms of this species may sometimes be confused with those of *Anacystis montana*. Where the cells are large, they may be distinguished from cells of *Coccochloris stagnina* by their more cylindrical shape and their apparently truncate ends.

Specimens examined:
GERMANY: Grunewaldsee [bei Berlin], am Ufer, *P. Hennings*, Mai 1888 (as *Coccochloris piscinalis* in Henn., Phyk. March. no. 44, FH, L; in Hauck & Richt., Phyk. Univ. no. 240, L); Schlachtensee [bei Berlin], *Hennings*, Aug. 1892 (as *C. piscinalis* in Henn., Phyk. March. no. 44, FH, L); Leipzig, *O. Bulnheim* (FH); Polenz bei Wurzen, *Bulnheim*, 1859 (FH, NY); Torfmoor, Wurzen bei Leipzig, *Bulnheim* (B); Leipzig, Altmanndorf, *P. Richter* (UC); Hanau, *Theobald* (Type of *Palmella pulchra* Kütz., L; isotype, UC). AUSTRIA: in fontis fundo prope Wagrain Austriae superioris, *de Mörl*, Aug. 1862 (Type of *Synechococcus violaceus* Grun., W); Tiergarten von Lainz unweit Wien, in Wassergefässen weiter cultiviert, *K. H. Rechinger*, Oct. 1904 (isotypes of *S. parvus* Mig., Crypt. Germ., Austr., & Helv. Exs., Alg. no. 123, MICH, NY, TA). SWITZERLAND: Zürich, im Katzensee, Überzug auf Schlamm am Ufer, *C. von Nägeli*, Jul. 1847 (Type of *S. elongatus* Näg. and *Protococcus elongatus* Näg. with *Leptothrix calcarea* Näg., ZT). FRANCE: Falaise, Calvados, *A. de Brébisson* (Type of *Micraloa elabens* Bréb., L [Fig. 168]; isotypes, B, FC, L, S; as *Polycystis elabens* in Rabenh. Alg. no. 2178, B, FH, L, S, UC; as *Microcystis elabens* in Desmaz., Pl. Crypt. Fr. no. 1952, FC), *Brébisson 726* (L); Aix en Provence, ex herb. Lenormand, *Castagne* (Type of *Palmella Castagnei* Kütz., L; isotypes, FC, UC); dans l'étang salé de la Valdue près Aix, *Castagne* (L, S; as *Palmella Castagnei* in Desmaz., Pl. Crypt. Fr. no. 1955, FI). BELGIAN CONGO: Mayya-moto, rivière Mokondo, Parc National Albert, *J. Lebrun* 9209, Dec. 1937 (Type of *Aphanothece Lebrunii* Duvign. & Sym., BR [Fig. 165]). UNION OF SOUTH AFRICA: Zeekoe Vlei, Cape Flats near Cape Town, *Mrs. G. E. Hutchinson 35*, 1928 (Type of *A. conglomerata* Rich, D[Fig. 166]).
NEW BRUNSWICK: Island lake, 30—50 miles due south of Dalhousie, *M. Le Mesurier 5*,

Aug. 1953 (FC); edge of lake, Blue Bell, Victoria county, *H. Habeeb 10349*, 7 Jul. 1948 (FC, HA); in lake 8 miles south of Grand Falls, *Habeeb 11650, 11655*, 22 Jun. 1951 (FC, HA). MASSA-CHUSETTS: Horn pond, Woburn, *F. S. Collins 5739*, Sept. 1908 (FH). RHODE ISLAND: Watch Hill pond, Watch Hill, *W. A. Setchell 6425*, 16 Jun. 1908 (FC, UC). CONNECTICUT: in ditches in salt marsh, Eastern point, Groton, *Setchell*, 13 Jul. 1888 (UC). MARYLAND: in a marsh pool at Chance, Somerset county, *P. W. Wolle & F. Drouet 2268*, 22 Aug. 1938 (FC). NORTH CAROLINA: on overhanging rocks below Dry falls, Highlands, Macon county, *H. C. Bold H220*, 29 Jun. 1939 (DA, FC); intertidal sand in turtle pen, Fish & Wildlife Laboratory, Beaufort, *H. J. Humm*, 5 Aug. 1949 (FC); intertidal sand inside beach, Shackleford Banks, Beaufort, *Humm*, 7 Aug. 1949 (FC). SOUTH CAROLINA: Alligator lake near Myrtle Beach, Horry county, *P. J. Philson SC1*, 2 Jul. 1932 (FC). FLORIDA: intertidal on sand in St. Andrews bay at Hathaway bridge, west of Panama city, Bay county, *F. Drouet & C. S. Nielsen 10928*, 15 Jan. 1949 (FC, T); Eastpoint, shore on U. S. highway no. 98, Franklin county, *G. C. Madsen, A. L. Pates, M. N. Hood, & L. Elias 1949*, 11 Sept. 1949 (FC, T); on Chara, Hernando county, *M. A. Brannon 557*, 23 Oct. 1948 (FC, PC); on pilings in the pond at 14th and Center streets, Leesburg, Lake county, *Drouet & Brannon 11077*, 19 Jan. 1949 (FC); saline pool, Southeast Hammock, Big Pine key, Monroe county, *E. P. Killip 42479*, 7 Jan. 1953 (FC, US); wet sand in the outlet of a sewer into Santa Rosa sound, Fort Walton, Okaloosa county, *Drouet, Nielsen, Madsen, D. Crowson, & Pates 10645*, 9 Jan. 1949 (FC, T); Buggs Springs, Okahumpka, Lake county, *J. Branham*, 10 Apr. 1955 (D, T).

MICHIGAN: pool along shore of Echo lake, Bois Blanc island, Mackinac county, *H. K. Phinney 34*, 8 Jul. 1942 (FC, PHI); submerged near shore, Carp lake, Emmet county, *W. A. Daily 43*, 2 Jun. 1935 (DA, FC); Sodon lake, Oakland county, *S. A. Cain*, Dec. 1947 (FC). INDIANA: lake 2 miles north of Columbus, Bartholomew county, *W. A. & F. K. Daily & C. Kenoyer 1162*, 19 Sept. 1943 (DA, FC); Little Eagle lake, Winona Lake, Kosciusko county, *C. M. Palmer B420*, 23 Aug. 1935 (DA, FC); shore of Messick lake north of Eddy, Lagrange county, *F. K. & W. A. Daily 2689*, 26 Aug. 1953 (DA, FC); in Lost lake, Culver, Marshall county, *F. K. & W. A. Daily 1510, 1516*, 3 Jul. 1946 (DA, FC); in Lake Maxinkuckee, Culver, *F. K. & W. A. Daily 1529*, 4 Jul. 1946 (DA, FC); in Bass lake, 5 miles south of Knox, Starke county, *F. K. & W. A. Daily 1562*, 6 Jul. 1946 (DA, FC); Crooked lake, Steuben county, *Palmer B49*, Sept. 1933 (DA, FC). WISCONSIN: Wandewoga Lake, *G. W. Prescott 3W57, 3W137*, 8 Aug. 1938 (FC). MISSOURI: in a muriatic spring, Kimmswick, Jefferson county, *J. A. Steyermark*, 1 May 1934 (D). LOUISIANA: in ponds, Little Lake Club, Plaquemines parish, *J. N. Gowanloch*, Sept. 1944 (FC).

NEBRASKA: culture from gravel pit northwest of Fremont, Dodge county, *W. Kiener 13769*, 16 Jul. 1944 (FC, KI); plankton in pool, Lincoln, *Kiener 16593, 16594*, 15 May 1944 (FC, KI), and culture, *Kiener 16351*, 13 Jul. 1944 (FC, KI); culture from soil in red oak woods 3 miles south of Lincoln, *Kiener 13741*, 16 Jul. 1944 (FC, KI); culture from roadside ditch 4 miles east of Maxwell, Lincoln county, *Kiener 16463c*, 1 Oct. 1944 (FC, KI); culture from Dismal river near Halsey, Thomas county, *Kiener 16750c*, 7 Oct. 1945 (FC, KI). WYOMING: Yellowstone National Park: border and bottom of Bath lake, Lower Geyser Basin, *W. A. Setchell 1932*, 27 Aug. 1898 (FC, UC); in stream from the spring under the main road and boardwalk, Mammoth hot springs, *Setchell 2021, 2022*, 1 Sept. 1898 (FC, UC); spring on Angel Terrace, Mammoth hot springs, *Setchell 2000*, 31 Aug. 1898 (FC, UC); in Bath lake, Mammoth hot springs, *Setchell 1995, 1997, 2004a*, 31 Aug. 1898 (FC, UC); in overflow and in pool, Cleopatra Terrace, Mammoth hot springs, *Setchell 1982, 1988*, 31 Aug. 1898 (FC, UC); borders of small spring above Devil's Kitchen, Mammoth hot springs, *Setchell 2002*, 31 Aug. 1898 (FC, UC); overflow pool from Minerva Terrace, Mammoth hot springs, *Setchell 2027*, 1 Sept. 1898 (FC, UC); pool from a very small spring, Pulpit Terrace, Mammoth hot springs, *Setchell 1999*, 31 Aug. 1898 (FC, UC); in a side pool of the spring (Laundry pool) nearest West Thumb station, *Setchell 1914*, 23 Aug. 1898 (FC, UC). Medicine Bow National Forest: bottom of Trail's Divide lake, *W. G. Solheim 14*, 22 Jun. 1933 (FC). NEW MEXICO: in the outlet of a hot spring, Montezuma (Hot Springs), San Miguel county, *F. Drouet & D. Richards 2671*, 19 Oct. 1939 (FC).

UTAH: Great Salt Lake: *A. S. Packard Jr.* on Hayden's Expedition (Type of *Polycystis Packardii* Farl., FH [Fig. 164]; isotype, D); *Mrs. E. A. McVicker* (FC, UC); *J. A. Harris*, Jul. 1929 (MIN); *E. R. Walker*, 1908 (FC, UC); *D. Stuard*, 17 Sept. 1928 (FC); Sta. Ac, *A. J. Eardley 3, 4*, 1934 (D); Black Rock, Salt Lake county, *A. O. Garrett 23, 27*, 27 Aug. 1929 (FC, UC); *R. P. Ehrhardt 1, 3*, 21 Oct. 1946 (FC); East bay, *Garrett 17*, 15 Jun. 1929 (FC, UC); Saltair, Salt Lake county, *Garrett 24*, 27 Aug. 1929 (FC, UC), *Drouet & H. B. Louderback 5734, 5763, 5766*, 21 Aug. 1946 (FC); West Point, *Garrett 11—14*, 22 Jun. 1929 (FC, UC); Garfield Beach, *J. E. Tilden*, 7 Jul. 1897 (Type of *Aphanothece utahensis* Tild. in Tild., Amer. Alg. no. 297, MIN; isotypes, FC [Fig. 169], L, UC; as *Polycystis Packardii* in Tild., Amer. Alg. no. 298, B, FC, L); near Stansbury island, *Garrett 3*, Aug. 1929 (FC, TA, UC). ARIZONA: south slope of Callville wash, Lake Mead, *E. U. Clover*, 12 Sept. 1940 (FC, MICH): NEVADA: sloughs near Fairbanks Springs house, Ash Meadows, Nye county, *I. La Rivers 1010*, 3 Apr. 1950 (FC); in hot water

30

just above level of Pyramid lake, west side of Pyramid, Washoe county, *La Rivers 1405*, 22 Sept. 1951 (FC). ALASKA: small pond, Milepost 42 (from Anchorage), Steese highway, *D. K. Hilliard 2*, 3 Sept. 1953 (FC); Mile 163 (from Anchorage), Glenallen highway, *Hilliard 17a*, 13 Sept. 1953 (FC). BRITISH COLUMBIA: Spillamacheen valley lakes near Bald mountain, Purcell range, Kootenay county, *W. R. Taylor*, Sept. 1925 (Type of *Aphanothece microscopica* var. *congesta* W. R. Taylor, TA). CALIFORNIA: culture from an aerating tank, West Berkeley, *N. L. Gardner 6926*, 21 Oct. 1930 (FC, UC); in a cold pool, Bridgeport, Mono county, *M. J. Groesbeck 42*, 5 Apr. 1940 (FC); small hot spring at Hot Creek, about 2 miles south of Whitmore Tubs, Mono county, *Groesbeck 61*, 16 Jun. 1940 (FC); in hot salt water near Key Route power house, Oakland, *Gardner 1557*, 6 Sept. 1905 (Type of *Synechococcus curtus* Setch., UC; isotypes, FC, and in Coll., Hold., & Setch., Phyc. Bor.-Amer. no. 1351, FC [Fig. 167], L, TA), *Gardner 1508*, Jul. 1905 (FC, UC); culture (from Woods Hole, Massachusetts, from E. H. Battley) in Hopkins Marine Station, Pacific Grove, *M. B. Allen*, 29 Jan. 1952 (FC).

BAHAMA ISLANDS: Nassau, *A. E. Wight*, Jan. 1905 (FH); Lake Cunningham, New Providence, *A. D. Peggs*, Jul. 1940, 1941, 10 May 1941 (FC). BRITISH WEST INDIES: in canals of salt works, Cockburn Harbor, South Caicos, *M. A. Howe 5581*, 16 Dec. 1907 (D, FH, L). PUERTO RICO: pools, College of Agriculture, Mayaguez, *L. R. Almodóvar 375, 412*, 22 Aug. 1954 (D, T).

JAMAICA: in a brackish pond, north cay, Morant cays, *V. J. Chapman 9a*, 1939 (FC); in the brackish pond, Gun cay, Kingston, *Chapman 3d*, 1939 (FC). DUTCH WEST INDIES: Bonaire: ten zuiden van Witte Pan, Pekelmeer, *P. Wagenaar Hummelinck*, Sept. 1930 (L); Salinja Tam, *Wagenaar Hummelinck*, Nov. 1930 (L); Salinja de Klein Bonaire, *Wagenaar Hummelinck*, Sept. 1930 (Type of *Aphanothece minor* Frémy, L). BRAZIL: in a freshwater pond in the dunes, Arpoadouros, Fortaleza, Ceará, *F. Drouet 1469*, 5 Oct. 1935 (D); praia da Sununga, Ubatuba, São Paulo, *A. B. Joly 9*, 3 July 1953 (D); rain pool on concrete dam of tanque, Anil, São Luiz, Maranhão, *Drouet 1325*, 24 Jul. 1935 (D). ECUADOR: salt lake south of Tagus cove, Albemarle island, Galapagos islands, *W. L. Schmitt 331—35*, 10 Dec. 1934 (FC). PERU: near Lima, *A. Maldonado 1*, Dec. 1940 (FC); salt lakes, Haucho, prov. Chancai, *Maldonado 88, 94, 95*, Feb., Mar. 1942 (FC). ARGENTINA: culture in Instituto Bacteriologico, Depto. Nacional de Higiene, Buenos Aires, *D. Rabinovich CP11*, Dec. 1943 (FC). GILBERT ISLANDS: fish-pond in south part of Beru, *Mrs. R. Catala 23*, Jul. 1951 (FC). WEST AUSTRALIA: Pink lake, Rottnest island, near Perth, *L. A. Doore*, 12 Oct. 1934 (in Tild., So. Pacific Pl., 2nd Ser., no. 18, FC). CHINA: plankton of West lake, Hangchow, Chekiang, *J. E. Nielsen*, 29 Sept. 1923 (FC). INDONESIA: plankton of Lake Ohoitiel, Kai islands, *H. Jensen & O. Hagerup*, 1922 (Type of *Microcystis elabens* var. *minor* Nyg., in the slide collection of G. Nygaard); Sitoe Bagendit, West Java, *A. Weber-van Bosse 755*, 1888 (D, FH, L).

4. COCCOCHLORIS PENIOCYSTIS Drouet & Daily, Lloydia 11: 78. 1948. *Gloeocapsa Peniocystis* Kützing, Tab. Phyc. 1: 25. 1846. *Peniocystis purpurea* Brébisson *pro synon.* in Kützing, loc. cit. 1846. *Gloeothece linearis* var. *purpurea* Brébisson in Rabenhorst, Alg. Eur. 218—220: 2193. 1870. *Bichatia Peniocystis* Kuntze, Rev. Gen. Pl. 2: 886. 1891. *Anacystis Peniocystis* Dr. & Daily, Amer. Midl. Nat. 27: 651. 1942. —Type from Arromanches, France (L). Wille in Nyt Mag. Naturvid. 62: 196 (1925) erroneously interpreted Kützing's original description as referring to *Anacystis thermalis*, plants of which are present in the original specimens.

Gloeocapsa purpurea Kützing, Tab. Phyc. 1: 18. 1846. *Bichatia purpurea* Kuntze, Rev. Gen. Pl. 2: 886. 1891. —Type from Arromanches, France (L). Wille in Nyt Mag. Naturvid. 62: 196 (1925) erroneously interpreted Kützing's original description as referring to *Anacystis thermalis*, plants of which are present in the original specimens.

Cryptococcus carneus Kützing, Sp. Algar., p. 146. 1849. —Type from St. Chamas, Bouches du Rhone, France (L).

Gloeothece linearis Nägeli, Gatt. Einzell. Alg., p. 58. 1849. —Type from Blickensdorf, Zug, Switzerland (ZT).

Aphanothece saxicola Nägeli, Gatt. Einzell. Alg., p. 59. 1849. *Palmogloea saxicola* Nägeli *pro synon.*, loc. cit. 1849. —Type from Küsnacht, Zürich, Switzerland (ZT).

Aphanothece saxicola var. *aquatica* Wittrock in Wittrock & Nordstedt, Alg. Exs. 6: 295. 1879. —Type from Upsala, Sweden (S). FIG. 170.

Aphanothece caldariorum var. *cavernarum* Hansgirg, Bot. Centralbl. 37: 38. 1889. —Type from Korno near Beroun, Bohemia (W).

Aphanothece subachroa Hansgirg, Sitzungsber. K. Böhm. Ges. Wiss., Math.-Nat. Cl., 1890(2):

31

98. 1891. *A. saxicola* f. *subachroa* Elenkin, Monogr. Algar. Cyanophyc., Pars Spec. 1: 151. 1938.
—Type from Smichov, Bohemia (W). FIG. 172.
   Gloeothece lunata W. & G. S. West, Journ. Linn. Soc. Bot. 30: 277. 1894. —Type from
St. Vincent, West Indies (BM).
   Synechococcus minutus West, Proc. Roy. Irish Acad. 31: 40. 1912. —Type from Clare island,
Ireland (BIRM).
   Synechocystis primigenia Gardner, Mem. New York Bot. Gard. 7: 2. 1927. —Type from
Arecibo, Puerto Rico (NY).
   Aphanothece bacilloidea Gardner, Mem. New York Bot. Gard. 7: 5. 1927. —Type from near
Arecibo, Puerto Rico (NY).
   Gloeocapsa cartilaginea var. *minor* Gardner, Mem. New York Bot. Gard. 7: 9. 1927. —Type
from Arecibo, Puerto Rico (NY).
   Gloeothece prototypa Gardner, Mem. New York Bot. Gard. 7: 14. 1927. —Type from
Arecibo, Puerto Rico (NY). FIG. 171.

   Original specimens have not been available to us for the following names;
their original descriptions are here designated as the Types until the specimens
can be found:
   Cylindrocystis coerulescens Brébisson in Meneghini, Mem. R. Accad. Sci. Torino, ser. 2, 5 (Sci.
Fis. & Mat.): 91. 1843. *Pleurococcus coerulescens* Brébisson in Rabenhorst, Fl. Eur. Algar. 3: 28.
1868. *Chroococcus coerulescens* Forti, Syll. Myxophyc., p. 25. 1907.
   Synechococcus racemosus Wolle, Bull. Torr. Bot. Club 8: 37. 1881.
   Bacillus muralis Tomaschek, Bot. Zeitung 45: 665. 1887. *Aphanothece caldariorum* b.
muralis Hansgirg, Prodr. Algenfl. Böhmen 2: 137. 1892. *A. caldariorum* var. *muralis* Hansgirg ex
Forti, Syll. Myxophyc., p. 80. 1907. *A. muralis* Lemmermann, Krypt.-Fl. Mark Brandenb. 3: 69.
1907.
   Aphanothece saxicola var. *violacea* W. & G. S. West, Trans. Roy. Microsc. Soc. 1894: 17.
1894.
   Rhabdoderma lineare Schmidle & Lauterborn, Ber. Deutsch. Bot. Ges. 18: 148. 1900.
   Microcystis orissica West, Journ. & Proc. Asiatic Soc. Bengal, N. S. 7: 84. 1911.
   Rhabdoderma Gorskii Woloszynskia, Bull. Acad. Sci. Cracovie, Cl. Math. & Nat., ser. B, 1917:
127. 1917.
   Rhabdogloea ellipsoidea Schröder, Ber. Deutsch. Bot. Ges. 35: 549. 1917. *Dactylococcopsis
ellipsoidea* Geitler in Engler & Prantl, Natürl. Pflanzenfam., ed. 2, 1b: 44. 1942.
   Aphanothece longior Naumann, K. Sv. Vet.-Akad. Handl. 62(4): 17. 1921. *A. saxicola* f.
longior Elenkin, Monogr. Algar. Cyanophyc., Pars Spec. 1: 149, 151. 1938.
   Spirillopsis irregularis Naumann, K. Sv. Vet.-Akad. Handl. 62(4): 18. 1921. *Rhabdoderma
irregulare* Geitler in Pascher, Süsswasserfl. 12: 113. 1925.
   Aphanothece clathrata var. *brevis* Bachmann, Verh. Naturf. Ges. Basel 35(1): 152, 165. 1923.
   Rhabdoderma sigmoideum Moore & Carter, Ann. Missouri Bot. Gard. 10: 398. 1923.
   Rhabdoderma sigmoideum f. *minus* Moore & Carter, Ann. Missouri Bot. Gard. 10: 398. 1923.
   Dactylococcopsis Smithii R. & F. Chodat, Veröffent. Geobot. Inst. Rübel Zürich 3: 455. 1925.
   Dactylococcopsis Smithii var. *sigmoidea* R. & F. Chodat, Veröffent. Geobot. Inst. Rübel
Zürich 3: 445. 1925.
   Aphanothece saxicola var. *clathratoides* Liebetanz, Bull. Int. Acad. Polon. Sci., Cl. Sci. Math.
& Nat., ser. B, 1925: 109. 1925.
   Aphanothece rubra Liebetanz, Bull. Int. Acad. Polon. Sci., Cl. Sci. Math. & Nat., ser. B, 1925:
109. 1925.
   Dactylococcopsis Elenkenii Roll, Ann. de Protistol. 1: 164. 1928.
   Gloeothece coerulea Geitler, Arch. f. Protistenk. 60: 440. 1928.
   Cyanocloster muscicola Kufferath, Ann. Crypt. Exot. 2: 49. 1929.
   Krkia croatica Pevalek, Acta Bot. Inst. Bot. Univ. Zagreb. 4: 55. 1929.
   Rhabdoderma Gorskii var. *spirale* Lundberg, Bot. Notiser 1931: 279. 1931.
   Tetrarcus Ilsteri Skuja, Acta Horti Bot. Univ. Latv. 7: 46. 1932.
   Gloeothece Kriegeri Frémy, Ann. Crypt. Exot. 5: 193. 1932.
   Dactylococcopsis alaskana Kol, Smiths. Misc. Coll. 101(16): 27. 1942.
   Dactylococcopsis planctonica Teiling, Bot. Notiser 1942: 64. 1942.
   Dactylococcopsis scenedesmoides Nygaard, K. Danske Videnskab. Selsk. Biol. Skr. 7(1): 185.
1949.
   Dactylococcopsis rhaphidioides var. *solitaria* Fjerdingstad, Dansk Bot. Ark. 14(3): 33. 1950.
   Rhabdoderma vermiculare Fott, Preslia 24: 192. 1952.

   Plantae violaceae, roseae, vel aerugineae, microscopicae vel macroscopicae, 1—
pluri—multi-cellulares, cellulis in divisione binis cylindraceis, aetate provecta longi-

cylindraceis et ad apices rotundis, haud raro attentuato-conicis, diametro 1—3μ crassis, usque ad 12-plo longiores, curvatis vel rectis; gelatino vaginale hyalino, homogeneo vel lamelloso, saepe omnino diffluente; protoplasmate violaceo, roseo, vel aerugineo, homogeneo, nonnumquam granuloso. FIGS. 170—172.

On wet rocks, wood, and soil, in seepage, in shallow and deep fresh water, and in the plankton. In the plankton there is sometimes encountered a growth-form with cells somewhat attenuated and bluntly conical at the tips. The large bacteria are often confused with this species, as are likewise certain smaller growth-forms of *Anacystis montana* and *Palmogloea protuberans* (Sm. & Sow.) Kütz., especially where the latter are poorly preserved. In hot springs the one- and two-celled hormogonia of *Phormidium laminosum* (Ag.) Gom., *P. tenue* (Menegh.) Gom., and *P. Treleasei* Gom. are reminiscent, in some respects, of the cells of *Coccochloris Peniocystis*. The short hormogonia of *Plectonema Nostocorum* Born. and of the organism passing under the name *Phormidium mucicola* Naum. & Hub. can also be mistaken as cells of this species.

Specimens examined:

SWEDEN: in piscina in Lassby backar prope Upsaliam, *V. Wittrock*, 20 Oct. 1878 (Type of *Aphanothece saxicola* var. *aquatica* Wittr. in Wittr. & Nordst., Alg. Exs. no. 295, S; isotypes, FC [Fig. 170], L); Kristineberg, Bohuslän, *G. Lagerheim*, 13 Jul. 1884 (D, S); in piscina ad Strömsberg in Smolandia, *O. Nordstedt*, Jul. 1890 (as *Coccochloris Trentepohlii* in Wittr., Nordst., & Lagerh., Alg. Exs. no. 1598, L, N); in spelunca Silfvergrotten montis Kullen Scaniae, *H. G. Simmons*, Jul. 1897 (S); Möja, Upland, *O. Borge*, Sept. 1932 (S). ROMANIA: ex palude Saltica, dist. Ilfov, *E. C. Teodorescu 1041*, 13 Nov. 1901 (FC, W). GERMANY: Dresden, hort. bot., *Lagerheim*, 6 Oct. 1882 (FC, D, L, S). CZECHOSLOVAKIA: Bohemia: in der Kalksteinhöhle unter dem Katarakte im Felsenthale unterhalb Korno nächst Beraun, *A. Hansgirg*, Sept. 1888 (Type of *Aphanothece caldariorum* var. *cavernarum* Hansg., W; isotype, FC); auf feuchten Hölzern in Warmhäusern in k. k. botanischen Garten am Smichow, *Hansgirg*, Dec. 1883 (Type of *Aphanothece subachroa* Hansg., W; isotype, FC); in caldariis horti in Tetschen, *Hansgirg*, Aug. 1890 (FC, W). AUSTRIA: Wien, Hort. bot., *Lagerheim*, 4 Oct. 1882 (D, S); ad parietes caldariorum horti Augarten Vindobonae, *Hansgirg*, 30 Dec. 1884 (as *A. caldariorum* in Wittr. & Nordst., Alg. Exs. no. 793, FC, L, NY, S; and in Kerner, Fl. Exs. Austro-Hung. no. 1998, L, MO, S); Vindobonae in tepidariis horti botanici universitatis, *K. Rechinger*, Apr. (as *Gloeothece tepidariorum* in Mus. Vindob. Krypt. Exs. no. 2138, FC, L). JUGOSLAVIA: Pola, Istria, *A. Hansgirg*, Apr. 1889 (FC, W). SWITZERLAND: Ct. Zug, Blickensdorf, an schattigen Felsen, *K. von Nägeli* (Type of *Gloeothece linearis* Näg., ZT); ad saxa, Zürich, *G. Winter*, Jul. 1878 (L); Ct. Zürich, Küsnacht, an feuchten Felsen, *Nägeli 300*, Sept. 1847 (Type of *Aphanothece saxicola* Näg., ZT). NETHERLANDS: Kortenhoefsche plashen, *A. Weber-van Bosse 144*, Jun. 1886 (FC, L); warme kas, Utrecht, *R. A. Maas Geesteranus*, Oct. 1942 (L); on artificial rocks, hortus botanicus, Leiden, *A. Meeuse & J. van Ooststrom 7658*, Oct. 1942 (L), *J. T. Koster 1032*, 16 Oct. 1946 (FC, L), *P. Leenhouts*, 14 Oct. 1948, 12 Oct. 1950 (FC, L). BRITISH ISLES: with Sphagnum, Mulranny, Clare island, Ireland, *W. West*, 7 Jun. 1911 (Type of *Synechococcus minutus* West, BIRM). FRANCE: Bouches du Rhone: St. Chamas, ex herb. Lenormand (Type of *Cryptococcus carneus* Kütz., L). Calvados: Arromanches, *A. de Brébisson* (Type of *Gloeocapsa Peniocystis* Kütz. and *G. purpurea* Kütz., L; isotypes, FC, NY, UC); dans une cascade des falaises d'Arromanches, *Brébisson* (as *G. purpurea* in Rabenh. Alg. no. 1596, FC, MIN, L), Sept. 1845 (L); sur des rochers calcaires arrosés par une source limpide, *Brébisson* (as *Gloeothece linearis* var. *purpurea* in Rabenh. Alg. no. 2193, L, MIN, S; in Roumeguère, Alg. de Fr. no. 519, L); rochers inondés d'une rivière, Falaise, *Brébisson* (L); Falaise, herb. A. Le Jolis, Aug. 1844 (NY); Falaise, herb. Lenormand (S); anciennes carrières d'Allemagne aux environs de Caen, *P. Frémy 739*, Apr. 1924 (NY); Bayeux, *Brébisson* (L, S); Caen, etc. (as *Palmella hyalina* in Chauvin, Alg. de Normand. no. 130, L). Cher: Bourges, *O. Ripart 265*, Apr. 1877 (S). Seine et Marne: environs de Melun, *Roussel* (as *Gloeocapsa Peniocystis* in Roumeguère, Alg. de Fr. no. 26, L). MASSACHUSETTS: quarry, West Quincy, *F. C. Seymour*, 25 Sept. 1938 (FC). RHODE ISLAND: Hope reservoir, Providence, *Appleton*, Dec. 1883 (FH). NEW YORK: Glen lake, Warren county, *W. R. Maxon* (NY). WEST VIRGINIA: on sandstone in a spring, Buckhannon river 5 miles below Buckhannon, Upshur county, *H. K. Phinney 529*, 30 Aug. 1938 (FC, PHI). NORTH CAROLINA: on wet rock near stream, rear of Duke University stadium, Durham, *F. Jones 3*, Apr. 1947 (FC). FLORIDA: on wet limestone about the entrance to the cave, Florida Caverns state park, Jackson county, *F. Drouet, C. S. Nielsen, G. C. Madsen, & D. Crowson 10384*, 4 Jan. 1949 (FC, T).

OHIO: on cement wall, Peterson's greenhouse, Cincinnati, *W. A. Daily & A. H. Blickle 286, 305*, 23 Apr. 1940 (DA, FC), *Daily 440, 441, 846*, 9 Jul. 1940, 11 Jan. 1941 (DA, FC). KEN-

TUCKY: on wet sandstone, Torrent, Wolfe county, *W. A. & F. K. Daily & R. Kosanke 465, 466*, 14 Jul. 1940 (DA, FC); on wet sandstone, Tight Holler, Pine Ridge, Wolfe county, *W. A. Daily, A. T. Cross, & J. Tucker 766, 810*, Sept. 1940 (DA, FC). INDIANA: on wall of cave, Elkhorn falls near Richmond, *L. J. King 201*, 8 Sept. 1940 (EAR, FC). WISCONSIN: sphagnum bogs near Trilby and Sand lakes, Vilas county, *G. W. Prescott 3W58, 3W278*, 15 Aug., 2 Sept. 1938 (FC).

MISSOURI: Sagamount, 2 miles south of Saginaw, Newton county, *E. J. Palmer M43*, 31 Jul. 1953 (D). ILLINOIS: in Thompson's lake at waterworks, Carbondale, *W. B. Welch CWT*, 23 Jul. 1941 (FC). ARIZONA: at waterfall, Bull Durham canyon, Lake Meade, Coconino county, *E. U. Clover 157*, 9 Sept. 1940 (FC, MICH). NEVADA: in tepid pool, Steamboat hot springs, Washoe county, *M. J. Groesbeck 114, 192*, 13 Jun., 4 Sept. 1940 (FC). CALIFORNIA: fresh water, *Lemmon*, 1877 (D, FH); in cold spring and pools, travertine quarry at Bridgeport, Mono county, *Groesbeck 175, 265, 472*, 3 Sept., 24 Nov. 1940, 8 Jul. 1941 (FC).

BAHAMA ISLANDS: on rocks in tide-pool, Eight Mile Rock, Great Bahama, *M. A. Howe 3694*, Feb. 1905 (NY). PUERTO RICO: on limestone, Arecibo to Hatillo, *N. Wille 1377* (Type of *Gloeocapsa cartilaginea* var. *minor* Gardn., NY), *1377a* (Type of *Gloeothece prototypa* Gardn., NY [Fig. 171]), *1377b* (Type of *Synechocystis primigenia* Gardn., NY), *1378, 1378a*, Feb. 1915 (NY); on limestone, Arecibo to Utuado, *Wille 1481a*, Mar. 1915 (NY); in depressions in limestone, Hato Arriba near Arecibo, *Wille 1407a* (Type of *Aphanothece bacilloidea* Gardn., NY), *1407b*, Mar. 1915 (NY); in a water pipe near a stream, Maricao, *Wille 1147, 1147c*, Feb. 1915 (FC, FH, NY). ST. VINCENT: [on damp wall of dam, Sharp's River], *W. R. Elliott 477* (Type of *Gloeothece lunata* W. & G. S. West, with *Cosmarium pseudopyramidatum* in the slide collection, BM). HONDURAS: on wet masonry, vicinity of El Zamorano, dept. Morazán, *P. C. Standley 4423*, Feb.-Mar. 1947 (FC). PERU: Lima, *A. Maldonado 2*, Dec. 1940 (FC); agua estanciada cerca mar, Lago Villa, *Maldonado 67*, Jan. 1942 (FC). NEW ZEALAND: caves, Piha, *V. J. Chapman*, Apr. 1947 (FC). PHILIPPINES: hot spring, College, Laguna, *G. T. Velasquez 2462*, 8 Aug. 1948 (FC, PUH). CHINA: on rock in rapid streams, Omei-shan and Shan-Shan-Kow, Omei, Szechwan, *H. Chu 1329, 1329a*, Aug. 1941, 21 Aug. 1942 (FC). INDONESIA: op de wand de grot bij Brabakan, *W. Docters van Leeuwen*, Nov. 1913 (L); in the caves where the edible swallows' nests are collected, Borneo, comm. Holmes (PC).

GENUS 2. ANACYSTIS

Meneghini, Consp. Algol. Eugan., p. 324. 1837. *Microcystis* Unterabtheilung *Anacystis* Kützing, Tab. Phyc. 1: 7. 1846. *Polycystis* Sectio *Anacystis* Hansgirg, Prodr. Algenfl. Böhmen 2: 144. 1892. —Type species: *Anacystis marginata* Menegh.

Pleurococcus Meneghini, Consp. Algol. Eugan., p. 337. 1837. —Type species: *P. communis* Menegh. Nägeli in Gatt. Einzell. Alg., p. 64 (1849) names "*P. vulgare* Menegh. part." as the Typus of "*Pleurococcus* Menegh. part.", but it seems to us that attaching part of a name to part of a type is an impossibility under the present International Rules of Nomenclature. We have designated as the Type species here one of the species described at the same time as the genus.

Gloeocapsa Kützing, Phyc. Gener., p. 173. 1843. *Gloeocapsa* Subgenus *Eugloeocapsa* Hansgirg, Notarisia 3: 589. 1888. *Gloeocapsa* Sectio *Eugloeocapsa* Hansgirg, Prodr. Algenfl. Böhmen 2: 152. 1892. *Gloeocapsa* Sectio *Hyalocapsa* Kirchner in Engler & Prantl, Natürl. Pflanzenfam., ed. 1, 1(1a): 54. 1898. —Type species: *Gloeocapsa atrata* Kütz. See Drouet in Doty, Lloydia 13: 9, footnote (1950).

Sorospora Hassall, Hist. Brit. Freshw. Alg. 1: 309. 1845. —Type species: *Ulva montana* Lightf.

Microcystis Unterabtheilung *Polycystis* Kützing, Tab. Phyc. 1: 7. 1846. *Polycystis* Kützing, Sp. Algar., p. 210. 1849; non Léveillé, 1846. *Diplocystis* Trevisan, Sagg. Monogr. Alg. Coccot., p. 40. 1848; nec Berkeley & Curtis, 1869; nec J. Agardh, 1896; nec Cleve, 1900. *Clathrocystis* Henfrey, Trans. Microsc. Soc. London, N. S., 4: 53. 1856. *Polycystis* Sectio *Clathrocystis* Hansgirg, Prodr. Algenfl. Böhmen 2: 146. 1892. —Type species: *Micraloa aeruginosa* Kütz.

Cagniardia Trevisan, Sagg. Monogr. Alg. Coccot., p. 47. 1848. —Type species: *Palmella cyanea* Kütz.

Chroococcus Nägeli, Gatt. Einzell. Alg., p. 45. 1849. *Chroococcus* Subgenus *Euchroococcus* Hansgirg, Notarisia 3: 589. 1888. *Chroococcus* Sectio *Euchroococcus* Hansgirg, Prodr. Algenfl. Böhmen 2: 161. 1892. —Type species: *Protococcus rufescens* Kütz. (= *Chroococcus rufescens* Näg.).

Monocapsa Itzigsohn in Rabenhorst, Alg. Sachs. 27—28: 263. 1853. —Type species: *M. stegophila* Itzigs.

Rhodococcus Sectio *Rhodocapsa* Hansgirg, Oesterr. Bot. Zeitschr. 34: 315. 1884. —Type species: *Gloeocapsa violacea* Kütz.

34

Eucapsis Clements & Shantz, Minn. Bot. Stud. 3: 134. 1909. *Chroococcus* Subgenus *Eucapsis* Elenkin, Not. Syst. Inst. Crypt. Horti Bot. Petropol. 2: 68. 1923. —Type species: *Eucapsis alpina* Clem. & Shantz.

Endospora Gardner, Mem. New York Bot. Gard. 7: 27. 1927. —Type species: *E. rubra* Gardn.

Chroococcidiopsis Geitler, Arch. f. Hydrobiol., Suppl. XII, 4: 625. 1933. —Type species: *C. thermalis* Geitl.

Radiocystis Skuja, Symbolae Bot. Upsal. 9(3): 43. 1948. —Type species: *R. geminata* Skuja.

Original specimens of the type species for the following generic and subgeneric names have not been available to us for study:

Aplococcus Roze, Journ. de Bot. 10: 321. 1896. —Type species: *A. natans* Roze.

Gloeocapsa Sectio *Cyanocapsa* Kirchner in Engler & Prantl, Natürl. Pflanzenfam., ed. 1, 1(1a): 54. 1898. *Gloeocapsa* Subgenus *Cyanocapsa* Kirchner ex Forti, Syll. Myxophyc., p. 39. 1907. —Type species: *Protococcus violasceus* Corda.

Coelosphaeriopsis Lemmermann, Abh. Nat. Ver. Bremen 16: 353. 1899. —Type species: *C. halophila* Lemm.

Planosphaerula Borzi, Nuova Notarisia 1905: 20. 1905. —Type species: *P. natans* Borzi.

Myxosarcina Printz, K. Norske Vidensk. Selsk. Skr. 1920(1): 35. 1920. —Type species: *M. concinna* Printz.

Sphaerodictyon Geitler, Beih. z. Bot. Centralbl., II, 41: 231. 1925. —Type species: *Polycystis reticulata* Lemm.

Plantae microscopicae vel macroscopicae, cellulis in divisione binis plus minusve hemisphaericis, aetate provecta globosis, per gelatinum vaginale irregulariter vel in seriebus in tribus planis unoquodque ad alia perpendicularibus distributis; divisionibus cellularum seriatim in tribus planis unoquodque ad alia perpendicularibus procedentibus; gelatino vaginale hyalino vel deinde (in *A. montana*) rubrescentes vel coerulescentes vel lutescentes.

In species of Anacystis the spherical cells divide successively in three planes perpendicular to each other. The cells thus formed become irregularly distributed through the gelatinous matrix or retain a regular cubical (eucapsoid) arrangement, distributed in series of rows in three planes perpendicular to each other. Bacteria, palmelloid Chlorophyceae and Rhodophyceae, and growth-forms of Nostocaceae and *Entophysalis* spp. are often mistaken as species of Anacystis.

Key to species of Anacystis:

1. Cells without pseudovacuoles, plants not developing as water blooms............ 3.

1. Cells containing pseudovacuoles, plants developing as water blooms .... 2.

2. Cells (2.5—)3—7(—10)μ in diameter .... 1. A. CYANEA

2. Cells 0.5—2μ in diameter .... 2. A. INCERTA

3. Cells 0.5—2μ in diameter ....3. A. MARINA

3. Cells larger .... 4.

4. Cells up to 6μ in diameter (larger where parasitized by fungi), the gelatinous matrix developing red, blue, or yellowish-brown pigments in aerial and subaerial habitats .... 4. A. MONTANA

4. Cells 6μ in diameter or more, gelatinous matrix hyaline .... 5.

5. Cells (8—)12—50μ in diameter, usually remaining angular after division .... 5. A. DIMIDIATA

5. Cells chiefly 6—12μ in diameter, soon becoming spherical after division.... 6.

6. Plants marine, usually macroscopic .... 6. A. AERUGINOSA

6. Plants of fresh water, usually microscopic .... 7. A. THERMALIS

1. ANACYSTIS CYANEA Drouet & Daily, Butler Univ. Bot. Stud. 10: 221. 1952. *Palmella cyanea* Kützing, Phyc. Gener., p. 172. 1843. *Cagniardia cyanea* Trevisan, Sagg. Monogr. Alg. Coccot., p. 47. 1848. —Type from Jever, Germany (L).

Micraloa aeruginosa Kützing, Linnaea 8: 371. 1833. *Sphaerothrombium aeruginosum* Kützing pro synon., ibid. p. 370. 1833. *Microcystis ichthyoblabe* Kützing, Phyc. Gener., p. 170. 1843. *Palmella ichthyoblabe* Kunze pro synon. in Kützing, loc. cit. 1843. *Micraloa ictyolabe* Meneghini, Mem. R. Accad. Sci. Torino, ser. 2, 5(Sci. Fis. & Mat.): 48, 104. 1843. *Palmella ictyolabe* Kunze pro synon. ex Meneghini, loc cit. 1843. *Microcystis aeruginosa* Kützing, Tab. Phyc. 1: 6. 1846. *Micraloa ichthyoblabe* Rabenhorst, Deutschl. Krypt.-Fl. 2(2): 13. 1847. *Diplocystis aeruginosa* Trevisan, Sagg. Monogr. Alg. Coccot., p. 40. 1848. *D. ichthyoblabe* Trevisan, loc. cit. 1848. *Polycystis ichthyoblabe* Kützing, Sp. Algar., p. 210. 1849. *P. aeruginosa* Kützing, loc. cit. 1849. *Clathrocystis aeruginosa* Henfrey, Trans. Microsc. Soc. London, N. S., 4: 53. 1856. *Microhaloa aeruginosa* Kützing ex Rabenhorst, Fl. Eur. Algar. 2: 54. 1865. *Polycystis elabens* var. *ichthyoblabe* Hansgirg, Prodr. Algenfl. Böhmen 2: 145. 1892. —Type from Stuttgart, Germany (L). FIG. 9.

Polycystis viridis A. Braun in Rabenhorst, Alg. Eur. 141 & 142: 1415. 1862. *Microcystis viridis* Lemmermann, Abh. Nat. Ver. Bremen 17: 342. 1903. *M. aeruginosa* f. *viridis* Elenkin, Monogr. Algar. Cyanophyc., Pars Spec. 1: 106. 1938. —Type from Salzungen, Germany (B).

Chroococcus indicus Zeller, Jour. Asiatic Soc. Bengal 42(2): 176. 1873; Hedwigia 12: 168. 1873. —Type from Prome, Burma (L). FIG. 8.

Polycystis prasina Wittrock in Wittrock & Nordstedt, Alg. Exs. 6: 297. 1879. *P. flos-aquae* var. *prasina* Wittrock ex Hansgirg, Prodr. Algenfl. Böhmen 2: 144. 1892. *Microcystis prasina* Lemmermann, Ark. f. Bot. 2(2): 146. 1904. *Clathrocystis prasina* Woronichin, Not. Syst. Sect. Crypt. Inst. Bot. Acad. Sci. U. R. S. S. 6(1—6): 39. 1950. —Type from Uppland, Sweden (S). FIG. 6.

Polycystis flos-aquae Wittrock in Wittrock & Nordstedt, Alg. Exs. 6: 298. 1879. *P. flos-aquae* f. *autumnalis* Wittrock ex Lemmermann, Krypt.-Fl. Mark Brandenburg 3: 75. 1907. *Microcystis flos-aquae* Kirchner ex Forti, Syll. Myxophyc., p. 86. 1907. *M. aeruginosa* f. *flos-aquae* Elenkin, Monogr. Algar. Cyanophyc., Pars Spec. 1: 103. 1938. —Type from Dalsland, Sweden (S). FIG. 4.

Microcystis caerulea Dickie, Journ. Linn. Soc. Bot. 18: 128. 1880. —Type from Rio Tapajoz, Brazil (BM).

Polycystis scripta Richter, Hedwigia 25: 254. 1886. *P. flos-aquae* var. *scripta* Hansgirg, Prodr. Algenfl. Böhmen 2: 144. 1892. *Microcystis scripta* Migula, Krypt.-Fl. Deutschl. II. Alg. 1a: 37. 1905. *M. aeruginosa* f. *scripta* Elenkin, Monogr. Algar. Cyanophyc., Pars Spec. 1: 103. 1938. —Type from Halle, Germany (B). FIG. 2.

Polycystis aeruginosa var. *major* Wittrock ex Hansgirg, Prodr. Algenfl. Böhmen 2: 146. 1892. *Clathrocystis aeruginosa* f. *major* Keissler, Oesterr. Bot. Zeitschr. 56: 56. 1906. *C. aeruginosa* var. *major* Wittrock ex Forti, Syll. Myxophyc., p. 95. 1907. *Microcystis aeruginosa* var. *major* Wittrock ex Lemmermann, Krypt.-Fl. Mark Brandenb. 3: 76. 1907. *M. aeruginosa* f. *major* Elenkin, Not. Syst. Inst. Crypt. Hort. Bot. Petropol. 3: 15. 1924. —Type from Libochowitz, Bohemia (W). FIG. 7.

Polycystis (Clathrocystis) insignis Beck, Mus. Vindob. Krypt. Exs. 2: 227. 1896. —Type from Bombay, India (W). FIG. 10.

Polycystis ochracea Brand, Ber. Deutsch. Bot. Ges. 16: 200. 1898. *Microcystis ochracea* Forti, Syll. Myxophyc., p. 86. 1907; Lemmermann, Krypt.-Fl. Mark Brandenb. 3: 76. 1907. *M. aeruginosa* f. *ochracea* Elenkin, Monogr. Algar. Cyanophyc., Pars Spec. 1: 103. 1938. —Type from Wurmsee, Bavaria, Germany (B). FIG. 3.

Clathrocystis robusta Clark, Proc. Biol. Soc. Washington 21: 94. 1908. *Microcystis robusta* Nygaard, Dansk Bot. Ark. 4(10): 8. 1925. *M. aeruginosa* var. *robusta* Nygaard and *M. flos-aquae* var. *robusta* Nygaard, K. Danske Videnskab. Selsk. Skr. 7(1): 185. 1949. —Type from Lake Amatitlan, Guatemala (FC). FIG. 11.

Microcystis protocystis Crow, New Phytol. 22: 62. 1923. *M. aeruginosa* f. *protocystis* Elenkin, Monogr. Algar. Cyanophyc., Pars Spec. 1: 106. 1938. —Type from Anuradhapoora, Ceylon (in the collection of F. E. Fritsch). FIG. 5.

Microcystis aeruginosa var. *major* Elenkin, Not. Syst. Crypt. Hort. Bot. Reipubl. Ross. 3: 15. 1924. —Isotypes from Uppland, Sweden (FC, L, MICH).

Microcystis aeruginosa f. *occidentalis* W. R. Taylor, Amer. Journ. Bot. 15: 606. 1928. —Type from Flagstaff, Arizona (TA).

Microcystis toxica Stephens, Trans. Roy. Soc. So. Africa 32(1): 107. 1949. —Type from Vaaldam, Transvaal (L). FIG. 1.

Eucapsis alpina f. *violacea* Thomasson, Svensk Bot. Tidskr. 46: 233. 1952. —Isotype from Bessvatn, Sweden (DA).

Original specimens have not been available to us for the following names;

36

their original descriptions are here designated as the Types until the specimens can be found:

Coelosphaerium dubium Grunow in Rabenhorst, Fl. Eur. Algar. 2: 55. 1865. —Only a sketch with descriptive notes is present in the herbarium of A. Grunow in the Naturhistorisches Museum, Vienna. The material discussed by Schmula from Graz, Styria, is cited among the specimens below (see Hedwigia 37, Beibl. 1: 47. 1898).

Microcystis pseudofilamentosa Crow, New Phytol. 22: 64. 1923. M. aeruginosa f. pseudofilamentosa Elenkin, Monogr. Algar. Cyanophyc., Pars Spec. 1: 105. 1938.

Microcystis aeruginosa f. minor Elenkin, Not. Syst. Inst. Crypt. Hort. Bot. Reipubl. Ross. 3: 15. 1924.

Microcystis fusca Zalessky, Rév. Gén. de Bot. 38: 33. 1926.

Microcystis ramosa Bharadwaja, Proc. Indian Acad. Sci. 2: 96. 1935.

Microcystis aeruginosa var. elongata B. Rao, Proc. Indian Acad. Sci. 6: 341. 1937.

Microcystis aeruginosa f. sphaerodictyoides Elenkin, Monogr. Algar. Cyanophyc., Pars Spec. 1: 100. 1938.

Microcystis botrys Teiling, Bot. Notiser 1942: 63. 1942.

Microcystis flos-aquae var. major Nygaard, K. Danske Videnskab. Selsk. Biol. Skr. 7(1): 179. 1949.

Microcystis lutescens Schiller, Sitzungsber. Oesterr. Akad. Wiss., Math.-Nat. Kl., 1, 163: 112. 1954.

Microcystis punctata Schiller, Sitzungsber. Oesterr. Akad. Wiss., Math.-Nat. Kl., 1, 163: 118. 1954.

Plantae smaragdinae vel laete aeruginosae, microscopicae, sphaericae, cylindraceae, vel ovoideae, varie et irregulariter lobatae et invaginatae vel clathratae, saepe conferte confluentes, planctonicae, cellulis in divisione binis depresso—sub-sphaericis, aetate provecta globosis diametro [2.5—]3—7[—10.5]μ crassis, sine ordine (aut in seriebus eucapsoideis) per gelatinum vaginale distributis; gelatino vaginale hyalino, homogeneo, ad superficiem distincte delimitato vel (saepe omnino) diffluente; protoplasmate dilute aerugineo, pseudovacuolis farcto. FIGS. 1—11.

In the plankton of freshwater lakes, ponds, and sluggish rivers, often appearing as conspicuous and heavy water blooms during the warm seasons of the year. The morphology, life history, and synonymy of this species have been discussed by Drouet & Daily in Field Mus. Bot. Ser. 20: 67—83 (1939). In pure cultures, the gelatinous matrix hydrolyzes completely and the cells become less pseudovacuolate than is characteristic of them in natural habitats. Often confused with this species are the various growth-forms of the bacterial Lamprocystis rosea (Kütz.) Dr. & Daily, the plants of which contain a red pigment in most living collections; this pigment is often lost in material preserved in liquid.

Specimens examined:

LATVIA: Riga, H. Skuja, 24 Aug. 1924 (DA, FC). SWEDEN: on lacu Kalungen Daliae, V. Wittrock, Aug. 1866 (Type of Polycystis flos-aquae Wittr. in Wittr. & Nordst., Alg. Exs. no. 298, S; isotypes, FC [Fig. 4], L, MIN, NY), 25 Aug. 1862 (D, S), Sept. 1882 (as P. flos-aquae in Wittr. & Nordst., Alg. Exs. no. 509, D, FC, L, MIN, NY), Jun. (as Aphanocapsa pulchra in Aresch., Alg. Scand. Exs., Ser. Nov., no. 388, FC); in lacu Mälaren ad Flottsund Uplandiae, Wittrock, 3 Nov. 1878 (Type of Polycystis prasina Wittr., S[Fig. 6]; isotypes, FC, L, MICH, TA, US); Bessvatn, K. Thomasson, 18 Jul. 1951 (isotype slide of Eucapsis alpina f. violacea Thomasson from the author, DA); in lacu Lötsjön par. Funbo Uplandiae, Wittrock, 6 Oct. 1878 (isotypes of Microcystis aeruginosa var. major Elenk. as Polycystis aeruginosa in Wittr. & Nordst., Alg. Exs. no. 296, FC, L, MICH), Oct. (as P. aeruginosa in Aresch., Alg. Scand. Exs., Ser. Nov., no. 429, FC, L); pond, Nötesjö, Malmöhus, B. Carlin-Nilsson 525, 17 Aug. 1937 (D, NY); Dagstorpssjön, Malmöhus, Carlin-Nilsson 140, 25 Aug. 1931 (D, NY); lake, Fiolen, Kronoberg, Carlin-Nilsson 516, 27 Jul. 1937 (D, NY); Hammarbysjö in Danviken, Stockholm, G. Lagerheim, 1882 (D, S); Ringsjön ad Bosjökloster, O. R. Holmberg, Sept. 1913 (S); in lacu ad Mullsjö Vestrogothiae, O. Nordstedt, Aug. 1900 (B, NY); Trehörningsjö, Upland, O. Borge, Sept. 1896 (NY); with Anacystis incerta in lacu Wombsjön Scaniae, Nordstedt, Jun. 1901 (B, NY, S, UC). DENMARK: in lacu prope Birkeröd Selandiae, C. Ostenfeld-Hansen, 30 Aug. 1896 (D, C); Sjön Naerum, C. Rasch, Jun. 1880 (C); Lilleröd, Sjaelland, T. Rosenvinge, Sept. 1879 (C); Hut-sö, J. Schmidt, Sept. 1898 (C); Hofmansgave, Fionen, Hofman-Bang (B). POLAND: Ockelsee bei Allenstein, Caspary, Aug. 1862

(B); Danzig, *Klinsmann*, 1858 (B); Sawadda-See (Westpreussen), *P. Hennings*, Sept. 1890 (B); Galgenberge bei Strehlen (Schlesien), *Hilse* (as *Anacystis marginata* in Rabenh. Alg. no. 1522, B, FC, L, NY), Jul. 1859 (as *Polycystis aeruginosa* in Rabenh. Alg. no. 1174, B, FC, MIN, NY); Strehlen, *Hilse* (B, L); bei Habendorf, *Hilse 8* (B); am Grossteiche von Habendorf bei Reichenbach, *Hilse*, Sept. 1862 (B); Rausern bei Breslau, *W. Migula*, Jul. 1877 (as *P. aeruginosa* in Hauck & Richt., Phyk. Univ. No. 296a, L, MIN), May 1887 (as *Microcystis aeruginosa* in Mig., Krypt. Germ., Austr., & Helv. Exs., Alg. no. 30, MICH, NY, TA); Breslau, *O. Kirchner*, 1874 (as *Clathrocystis aeruginosa* in Rabenh. Alg. no. 2424, FC, L, MIN, NY, TA); im Teiche bei Schoffschütz in Oberschlesien, *A. Utgenannt & S. Schmula*, Sept. 1895 (as *Polycystis elabens* in Hauck & Richt., Phyk. Univ. no. 684, L, MIN, NY); Teich in Buchwald bei Schmiedeberg, *G. Hieronymus*, Sept. 1887 (B).

GERMANY: Baden: aus dem Grünloch in Ichenheim, *L. Leiner*, Sept. 1858 (as *P. aeruginosa* in Jack, Leiner, & Stizenb., Krypt. Badens no. 461, FC, YU). Bavaria: Nürnberg, *P. Reinsch* (MICH); in Wassertümpeln bei Erlangen, *Glück*, Sept. 1895 (L, US); Erlangen, *Reinsch* (L, NY, US); Wurmsee, *F. Brand*, Sept. 1897 (Type of *P. ochracea* Brand, B [Fig. 3]). Brandenburg: Wannensee, *P. Hennings*, Sept. 1892 (Henn., Phyk. March. no. 47c, B, FH); Halensee, *Hennings*, Nov. 1890 (Henn., Phyk. March. no. 47b, B, FH); Wilmersdorfer See, Berlin, *Hennings*, Jul. 1882 (Henn., Phyk. March. no. 47a, B, FH); Müggel-See, Friedrichshagen, *Hennings*, Aug. 1892 (B), Sept. 1892 (Henn., Phyk. March. no. 47d, B, FH), *A. Braun* (B); in der Havel bei der Pfaueninsel, *Braun*, Aug. 1863 (B), bei Potsdam, *Bauer* (B, D, FH, L, NY), *L. Rabenhorst*, Jul. 1857 (L); Berlin, *Braun*, Aug. 1854 (B, L), *Jahn*, 1857 (B); Hort. Berol., *Braun*, Aug. 1852 (B); Golssen, *Schumann*, Jul. 1865, 1866 (B); Grunewald-See bei Berlin, *G. Hieronymus*, Aug. 1891 (B); in einem Graben, Neudamm, *Itzigsohn*, May 1855 (B); im Tempelhofer Parkteiche bei Berlin, *Hennings*, Jul. 1883 (B; Henn., Phyk. March. no. 46, B, FH). Hamburg: Aussenalster, *Fitschen*, Jul. 1906 (S). Mecklenburg-Pomerania: Wolgast-See bei Haeringsdorf, *A. Braun*, Sept. 1864 (B, L). Oldenburg: Jever, *Jürgens* (Type of *Palmella cyanea* Kütz., L). Saxony: Teiche in Bernbruch bei Lausigk, *P. Richter*, Jul. 1864 (as *Coelosphaerium Kuetzingianum* in Rabenh. Alg. no. 1791, FC); Mannsfeld, *M. Marsson*, Jul. 1896 (NY); Leipzig, Lindenau, *Marsson*, Sept. 1897 (NY); Teich in Collau am Mulde, *Marsson*, Sept. 1897 (NY); Gohlis bei Leipzig, *Richter*, Sept. 1878 (NY), *Auerswald*, Sept. 1852 (B); in lacu salso Mansfeldensis prope Halam, *P. Richter* (Type of *Polycystis scripta* Richt. in Hauck & Richt., Phyk. Univ. no. 92, B [Fig. 2]; isotypes, L, MIN); Brösen bei Grimma, *Richter*, Aug. 1894 (in Hauck & Richt., Phyk. Univ. no. 748a, L, MIN, NY); in einem Teiche bei Anger, Leipzig, *Richter*, Aug. 1879 (L, PH, TA); auf Teichen, Mortizburg (L); auf einem Fischteiche, Ponickau, *Auerswald* (B, FC, FH, NY), 27 Jul. 1852 (as *P. ichthyoblabe* in Rabenh. Alg. no. 210, B, FC, L, MIN, NY); Leipzig (Lipsiae), *Kunze* (B, FC, NY, PH), *Auerswald* (FC), *Richter* (FC); in einem Graben am Grossen Garten in Dresden, *C. Schiller*, Jul. 1888 (in Hauck & Richt., Phyk. Univ. no. 296b, L, MIN); Schadebach bei Makranstadt (Leipzig), *H. Reichelt* (in Hauck & Richt., Phyk. Univ. no. 297, L, MIN); Burgsee bei Salzungen, *A. Braun*, Sept. 1862 (Type of *Polycystis viridis* A. Br. in Rabenh. Alg. no. 1415, B; isotypes, FC, L, MIN, NY); Salzunger See, *A. Röse* (as *Microhaloa firma* in Rabenh. Alg. no. 453, FC, FH, L, NY, UC); Salzsee bei Halle, *O. Kuntze* (NY). Thuringia: Georgental, in einem Teich am Bahnhof, *W. Migula*, 1 Sept. 1928 (as *Microcystis aeruginosa* in Mig., Krypt. Germ., Austr., & Helv. Exs., Alg. no. 242, FC, TA). Württemberg: Stuttgart, *G. von Martens* (Type of *Micraloa aeruginosa* Kütz., L [Fig. 9]; isotypes, B, L, NY, UC), Jun. 1827, Aug. 1830, Aug. 1947 (B).

CZECHOSLOVAKIA: Moravia: près Eisgrub (L). Bohemia: Bystrice u Benesova, rybníky u drahy, *A. Hansgirg*, Aug. 1884 (FC, W); piscina Velky Pálenec prope Blatná, *B. Fott*, 4 Aug. 1948 (FC); Chlumec—Pilár, blíz stanice, *Hansgirg*, Aug. 1887 (FC, W); Schlossteich in Cimelic, *Hansgirg*, Aug. 1887 (FC, W); Dobrís, *Hansgirg*, Aug. 1886 (FC, W); the Great Pond of Doksy, *Fott*, 12 Jul. 1947 (FC); Krivoklát, *Hansgirg*, Sept. 1884 (FC, W); Amalienhof u Krivoklátu, *Hansgirg*, Sept. 1884 (FC, W); Gumnitzsee, Joachimstal, *W. Panknin*, Aug. 1937 (B); bei Libochowitz, *Hansgirg*, Jul. 1888 (Type of *Polycystis aeruginosa* var. *major* Wittr., W; isotype, FC [Fig. 7]); in Dolejsi pond near Lnáre, *Fott*, 15 Aug. 1947 (FC); pond Mlýnský near Lnáre, *Fott*, 3 Aug. 1947 (FC); Nepomuk, *Hansgirg*, Aug. 1887 (FC, W); Stráncice, *Hansgirg*, Sept. 1885 (FC, W); Teplice, *Hansgirg*, Jul. 1883 (FC, W); pond Starý Hospidár near Trebon, *Fott*, 19 Jun. 1946 (FC); pond Velká Kus, *Fott*, 2 Aug. 1947 (FC); Planina a Zíc u Chlumec blíz Treboné, *Hansgirg*, Sept. 1887 (FC, W). AUSTRIA: Teich im Schwarzenberggarten, Wien, *K. Rechinger*, Jun. 1907 (as *Coelosphaerium dubium* in Mig., Krypt. Germ., Austr., & Helv. Exs., Alg. no. 154, MICH, TA); Vindobonae in piscinis horti Caesarei Schönbrunn, *C. de Keissler* (as *Clathrocystis aeruginosa* in Mus. Vindob. Krypt. Exs. no. 1517, B, FC, L, NY, US); Vindobonae in piscinis hortorum publicorum, *Rechinger* (as *Microcystis flos-aquae* in Mus. Vindob. Krypt. Exs. no. 2335, B, FC, L, NY); in lacu Hilmteich ad Graz, *Schmula* (as *Coelosphaerium dubium* in Kerner, Fl. Exs. Austro-Hung. no. 3196, FC, L, MO); Ober-Premstätten—Liebach prope Graz, *Hansgirg*, Sept. 1890 (FC, W). HUNGARY: Budapest in lacu Városligeti tó, *F. Filárszky* (as *Polycystis aeruginosa* in Mus. Vindob. Krypt. Exs. no. 226, B, FC, L, NY); Budapest in excavationibus Lágymányosi holt Dunaág, *Filarszky*, May 1911 (Fl. Hungar. Exs. Alg. no. 21, FC, L, US). TRIESTE: ex herb. Hauck (D, FH, NY). SWITZER-

LAND: Zürich, Katzensee, auf dem Grunde am Strande liegend, *K. von Nägeli* (ZT). ITALY: in una vasca all'Orto Botanico di Pisa, *Savi,* 1865 (UC). NETHERLANDS: Haague, *van den Bosch* (L); Haagsche Bosch, den Haag, *W. F. R. Suringar D18,* May 1857 (D, L); Witte Singel, Leiden, *J. T. Koster 233,* 11 Aug. 1938 (FC, L); in aq. dulc. stagn. Ludg.-Batav. (Leiden), *van den Bosch 376,* Aug. 1846 (B, L); Holland, *A. Weber-van Bosse,* Oct. 1891 (L); forest, The Hague, *W. Trelease,* Jun. 1884 (MO); de Kaag, op de plas, *Koster 953* (L); Rotterdam, Kralingse plas, *J. Poolman,* Jul. 1944 (L); Noorden, sloot langs de weg, *Koster 501,* Jun. 1940 (L); Kolkven, Oisterwijk, *G. P. H. van Heusden,* 17 May 1947 (FC, L). FRANCE: Falaise, Calvados, *A. de Brébisson* (B), *Brébisson 528, 594* (L); Carentan, Manche, à la surface du canal de l'hospice, herb. Lebel (B, FC); dans un étang du parc du Château de Flamanville, Manche, *P. Frémy,* 10 Aug. 1932 (as *Microcystis aeruginosa* in Hamel, Alg. de Fr. no 151, FC); hortus, Clermont-Ferrand, *W. F. R. Suringar* (FC, L); Angers, Maine et Loire, *F. Hy* (B).

PORTUGUESE EAST AFRICA: Maloti lake near Masiyeni, S. Chopiland, *E. L. Stephens 38,* 27 Jun. 1928 (D, NY). UNION OF SOUTH AFRICA: Transvaal: Hartebeestpoort dam, *Mrs. G. E. Hutchinson 15,* 29 Apr. 1928 (D); Rietkuil, Bethal district, *M. E. Blenkiron & D. Weintroub 24,* 25 Feb. 1928 (D); Vaaldam, *E. L. Stephens,* Apr. 1945 (Type of *Microcystis toxica* Steph., L; isotype, FC [Fig. 1]). Cape of Good Hope: Zeekoe Vlei near Cape Town, *Mrs. Hutchinson 35,* 5 Feb. 1928 (D).

MASSACHUSETTS: Arlington, *E. Dewart,* 27 Jul. 1891 (FC, NY); cemetery pond, Belmont, *A. B. Seymour,* Sept. 1918 (FC); Herring Run, Brewster, *L. Walp,* Aug. 1940 (FC); Fresh pond, Cambridge, *W. G. Farlow,* Oct. 1882 (FH, MO), 25 Oct. 1879 (FC), *G. T. Moore,* Oct. 1893 (B), *H. H. Bartlett 1179,* Oct. 1907 (MICH), *A. H. Moore 4090,* 11 Oct. 1907 (FC), *A. B. Seymour,* Jul. 1913 (FC); pond, Cuttyhunk island, *F. S. Collins 5723,* Aug. 1907 (NY); pond south of Nonquitt, Dartmouth, *E. T. Rose & F. Drouet 1891,* 18 Jul. 1936 (D); Fresh pond, Falmouth, *W. R. Taylor,* Jul. 1921 (TA), *T. E. Hazen 5,* 6 Jul. 1926 (DA, FC); Oyster pond, Falmouth, *Taylor,* 12 Aug. 1922 (FC, TA), *Drouet 2118,* 30 Jul. 1937 (D, FH, NY, S), *J. Bader 84* (FC), *129* (D), Jul. 1938; Episcopal Ocean, Falmouth, *Rose & Drouet 1867,* 4 Jul. 1936 (D, NY, S), *C. M. Palmer,* 2 Sept. 1937 (DA, FC), *R. N. Webster,* 20 Jun. 1938 (D, FH, NY), *Bader 127,* 30 Jul. 1938 (D); Shivericks pond, Falmouth, *E. T. Moul 6409,* 2 Aug. 1949 (FC); basin no. 3, Framingham, *Farlow,* Nov. 1881 (FH); Ashumet lake northeast of Hatchville, *Moul 6404,* 27 Jun. 1949 (FC); Hummock pond, Nantucket island, *W. R. Taylor,* 1920 (FC, TA); Long pond, Nantucket, *Taylor & B. F. D. Runk,* 17 Jul. 1938 (D); Hammond pond, Newton, *H. M. Richards,* Oct. 1889 (NY); Horn pond, Woburn, *Farlow,* 2 Aug. 1879 (FC, D, FH). RHODE ISLAND: Mashapaug pond, Providence, *W. J. V. Osterhout,* 15 Oct. 1892 (as *Clathrocystis aeruginosa* in Coll., Hold., & Setch., Phyc. Bor.-Amer. no. 51, FC, L, MICH, MIN, NY); Providence, *Nichols,* 1877 (D, FH); Haley's pond, Cranston, *F. S. Collins 6444,* Sept. 1911 (NY). CONNECTICUT: plankton of Smith Brothers pond, central Connecticut, *G. E. Hutchinson,* 22 Dec. 1942 (FC). NEW YORK: Central park, New York City, Jul. 1865 (B); in Basic reservoir, Rensselaerville, *M. S. Markle 25,* Jul. 1947 (EAR, FC). NEW JERSEY: small woodland lake 2 miles north of Glenwood, Sussex county, *E. T. Moul 7169,* 18 Oct. 1950 (FC). PENNSYLVANIA: pond north of Kennett Square, *F. W. Pennell & W. S. May,* Oct. 1921 (TA); plankton of Shenango river, Pymnatuning Dam, Greenville, Clarksville, Sharon, and Wheatland, *L. Walp & W. Murchie 6—15,* Jul. & Sept. 1941 (FC).

VIRGINIA: in Queens Creek (Haw Tree Landing), York county, *A. F. Chestnut,* 8 Sept. 1942 (FC, ST). SOUTH CAROLINA: in Lake Chapin near Myrtle Beach, Horry county, *P. J. Philson SC72,* 21 Sept. 1938 (FC). FLORIDA: Gainesville, *M. A. Brannon 11, 12, 25, 27, 227b, 365, 419, 422—426,* Sept. 1941—May 1947 (FC, PC); Lake City, *R. K. Strawn 589,* 2 Aug. 1947 (FC, PC); in Lake Harris, Leesburg, Lake county, *F. Drouet & Brannon 11049, 11053, 11056, 11062,* 19 Jan. 1949 (FC); Henderson camp (Lee county?), *O. Tollin,* May 1886 (S). ONTARIO: Bay of Quinte, Lake Ontario, Prince Edward county, *A. Tucker,* 1946 (DA, FC); Clear lake, Kawartha lakes, Peterborough county, *W. A. Strow,* 26 Aug. 1951 (DA, FC). OHIO: Auglaize county: St. Mary's lake, St. Marys, *W. A. & F. K. Daily 633, 634,* 22 Jul. 1940 (DA, FC). Franklin county: Goodale park, Columbus, *E. H. Ahlstrom,* 17 Oct. 1933 (DA, FC). Hamilton county: pure culture no. 15 (from G. C. Gerloff, culture no. 1036) in Environmental Health Center, Cincinnati, *C. M. Palmer* (FC); pond in Arlington cemetery, Cincinnati, *W. A. Daily & J. H. Hoskins 109,* 4 Oct. 1939 (DA, FC), *Daily 418,* 5 Jul. 1940 (DA, FC), *W. A. & F. K. Daily 741,* 2 Sept. 1940 (DA, FC), *E. Ward,* 18 Jul. 1938 (DA, FC); lake in Burnet Woods, Cincinnati, *Daily & A. Blickle 104,* 1 Oct. 1939 (DA, FC); Sharon Woods lake, Cincinnati, *W. A. & F. K. Daily 743,* 2 Sept. 1940 (DA, FC). Defiance county: in canal, Independence state park, east of Defiance, *W. A. & F. K. Daily 642,* 23 Jul. 1940 (DA, FC). Ottawa county: Lake Erie, Put-in-Bay, *C. E. Taft,* 1935, 3 Aug. 1938 (DA, FC), near Gibraltar Isle, *Daily & Taft 649A,* 24 Jul. 1940 (DA, FC); pond, Port Clinton, *F. K. & W. A Daily 648,* 23 Jul. 1940 (DA, FC). Putnam county: in ponds, Erie-Miami canal, Delphos, *W. A. & F. K. Daily 635, 636,* 22 Jul. 1940 (DA, FC); in Auglaise river at Cascade park near Dupont, *W. A. & F. K. Daily 640,* 22 Jul. 1940 (DA,

FC). Stark county: in reservoir, Beach City, *W. A. & F. K. Daily 576*, 26 Jul. 1940 (DA, FC). Summit county: in Portage lake south of Akron, *W. A. & F. K. Daily 575*, 26 Jul. 1940 (DA, FC). KENTUCKY: pond near Walton, Boone county, *B. B. McInteer 13, 652*, 22 Aug. 1929 (DA, FC), *J. B. Lackey*, 10 Oct. 1940 (DA, FC); Blue pond, Fulton county, *McInteer 545*, 13 Aug. 1929 (DA, FC); pond 12 miles east of Hickman, Fulton county, *McInteer 529*, 13 Aug. 1929 (DA, FC); pond 2 miles south of Glasgow, Barren county, *McInteer 516*, 10 Aug. 1929 (DA, FC); in reservoir, Lexington, *Daily, McInteer, & A. Harvill 845*, 30 Oct. 1940 (DA, FC); pond in fish hatchery, Fort Mitchell, Kenton county, *J. H. Hoskins & Daily 251*, 2 Dec. 1939 (DA, FC); sandpit, Greenwood road near Louisville, *H. Bishop*, 30 Oct. 1931 (DA, FC).

TENNESSEE: Radnor Lake, *H. C. Bold*, 20 Oct. 1936 (FC); Percy Warner lake, Nashville, *Bold 13*, 1933 (FC); pond on M. Hayes farm, Obion county, *H. Silva 1316, 1317, 1319*, 12 Jul. 1949 (FC, TENN). ALABAMA: pond no. 9, Marion Fish Cultural Station, Marion, *J. R. Snow 7*, 9 Jul. 1953 (DA,FC). MICHIGAN: Thayer's lake, Torch lake, and Elk lake, Antrim county, *J. H. Hoskins*, 24 Aug. 1934 (DA, FC); McDonald lake, Yankee Springs, Hastings, Barry county, *G. T. Velasquez 16*, 4 Aug. 1936 (D, TA); French Farm lake, Cheboygan county, *G. W. Prescott 3M6*, Aug. 1941 (FC); Interlochen, Grand Traverse county, *J. H. Hoskins*, 29 Aug. 1934 (DA, FC); Lake Leelanau, Leelanau county, *Hoskins*, 28 Aug. 1934 (DA, FC); Pasinski pond, Genoa township, Livingston county, *W. F. Carbine*, 1938 (FC, TA); Mud lake west of Curtis, Mackinac county, *H. K. Phinney 4M41/15-1*, 19 Jul. 1941 (FC, PHI); Lake George, Oakland county, *C. E. Taft 155*, Aug. 1936 (DA, FC); Anne Louise lake, Schoolcraft county, *Phinney 9M41-16*, 19 Jul. 1941 (FC, PHI); Three Lakes, Ann Arbor, *L. N. Johnson 5*, 27 Sept. 1892 (FC, NY).

INDIANA: Bartholomew county: Columbus By-pass pond, Columbus, *F. K. & W. A. Daily 1581B*, 20 Apr. 1947 (DA, FC), *F. K. & W. A. Daily & M. F. Cundiff 2104*, 13 May 1949 (DA, FC). Carroll county: Lake Freeman near Monticello, *Daily 5, 15, 877*, Jul. 1938, 1941 (DA, FC). Decatur county: in Cupp's pond 5 miles east of Greensburg and in limestone pond between Hartsville and Greensburg, *W. A. & F. K. Daily & C. Kenoyer 1166, 1168*, 19 Sept. 1943 (DA, FC). Elkhart county: Simonton lake 4 miles north of Elkhart, *F. K. & W. A. Daily 2616, 2618*, 26 Aug. 1952 (DA, FC). Fulton county: Nyona (North Mud) lake 4 miles east of Fulton, *F. K. & W. A. Daily 1476, 1477, 1479, 1480*, 29 Jun. 1946 (DA, FC). Kosciusko county: Big Barbee lake 3 miles south of North Webster, *F. K. & W. A. Daily 2271*, 27 Jul. 1950 (DA, FC); Dewart lake, *F. K. & W. A. Daily 2115*, 9 Jul. 1949 (DA, FC); Eagle lake, *H. W. Clark & B. W. Evermann 255*, Aug. 1906 (FC, US); Kuhn lake, Tippecanoe township, *F. K. & W. A. Daily 2268*, 27 Jul. 1950 (DA, FC); Pappakeechie lake south of Lake Wawasee, *F. K. & W. A. Daily 2273*, 28 Jul. 1950 (DA, FC), *B. H. Smith 145*, 15 Sept. 1929 (DA, FC); Ridinger lake, Washington township, *F. K. & W. A. Daily 2270*, 27 Jul. 1950 (DA, FC); Beaver Dam lake 4 miles northwest of Silver Lake, *Daily 82*, 10 Jun. 1939 (DA, FC); Silver lake north of Silver Lake, *F. K. & W. A. Daily 2168, 2169*, 22 Jul. 1950 (DA, FC); Lake Wawasee, *F. K. &W. A. Daily 2190, 2209, 2232, 2258, 2274, 23—28* Jul. 1950 (DA, FC); Lake Winona, *B. H. Smith 199*, 15 Sept. 1929 (DA, FC), *C. M. Palmer*, 30 Aug. 1935 (D, DA, FC), *Palmer 168*, Summer 1936 (DA, FC). Lake county: Cedar lake at Cedar Lake, *F. K. & W. A. Daily 2561, 2562*, 28 Aug. 1951 (DA, FC); Calumet river north of Miller station, *P. D. Voth & F. Drouet 2367*, 28 Sept. 1938 (FC, NY). Marion county: in Bacon's Bog pond, Indianapolis, *W. A., F. K., J. H., & M. Daily 861, 862*, 27 Jul. 1941 (DA, FC). Marshall county: Lake of the Woods northeast of Plymouth, *Clark & Evermann 27*, 12 Jul. 1909 (FC, US); Lost lake, Culver, *F. K. & W. A. Daily 1518*, 3 Jul. 1946 (DA, FC); in spring by Culver depot, *Clark & Evermann 272*, 27 Sept. 1900 (FC, US); Lake Maxinkuckee, *Clark & Evermann s. n. and 52*, 18 Oct., 15 Nov. 1906 (FC, US). Monroe county: in a limestone sinkhole pond 3 miles south of Bloomington, *F. K. & W. A. Daily 1004*, 7 Sept. 1942 (DA, FC); Wiemer's lake 4 miles northwest of Bloomington, *F. K. & W. A. Daily 1005*, 7 Sept. 1942 (DA, FC). Montgomery county: in lagoon in Shades state park, Waveland, *F. K. & W. A. Daily 965*, 6 Jun. 1942 (DA, FC), *L. Lee*, 17 Jun. 1942 (DA, FC). Morgan county: Brickyard pond adjoining Bethany Park, *F. K. & W. A. Daily 1031*, 20 Sept. 1942 (DA, FC). Porter county: in Wahob lake and in a pond connected with Spectacle lake, north of Valparaiso, *F. K. & W. A. Daily 2572, 2575*, 30 Aug. 1951 (DA, FC). Steuben county: Clear lake 6 miles east of Fremont, *F. K. & W. A. Daily 1933*, 13 Sept. 1947 (DA, FC); Crooked lake at Crooked Lake, *Palmer B50*, 30 Sept. 1933 (DA, FC), *F. K. & W. A. Daily 2459, 2461*, 18 Jun. 1951 (DA, FC); Fox lake southwest of Angola, *F. K. & W. A. Daily 2487*, 19 Jun. 1951 (DA, FC); Lake George 6 miles north of Angola, *F. K. & W. A. Daily 1934*, 13 Sept. 1947 (DA, FC); Green lake west of Fremont, *F. K. & W. A. Daily 2450*, 17 Jun. 1951 (DA, FC); Hamilton lake, Circle park, Hamilton, *F. K. & W. A. Daily 2443, 2445*, 17 Jun. 1951 (DA, FC); Lake James, Pokagon state park, *F. K. & W. A. Daily 1918*, 12 Sept. 1947 (DA, FC); Mill Pond lake at Nevada Mills, *F. K & W. A. Daily 1936*, 14 Sept. 1947 (DA, FC); Lake Pleasant 4 miles east of Orland, *F. K. & W. A. Daily 2470*, 18 Jun. 1951 (DA, FC); Pleasant lake at Pleasant Lake, *F. K. & W. A. Daily 2458*, 18 Jun. 1951 (DA, FC); Silver lake 4 miles west of Angola, *F. K. & W. A. Daily 2488, 2490*, 19 Jun. 1951 (DA, FC). Vigo county: *B. H. Smith 647*, 4 Aug. 1930 (DA, FC); St. Mary's lake, *Smith*

393, 13 Aug. 1930 (DA, FC). Wells county: in Kunkel lake 2 miles east of Bluffton, *F. K. & W. A. Daily 2433*, 16 Jun. 1951 (DA, FC).
WISCONSIN: Pleasant lake near Lauderdale lakes, *G. W. Prescott 3W28*, Aug. 1938 (FC); Razorback lake, *E. J. Kryski 49*, 26 Jul. 1938 (DA, FC). Burnett county: Fish lake, *G. M. Smith 64* (FC). Dane county: laboratory cultures, *R. Evans, G. Fitzgerald*, 8 Jun., 25 Oct. 1949 (FC); Fourth lake, Madison, *W. Trelease*, 1882 (MO); Lake Mendota, Madison, *Trelease*, 1882 (FC, D, FH, MO), *R. E. Blount*, 2 Oct. 1886 (FC). Kenosha county: Silver lake in Silver Lake, *E. H. Ahlstrom*, 5 Jun. 1932 (DA, FC). Polk county: North Twin lake, *G. M. Smith 61*, 22 Aug. 1917 (FC). Rock county: in small lake near Clear lake, *J. B. Lackey*, Aug. 1943 (FC). Vilas county: Alequash lake, *Prescott W155, 2W268*, 13 Aug. 1936, 11 Jul. 1937 (FC); High lake, *Prescott 2W59*, 22 Jun. 1937 (FC). Washburn county: Spooner lake, *W. Trelease* (MO). Waushara county: pond east of Wild Rose near Silver lake, *Prescott 2W330*, 19 Jul. 1937 (FC). ILLINOIS: Lake county: Diamond lake, Fox lake, Lake Marie, Petit lake, Pistakee lake, Slocum lake, and Lake Zurich, *E. H. Ahlstrom*, 5 Jun. 1932 (DA, FC); Grass lake and Slocum lake, *M. E. Britton 924, 926*, 24-25 Aug. 1939 (DA, FC). MINNESOTA: Fountain and Iowa lakes (western prairie lakes), *C. B. Reif*, 28 Jun. 1936, 26 Aug. 1938 (DA, FC). Anoka county: Rice lake, *S. Eddy*, 16 Aug. 1932 (DA, FC). Cass county: Cass lake, *S. Eddy*, 10 Sept. 1936 (DA, FC). Clearwater county: Lake Itasca and De Soto lake, *Eddy*, 29 Aug. 1935, 20 Aug. 1940 (DA, FC). Hennepin county: Bass pond, *Eddy*, 22 Sept. 1931 (DA, FC); Cedar lake, *Eddy*, 14 Aug. 1931 (DA, FC); Cross lake just north of Minneapolis, *Reif*, 22 Sept. 1936 (DA, FC); Long lake, *B. T. Shaver & J. E. Tilden*, 5 Sept. 1895 (as *Clathrocystis aeruginosa* in Tild., Amer. Alg. no. 194A, B, FC, MIN, NY, US); Lake of the Isles, Minneapolis, *K. Damann*, 18 Aug. 1936 (FC), *F. Drouet 5594*, 19 Aug. 1944 (FC); Lake Minnetonka at Spring Park, *Drouet & A. A. & E. Cohen 5007, 5016*, 8 Aug. 1943 (FC); Lake Nokomis, *Eddy*, 14 Aug. 1931 (DA, FC). Itasca county: Bowstring lake, *Reif*, 29 Jul. 1936 (DA, FC). Jackson county: Heron lake, *Reif*, 11 Jul. 1938 (DA, FC). Kandiyohi county: Kandiyohi lake, *Reif*, 7 Sept. 1938 (DA, FC). Ottertail county: Crystal, Little Pelican, Sand, Sauer, Sear, Spirit, and Wimer lakes, *Eddy*, Jul.—Aug. 1940 (DA, FC); Ottertail lake, *Reif*, 13 Sept. 1938 (DA, FC). Ramsey county: in lake, Como park, St. Paul, *J. E. Tilden*, 10 Aug. 1895 (as *Clathrocystis aeruginosa* in Tild., Amer. Alg. no. 194B, B, FC, MIN, NY, US); Lake Josephine, *Eddy*, 18 Jul. 1931 (DA, FC); in Keller lake north of St. Paul, *Drouet & E. Cohen 5045, 5050*, 9 Aug. 1943 (FC); McCarron's pond, *Reif*, 15 Aug. 1931 (DA, FC); Lake Owasso, *Eddy*, 15 Aug. 1931 (DA, FC); Round lake, *Reif*, 13 Jul. 1938 (DA, FC).
IOWA: Center lake, *G. W. Prescott 317*, 10 Jul. 1925 (D, NY); Des Moines river, Boone county, *W. C. Starrett P34, P38, P47, P165*, 18 & 29 Jul. 1946, 9 Aug. 1947 (FC); Lake Okoboji, *M. S. Markle*, Summer 1926 (DA, FC), *Prescott 316, 338*, 30 Jun. 1925, 24 Jun. 1926 (D), *Prescott*, 23 Aug. 1930 (FC). MISSOURI: in lake, Pertle Springs, Warrensburg, *Drouet 778*, 26 Oct. 1930 (D); St. Louis, *G. T. Moore*, Aug. 1913 (D); Agriculture pond, Columbia, *Drouet 990*, 27 Jul. 1932 (D); laboratory culture (no. 1036 from Dr. Gerloff), University of Missouri, Columbia, *V. W. Proctor*, 31 Oct. 1952 (FC). ARKANSAS: Monticello, Drew county, *D. Demaree 24557*, 16 Jul. 1943 (FC); lake in Malvern, Hot Spring county, *N. E. Gray 284*, 9 Sept. 1939 (FC); artificial lake east of 1000 Drippings, Garland county, *Gray 659*, 3 Sept. 1940 (FC); Lake Hamilton, Garland county, *Gray 134-35*, 18 Aug. 1939 (FC). LOUISIANA: pond in pasture 1 mile northwest of Baton Rouge, *G. W. Prescott La17*, 20 Jun. 1938 (D). SASKATCHE-WAN: lake near Prince Albert, *F. Rising-Moore & O. H. Stiffs Brancerethe*, 2 Aug. 1951 (FC). NORTH DAKOTA: Devils Lake, *M. A. Brannon 1*, Aug. 1913 (FC, UC). NEBRASKA: Arthur county: Haythorn lake, *W. Kiener 15457—15460*, 10 Sept. 1943 (FC, KI). Dawson county: Johnson lake south of Lexington, *Kiener 24671, 24672a*, 7 Jul. 1949 (FC, KI). Dodge county: sandpit lakes, Fremont, *Kiener 19700, 23934*, 15 Sept. 1945, 1 Jul. 1948 (FC, KI). Garden county: Crescent lake, Oshkosh, *Kiener 16183, 16190*, 4 Nov. 1943 (FC, KI). Cherry county: Hackberry, Dewey and Watts lakes, *E. R. Walker & E. N. Anderson*, Jul. 1912 (NEB); Hackberry lake, *E. Palmatier*, Jul. 1936 (NEB). Hall county: sandpit pond, Grand Island, *Kiener 15176, 15177*, 29 Aug. 1943 (FC, KI); gravel-pit pond southeast of Grand Island, *Kiener 17176*, 17 Aug. 1944 (FC, KI). Hitchcock county: lake 3 miles east of Stratton, *Kiener 10470, 10477*, 24 Jul. 1941 (FC, KI). Holt county: shore of Goose lake 25 miles south of O'Neill, *Kiener 21390*, 8 Aug. 1946 (FC, KI). Kimball county: Bennett reservoir east of Kimball, *Kiener 22706*, 18 Aug. 1947 (FC, KI). Lancaster county: Du Teau farm pond, Lincoln, *Kiener 21431*, 21 Aug. 1946 (FC, KI). Lincoln county: Sutherland reservoir, Sutherland, *Kiener 24693, 24695*. 11 Jul. 1949 (FC, KI); fish-hatchery ponds south of North Platte, *Kiener 17261*, 22 Aug. 1944 (FC, KI); Lake Maloney, North Platte, *Kiener 24685, 24687, 24689—24691*, 10 Jul. 1949 (FC, KI), *Kiener 24029, 24031*, 28 Jul. 1949 (FC, KI).
KANSAS: stagnant pond, Pittsburg, *R. Patrick*, 5 Sept. 1938 (FC). TEXAS: sewage disposal lakes at Camp Barkeley, Taylor county, *W. L. Tolstead 7750*, 29 Jan. 1944 (FC). COLORADO: Barr lake, Adams county, *R. Prettyman*, 30 Jul. 1939 (DA, FC). UTAH: Strawberry reservoir, Wasatch county, *Utah Party F19*, 17 Mar. 1934 (FC), *A. S. Hazzard F16*, 24 Aug. 1933 (FC).

41

ARIZONA: station 13, Mary's lake, Flagstaff, *H. S. Colton*, 28 Aug. 1923 (Type of *Microcystis aeruginosa* f. *occidentalis* W. R. Taylor, TA). BRITISH COLUMBIA: Deer lake, Burnaby, New Westminster county, *J. E. Nielsen 84, 87*, 7 Sept. 1924, 24 May 1925 (FC). WASHINGTON: Green lake, Seattle, *N. L. Gardner*, Dec. 1903 (as *Clathrocystis aeruginosa* in Coll., Hold., & Setch., Phyc. Bor.-Amer. no. 1153, FC, L, MICH, MIN, NEB, NY, TA, US); lake on Fidalgo island, *L. E. Griffin*, Summer 1938 (FC). CALIFORNIA: in Clear lake, 3 miles north of Lower Lake, Lake county, *Gardner 7996*, 25 Oct. 1936 (FC, UC); in Puddingstone lake near La Verne, Los Angeles county, *G. J. Hollenberg 1527, 1646*, 30 Sept. 1924, 13 Sept. 1936 (FC, UC); Sweet-water dam, San Diego county, *B. Bostwick 72* (FC, UC); Laguna Puerca, San Francisco, *K. Brandegee*, Mar. 1913 (UC). MEXICO: in Lake Chapultepec, Mexico City, *R. Patrick 148*, 19 Jul. 1947 (FC, PH). GUATEMALA: Lake Amatitlan, *W. A. Kellerman 5034*, Jan. 1906 (FC), *S. E. Meek 1* (Type of *Clathrocystis robusta* Clark, FC [Fig. 11]), *9, 11, 12, 16, 21—23, 25, 26, 40*, Jan., Feb. 1906 (FC); La Laguna one mile north of Jalapa, *J. A. Steyermark 32246*, 30 Nov. 1939 (FC); Lago Atescatempa south of Asunción Mita, dept. Jutiapa, *Steyermark 31841*, 17 Nov. 1939 (FC). COSTA RICA: Laguna del Misterio, al norte de San José, *J. M. Orozco C. 35—37, 48—50, 52*, 1935 (FC). PANA-MA: Barro Colorado island, Canal Zone, *A. M. Chickering*, 5 Jul., 4 Aug. 1936 (D, FH, L, NY, S). BRAZIL: Ceará: Açude São Francisco, São Francisco, *S. Wright*, 2 Nov. 1937 (D, FH, NY); Açude Cedro near Quixadá, *Wright & P. de Azevedo 1539*, 31 Aug. 1935 (D, FH, MICH, NY); Lagôa Porangabuçú, Fortaleza, *F. Drouet 1504*, 3 Dec. 1935 (D, DT, NY). Pará: in small bays in Rio Tapajoz, *J. W. H. Trail 165* (Type of *Microcystis caerulea* Dickie, BM). Paraíba: Açudes Baixo de Pão, Lapa, Puxinana, Simão, and Velho, Campina Grande, *Wright 1558, 1559, 1561, 1565, 1567, 1568, 1572—1574, 1578, 1587, 1589, 1592, 1606, 1608, 1967, 1985, 1992, 1998—2000, 2004, 2017, 2019, 2041* (D), *4258* (DA, FC), 1933—35; Açude Esperança near Esperança, *Wright 1595, 1974, 2043*, 12 Dec. 1933, 19 Sept. 1934 (D); Açude Linda Flôr near Mojeiro de Baixo, *Wright 1562*, 24 Nov. 1934 (D, MICH, NY); açude near Serra Branca, *Wright 1995*, 6 Apr. 1934 (D). Rio Grande do Sul: Lagôa dos Quadros, *H. Kleerekoper*, Oct. 1941 (FC). ARGENTINA: Laguna Los Chilenos near Dufaur, Laguna Chascomus, and Laguna Blanca Grande, prov. Buenos Aires, *S. Wright 2098, 2101, 2113*, Nov. 1936, Jan. 1937 (D); Lagunas Garcia, Tala, and La China, prov. San Luis, *Wright 2096, 2111, 2112*, Dec. 1936 (D). PERU: Cocha Zapote, Rio Pacaya, Bajo Ucayali, dept. Loreto, *F. Anceita*, July, 1947 (FC).

FORMOSA: fish pond, Tainan, *C. C. Chou*, 11 May 1951 (D). PHILIPPINES: Pasig river, Manila, *W. R. Shaw 347*, Mar. 1909 (B, L, NY, US), *E. Quisumbing 9*, 1929 (TA); Laguna Bay, Laguna, *R. Medina & G. T. Velasquez 375*, 15 Jan. 1940 (FC, S). CHINA: Woo-Sung, Chekiang, *J. E. Nielsen 63*, 29 Jul. 1923 (FC); Tientsin, *M. S. Clemens 6014*, 22 Jul. 1913 (FC, UC); Tali lake, Yunnan, *S. C. Hsiao*, Nov. 1945—Feb. 1946 (FC). AUSTRALIA: Somerset dam, Queensland, *A. B. Cribb*, Jul. 1947 (FC). BURMA: Prome *{S. Kurz}* (Type of *Chroococcus indicus* Zell., L [Fig. 8]); Rangoon, *L. P. Khanna 576, 589*, Spring 1936 (FC); Cantonment Gardens, Rangoon, *Khanna 332, 334*, 9 May 1935 (FC). CAMBODIA: im Bien-ho [Toulé-Sap], *O. Kuntze*, 1875 (NY). INDIA: in einem Bassin der Victoria-Garden, *A. Hansgirg*, Nov. 1895 (in Hauck & Richt., Phyk. Univ. no. 748B, L, MIN, NY); Bombay, in horto Victoria Garden, *Hansgirg*, Sept. (Type of *Polycystis insignis* Beck in Mus. Vindob. Krypt. Exs. no. 227, W; isotypes, DT, FC [Fig. 10], L, NY). CEYLON: Perithpan-pokuna near Isurumunija temple at Anuradhapoora, *F. E. Fritsch*, 3 Oct. 1903 (Type of *Microcystis protocystis* Crow in the collection of F. E. Fritsch; isotype D [Fig. 5]); Colombo, *P. J. Philson & I. B. Warnock*, Sept. 1934 (as *M. aeruginosa* in Tild., So. Pacific Pl., 2nd Ser., no 3, FC).

2. ANACYSTIS INCERTA Drouet & Daily, Butler Univ. Bot. Stud. 10: 221. 1952. *Polycystis incerta* Lemmermann, Forschungsber. Biol. Sta. Plön 7: 132. 1899. *Microcystis incerta* Lemmermann, Krypt.-Fl. Mark Brandenb. 3: 76. 1907; Forti, Syll. Myxophyc., p. 86. 1907. *M. pulverea* var. *incerta* Crow, New Phytol. 22: 66. 1923. *M. pulverea* f. *incerta* Elenkin, Monogr. Algar. Cyanophyc., Pars Spec. 1: 124. 1938. *Diplocystis incerta* Drouet & Daily, Lloydia 11: 78. 1948. —Specimen from Lake Wombsjön, Sweden, determined by the author, designated here as the Type (S).

Polycystis elongata W. & G. S. West, Trans. Roy. Irish Acad. 32(B): 79. 1902. *Clathrocystis elongata* Forti, Syll. Myxophyc., p. 96. 1907. —Type from Lough Neagh, Ireland (BIRM).

Clathrocystis holsatica Lemmermann, Forschungsber. Biol. Sta. Plön 10: 150. 1903. *Microcystis holsatica* Lemmermann, Krypt.-Fl. Mark Brandenb. 3: 77. 1907. *M. pulverea* f. *holsatica* Elenkin, Monogr. Algar. Cyanophyc., Pars Spec. 1: 124. 1938. —Specimen from Neudamm, Germany, de-termined by the author, designated here as the Type (B).

Microcystis minutissima West, Proc. Roy. Irish Acad. 31: 41. 1912. *Aphanothece saxicola* f.

minutissima Elenkin, Monogr. Algar. Cyanophyc., Pars Spec. 1: 151. 1938. —Type from Achill island, Ireland (BIRM).

Aphanocapsa elachista var. *conferta* W. & G. S. West, Journ. Linn. Soc. Bot. 40: 432. 1912. *Microcystis pulverea* f. *conferta* Elenkin, Monogr. Algar. Cyanophyc., Pars Spec. 1: 124. 1938. —Type from Perthshire, Scotland (BIRM).

Microcystis pallida Migula, Crypt. Germ., Austr. & Helvet. Exs. 52(Alg.): 264. 1931. *M. Migulae* J. de Toni, Archivio Bot. 15: 291. 1939. —Isotypes from Eisenach, Germany (B, FC, FH, NY, S, TA). FIG. 13.

Aphanocapsa minima Migula, Crypt. Germ., Austr. & Helvet. Exs. 58(Alg.): 277. 1933. —Isotype from Marksuhl, Germany (FC, FH, NY, S, TA). FIG. 14.

Aphanothece clathrata var. *rosea* Skuja, Symbolae Bot. Upsal. 9(3): 36. 1948. —Paratype received from the author from Gr. Plöner See, Germany (D). FIG. 12.

Original specimens have not been available to us for the following names; their original descriptions are here designated as the Types until the specimens can be found:

Polycystis (*Clathrocystis*) *reticulata* Lemmermann, Bot. Centralbl. 76: 153. 1898. *Coelosphaerium reticulatum* Lemmermann, Krypt. Fl. Mark Brandenb. 3: 84. 1907. *Cyanodictyon reticulatum* Geitler in Pascher, Süsswasserfl. 12: 103. 1925. *Sphaerodictyon reticulatum* Geitler, Beih. z. Bot. Centralbl., II, 41: 231. 1925.

Coelosphaerium natans Lemmermann, Ber. Deutsch. Bot. Ges. 18: 309. 1900. *Microcystis incerta* var. *elegans* Lemmermann, Forschungsber. Biol. Sta. Plön 10: 150. 1903.

Aphanocapsa delicatissima W. & G. S. West, Journ. Linn. Soc. Bot. 40: 431. 1912. *Microcystis pulverea* f. *delicatissima* Elenkin, Monogr. Algar. Cyanophyc., Pars Spec. 1: 124. 1938. —The original sketches and notes are reproduced here from the West notes in the British Museum (Natural History): FIG. 15.

Microcystis pulverea f. *elongata* Crow, New Phytol. 22: 66. 1923.

Microcystis natans Lemmermann ex Skuja, Acta Horti Bot. Univ. Latv. 7: 45. 1934.

Plantae smaragdinae vel laete aerugineae, microscopicae, sphaericae, ovoideae, vel cylindricae, varie et irregulariter lobatae vel clathratae, saepe conferte confluentes, cellulis in divisione binis depresso—sub-sphaericis, aetate provecta globosis diametro 0.5—2μ crassis, sine ordine per gelatinum vaginale distributis; gelatino vaginale hyalino, homogeneo, diffluente; protoplasmate dilute aerugineo, pseudovacuolas continente. FIGS. 12—15.

Developing sparsely in freshwater plankton, seldom the dominant species, and rarely forming conspicuous water blooms. One of the smallest of the Myxophyceae, this species may often be mistaken for coccoid bacteria. Its morphology, life-history, and synonymy have been discussed by Drouet & Daily in Field Mus. Bot. Ser. 20: 67—83 (1939).

Specimens examined:

SWEDEN: in lacu Wombsjön Scaniae, *O. Nordstedt*, Jun. 1901 (designated as Type of *Polycystis incerta* Lemm., S; isotypes, B, NY, UC); duck pond, Lovestad, *B. Carlin-Nilsson*, 15 Jul. 1934 (FC). POLAND: auf dem feuchten Schlamme ausgetrockneter Lachen auf dem Galgenberge, *Hilse*, 1862 (as *Palmella hyalina* in Rabenh. Alg. no. 1525, FC, L, UC). GERMANY: Gr. Plöner See, *W. Rohde*, 15 Sept. 1950 (paratype—from the author—of *Aphanothece clathrata* var. *rosea* Skuja, D [Fig. 12]); Teich bei Fürstenfelde bei Neudamm, *Itzigsohn & de Bary*, Sept. 1852 (designated as Type of *Clathrocystis holsatica* Lemm., B); Salzunger See, *A. Röse* (as *Microhaloa firma* in Rabenh. Alg. no 453, NY); Marksuhl (Thüringen), in einem kleinen Teich, *W. Migula*, Sept. 1927 (isotypes of *Aphanocapsa minima* Mig. in Mig., Crypt. Germ., Austr., & Helv. Exs. Alg. no. 277, FC [Fig. 14], FH, NY, TA, S); Eisenach, im Plankton des Prinzenteiches, *Migula*, 24 Aug. 1931 (FH, S; isotypes of *Microcystis pallida* Mig. in Mig., Crypt. Germ. Austr., & Helv. Exs. Alg. no. 264, B, FC [Fig. 13], FH, NY, S, TA); Altenburger See, Thüringen, *Migula*, Sept. 1927 (as *Aphanothece nidulans* in Mig., Krypt. Germ., Austr., & Helv. Exs. Alg. no. 228, TA). CZECHOSLOVAKIA: Starý Hospidár near Trebon, *B. Fott*, 10 Jun. 1946 (FC). BRITISH ISLES: Loch Katrine, Perthshire, Scotland, *W. & G. S. West*, Mar. 1909 (Type of *Aphanocapsa elachista* var. *conferta* W. & G. S. West, BIRM); plankton, Lough Straheen, Achill island, county Mayo, Ireland, 11 June (Type of *Microcystis minutissima* West, BIRM); Lough Neagh, North Ireland, *W. West*, May 1900 (Type of *Polycystis elongata* W. & G. S. West, BIRM; isotypes, FC, UC). UNION OF SOUTH AFRICA: tow-netting in channel, Barberspan, southwest Transvaal, *Mrs. G. E. Hutchinson 4*, 6 Apr. 1928 (D).

OHIO: Wehrle's pond on Middle Bass island, Smith's pond on North Bass island, and East Harbor near Catawba island, Ottawa county, *E. H. Ahlstrom*, June, Jul. 1931—32 (DA, FC); in pond of fish hatchery, Newtown, Hamilton county, *F. Brinley*, 29 Oct. 1941 (DA, FC). KENTUCKY: reservoir of Walton, *J. B. Lackey*, Summer 1940 (FC). INDIANA: pond on Wisniewski farm east of Cook, Lake county, *F. K. & W. A. Daily 2565*, 29 Aug. 1951 (DA, FC). WISCONSIN: laboratory culture, University of Wisconsin, Madison, *R. Evans*, 8 Jun. 1949 (FC). ILLINOIS: Goose lake northwest of Quincy, Adams county, *T. C. Dorris 59*, 6 Sept. 1950 (FC). MINNESOTA: with *Anacystis cyanea*, Heron lake, Jackson county, *Minnesota Fish Commission*, Jul. 1938 (FC). NEBRASKA: floating in lake, Antioch, Sheridan county, *W. Kiener 20613*, 23 May 1946 (FC, KI). BRAZIL: with *A. cyanea*, Açude Velho near Campina Grande, Paraíba, *S. Wright 1570*, Mar. 1935 (D). PERU: Cocha Zapote, Rio Pacaya, Bajo Ucayali, dept. Loreto, *F. Anceita*, Jul. 1947 (FC).

3. ANACYSTIS MARINA Drouet & Daily, Lloydia 11: 77. 1948. *Aphanocapsa marina* Hansgirg in Foslie, Contrib. Knowl. Mar. Alg. Norway 1: 169. 1890. —Type from Pasvig, Norway (W). FIG. 376.

 Aphanothece nidulans Richter in Wittrock & Nordstedt, Alg. Exs. 14: 694. 1884; Bot. Not. 1884: 128. 1884; Hedwigia 23: 66. 1884. *A. saxicola* f. *nidulans* Elenkin, Monogr. Algar. Cyanophyc., Pars Spec. 1: 151. 1938. *Coccochloris nidulans* Drouet & Daily, Lloydia 11: 78. 1948. *Anacystis nidulans* Drouet & Daily, Butler Univ. Bot. Stud. 10: 221. 1952. —Type from Anger near Leipzig, Germany (FC).

 Microcystis parasitica var. *glacialis* Fritsch, Nat. Antarctic Exped. (1901—04), Nat. Hist. 6(Freshw. Alg.): 15. 1912. —Type from McMurdo bay, Antarctica (in the collection of F. E. Fritsch).

 Aphanocapsa Lewisii Keefe, Rhodora 29: 41. 1927. —Type from Falmouth, Massachusetts (FH). FIG. 377.

 Original specimens have not been available to us for the following names; their original descriptions are here designated as the Types until the specimens can be found:

 Aphanothece pulverulenta Bachmann, Mitt. Naturf. Ges. Luzern 8: 10. 1921.
 Aphanothece Protohydrae Häyrén in Luther, Acta pro Fauna & Flora Fennica 52(3): 24. 1923.
 Synechocystis parvula Perfiljev, Bull. Kom. Sapropel. Leningrad 1: 61. 1923.
 Aphanocapsa salina Woronichin, Bull. Jard. Bot. Princip. U. R. S. S. 28: 30. 1929.

Plantae laete aerugineae, praecipue sphaericae vel ovoideae microscopicae, aut in strata gelatinosa macroscopica evolventes, cellulis in divisione binis depresso—subsphaericis, aetate provecta globosis diametro 0.5—2μ crassis sine ordine (frequenter conferte) per gelatinum vaginale distributis; gelatino vaginale hyalino, homogeneo, (saepe omnino) diffluente; protoplasmate aerugineo, homogeneo. FIGS. 376, 377.

In old pools of standing fresh or brackish water, and among other algae on wet rocks, wood, and soil; often developing in cultures and in old aquaria. Plants of this species sometimes give the impression of having originated as growth-forms of *Anacystis montana* or *Entophysalis* spp. Coccoid bacteria may be easily confused with *Anacystis marina*.

Specimens examined:
 NORWAY: [in a rock pool at or a little above high water mark,] Pasvig, *M. Foslie*, 3 Aug. 1889 (Type of *Aphanocapsa marina* Hansg. in Hauck & Richt., Phyk. Univ. no. 468, W; isotypes, D [Fig. 376], FH, L, NY, UC). GERMANY: ad parietes caldarii in Anger prope Lipsiam, *P. Richter*, 2 Mar. 1882 (Type of *Aphanothece nidulans* Richt. in Wittr. & Nordst., Alg. Exs. no 694, FC; isotypes, D, L, NY). NETHERLANDS: ten Noorden van Harlingen, Friesland, *J. T. Koster 746*, Aug. 1942 (L). MASSACHUSETTS: Salt pond, Falmouth, *A. M. Keefe* (Type of *Aphanocapsa Lewisii* Keefe, FH; isotype, D [Fig. 377]); freshwater pools, Angelica point near New Bedford, *E. H. Battley*, culture from A. W. Frenkel (FC). CONNECTICUT: laboratory culture, New Haven, *R. A. Lewin*, May 1948 (FC). QUEBEC: in a peat bog, St. Hubert, Chambly county, *Bro. Irénée-Marie 469*, 30 Oct. 1937 (FC). WISCONSIN: laboratory culture, Madison, *R. Evans*, 23 May 1949 (FC). ILLINOIS: laboratory culture, Urbana, *R. C. Hecker*, 29 Dec. 1947 (FC). NEBRASKA: culture from Whitetail creek near Keystone, Keith county, *W. Kiener 15621*, 16 Jul. 1944 (FC, KI); culture from creek south of Gretna, Sarpy county, *Kiener 13683a*, 16 July 1944 (FC, KI); culture from floodplain pond southeast of Grand Island, Hall county, *Kiener 16396b*, 11 July 1944

(FC, KI); culture from roadside ditch 9 miles east of North Platte, Lincoln county, *Kiener 16870*, 13 Jul. 1944 (FC, KI). ANTARCTICA: brick-colored ice 4 feet above frozen watercourse through "Penknife" ice, McMurdo Bay, *National Antarctic Expedition*, 13 Sept. 1902 (Type of *Microcystis parasitica* var. *glacialis* Fritsch in the collection of F. E. Fritsch; isotype, D). PHILIP-PINES: on moist wall under leaking pipe, Iloilo Normal School, Iloilo City, *J. D. Soriano 1559*, 4 Jan. 1953 (FC, PUH).

4. ANACYSTIS MONTANA (Lightf.) Dr. & Daily.

Plantae aerugineae, rubrae, roseae, violaceae, nigrae, vel decoloratae, microscopicae vel macroscopicae, libere natantes vel in strata invenientes, cellulis in divisione binis depresso—sub-sphaericis (raro truncato-hemisphaericis), aetate provecta globosis diametro ad 6μ (ad 20μ ubi a fungis parasitatis) crassis, sine ordine aut in seriebus eucapsoideis per gelatinum vaginale distributis; gelatino vaginale homogeneo vel lamelloso, primum hyalino deinde (in f. *montana*) saepe partim vel omnino rubrescente vel coerulescente vel raro lutescente; protoplasmate aerugineo, homogeneo passim granuloso. FIGS. 16—99.

Key to forms of *Anacystis montana*:

Cells 2—6μ in diameter, larger where parasitized by fungi; gelatinous matrix lamellose or homogeneous, becoming red, blue, or rarely yellowish-brown in subaerial and aerial situations .... 4a. f. MONTANA

Cells 2—3μ in diameter; gelatinous matrix hyaline and homogeneous.. 4b. f. MINOR

Cells 5—6μ in diameter; gelatinous matrix hyaline and homogeneous
.... 4c. f. GELATINOSA

4a. ANACYSTIS MONTANA f. MONTANA. *A. montana* Drouet & Daily, Butler Univ. Bot. Stud. 10: 221. 1952. *Ulva montana* Lightfoot, Fl. Scot. 2: 973. 1777. *Palmella alpicola* Lyngbye, Tent. Hydrophyt. Dan., p. 206. 1819. *Conferva alpina* Lyngbye *pro synon.*, loc. cit. 1819. *Merrettia alpicola* S. F. Gray, Nat. Arr. Brit. Pl., p. 349. 1821. *Palmella montana* Agardh, Syst. Algar., p. 15. 1824. *Coccochloris alpicola* Sprengel, Linn. Syst. Vegetab., ed. 16, 4(1): 373. 1827. *Phytoconis alpicola* Meneghini, Mem. R. Accad. Sci. Torino, ser. 2, 5(Sci. Fis. & Mat.): 53. 1843. *Sorospora montana* Hassall, Hist. Brit. Freshw. Alg. 1: 309. 1845. *Bichatia montana* Trevisan, Sagg. Monogr. Alg. Coccot., p. 60. 1848. *Chaos alpicola* Bory *pro synon.* ex Trevisan, loc. cit. 1848. *Bichatia alpicola* Kuntze, Rev. Gen. Pl. 2: 886. 1891. *Gloeocapsa alpicola* Bornet in Wittrock, Nordstedt & Lagerheim, Alg. Exs. 32: 1540. 1903; Bot. Not. 1903: 140. 1903. *Gloeocapsa montana* Wille, Nyt Mag. Naturvid. 56(1): 45. 1918. —Type from Isle of Skye, Scotland (K).

Palmella sanguinea Agardh, Syst. Algar., p. 15. 1824. *Haematococcus sanguineus* Agardh, Icon. Algar. Eur., no. 24. 1828—35. *Microcystis sanguinea* Kützing, Linnaea 8: 372. 1833. *Palmella cryptophylla* Carmichael *pro synon.* in Hooker, Engl. Fl. 5(1): 395. 1833; Hooker, Brit. Fl. 2: 395. 1833. *Protococcus sanguineus* Brébisson & Godey, Mém. Soc. Acad. Sci., Arts & Belles-Lettres Falaise 1835: 40. 1836. *Gloeocapsa sanguinea* Kützing, Phyc. Gener., p. 175. 1843. *Bichatia sanguinea* Trevisan, Nomencl. Algar. 1: 62. 1845. *Ephebe pubescens* δ *versicolor* Flotow, Bot. Zeit. 8: 76. 1850. —Type from near Stockholm, Sweden (LD). FIG. 80.

Haematococcus frustulosus Harvey in Hooker, Engl. Fl 5(1): 395. 1833; in Hooker, Brit. Fl. 2: 395. 1833. *Palmella frustulosa* Carmichael *pro synon.* in Hooker, loc. cit. 1833. *Bichatia frustulosa* Trevisan, Nomencl. Algar. 1: 60. 1845. *Chlorococcum frustulosum* Rabenhorst, Fl. Eur. Algar. 3: 59. 1868. *Gloeocystis frustulosa* W. & G. S. West, Trans. Yorks. Nat. Union 4: 133. 1901. —Type from Scotland (K). FIG. 38.

Palmella livida Carmichael in Hooker, Engl. Fl. 5(1): 397. 1833; in Hooker, Brit. Fl. 2: 397. 1833. *Microcystis livida* Meneghini, Mem. R. Accad. Sci. Torino, ser. 2, 5(Sci. Fis. & Mat.): 47. 1843. *Phytoconis livida* Brébisson *pro synon.* in Meneghini, loc. cit. 1843. *Haematococcus lividus*

Hassall, Hist. Brit. Freshw. Alg. 1: 332. 1845. *Bichatia livida* Trevisan, Nomencl. Algar. 1: 61. 1845. *Phytoconis sordida* Brébisson *pro synon.* in Trevisan, Sagg. Monogr. Alg. Coccot., p. 65. 1848. —Type from Scotland (K). FIG. 65.

Protococcus magma Brébisson in Brébisson & Godey, Mém. Soc. Acad. Sci., Arts & Belles-Lettres Falaise 1835: 40. 1836. *Pleurococcus magma* Meneghini, Mem. R. Accad. Sci. Torino, ser. 2, 5(Sci. Fis. & Mat.): 10, 43. 1843. *Gloeocapsa magma* Kützing, Tab. Phyc. 1: 17. 1846. —Type from Falaise, France (L).

Anacystis marginata Meneghini, Consp. Algol. Eugan., p. 324. 1837. *Palmella marginata* Kützing, Phyc. Gener., p. 172. 1843. *Microcystis marginata* Kützing, Tab. Phyc. 1: 6. 1846. *Polycystis marginata* Richter ex Hansgirg, Prodr. Algenfl. Böhmen 2: 145. 1892. *Microcystis aeruginosa* f. *marginata* Elenkin, Monogr. Algar. Cyanophyc., Pars Spec. 1: 106. 1938. —Type from the Euganean springs, Italy (L). FIG. 19.

Pleurococcus communis Meneghini, Consp. Algol. Eugan., p. 338. 1837. —Type from Padua, Italy (L). FIG. 83.

Palmella Ralfsii Harvey, Man. Brit. Alg., p. 179. 1841. *Sorospora Ralfsii* Hassall, Hist. Brit. Freshw. Alg. 1: 310. 1845. *Gloeocapsa Ralfsiana* Kützing, Tab. Phyc. 1: 18. 1846. *Bichatia Ralfsii* Trevisan, Sagg. Monogr. Alg. Coccot., p. 59. 1848. *Gloeocapsa Ralfsii* Lemmermann, Krypt.-Fl. Mark Brandenb. 3: 65. 1907. —Type from Dolgelley, Wales (BM).

Coccochloris muscicola Meneghini, Mem. R. Accad. Sci. Torino, ser. 2, 5(Sci. Fis. & Mat.): 60. 1843. *Palmella muscicola* Kützing, Phyc. Germ., p. 149. 1845. *P. botryoides* var. *muscicola* Hansgirg, Prodr. Algenfl. Böhmen 1: 138. 1886. *Aphanocapsa muscicola* Wille, Nyt Mag. Naturvid. 56(1): 43. 1918. *Microcystis muscicola* Elenkin, Monogr. Algar. Cyanophyc., Pars Spec. 1: 132. 1938. —Type from the Euganean springs, Italy (FI). FIG. 79.

Coccochloris parietina Meneghini, Mem. R. Accad. Sci. Torino, ser. 2, 5(Sci. Fis. & Mat.): 61. 1843. —Type from the Euganean mountains, Italy (FI). FIG. 77.

Microcystis nigra Meneghini, Mem. R. Accad. Sci. Torino, ser. 2, 5(Sci. Fis. & Mat.): 75. 1843. *Bichatia nigra* Trevisan, Nomencl. Algar. 1: 61. 1845. *Gloeocapsa nigra* Grunow in Rabenhorst, Fl. Eur. Algar. 2: 36. 1865. —Type from the Euganean mountains, Italy (FI). FIG. 64.

Protococcus fusco-ater Kützing, Phyc. Gener., p. 168. 1843. *Chroococcus fusco-ater* Rabenhorst, Fl. Eur. Algar. 2: 34. 1865. —Type from Nordhausen, Germany (L). FIG. 46.

Palmella didyma Kützing, Phyc. Gener., p. 172. 1843. *Gloeocapsa didyma* Kützing, Tab. Phyc. 1: 14. 1846. *Bichatia didyma* Trevisan, Sagg. Monogr. Alg. Coccot., p. 62. 1848. —Type from Padua, Italy (L).

Gloeocapsa aeruginosa Kützing, Phyc. Gener., p. 174. 1843. *G. aeruginea* Kützing, Phyc. Germ., p. 151. 1845. *Bichatia aeruginosa* Trevisan, Nomencl. Algar. 1: 59. 1845. —Type from Nordhausen, Germany (L). FIG. 90.

Gloeocapsa atrata Kützing, Phyc. Gener., p. 174. 1843. *Ephebe pubescens* δ *atra* Flotow, Bot. Zeit. 8: 76. 1850. —Type from Berne, Switzerland (L).

Gloeocapsa sanguinolenta Kützing, Phyc. Gener., p. 174. 1843. *Bichatia sanguinolenta* Trevisan, Nomencl. Algar. 1: 62. 1845. —Type from Sachswerfen, Thuringia, Germany (L). FIG. 61.

Gloeocapsa coracina Kützing, Phyc. Gener, p. 174. 1843. *Palmella placentaris* Wallroth *pro synon.* in Kützing, loc. cit. 1843. *Bichatia coracina* Trevisan, Nomencl. Algar. 1: 60. 1845. *Gloeocapsa coracina* β *placentaris* Kützing, Spec. Algar., p. 219. 1849. *G. coracina* var. *placentaris* Kützing ex Forti, Syll. Myxophyc., p. 56. 1907. —Type from Steigerthal, Thuringia, Germany (L).

Gloeocapsa rosea Kützing, Phyc. Gener., p. 174. 1843. *Bichatia rosea* Trevisan, Nomencl. Algar. 1: 62. 1845. —Type from Trieste (L).

Gloeocapsa Shuttleworthiana Kützing, Phyc. Gener., p. 175. 1843. *Bichatia Shuttleworthiana* Trevisan, Nomencl. Algar. 1: 62. 1845. —Type from M. Ben Gower, Ireland (L). FIG. 37.

Gloeocapsa rubicunda Kützing, Phyc. Gener., p. 175. 1843. *Bichatia rubicunda* Trevisan, Nomencl. Algar. 1: 62. 1845. —Type from Scandinavia (L).

Protococcus pygmaeus Kützing, Phyc. Germ., p. 144. 1845. —Type from Battaglia, Italy (L). FIG. 27.

Haematococcus binalis Hassall, Hist. Brit. Freshw. Alg. 1: 331. 1845. *Pleurococcus binalis* Trevisan, Sagg. Monogr. Alg. Coccot., p. 32. 1848. —Specimen from Erigal, county Donegal, Ireland, named by the author, designated here as the Type (BM).

Haematococcus aeruginosus Hassall, Hist. Brit. Freshw. Alg. 1: 333. 1845. *Palmella aeruginosa* Carmichael *pro synon.* in Hassall, loc. cit. 1845. *Bichatia Carmichaelii* Trevisan, Sagg. Monogr. Alg. Coccot., p. 65. 1848. —Type from Scotland (K). FIG. 84.

Gloeocapsa conglomerata Kützing, Tab. Phyc. 1:˙16. 1846. *Bichatia conglomerata* Trevisan, Sagg. Monogr. Alg. Coccot., p. 64. 1848. *Gloeocapsa glomerata* Kützing ex Roumeguère, Alg. des Eaux Douces France, no. 1016. —Type from the Harz mountains, Germany (L). FIG. 81.

Gloeocapsa cryptococcoides Kützing, Tab. Phyc. 1: 16. 1846. *Bichatia cryptococcoides* Trevisan, Sagg. Monogr. Alg. Coccot., p. 64. 1848. *Gloeocapsa gelatinosa* f. *cryptococcoides*

Rabenhorst, Fl. Eur. Algar. 2: 39. 1865. *G. gelatinosa* var. *cryptococcoides* Kützing ex Bizzozero, Fl. Veneta Crittog. 2: 63. 1885. —Type from Abano, Italy (L).

Protococcus haematodes Kützing, Tab. Phyc. 1: 5. 1846. *Gloeocapsa haematodes* Kützing, Sp. Algar., p. 222. 1849. *Bichatia haematodes* Kuntze, Rev. Gen. Pl. 2: 886. 1891. —Type from Domfront, Orne, France (L). FIG. 88.

Gloeocapsa rupestris Kützing, Tab. Phyc. 1: 17. 1846. *Bichatia rupestris* Kuntze, Rev. Gen. Pl. 2: 886. 1891. *Bichatia fuscescens* Lagerheim, Ber. Deutsch. Bot. Ges. 10: 526. 1892. *Gloeocapsa fuscescens* Lagerheim ex Forti, Syll. Myxophyc., p. 58. 1907. —Type from Calvados, France (L). Kützing's note, "an Felsen im Harze" with the original description is surely in error, since only de Brébisson's specimens are labeled thus in the Kützing herbarium.

Gloeocapsa compacta Kützing, Tab. Phyc. 1: 24. 1846. *Rhodocapsa violacea* β *compacta* Richter in Hansgirg, Österr. Bot. Zeitschr. 34: 315. 1884. *Bichatia compacta* Kuntze, Rev. Gen. Pl. 2: 886. 1891. *Gloeocapsa violascea* β *compacta* Richter in Hansgirg, Prodr. Algenfl. Böhmen 2: 149. 1892. *G. violascea* var. *compacta* Richter ex Forti, Syll. Myxophyc., p. 40. 1907. —Type from Vire, France (L). FIG. 49.

Gloeocapsa violacea Kützing, Tab. Phyc. 1: 25. 1846. *G. lignicola* Rabenhorst, Fl. Eur. Algar. 2: 41. 1865. *Rhodocapsa violacea* Hansgirg, Österr. Bot. Zeitschr. 34: 315. 1884. *Bichatia lignicola* Kuntze, Rev. Gen. Pl. 2: 886. 1891. *Gloeocapsa alpina* f. *lignicola* Hollerbach in Elenkin, Monogr. Algar. Cyanophyc., Pars Spec. 1: 189. 1938. —Type from Paris, France (L). FIG. 42.

Palmella brunnea A. Braun in Kützing, Sp. Algar., p. 212. 1849. *Aphanocapsa brunnea* Nägeli, Gatt. Einzell. Alg., p. 52. 1849. *Anacystis brunnea* Wolle, Fresh-w. Alg. U. S., p. 329. 1887. —Type from Freiburg, Baden, Germany (L). FIG. 87.

Gloeocapsa geminata Kützing, Sp. Algar., p. 219. 1849. *Oocystis geminata* Nägeli *pro synon.* in Kützing, loc. cit. 1849. *Oocystis minor* (*micrococca*) Itzigsohn *pro synon.* in Rabenhorst, Fl. Eur. Algar. 3: 53. 1868. *Bichatia geminata* Kuntze, Rev. Gen. Pl. 2: 886. 1891. —Type from Zürich, Switzerland (L). FIG. 33.

Gloeocapsa scopulorum Nägeli in Kützing, Sp. Algar., p. 220. 1849. *Bichatia scopulorum* Kuntze, Rev. Gen. Pl. 2: 886. 1891. *Microcystis scopulorum* Nägeli ex Wille, Nyt Mag. Naturvid. 62: 198. 1925. —Type from Schaffhausen, Switzerland (L). FIG. 16.

Gloeocapsa ianthina Nägeli in Kützing, Sp. Algar., p. 220. 1849. *Bichatia ianthina* Kuntze, Rev. Gen. Pl. 2: 886. 1891. —Type from Zürich, Switzerland (L). FIG. 23.

Gloeocapsa versicolor Nägeli in Kützing, Sp. Algar., p. 220. 1849. *Bichatia versicolor* Kuntze, Rev. Gen. Pl. 2: 886. 1891. —Type from Zürich, Switzerland (L). FIG. 50.

Gloeocapsa ambigua Nägeli in Kützing, Sp. Algar., p. 220. 1849; Nägeli, Gatt. Einzell. Alg., p. 51. 1849. *G. ambigua* a. *fuscolutea* Nägeli, loc. cit. 1849. *G. fuscolutea* Kirchner, Krypt.-Fl. Schlesien 2(1): 260. 1878. *Bichatia ambigua* Kuntze, Rev. Gen. Pl. 2: 886. 1891. *B. fuscolutea* Lagerheim, Ber. Deutsch. Bot. Ges. 10: 526. 1892. *Gloeocapsa ambigua* f. *fuscolutea* Nägeli ex Setchell & Gardner, Univ. Calif. Publ. Bot. 1: 179. 1903. *G. alpina* f. *ambigua* Hollerbach in Elenkin, Monogr. Alg. Cyanophyc., Pars Spec. 1: 189. 1938. —Type from Zürich, Switzerland (L).

Gloeocapsa ambigua β *rivularis* Nägeli in Kützing, Sp. Algar., p. 221. 1849. —Type from Zürich, Switzerland (L).

Gloeocapsa ambigua b. *violacea* Nägeli, Gatt. Einzell. Alg., p. 50. 1849. *G. ambigua* var. *violacea* Nägeli ex Rabenhorst, Alg. Sachs. 61 & 62: 607. 1857. —Type from Zürich, Switzerland (L).

Gloeocapsa ambigua var. *pellucida* Nägeli, Gatt. Einzell. Alg., p. 50. 1849. *G. ambigua* b. *pellucida* Nägeli ex Rabenhorst, Fl. Eur. Algar. 2: 42. 1865. *G. ambigua* f. *pellucida* Nägeli in Roumeguère, Alg. des Eaux Douces France, no. 825. —Type from Zürich, Switzerland (L).

Gloeocapsa opaca Nägeli in Kützing, Sp. Algar., p. 221. 1849; Nägeli, Gatt. Einzell. Alg., p. 50. 1849. *G. magma* b. *opaca* Nägeli in Kirchner, Krypt.-Fl. Schlesien 2(1): 251. 1878. *G. magma* var. *opaca* Kirchner ex Hansgirg, Prodr. Algenfl. Böhmen 2: 147. 1892. *G. magma* f. *opaca* Hollerbach in Elenkin, Monogr. Algar. Cyanophyc., Pars Spec. 1: 176. 1938. —Type from Zürich, Switzerland (L). Nägeli *no.* 378 as cited by Kützing with the original description is obviously a typographical error.

Gloeocapsa opaca β *pellucida* Kützing, Sp. Algar., p. 221. 1849. *G. opaca* var. *pellucida* Nägeli, Gatt. Einzell. Alg., p. 50. 1849. *G. magma* b. *pellucida* Nägeli in Rabenhorst, Fl. Eur. Algar. 2: 43. 1865. *G. pellucida* Brébisson and *Protococcus pellucidus* Nägeli *pro synon.* in Roumeguère, Alg. des Eaux Douces France no. 410. *G. magma* var. *pellucida* Nägeli ex Hansgirg, Prodr. Algenfl. Böhmen 2: 147. 1892. —Type from Zürich, Switzerland (L).

Gloeocapsa punctata Nägeli in Kützing, Sp. Algar., p. 222. 1849; Nägeli, Gatt. Einzell. Alg., p. 51. 1849. *Bichatia punctata* Kuntze, Rev. Gen. Pl. 2: 886. 1891. *Coelosphaerium Kuetzingianum* var. *punctatum* Playfair, Proc. Linn. Soc. New South Wales 39: 135. 1914. *C. punctatum* Playfair, Census New South Wales Pl. Suppl. 1: 260. 1917. —Type from Zürich, Switzerland (L).

Ephebe pubescens d. *cruenta* Flotow, Bot. Zeit. 8: 75. 1850. —Type from the Riesengebirge, Bohemia (S).

Monocapsa stegophila Itzigsohn in Rabenhorst, Alg. Sachs. 27 & 28: 263. 1853. *Gloeocapsa stegophila* Rabenhorst, Krypt.-Fl. Sachsen 1: 72. 1863. *Bichatia stegophila* Kuntze, Rev. Gen. Pl. 2: 886. 1891. —Type from Neudamm, Germany (FC). FIG. 48.

Chroococcus cohaerens Nägeli in Rabenhorst, Alg. Sachs. 45 & 46: 446. 1855. —Type from Freiburg, Baden, Germany (FC). FIG. 78.

Gloeocapsa nigrescens Nägeli in Rabenhorst, Alg. Sachs. 63 & 63: 629. 1857. *Bichatia nigrescens* Kuntze, Rev. Gen. Pl. 2: 886. 1891. *Gloeocapsa Corrensii* J. de Toni, Noter. Nomencl. Algol. 3. 1936. —Type from the Berner Oberland, Switzerland (ZT).

Gloeocapsa saxicola Wartmann in Rabenhorst, Alg. Sachs. 81 & 82: 813. 1859. *G. alpina* f. *saxicola* Rabenhorst, Fl. Eur. Algar. 2: 40. 1865. *Bichatia saxicola* Kuntze, Rev. Gen. Pl. 2: 886. 1891. *Gloeocapsa alpina* var. *saxicola* Rabenhorst in Hansgirg, Prodr. Algenfl. Böhmen 2: 150. 1892. —Type from St. Gallen, Switzerland (FC). FIG. 22.

Gloeocapsa alpina Nägeli ex Cramer in Rabenhorst, Alg. Sachs. 87 & 88: 869. 1859. —Type from Engelberg, Unterwalden, Switzerland (FC). See Brand, Bot. Centralbl. 83: 224. 1900. FIG. 17.

Aphanocapsa montana Cramer in Wartmann & Schenk, Schweiz. Krypt. no. 134. 1862. *A. montana* var. *micrococca* Cramer *pro synon.*, loc. cit. 1862. *A. montana* f. *micrococca* Cramer in Brügger, Jahresber. Naturf. Ges. Graubündens, N. F. 8: 244. 1863. *A. montana* b. *micrococca* Cramer ex Rabenhorst, Fl. Eur. Algar. 2: 50. 1865. —Type from Rigi, Switzerland (ZT).

Aphanocapsa montana var. *macrococca* Cramer in Wartmann & Schenk, Schweiz. Krypt. no. 134. 1862; Cramer in Brügger, Jahresber. Naturf. Ges. Graubündens, N. F. 8: 244. 1863. *A. montana* a. *macrococca* Cramer ex Rabenhorst, Fl. Eur. Algar. 2: 50. 1865. —Type from Trinser See, Switzerland (ZT).

Chroococcus violaceus Rabenhorst, Fl. Eur. Alg. 2:32. 1865. —Type from Baden-Baden, Germany (L). FIG. 86.

Aphanocapsa violacea Grunow in Rabenhorst, Fl. Eur. Algar. 2: 51. 1865. —Type from Neehaus, Lower Austria (W).

Gloeocapsa Kalchbrenneri Grunow in Rabenhorst, Fl. Eur. Algar. 2: 41. 1865. *Bichatia Kalchbrenneri* Kuntze, Rev. Gen. Pl. 2: 886. 1891. —Type from Badoczer Wiesen, Hungary (W).

Gloeocapsa Titiana Grunow in Rabenhorst, Fl. Eur. Algar. 2: 47. 1865. *Bichatia Titiana* Kuntze, Rev. Gen. Pl. 2: 866. 1891. —Type from the Euganean mountains, Italy (W).

Gloeocapsa tropica Crouan fr. in Schramm & Mazé, Essai Class. Alg. Guadeloupe, p. 29. 1865. —Type from La Guadeloupe (PC). FIG. 36.

Coccochloris tuberculosa Areschoug, Alg. Scand. Exs., Ser. Nov., 6: 292. 1866. *Aphanothece tuberculosa* Forti, Syll. Myxophyc., p. 79. 1907. —Type from Bursjön, Uppland, Sweden (S).

Chroococcus varius A. Braun in Rabenhorst, Alg. Eur. 246—248: 2451. 1876. *C. varius* f. *fuscoluteus* A. Braun ex Wille, Nyt Mag. Naturvid. 62: 181. 1925. *Gloeocapsa varia* Hollerbach in Elenkin, Monogr. Algar. Cyanophyc., Pars Spec. 1: 208. 1938. *Pleurocapsa varia* Drouet & Daily in Daily, Amer. Midl. Nat. 27: 644. 1942. —Type from Poppelsdorf near Bonn, Germany (FC).

Aphanocapsa biformis A. Braun in Rabenhorst, Alg. Eur. 246—247: 2453b. 1876; A. Braun, Hedwigia 1879: 194. 1879. —Type from Berlin, Germany (FC). FIG. 30.

Aphanocapsa nebulosa A. Braun in Rabenhorst, Alg. Eur. 246—248: 2454. 1876. *Aphanothece caldariorum* Richter ex Hansgirg, Prodr. Algenfl. Böhmen 2: 136. 1892. *Gloeothece caldariorum* Hollerbach in Elenkin, Monogr. Algar. Cyanophyc., Pars Spec. 1: 255. 1938. —Type from Berlin, Germany (FC). FIG. 69.

Gloeothece inconspicua A. Braun in Rabenhorst, Alg. Eur. 246—248: 2455. 1876. —Type from Berlin, Germany (FC).

Entophysalis magnoliae Farlow, Mar. Alg. New England, p. 29. 1881. —Type from Magnolia, Massachusetts (FH).

Gloeocapsa Itzigsohnii Bornet in Zopf, Spaltpfl., p. 60. 1882. *G. magma* var. *Itzigsohnii* Hansgirg, Sitzungsber. K. Böhm. Ges. Wiss., Math.-Nat. Cl., 1890(2): 137. 1891. *G. magma* f. *Itzigsohnii* Hollerbach in Elenkin, Monogr. Algar. Cyanophyc., Pars Spec. 1: 176. 1938. —Type from Versailles, France (FH).

Aphanocapsa Naegelii Richter, Hedwigia 23: 66. 1884. —Type from Anger near Leipzig, Germany (L).

Gloeothece tepidariorum var. *cavernarum* Hansgirg, Bot. Centralbl. 37: 38. 1889. *G. rupestris* var. *cavernarum* Hansgirg, Prodr. Algenfl. Böhmen 2: 136. 1892. *G. palea* var. *cavernarum* Lemmermann, Krypt.-Fl. Mark Brandenb. 3: 49. 1907. —Type from Korno near Beroun, Bohemia (W). FIG. 68.

Anacystis Reinboldii Richter in Reinbold, Schrift. Naturw. Ver. Schleswig-Holst. 7(2): 180. 1889. *Microcystis Reinboldii* Forti, Syll. Myxophyc., p. 91. 1907. —Type from Friedrichsort, Germany (NY).

Coelosphaerium anomalum var. *minus* Hansgirg, Sitzungsber. K. Böhm. Ges. Wiss., Math.-Nat. Cl., 1890(1): 19. 1890. —Type from Kuchelbad near Prague (W).

Aphanocapsa fonticola Hansgirg, Sitzungsber. K. Böhm. Ges. Wiss., Math.-Nat. Cl., 1890(1): 19. 1890. —Type from Hostin near Beroun, Bohemia (W).

Gloeocapsa alpina var. *mediterranea* Hansgirg, Sitzungsber. K. Böhm. Ges. Wiss., Math.-Nat. Cl., 1891(1): 356. 1891. —Type from Pinguente, Istria, Jugoslavia (W). FIG. 45.

Aphanocapsa Richteriana Hieronymus in Hauck & Richter, Phyk. Univ. 10: 485. 1892. —Type from Nimptsch in Silesia, Poland (L).

Aphanothece conferta Richter in Hauck & Richter, Phyk. Univ. 10: 487. 1892. —Type from Oschatz, Saxony, Germany (L).

Aphanocapsa thermalis var. *minor* Hansgirg, Prodr. Algenfl. Böhmen 2: 159. 1892. *A. violacea* var. *minor* Hansgirg ex Forti, Syll. Myxophyc., p. 75. 1907. —Type from Karlsbad, Bohemia (W). FIG. 71.

Chroococcus varius var. *luteolus* Hansgirg, Prodr. Algenfl. Böhmen 2: 164. 1892. —Type from Smichov, Bohemia (W). FIG. 74.

Aphanocapsa fuscolutea Hansgirg, Prodr. Algenfl. Böhmen 2: 156. 1892. *Microcystis Hansgirgiana* Elenkin, Monogr. Algar. Cyanophyc., Pars Spec. 1: 130. 1938. —Type from Prague, Bohemia (W).

Chroococcus schizodermaticus West, Trans. Roy. Microsc. Soc. 1892: 742. 1892. —Type from Cantley, Yorkshire, England (BIRM). In the Type, the cells are much enlarged and parasitized.

Pleurocapsa muralis Lagerheim in Wittrock, Nordstedt & Lagerheim, Alg. Exs. 23: 1097. 1893. —Isotypes from Quito, Ecuador (L, MIN, NY).

Aphanocapsa nivalis Lagerheim, Nuova Notarisia 1894: 652. 1894. —Type from Spitzbergen, Norway (DT).

Aphanothece saxicola var. *sphaerica* W. & G. S. West, Trans. Roy. Microsc. Soc. 1894: 17. 1894. *A. saxicola* f. *sphaerica* Elenkin, Monogr. Algar. Cyanophyc., Pars Spec. 1: 151. 1938. —Type from Snowdon, Waies (BIRM). FIG. 55.

Gloeocapsa gigas W. & G. S. West, Journ. Linn. Soc. Bot. 30: 276. 1894. *Anacystis gigas* Gardner, Mem. New York Bot. Gard. 7: 15. 1927. —Type from St. Vincent, West Indies (BM).

Aphanocapsa elachista W. & G. S. West, Journ. Linn. Soc. Bot. 30: 276. 1894. *Microcystis pulverea* f. *elachista* Elenkin, Monogr. Algar. Cyanophyc., Pars Spec. 1: 124. 1938. —Type from Dominica, West Indies (BIRM).

Chroococcus minor f. *minimus* W. & G. S. West, Journ. Linn. Soc. Bot. 30: 275. 1894. —Type from Dominica, West Indies (FC).

Gloeocapsa Holstii Hieronymus in Engler, Pflanzenwelt Ostafr. C: 8. 1895. —Type from Usumbura, Ruanda Urundi (O).

Gloeocapsa Reicheltii Richter in Hauck & Richter, Phyk. Univ. 13: 647. 1895; Hedwigia 34: 25. 1895; Sitzungsber. Naturf. Ges. Leipzig 19—21: 152. 1895. —Type from Leipzig, Germany (L).

Chroococcus alpinus Schmidle in Simmer, Allgem. Bot. Zeitschr. 1899:194. 1900. —Type from the Kolbitsch, Carinthia, Austria (Z).

Chroococcus decolorans Migula, Crypt. Germ., Austr. & Helvet. Exs. 26 & 27(Alg.): 83. 1906. —Isotypes from Eisenach, Germany (MICH, NY, TA).

Gloeocapsa haematodes var. *violascens* Grunow ex Forti, Syll. Myxophyc., p. 38. 1907. —Type from Kufstein, Austria (W). FIG. 70.

Microcystis densa G. S. West, Journ. of Bot. 47: 246. 1909. —Type from Lake Albert Nyanza, Uganda (BIRM).

Eucapsis alpina Clements & Shantz, Minn. Bot. Stud. 4: 134. 1909. *Merismopoedia cubica* West, Proc. Roy. Irish Acad. 31(16): 42. 1912. —Type from near Pikes Peak, Colorado (US).

Gloeothece samoensis Wille in Rechinger, Bot. & Zool. Ergebn. Samoa- & Salomoninseln, Süsswasseralg., p. 6. 1914; Hedwigia 53: 144. 1913. —Type from Savaii, Samoa (O).

Gloeocapsa aeruginosa f. *lignicola* Wille in Rechinger, Bot. & Zool. Ergebn. Samoa & Salomoninseln, Süsswasseralg., p. 7. 1914. —Type from Upolu, Samoa (O).

Gloeocapsa rupestris Kütz. ex Wille, Nyt Mag. Naturvid. 62: 197. 1925. —Type from Falaise, France (L).

Pleurocapsa Mayorii Setchell, Publ. Carnegie Inst. Wash. Dept. Mar. Biol. 20: 258. 1924. —Type from Rose atoll, Samoa (UC). FIG. 40.

Chondrocystis Bracei Howe, Bull. Torr. Bot. Club 51: 355. 1924. *Gloeocapsa Bracei* J. de Toni, Not. Nomencl. Algol. 4. 1936. —Type from Long cay, Bahama islands (NY). FIG. 82.

Chroococcus varius f. *atrovirens* A. Braun ex Wille, Nyt Mag. Naturvid. 62: 181. 1925. —Type from Poppelsdorf near Bonn, Germany (FC). FIG. 85.

Gloeocapsa multisphaerica Gardner, Univ. Calif. Publ. Bot. 14: 1. 1927. —Type from near Foochow, China (FH). FIG. 21.

Gloeocapsa minutula Gardner, Univ. Calif. Publ. Bot. 14: 2. 1927. —Type from Huangchun, Foochow, China (FH).

49

Aphanocapsa intertexta Gardner, Mem. New York Bot. Gard. 7: 4. 1927. —Type from near Adjuntas, Puerto Rico (NY). FIG. 32.

Chroococcus cubicus Gardner, Mem. New York Bot. Gard. 7: 5. 1927. —Type from Santurce, Puerto Rico (NY). FIG. 31.

Aphanothece opalescens Gardner, Mem. New York Bot. Gard. 7: 5. 1927. —Type from Maricao, Puerto Rico (NY). FIG. 28.

Chroococcus muralis Gardner, Mem. New York Bot. Gard. 7: 7. 1927. —Type from Coamo Springs, Puerto Rico (NY). FIG. 26.

Gloeocapsa livida var. *minor* Gardner, Mem. New York Bot. Gard. 7: 10. 1927. *G. livida minor* Gardner, New York Acad. Sci. Sci. Surv. Porto Rico 8: 258. 1932. —Type from Hato Arriba near Arecibo, Puerto Rico (NY). FIG. 52.

Gloeocapsa calcicola Gardner, Mem. New York Bot. Gard. 7: 11. 1927. —Type from Laguna Tortuguero, Puerto Rico (NY). FIG. 39.

Gloeothece endochromatica Gardner, Mem. New York Bot. Gard. 7: 13. 1927. —Type from near Arecibo, Puerto Rico (NY). FIG. 53.

Gloeothece parvula Gardner, Mem. New York Bot. Gard. 7: 14. 1927. —Type from Hato Arriba near Arecibo, Puerto Rico (NY). FIG. 35.

Anacystis nigropurpurea Gardner, Mem. New York Bot. Gard. 7: 18. 1927. —Type from near Arecibo, Puerto Rico (NY). FIG. 75.

Anacystis nigroviolacea Gardner, Mem. New York Bot. Gard. 7: 19. 1927. —Type from near Utuado, Puerto Rico (NY).

Anacystis compacta Gardner, Mem. New York Bot. Gard. 7: 20. 1927. —Type from Caguas, Puerto Rico (NY). FIG. 25.

Anacystis magnifica Gardner, Mem. New York Bot. Gard. 7: 21. 1927. —Type from San Juan, Puerto Rico (NY). FIG. 58.

Anacystis distans Gardner, Mem. New York Bot. Gard. 7: 21. 1927. —Type from Maricao, Puerto Rico (NY).

Anacystis amplivesiculata Gardner, Mem. New York Bot. Gard. 7: 22. 1927. —Type from Maricao, Puerto Rico (NY). FIG. 20.

Anacystis gloeocapsoides Gardner, Mem. New York Bot. Gard. 7: 22. 1927. —Type from Mayaguez, Puerto Rico (NY). FIG. 44.

Anacystis microsphaeria Gardner, Mem. New York Bot. Gard. 7: 22. 1927. —Type from Coamo Springs, Puerto Rico (NY).

Anacystis pulchra Gardner, Mem. New York Bot. Gard. 7: 23. 1927. —Type from near Maricao, Puerto Rico (NY). FIG. 63.

Anacystis nidulans Gardner, Mem. New York Bot. Gard. 7: 23. 1927. —Type from Mayaguez, Puerto Rico (NY). FIG. 62.

Anacystis Willei Gardner, Mem. New York Bot. Gard. 7: 24. 1927. —Type from Mayaguez, Puerto Rico (NY).

Anacystis irregularis Gardner, Mem. New York Bot. Gard. 7: 24. 1927. —Type from Coamo Springs, Puerto Rico (NY).

Anacystis consociata Gardner, Mem. New York Bot. Gard. 7: 25. 1927. —Type from Santurce, Puerto Rico (NY).

Anacystis minutissima Gardner, Mem. New York Bot. Gard. 7: 25. 1927. —Type from near Cabo Rojo, Puerto Rico (NY). FIG. 73.

Anacystis radiata Gardner, Mem. New York Bot. Gard. 7: 26. 1927. —Type from Coamo Springs, Puerto Rico (NY). FIG. 60.

Anacystis radiata var. *major* Gardner, Mem. New York Bot. Gard. 7: 26. 1927. *A. radiata major* Gardner, New York Acad. Sci. Sci. Surv. Porto Rico 8: 263. 1932. —Type from Coamo Springs, Puerto Rico (NY). FIGS. 51 and 56.

Anacystis anomala Gardner, Mem. New York Bot. Gard. 7: 26. 1927. —Type from Fajardo, Puerto Rico (NY).

Endospora rubra Gardner, Mem. New York Bot. Gard. 7: 28. 1927. —Type from Coamo Springs, Puerto Rico (NY). FIG. 57.

Endospora olivacea Gardner, Mem. New York Bot. Gard. 7: 29. 1927. —Type from Caguas, Puerto Rico (NY). FIG. 41.

Endospora bicoccus Gardner, Mem. New York Bot. Gard. 7: 28. 1927. —Type from near Maricao, Puerto Rico (NY). FIG. 34.

Endospora mellea Gardner, Mem. New York Bot. Gard. 7: 28. 1927. —Type from near Utuado, Puerto Rico (NY). FIG. 29.

Endospora nigra Gardner, Mem. New York Bot. Gard. 7: 29. 1927. —Type from near Sabana Grande, Puerto Rico (NY). FIG. 24.

Placoma Willei Gardner, Mem. New York Bot. Gard. 7: 29. 1927. —Type from Maricao, Puerto Rico (NY). FIG. 66.

Entophysalis violacea Gardner, Mem. New York Bot. Gard. 7: 30. 1927. *E. Willei* Gardner, New York Acad. Sci. Sci. Surv. Porto Rico 8(2): 260. 1932. —Type from Hato Arriba near Arecibo, Puerto Rico (NY).

Aphanothece uliginosa W. R. Taylor, Proc. Acad. Nat. Sci. Philadelphia 80: 86. 1928. —Type from Purcell range, British Columbia (TA).

Gloeocapsa Dvorakii Novácek, Zpravy Kom. Prirodov. Moravy a Slezka, Oddel. Bot., 7: 1. 1929. —Type from Mohelno, Moravia (D). FIG. 59.

Entophysalis atroviolacea Novácek, Prace Moravské Prirodov. Spolecn. 7(3): 1. 1932. —Type from Mohelno, Moravia (D). FIG. 18.

Gloeocapsa compacta var. *coeruleoatra* Novácek in Dvorák & Novacek, Sborn. Klub. Prirodov. Brne 15: 5. 1933. —Type from Mohelno, Moravia (D). FIG. 43.

Gloeocapsa pleurocapsoides Novácek in Dvorak & Novácek, Sborn. Klub. Prirodov. Brne 15: 5. 1933. —Type from Mohelno, Moravia (D). FIG. 89.

Chroococcidiopsis thermalis Geitler, Arch. f. Hydrobiol., Suppl. XII, 4: 625. 1933. —Type from Kadjaj, Sumatra, Indonesia (in the collection of L. Geitler).

Gloeocapsa nigrescens f. *vitrea* Novácek, Mohelno 1934: 110, 144. 1934. —Type from Mohelno, Moravia (D). FIG. 47.

Gloeocapsa chroococcoides Novácek, Mohelno 1934: 100, 139. 1934. —Type from Mohelno, Moravia (D). FIG 54.

Gloeothece tophacea Skuja in Handel-Mazzetti, Symb. Sinicae 1: 15. 1937. —Type from Dschungdien, Yunnan, China (W).

Synechococcus ambiguus Skuja in Handel-Mazzetti, Symb. Sinicae 1: 16. 1937. —Type from Dschungdien, Yunnan, China (W). FIG. 67.

Coelosphaerium limnicola Lund, Journ. of Bot. 1942: 66. 1942. —Type from Richmond Park, Surrey, England (in the collection of J. W. G. Lund). FIG. 76.

Anacystis hyalina Chu, Sinensia 15: 153. 1944. *Asterocapsa hyalina* Chu, Ohio Journ. Sci. 52: 100. 1952. —Type from Chin-Chen-Shan, Szechwan, China (FC).

Gloeocapsa incrustata Chu, Sinensia 15: 154. 1944. —Type from Chin-Chen-Shan, Szechwan, China (FC).

Entophysalis sinensis Chu, Sinensia 15: 154. 1944. —Type from Chin-Chen-Shan, Szechwan (FC).

Gloeothece rhodochlamys Skuja, Nova Acta Reg. Soc. Sci. Uppsal., ser. 4, 14(5): 18. 1949. —Type from Kyauktan, Rangoon, Burma (FC). FIG. 72.

Original specimens have not been seen for the following names; their original descriptions are here designated as the Types until the specimens can be found:

Lepraria atra Acharius, Lich. Univ., p. 668. 1810. *Globulina atra* Turpin, Végét. Acot., pl. 6, fig. 2, in Dict. Sci. Nat., vol. 65. 1816—29; Turpin, Mém. Mus. d'Hist. Nat. Paris 14: 26, 63. 1827. *Protococcus ater* Kützing, Linnaea 8: 368. 1833. *Microcystis atra* Kützing, ibid. 8: 375. 1833. *Bichatia atra* Trevisan, Nomencl. Alg. 1: 60. 1845.

Protococcus violasceus Corda in Sturm, Deutschl. Fl., II, 6: 37. 1833. *Microcystis violascea* Kützing, Linnaea 8: 373. 1833. *Haematococcus violasceus* Meneghini, Mem. R. Accad. Sci. Torino, ser. 2, 5(Sci. Fis. & Mat.): 9, 24. 1843. *Protosphaeria violascea* Trevisan, Sagg. Monogr. Alg. Coccot., p. 28. 1848. *Gloeocapsa violascea* Rabenhorst, Fl. Eur. Algar. 2: 41. 1865. *Bichatia violascea* Kuntze, Rev. Gen. Pl. 2: 886. 1891.

Haematococcus alpestris Hassall, Hist. Brit. Freshw. Alg. 1: 328. 1845. *Pleurococcus alpestris* Trevisan, Sagg. Monogr. Alg. Coccot., p. 34. 1848.

Haematococcus theriacus Hassall, Hist. Brit. Freshw. Alg. 1: 333. 1845. *Bichatia theriaca* Trevisan, Sagg. Monogr. Alg. Coccot., p. 58. 1848.

Synechococcus parvulus Nägeli, Gatt. Einzell. Alg., p. 56. 1849. —Nägeli's sketch and notes, in the Pflanzenphysiologisches Institut, Eidgenossische Technische Hochschule, Zürich, are reproduced here: FIG. 90A.

Chroococcus refractus Wood, Proc. Amer. Philos. Soc. 11: 122. 1869.

Palmella fuscata Zanardini, N. Giorn. Bot. Ital. 10: 40. 1878.

Sphaerogonium nigrum Rostafinski, Rozpr. Akad. Umiej. Krakow., Wydz. Mat.-Przyr., 10: 300. 1883.

Polycystis amethystina Filarszky, Hedwigia 39: 139. 1900. *Microcystis amethystina* Forti, Syll. Myxophyc., p. 89. 1907.

Planosphaerula natans Borzi, Nuova Notarisia 1905: 20. 1905.

Gloeocapsa thermalis Lemmermann, Engler Bot. Jahrb. 34: 614. 1905.

Gloeocapsa haematodes var. *brunneola* Kirchner ex Forti, Syll. Myxophyc., p. 38. 1907.

Clathrocystis montana Teodorescu, Beih. z. Bot. Centralbl. 21(2): 106. 1907. *Microcystis montana* J. de Toni, Diagn. Algar. Nov. I., Myxophyc. 6: 570. 1939.

51

Chroococcus aurantiacus Bernard, Protococc. & Desmid. d'Eau Douce, p. 48. 1908. *C. Detonii* Bernard, Nova Guinea, Rés. de l'Exped. Sci. Néerl. à la Nouv. Guinée, 1907 & 1909, 8(2, Bot.): 259. 1910.
Chroococcus Rechingeri Wille in Rechinger, Denkschr. K. Akad. Wiss. Wien, Math.-Nat. Kl., 91: 160. 1915.
Microcystis pulverea var. *major* Elenkin, Estest.-Istorich. Koll. Gr. E. P. Sheremetevoi S. E. Mikhailovsk. Moskovskoi Gub. VI, Vodor. Okt. 1, Schizophyc, p. 6. 1915.
Microcystis amethystina var. *vinea* Printz, K. Norske Videnskab. Selsk. Skr. 1920(1): 33. 1920.
Myxosarcina concinna Printz, K. Norske Videnskab. Selsk. Skr. 1920(1): 35. 1920.
Chroococcus minor f. *violaceus* Wille in Hedin, Southern Tibet 6(3, Bot.): 165. 1922.
Chroococcus cohaerens var. *antarcticus* Wille, Deutsche Südpolar-Exped. 1901—02, 8: 382. 1924.
Synechocystis Pevalekii Ercegovic, Acta Bot. Inst. Bot. Univ. Zagreb. 1: 77. 1925.
Gloeocapsa alpina var. *polyedrica* Ercegovic, Acta Bot. Inst. Bot. Univ. Zagreb. 1: 80. 1925.
Gloeocapsa alpina f. *violacescens* Ercegovic, Acta Bot. Inst. Bot. Univ. Zagreb. 1: 80. 1925.
Gloeocapsa biformis Ercegovic, Acta Bot. Inst. Bot. Univ. Zagreb. 1: 80. 1925.
Gloeocapsa biformis f. *dermochroa* Ercegovic, Acta Bot. Inst. Bot. Univ. Zagreb. 1: 81. 1925.
Gloeocapsa biformis f. *punctata* Ercegovic, Acta Bot. Inst. Bot. Univ. Zagreb. 1: 81. 1925.
Synechocystis minuscula Woronichin, Arch. f. Hydrobiol. & Planktonk. 17: 642. 1926.
Gloeocapsa lacustris Huber-Pestalozzi, Arch. f. Hydrobiol. & Planktonk. 16: 162. 1926.
Eucapsis alpina var. *minor* Skuja, Acta Horti Bot. Univ. Latv. 1(3): 155. 1926. *E. minor* Elenkin, Acta Inst. Bot. Acad. Sci. U. R. P. S. S., ser. 2, Pl. Crypt. 1: 15. 1933.
Aphanocapsa Grevillei f. *densa* Strφm, Norweg. Mountain Alg., p. 148. 1926.
Chroococcus sarcinoides Huber-Pestalozzi, Verh. Int. Ver. Theor. & Ang. Limnol. 4: 366. 1929. *C. Huberi* J. de Toni, Noter. Nomencl. Algol. 3. 1936.
Gloeocapsa dirumpens Beck-Mannagetta, Arch. f. Protistenk. 66: 9. 1929.
Gloeothece ustulata Beck-Mannagetta, Arch. f. Protistenk. 66: 9. 1929.
Entophysalis Perrieri Frémy, Ann. Crypt. Exot. 3: 207. 1930.
Chroococcus minor f. *glomeratus* Frémy, Bull. Soc. Bot. France 77: 673. 1930. *Gloeocapsa minor* f. *glomerata* Frémy ex Hollerbach in Elenkin, Monogr. Algar. Cyanophyc., Pars Spec. 1: 239. 1938.
Chroococcus verrucosus Krieger, Hedwigia 70: 151. 1931.
Chroococcus niger Starmach, Acta Soc. Bot. Polon. 11: 293. 1934.
Microcystis thermalis Yoneda, Acta Phytotax. & Geobot. 7: 139. 1938.
Aphanothece cretaria Frémy, Bull. Soc. Linn. Normandie, ser. 9, 3: 21. 1943.
Anacystis trochiscioides Jao, Sinensia 15: 75. 1944. *Asterocapsa trochiscioides* Chu, Ohio Journ. Sci. 52(2): 100. 1952.
Anacystis purpurea Jao, Sinensia 15: 76. 1944.
Chroococcus limneticus var. *multicellularis* Chu, Ohio Journ. Sci. 52(2): 96. 1952.
Gloeocapsa attingens Schiller, Sitzungsber. Oesterr. Akad. Wiss., Math.-Nat. Kl., 1, 163: 125. 1954.

Plantae aerugineae, rubrae, nigrae, violaceae, vel olivaceae, in stratum gelatinosum vel libere inter alias algas evolventes, cellulis in divisione binis depresso—subsphaericis (raro truncato-hemisphaericis) aetate provecta globosis diametro 2—6μ (ad 20μ ubi a fungis parasitatis) crassis, per gelatinum vaginale sine ordine aut in seriebus eucapsoideis distributis; gelatino vaginale homogeneo vel lamelloso, primum hyalino deinde vulgo partim vel omnino rubrescente vel coerulescente vel lutescente; protoplasmate aerugineae, olivaceae, vel violaceae, homogeneo vel granuloso. FIGS. 16—90A.

In strata and among other algae on rocks, wood, and soil wet by rains, and especially in seepage; often found in temporary pools and on emergent shores. The red and blue pigments in the gelatinous matrix, apparently developing in response to desiccation and exposure to direct sunlight, seem to be influenced by the acidity or alkalinity of the substratum. According to Cramer in Rabenh. Alg. no. 869 (1859), Nägeli and Wartmann noted that mounting in acid solutions changes the color of the matrix from blue to red, and that this change can be reversed in alkaline solutions. The same was remarked by Wartmann in Rabenh. Alg. no. 813 (1859). It had long been our own observation that formalin-preserved material almost invariably exhibits red sheaths. O. Jaag & N. Gemsch in Verh.

Schweiz. Naturf. Ges. Locarno 1940: 158—159 (1940) and Gemsch in Beih. Schweiz. Bot. Ges. 53: 121—192 (1943) report careful observational and experimental work on this matter. Yellow and brown pigments are also produced in the gelatinous matrix; perhaps fungi are responsible for these.

Fungus hyphae often enter the protoplasts themselves, causing the cells to enlarge and to cease the secretion of gelatinous material. The hyphae of some fungi form conspicuous (often ornamented) walls about the protoplasts; this phenomenon is the basis for the reports of thick-walled spores in *Anacystis montana*. It has been observed that such parasitization results in the death of cells of *Anacystis montana;* future observations may disclose that there exist fungi which, parasitizing this alga, do not destroy it but continue to grow with it as a lichen.

Bacteria, palmelloid Chlorophyceae, *Entophysalis deusta,* and dissociated cells of *Agmenellum* spp. and *Gomphosphaeria lacustris* may sometimes be confused with *Anacystis montana*.

Specimens examined:
SPITZBERGEN: "Chlamydococcus nivalis. Spitzberg. Ex herb. Zanardini." (Type of *Aphanocapsa nivalis* Lagerh., DT); King's bay, *T. M. Fries,* Aug. 1868 (S). FINLAND: Karis, *R. Grönblad,* Apr. 1935 (S). SCANDINAVIA?: with *Stigonema* (*Sirosiphon*) *panniforme,* Kunze in herb. Kützing (Type of *Gloeocapsa rubicunda* Kütz., L). NORWAY: *K. Sprengel,* as *G. rupicola* in herb. Kützing (L); in *Neckera oligocarpa* Schimp, ad latera saxorum prope pontem Stuelsboren, Ringeboe, Gudbrandsdalen, *J. E. Zetterstedt & J. A. O. Wickbom,* 22 Jun. 1870 (FC). SWEDEN: Hessingen ved Stockholm, 1823, in herb. Agardh (Type of *Palmella? sanguinea* Ag., LD [Fig. 80]; isotypes, K, S, UC); in lacu Bursjön in Uplandia, *V. B. Wittrock,* Aug. (Type of *Coccochloris tuberculosa* Aresch. in Areschoug, Alg. Scand. Exs., Ser. Nov., no. 292, S; isotypes, FC, L, NY, UC); ad lapides in Lassby backar prope Upsaliam, *G. Lagerheim,* 28 Mar. 1882 (as *Gloeocapsa opaca* in Wittr. & Nordst., Alg. Exs., no. 597, FC, L, S); in rupibus ad Danviken prope Holmiam, *Lagerheim,* Nov. 1881 (as *G. magma* in Wittr. & Nordst., Alg. Exs. no. 500b, FC, L, S); ad Nacka qvarnar prope Stockholmiam, *Lagerheim,* Jun. 1883 (as *G. versicolor* in Wittr., Nordst., & Lagerh., Alg. Exs. no. 1542, L, S); in aquaeducta ad Nacka pr. Stockholmiam, *N. Wille,* Jun. 1884 (D, S); ad saxa humida prope Qvickjock in Lapponia Lulensi, *Lagerheim,* Jul. 1883 (S; as *G. alpicola* in Wittr., Nordst., & Lagerh., Alg. Exs. no. 1540, L, NY); Scania, *J. Agardh,* 1843 (L); Hammarbysjön pr. Stockholmiam, *Lagerheim,* Jun. 1881 (S); Vassijokk, Torne Lappmark, *O. Borge 105, 116,* Jul. 1909 (S); Drottningholm, *Borge,* Mai 1920 (S); Möja, *Borge 33/43,* Aug. 1933 (S); Kullaborge, Skåne, *Borge,* Oct. 1898 (S); Arboga, *O. F. Andersson,* Nov. 1888 (S); Hederviken, Uppsala, *Areschoug,* Jul. 1857 (S), Aug. 1863 (S).

POLAND: Breslau, im Warmhaus des bot. Gartens, *W. Migula,* Feb. 1888 (S; as *Chroococcus varius* in Hauck & Richt., Phyk. Univ. no. 484, L, UC); Riesengebirge, *Flotow,* Sept. 1851 (S); Silesia, in herb. Kützing (L); feuchte Steinwand auf Burg Kynast, *G. Hieronymus 746,* Sept. 1887 (MICH; as *Chroococcus cohaerens* in Hauck & Richt., Phyk. Univ. no. 243, L); in caldario horti botanici, Breslau, *B. Schröder,* Nov. 1894 (S); bei Strehlen, *Hilse* (as *Palmella testacea* in Rabenh. Alg. no. 1524, UC); an den Felswänden am Galgenberge bei Strehlen in Schlesien, *Hilse* (as *Gloeocapsa opaca* in Rabenh. Alg. 1169, FC); an überrieselten Felswänden in Steinbrüchen bei Gross-Willkau, Kreis Nimptsch in Schlesien, *Hieronymus,* 2 Okt. 1891 (TA; Type of *Aphanocapsa Richteriana* Hieron. in Hauck & Richt., Phyk. Univ. no. 485, L; isotypes, MIN, UC). ROMANIA: Bucuresti-Cotroceni, in caldario horti botanici ad muros, *E. C. Teodorescu 1384,* 23 Dec. 1903 (FC, W).

GERMANY: Schleswig-Holstein: auf Sand am Strande nördlich Friedrichsort, *T. Reinbold,* Sept. 1890 (Type of *Anacystis Reinboldii* Richt. in Hauck & Richt, Phyk. Univ. no. 447, NY; isotype, MIN); Kiel, *Reinbold* (DT). Brandenburg: Berliner bot. Garten, Palmhaus, *G. Lagerheim,* Oct. 1882 (S), ad parietem caldariorum, *Lagerheim,* 9 Oct. 1882 (as *Gloeocapsa granosa* in Wittr. & Nordst., Alg. Exs. no. 595, FC, D, L, S), im Orchideenhause, *A. Braun,* Nov. 1875 (Type of *Aphanocapsa nebulosa* A. Br. in Rabenh. Alg. no. 2454, FC [Fig. 69]; isotypes, S, TA; Type of *Gloeothece inconspicua* A. Br. in Rabenh. Alg. 2455, FC; isotype, D), Oct. 1875 (as *A. nebulosa* in Rabenh. Alg. 2454b, FC), Mai 1875 (Type of *A. biformis* A. Br. in Rabenh. Alg. no. 2453, FC, [Fig. 30]; isotypes, L, NY, TA, S; in Roumeguère, Alg. de Fr. no. 616, L); an der Hinterwand des Warmhauses in Universitätsgarten zu Berlin, *Braun,* Dec. 1875 (as *A. biformis* in Rabenh. Alg. no. 2453b, FC, L, NY, S, TA); Neudamm, *Itzigsohn,* Dec. 1852 (Type of *Monocapsa stegophila* Itzigs. in Rabenh. Alg. no. 263, FC [Fig. 48]; isotype, L). Saxony: im Taubenbachthale bei der Schweizermühle, *Biene,* Mai 1862 (as *Gloeocapsa magma* in Rabenh. Alg. no. 1334, UC); Porphyrfelsen bei Grimma, *G. Winter,* 5 Apr. 1868 (UC); auf den steinernen Stufen zum Altarraum in der Kirche zu Döben bei Grimma, *P. Richter* (as *G. bituminosa* in Hauck & Richt.,

Phyk. Univ. no. 749, L, NY, S); an einem Brunnen in Königstein, *Richter*, Sept. 1880 (TA); Utewalde, *C. Cramer* (as *G. opaca* in Rabenh. Alg. no. 544, FC, FH, L), Aug. 1854 (FH), *Richter*, 26 Mai 1880 (as *G. sanguinea* in Wittr. & Nordst., Alg. Exs. no. 499, FC, L, S), Jun. 1881 (TA), 29 Mai 1882 (as *G. quaternata* in Wittr. & Nordst., Alg. Exs. no. 598, FC, D, L, S), 27 Mar. 1884 (as *G. sabulosa* in Wittr. & Nordst., Alg. Exs. no. 698, FC, L, PC, S), without date (FC, UC); in der sächsischen Schweiz, an Felsen, *Hantzsch* (UC), an feuchten Felswänden (as *G. magma* in Rabenh. Alg. no. 84, FC, L, NY, TA; in Breutel, Alg. Exs. no. 328, L; as *G. ambigua* var. in Rabenh. Alg. no. 32, FC, L, NY, TA); in rupibus humectis ad Hohnstein prope Dresdam, *Richter*, 28 Mai 1881 (as *G. magma* in Wittr. & Nordst., Alg. Exs. no. 500, FC, L, S); an einer schwach überrieselten Steinwand in Steinbruche zu Beucha bei Brandis um Leipzig, *H. Reichelt*, Sept. 1890 (Type of *G. Reicheltii* Richt. in Hauck & Richt., Phyk. Univ. no. 647, L; isotype, MIN); Moritzburg, auf Wasser schwimmend (as *Microcystis ichthyoblabe* in Rabenh. Alg. no. 16, FC); in einem Gewächshause, Oschatz, *E. May*, Feb. 1892 (Type of *Aphanothece conferta* Richt. in Hauck & Richt., Phyk. Univ. no. 487, L; isotypes, MIN, UC); Leipzig, an den Wänden des Gewächshauses im bot. Garten, *Richter* (UC); an einer feuchten Wand im Gewächshause der Dreyzehnerschen Gärtnerei in Anger bei Leipzig, *Richter*, Apr. 1888 (Type of *Aphanocapsa Naegelii* Richt. in Hauck & Richt., Phyk. Univ. no. 195, L; isotype, MIN); in einem Warmhause in Anger, *Richter*, Feb. 1879 (L, S), Apr. 1879 (TA); in parietes caldarii ad Anger, *Richter*, 12 Apr. 1881 (as *Chroococcus varius* and *Aphanocapsa biformis* in Wittr. & Nordst., Alg. Exs. no. 600, FC, D, L, S; with *Protococcus grumosus* Richt. in Wittr. & Nordst., Alg. Exs. no. 1090, FI). Thuringia: bei Gotha, *A. Braun* (NY); Kohnstein, herb. Kützing (L); an Steinen im Gehege, Nordhausen, *F. T. Kützing* (Type of *Protococcus fusco-ater* Kütz., L[Fig. 46]); Gypsberg, *Wallroth* in herb. Kützing (L); an Gypsfelsen bei Sachswerfen, *Kützing*, Aug. 1839 (Type of *Gloeocapsa sanguinolenta* Kütz., L [Fig. 61]; isotype, UC); "Palmella placentaris Wallr.", auf Gipsberge bei Steigerthal, *Wallroth*, Apr. 1839 (Type of *G. coracina* Kütz., L); Hercynia, in herb. Kützing (Type of *G. conglomerata* Kütz., L [Fig. 81]; isotype, UC); Nordhausen, *Kützing* (Type of *G. aeruginosa Kütz.*, L [Fig. 90)]; isotype, UC); an Felsen beim Weissenfelser-See, *F. Hauck*, 16 Aug. 1881 (L); Eisenach, Betonmauer des Nessewehres, *W. Migula*, Sept. 1905 (Isotypes of *Chroococcus decolorans* Mig. in Migula, Crypt. Germ. Austr. & Helv. Exs., Alg. no. 83, MICH, NY, TA); Tambach, an feuchten Felsen der Talsperre, *Migula*, 7 Sept. 1928 (as *Gloeocapsa dermochroa* in Migula, Crypt. Germ. Austr. & Helv. Exs., Alg. no. 235, FC, NY, S, TA, Z). Bavaria: Berchtesgaden, *P. Magnus*, Sept. 1874 (FH); Weg nach dem Königssee, Berchtesgaden, *Magnus*, 1874 (NY); an der Wand eines Gewölbes der über den Rödelheim führenden Brücke des Donau-Mainkanales bei Erlangen, *P. Reinsch*, März 1864 (as *Gloeocapsa atrata* with *Coccochloris stagnina* Spreng. in Rabenh. Alg. no. 1914, FC); Oberstdorf, an feuchten Kalkfelsen, *Migula*, Juli 1926 (S; as *Gloeocapsa sanguinea* in Migula, Crypt. Germ. Austr. & Helv. Exs. Alg. no. 258, FC, TA). Württemberg: Canstatt, am Rande einer Mineralwasser-Bassin, *G. Zeller* (as *Chroococcus minor* in Rabenh. Alg. no. 1143, FC). Rhineland: an der Wand eines Warmhauses zu Poppelsdorf bei Bonn, *C. Bouché*, Sept. 1875 (Type of *Chroococcus varius* A. Br. in Rabenh. Alg. no. 2451, FC; isotypes, L, S, TA), *A. Braun*, Sept. 1875 (Type of *C. varius* f. *atrovirens* A. Br. in Rabenh. Alg. no. 2452, FC [Fig. 85]; isotypes, L, S, TA, UC). Baden: Hirschensprung im Höllenthal, *B. Wartmann*, Jun. 1854 (as *Palmella microspora* in Rabenh. Alg. no. 406, FC, L); Freiburg, *A. Braun*, Oct. 1847 (Type of *Palmella brunnea* A. Br., L [Fig. 87]; isotype, NY), an Mühleradern, *A. Braun*, Apr. 1848 (FH, ZT), auf dem Mörtel alter pfutzigen Mauern, *Braun 117*, Oct. 1848 (L); im Warmhause des bot. Gartens zu Freiburg, *Wartmann*, Oct. 1854 (Type of *Chroococcus cohaerens* Näg. in Rabenh. Alg. no. 446, FC [Fig. 78]; isotypes, FH, L, NY); in den Thermalbrunnen, Badensweiler, *Braun 129*, Apr. 1848 (L), Aug. 1848 (L); Carlsruhe, *Braun*, Nov. 1845 (L), Apr. 1848 (UC); Karlsruhe, an den Mauern eines Gewächshauses, *W. Migula*, Feb. 1905 (as *Gloeocapsa granosa* in Migula, Crypt. Germ. Austr. & Helv. Exs., Alg. no. 98, MICH, TA, Z); an den Speichen eines Wasserrades bei Constanz, *E. Stizenberger*, Aug. 1857 (as *G. ambigua* var. *violacea* with *Coccochloris stagnina* Spreng. in Rabenh. Alg. no. 607, FC, MIN, L, PC; in Jack, Leiner & Stizenb., Kryptog. Badens no. 1, TA, UC); Baden, *W. F. R. Suringar 1072*, Sept. 1862 (Type of *Chroococcus violaceus* Rabenh., L [Fig. 86]).

CZECHOSLOVAKIA: Moravia: Mohelno, *J. Suza*, Jul. 1931 (Type of *Gloeocapsa compacta* var. *coeruleoatra* Nov., D [Fig. 43]; isotype, FC), *F. Nováček*, 10 Mar. 1926 (Type of *G. nigrescens* f. *vitrea* Nov., D [Fig. 47]; isotype, FC), 21 Apr. 1926 (D, FC), May 1926 (D, FC), 31 Jul. 1926 (Type of *Entophysalis atroviolacea* Nov., D [Fig. 18]; isotype, FC), 9 Aug. 1926 (Type of *Gloeocapsa chroococcoides* Nov., D [Fig. 54]; isotype, FC), 24 Sept. 1926 (D, FC), 21 Oct. 1926 (Type of *G. pleurocapsoides* Nov., D [Fig. 89]; isotype, FC), 22 Oct. 1926 (D, FC), 19 Sept. 1927 (D, FC), 22 Sept. 1927 (D, FC), 4 Oct. 1928 (D, FC), 20 Oct. 1928 (Type of *G. Dvorakii* Nov., D [Fig. 59]; isotype, FC), 20 Oct. 1930 (D, FC), 17 Apr. 1932 (D, FC), 8 Mar. 1933 (D, FC), 11 Mar. 1933 (D, FC), 12 Mar. 1933 (D, FC). Slovakia: Felkenthal, an der Granaten-Wand, Tatra, *G. Hieronymus 736*, Aug. 1888 (MICH; as *G. Ralfsiana* in Hauck & Richt., Phyk. Univ. no. 334, L). Bohemia: Belá, *A. Hansgirg*, Aug. 1883 (FC, W); Brenn-Schiessnitz prope Böhmisch-Leipa, *Hansgirg*, Sept. 1890 (FC, W); Budnany, na klouzavce, *Hansgirg*, Jul. 1884 (FC, W); Chuchel, *Hansgirg*, Apr. 1884 (FC, W), Nov. 1884 (FC, W); Chvateruby u Kralup, *Hansgirg*, Apr. 1888 (FC, W); Dittersbach—Hinter-Dittersbach, *Hansgirg*, Jul. 1889 (FC, W),

Dux, *Hansgirg*, Aug. 1883 (FC, W); Edmundsklamm prope Herrnskretchen, *Hansgirg*, Sept. 1890 (FC, W); Pampferhütte bei Eisenstein, *Hansgirg*, Aug. 1887 (FC, W); Friedland prope Reichenberg, *Hansgirg*, 1891 (FC, W); Habstein, *Hansgirg*, Jul. 1883 (FC, W); in lapidibus calcareis viaductus magni ad Hlubocep in agro Pragensi, *Hansgirg* (as *Gloeocapsa rupicola* in Fl. Exs. Austro-Hung. no. 3199, L, MO, S); Hohenfurth, *Hansgirg*, Aug. 1884 (FC, W); an Kalksteinen in offenen Felsenbrunnen bei Hostin nächst Beroun, *Hansgirg*, Jul. 1888 (Type of *Aphanocapsa fonticola* Hansg., W; isotype, FC); teplá voda, Hradec Kralove, *Hansgirg*, Sept. 1883 (FC, W); horní cást udolí u Karlíka pod Roblínem, *Hansgirg*, Jul. 1888 (FC, W); Mauern unter dem Sprudel, Carlsbad, *Hansgirg*, Jul. 1886 (FC, W); Karlsbad, am sog. kleinen Sprudel u.s.w. im Bette der Tepl unter der Sprudelkolonnade, *Hansgirg*, Aug. 1883 (Type of *A. thermalis* var. *minor* Hansg., W [Fig. 71]; isotype, FC); Karlstein, u hradu, *Hansgirg*, Jul. 1884 (FC, W); Koda prope Karlstein, Sedlec prope Lodenic, *Hansgirg*, May 1890 (FC, W); auf mässig beleuchteten Stellen der Kaskadenhöhle unterhalb Korno nächst Beraun, *Hansgirg*, Sept. 1888 (Type of *Gloeothece tepidariorum* var *cavernarum* Hansg. and *G. rupestris* var. *cavernarum* Hansg., W; isotype, FC [Fig. 68]); Krc, *Hansgirg*, Sept. 1886 (FC, W); Krumlov, *Hansgirg*, Aug. 1883 (FC, W), Aug. 1884 (FC, W); in den Sandgruben oberhalb Kuchelbad nächst Prag, *Hansgirg*, Jul. 1883 (Type of *Coelosphaerium anomalum* var. *minus* Hansg., W; isotype, FC); plovárna, Litomerice, *Hansgirg*, Jul. 1884 (FC, W); mlýny v Labi, Lovosice, *Hansgirg*, Jul. 1884 (FC, W); Peiperz-Maxdorf prope Bodenbach, *Hansgirg*, Sept. 1890 (FC, W); Vereinsgarten, Prag, *Hansgirg*, Jan. 1885 (FC, W); auf feuchten Fensterscheiben im Vermehrungshause des Prager Vereinsgartens, *Hansgirg*, Oct. 1884 (Type of *Aphanocapsa fuscolutea* Hansg., W; isotype, FC); Krumlov, Pravápmo, *Hansgirg*, Sept. 1884 (FC, W); kameny studánka, Rovné pod Pipem u Roudnice, *Hansgirg*, Jul. 1884 (FC, W); in rupibus madidis ad Selc in agro Pragensi, *Hansgirg*, Jun. 1886 (as *Gloeocapsa ambigua* in Wittr. & Nordst., Alg. Exs. no. 1099, L, S); Selc u Prahy, *Hansgirg*, Sept. 1885 (FC, W), Nov. 1886 (FC, W); Kinski Zahrada, Smichov, *Hansgirg*, Dec. 1883 FC, W); im Vermehrungshause des k. botanischen Gartens am Smichow, *Hansgirg*, Jan. 1884 (Type of *Chroococcus varius* var. *luteolus* Hansg., W [Fig. 74]; isotype, FC); Spindelmühle im Riesengebirge, *Hansgirg*, Nov. 1883 (FC, W); Spitzberg im Böhmerwalde (u tunelu), *Hansgirg*, Aug. 1883 (FC, W); Stríbro, *Hansgirg*, Aug. 1883 (FC, W); Stupcic prope Tábor, *Hansgirg*, Aug. 1888 (FC, W); Tábor, *Hansgirg*, Aug. 1883 (FC, W); Riesengebirge, *v. Flotow* (Type of *Ephebe pubescens* d. *cruenta* Flot., S); skály nad Tetínem nad Berounkou, *Hansgirg*, Sept. 1888 (FC, W); viadukt na Stvanice, *Hansgirg*, Mar. 1883 (FC, W); Volsany u Nepomuk, *Hansgirg*, Aug. 1887 (FC, W); Zel. Brod, *Hansgirg*, Jul. 1885 (FC, W); Zleby prope Caslov, *Hansgirg*, 1891 (FC, W). HUNGARY: in terra muscosa calce incrustata et aqua minerali humectata, Badoczer Wiesen, *Kalchbrenner* (Type of *Gloeocapsa Kalchbrenneri* Grun., W).

AUSTRIA: Lower Austria: in horto botanico Vindobonensi, *A. Kerner* (as *G. sanguinea* in Fl. Exs. Austro-Hung. no. 1196, L, S), *G. Lagerheim*, 4 Oct. 1882 (as *G. compacta* in Wittr. & Nordst., Alg. Exs. no. 596, FC, L, S); auf Leucodon bei Wien, ex herb. Grunow. (FH); in rupibus prope pagum Purkersdorf (in silva Wiener Wald), *K. de Keissler*, Mar. (as *G. sanguinea* in Mus. Vindob. Krypt. Exs. no. 2049, FC, L, S, Z); in rupibus vallis Krummbachgraben in monte Schneeberg, *G. de Beck*, Jul. (as *Aphanocapsa montana* in Mus. Vindob. Krypt. Exs. no. 146b, FC, L, NY, S); auf Holz in einem Brünn in Neehaus, *A. Grunow*, Aug. 1858 (Type of *Aphanocapsa violacea* Grun., W). Styria: Hortus Sackeli, Graz, *A. Hansgirg*, Sept. 1890 (FC, W); "äussere Wand der Wasserarche" prope Mürzzuschlag, *F. Pfeiffer & Schmula* (as *Gloeocapsa nigrescens* in Fl. Exs. Austro-Hung. no. 3198, FC, FH, L, MO, S, Z). Salzburg: in cortice arborum ad torrentes Gollinger Wasserfall, *C. de Keissler*, Jun. (as *G. alpina* in Mus. Vindob. Krypt. Exs. no. 228b, FC, L, NY, S, Z); Gastein, *Diesing* (FH); ad saxa irrorata et humida prope Wildbad-Gastein, *G. de Beck*, Jun. (as *Aphanocapsa montana* in Mus. Vindob. Krypt. Exs. no. 146, FC, L, S). Carinthia: St. Veit-an-der-Glan, *Hansgirg*, Sept. 1889 (FC, W); Pontafel, *Hansgirg*, Aug. 1885 (FC, W); Kalk und Tuff am Kolbitsch, *H. Simmer 1007*, 29 Aug. 1898 (Type of *Chroococcus alpinus* Schmidle, Z; isotype, S); Kreuzeckgebiet, *Simmer*, Aug. 1898 (S), bei Rittersdorf, *Simmer*, Oct. 1898 (S), auf Zannhölzern, *Simmer 1058*, 29 Aug. 1898 (Z), am Knoten, *Simmer*, Aug. 1897 (S), in Ameisgraben, *Simmer 1056*, 14 Apr. 1898 (Z), bei Zwickenberg auf feuchtem Sande, *Simmer 1059*, 31 Aug. 1898 (Z). Tyrol: Brixlegg, *A. Hansgirg*, 1891 (FC, W); Hall, *Hansgirg*, 1891 (FC, W); zwischen Hall und St. Magdalene, *Hansgirg*, 1891 (FC, W); Rothholz nächst Jenbach, *Hansgirg*, 1891 (FC, W); Kufstein, *Hansgirg*, 1891 (FC, W); Thierberg, Hechtsee, etc. bei Kufstein, *Hansgirg*, 1891 (FC, W); feuchte Felsen an Langensee bei Kufstein, *Heufler*, Aug. 1860 (Type of *Gloeocapsa haematodes* var. *violascens* Grun., W [Fig. 70]); auf nassen Kalkfelsen bei Mühlau, *Kerner* (S).

JUGOSLAVIA: Slovenia: Bischoflack, *A. Hansgirg*, Aug. 1889 (FC, W); Brunndorf et Draga prope Laibach, *Hansgirg*, Sept. 1889 (FC, W); Franzdorf, *Hansgirg*, Aug. 1889 (FC, W); ex caverna (Jelenska Jerna), Franzdorf, *Hansgirg*, Aug. 1889 (FC, W); Marburg, *Hansgirg*, Aug. 1890 (FC, W); Podnart, *Hansgirg*, Aug. 1889 (FC, W); Tüffer, *Hansgirg*, Aug. 1890 (FC, W). Istria: in saxis calcareis ad fl. Quieto prope Pinguente, *Hansgirg*, 1891 (Type of *G. alpina* var. *mediterranea* Hansg., W; isotype, FC [Fig. 45]). Dalmatia: Cannosa—Valdinoce prope Ragusa, *Hansgirg*, 1891 (FC, W); Topla—Gruda prope Castelnuovo, *Hansgirg*, 1891 (FC, W); Lesina, herb. Kützing (L); St. Dominica, Lesina, 1857, herb. Kützing (L). TRIESTE: *F. Hauck*, Sept. 1874 (S); in saxis in ostio fluminis Quieto ad Cittanova, *Hansgirg*, 1891 (FC, W); Pirano, *Hansgirg*, Apr. 1889 (FC, W);

Pisino, *Hansgirg*, Aug. 1889 (FC, W); in saxis prope Prosecco—Miramar, *Hansgirg*, 1891 (FC, W); with *Scytonema crustaceum* Ag., Monte Spaccato, Triest, *F. T. Kützing*, Apr. 1835 (Type of *Gloeocapsa rosea* Kütz., L); Boschetto, Triest, *F. Hauck*, Mar. 1884 (L). ALBANIA: Radhima, acqua di una fonte, *G. de Toni 65a*, 16 Jun. 1941 (FC, DT); Fushes-Dukati, roccia bagnata presso una sorgente, *de Toni 75*, 23 Jun. 1941 (FC, DT), pendici del Shendelliut, *de Toni & L. Boldori 82*, 6 Jul. 1941 (FC, DT). LIECHTENSTEIN: ad saxa calcarea supra Vaduz, *G. de Beck* (as *G. alpina* in Mus. Vindob. Krypt. Exs. no. 228, FC, L, NY, Z).

SWITZERLAND: an Felsen, Bodenhaus, Berner-Oberland, *C. von Nägeli*, 26 Jul. 1850 (ZT), an bespritzen Balken, with *Schizosiphon parietinus* var. *alpinus* Näg., *von Nägeli*, 20 Jul. 1850 (Type of *G. nigrescens* Näg., ZT); Bern, in herb. Kützing (Type of *G. atrata* Kütz., L; isotypes, FI, UC); an feuchten Felswänden, Engelberg, Kt. Unterwalden, *C. Cramer*, Jul. 1859 (Type of *G. alpina* Näg., FC [Fig. 17]); with *Scytonema incrustans* Kütz., Erlenbacher Tobel, auf Felsen, *von Nägeli*, Oct. 1850 (ZT); Moose überziehend, Gäbris, Appenzell, *B. Wartmann*, Jul. 1862 (as *Aphanocapsa montana* var. *micrococca* in Wartm. & Schenk, Schweiz. Krypt. no. (134), MIN, ZT); St. Gotthardstrasse oberhalb Göschenen, Uri, *Cramer*, 4 Jul. 1862 (as *Gloeocapsa opaca* in Wartm. & Schenk, Schweiz. Krypt. no. 235, Z); im stehenden Wasser vom schmelzenden Schnee auf der Gemmi, Wallis, *R. J. Shuttleworth 232*, 22 Aug. 1841 (BM), *216*, Aug. 1841 (BM); an Felsen, Maderanerthal, Uri, *Cramer*, Aug. 1876 (FC, FH); Maderanertal along the Kerstelenbach, *J. T. Koster 1297*, 28 Jun. 1948 (FC); on roots of *Hedera helix*, Meride-Serpiano, Ticino, *R. A. Maas Geesteranus 3403*, 30 May 1946 (FC); rochers calcaires, Meride-Tremona, Tessino, *Maas Geesteranus 3261*, 2 Jun. 1946 (FC); Rigi, *Cramer*, Sept. 1856 (Type of *Aphanocapsa montana* and var. *micrococca* Cramer in Wartm. & Schenk, Schweiz. Krypt. no. 134, ZT; isotype, MIN); Schaffhausen, in herb. Kützing (Type of *Gloeocapsa scopulorum* Näg., L [Fig. 16]; isotypes bearing variant labels: "Rheinfall, *von Nägeli*, Jul. 1847", M, and "ad Rheni cataractam, *Nägeli*", FI); am Nagelfluhfelsen der Bernegg bei St. Gallen, *Wartmann*, 1858, 1861 (as *G. coracina* in Rabenh. Alg. no. 814, FC, L; as *G. coracina* in Wartm. & Schenk, Schweiz. Krypt. no. 35, Z), 21 Dec. 1858 (Type of *G. saxicola* Wartm. in Rabenh. Alg. no. 813, FC [Fig. 22]; isotypes, L, and in Wartm. & Schenk, Schweiz. Krypt no. 345, Z); Trinser See auf Kalkfelsen, *G. C. Brügger* (Type of *Aphanocapsa montana* var. *macrococca* Cramer, ZT); ad *Populi pyranii* corticem, Zürich, *G. Winter*, Mar., Apr. 1877 (FC, L, M, S), Jul. 1878 (FC, S); ad terram arenosam humidam, Zürich, *Winter*, May 1878 (FC), Jan. 1879 (L); auf feuchtem Holz, Zürich, *Nägeli 108* (Type of *Gloeocapsa versicolor* Näg., L [Fig. 50]; isotype, UC); an nassen Felsen, Zürich, *Nägeli 216* (Type of *G. geminata* Kütz., L [Fig. 33]); Zürich, *Nägeli 299* (Type of *G. ianthina* Näg., L [Fig. 23]; isotype, UC), *Nägeli 301* (Type of *G. ambigua* Näg., var *pellucida* Näg., and b. *violacea* Näg., L); an nassen Felsen, Küssnacher Tobel, *Nägeli 302*, Sept. 1847 (M); Zürich, *Nägeli 304* (Type of *G. punctata* Näg., L; isotype, UC), *Nägeli 364* (L), *Nägeli 365* (Type of *G. ambigua β rivularis* Näg., L; isotype, UC), *Nägeli 387* (Type of *G. opaca* Näg. and var. *pellucida* Näg., L; isotype, UC); an einem Felsen unten der Sihl, Zürich, *von Nägeli*, 27 Mai 1849 (ZT); an einem Brunnen bei Küssnacht unweit Zürich, *Wartmann* (as *G. nigrescens* in Rabenh. Alg. no. 629, FC, FH, L, TA).

ITALY: Venetia: Aquilegia, *Biasoletto*, 1850 (UC); Isonzo, *A. Hansgirg*, 1891 (FC, W); Görz, *Hansgirg*, 1891 (FC, W); Sagrado prope Görz, *Hansgirg*, 1891 (FC, W); Oliero, *E. de Toni* (DT); Conegliano, *G. B. de Toni & D. Levi*, 1887 (as *G. versicolor* in de Toni & Levi, Phyc. Ital. no. 85, UC; in Roumeguère, Alg. de Fr., L); Nostoc no. 2, Padova, *Meneghini* (Type of *Palmella didyma* Kütz., L; isotype, UC); S. Pietro in thermis Euganeis, *O. Beccari* (B); with *Coccochloris stagnina* Spreng., Abano, therm. Eugan., *Meneghini* (Type of *Gloeocapsa cryptococcoides* Kütz., L; isotypes, UC, S); Battaglia, *Meneghini* (Type of *Protococcus pygmaeus* Kütz., L [Fig. 27]); ubique super muros Patavii, *Meneghini* (Type of *Pleurococcus communis* Menegh., FI [Fig. 83]; isotypes or cotypes with variant labels, "ad muros humidos", PC, and "ad parietes humidas", W, FI); ad muscos ad rupes in udis [Euganeorum], S. Elena, *Meneghini*, Sept. 1837 (Type of *Coccochloris muscicola* Menegh., FI [Fig. 79]; isotypes, L, UC); in Euganeis, *Meneghini* (Type of *C. parietina* Menegh., FI [Fig. 77]; isotypes, L, PC); in thermis Euganeis, *Meneghini* (Type of *Anacystis marginata* Menegh., L [Fig. 19]; isotypes, FC, B, S), in Euganeis, *Meneghini* (Type of *Microcystis nigra* Menegh., FI [Fig. 64]; isotypes, L, LD, S, UC); in rupibus humidis in mont. eugan., *Titius*, 1853 (Type of *Gloeocapsa Titiana* Grun., W); Sterzing—Gossensass, *Hansgirg*, 1891 (FC, W); Ala, *Hansgirg*, 1891 (FC, W); S. Margarita prope Ala, *Hansgirg*, 1891 (FC, W); Auer, *Hansgirg*, 1891 (FC, W); Gossensass—Pflersch, *Hansgirg*, 1891 (FC, W); Kardaun—Blumau prope Bozen, *Hansgirg*, 1891 (FC, W). Lombardy: Bormio, su rupi umide presso la sorgente Pliniana, *E. Levier*, Aug. 1871 (as *Gloeocapsa rupestris* in Erb. Critt. Ital., Ser. II, no. 663, FC, L); in muris humidis, Bellagio, *N. & G. Lagerheim*, 17 Sept. 1882 (D, S). Piedmont: su muschi rupicoli lungo i torrenti del Val Intrasca al Lago Maggiore, *De Notaris*, Aug. 1865 (as *G. livida* in Erb. Critt. Ital., Ser. II, no. 340, UC). Liguria: ad Oleam vetustam, Varigotti, *C. Sbarbaro*, 28 Mar. 1950 (FC); ad rupes inter Porto Venere et Spezia, *U. Martelli*, Nov. 1890 (DT). Emilia: Ballata Bagnatori prope Bologna, *C. Zanfrognini*, Aug. 1896 (DT). Tuscany: sulla scorza del *Diospyros virginiana* nell'Orto Botanico Pisano, *Archangeli*, Jan. 1883 (as *Gloeocapsa versicolor* in Erb. Critt. Ital., Ser. II, no. 1323, FC); Firenze (FH), sulla corteccia dei cipressi al Giardino di Boboli, *G. Archangeli*, Jan. 1878 (as *G. livida* in Erb. Critt. Ital., Ser. II, no.

664, FC, L, NY, UC, Z); Vallombrosa, *A. Borzi* (FH), comm. F. Ardissone (NY); sui muri umidi di un ponte presso Vallombrosa, *Borzi*, Oct. 1878 (as *Aphanocapsa pulchra* in Erb. Critt. Ital., Ser. II, no. 784, FC, NY, S). Roma: Roma, *E. Fiorini-Manzatti* (PC). Campania: amphitheater, Pozzuoli, *B. M. Davis*, May 1904 (MICH). Apulia: serra di Otranto, in herb. Kützing (L).

FAEROES ISLANDS: Faeroa, *Lyngbye*, as *Palmella alpicola* (FC, MO). SCOTLAND: "*Palmella frustulosa Car.*" [Appin?], *Carmichael* (Type of *Haematococcus frustulosus* Harv., K [Fig. 38]; isotype?, "ex herb. Thwaites", BM); "Palmella livida C." [Appin?], *Carmichael* (Type of *Palmella livida* Carm., K [Fig. 65]); "Palmella aeruginosa Car." [Appin], *Carmichael* (Type of *Haematococcus aeruginosus* Hass., K [Fig. 84]); Isle of Skye, *Lightfoot* (Type of *Ulva montana* Lightf., K); summit of Ben Arthur, Arroquhar, *W. Borrer* (BM). ENGLAND: Cantley, Yorkshire, *W. West*, Aug. 1887 (Type of *Chroococcus schizodermaticus* West, BIRM); Brant Fell, English Lake District, *G. S. West*, Aug. 1891 (UC); wet limestone rocks, Helln Pt., west Yorkshire, *G. S. West*, May 1897 (UC); Linn Gill, west Yorkshire, 12 Apr. 1895 (FC, UC); in river at Fairford, Gloucestershire, *W. L. Tolstead 8496*, 3 Oct. 1944 (FC); Upper Pen pond, Richmond Park, Surrey, *J. W. G. Lund* (Type of *Coelosphaerium limnicola* Lund in the private collection of J. W. G. Lund; isotype D [Fig. 76]). WALES: Llyn Teyrn, Snowdon, *G. S. West*, Aug. 1891 (Type of *Aphanothece saxicola* var. *sphaerica* W. & G. S. West, BIRM [Fig. 55]); Dolgelley, *J. Ralfs* (Type of *Palmella Ralfsii* Harv., BM; isotypes with variant label, "Welsh Mountains", L, UC). IRELAND: Ballintay, *D. Moore* (K); cavy white rocks, Portrush, county Antrim (FC, L); M. Ben Gower, *Shuttleworth 65* (Type of *Gloeocapsa Shuttleworthiana* Kütz., L [Fig. 37]; isotype, C); wet rocks, Erigal, county Donegal, *G. Dickie*, Jul. 1852 (designated as Type of *Haematococcus binalis* Hass., BM). NETHERLANDS: Nieuwkoopsche plas, *J. T. Koster 562*, Jun. 1940 (L); on wall in hothouse, Hortus Botanicus, Leiden, *Koster 220*, 5 May 1938 (FC, S), *Koster 856*, Nov. 1943 (L); in a small cement trench, hothouse, Hortus Botanicus, Leiden, *P. Leenhouts*, 12 Oct. 1950 (FC); tegen rotswanden, Valckenburg, *W. F. R. Suringar*, Jul. 1861 (L).

FRANCE: ad parietes cryptarum (as *Pleurococcus bituminosus* in Rabenh. Alg. no. 2188, L). Aisne: sur les casemates, Berthenicourt-par-Moy, *J. Mabille 1, 3*, May 1952 (FC). Alpes Maritimes: Antibes, *G. Thuret*, 9 Feb. 1869 (FH, L), 31 Jan. 1869 (PC), 9 Mar. 1870 (FC, NY), 21 Mar. 1870 (FC); grottes de Menton, *C. Flahault*, Apr. 1882 (PC), Aug. 1882 (S); rochers de la Croupelastière, *E. Bornet*, 20 Jan. 1872 (PC). Alsace: stagnes d'eau saumâtre, environs de Strasbourg, *C. Roumeguère* (as *Chroococcus varius* in Roumeguère, Alg. de Fr. no. 507, L). Ardennes: sur rocher humide, Monthermé, *J. Mabille*, 1952 (FC). Ariège: *H. Filhol* (as "Gloeosapsa rufescens" in Roumeguère, Alg. de Fr. no. 517, L). Bouches du Rhone: Marseille, falaises d'Arene, *Thuret*, Nov. 1854 (PC). Calvados: Falaise, *A. de Brébisson* (D, FC, FH, L, NY, S, UC, ZT), *557* (L, UC), *815* (L), *979* (L), sur les rochers humides (Type of *Protococcus magma* Bréb. in Desmaz., Pl. Crypt. de Fr. no. 213, L; isotype, FC); Falaise, *de Brébisson* (as *Gloeocapsa livida* in Desmaz., Pl. Crypt. de Fr. no. 1957, K, NY; in Rabenh. Alg. no. 2029, FC, K, L, UC); rochers granitiques et humides, *Brébisson 172* (Type of *G. rupestris* Kütz., L); Falaise, *A. de Brébisson* (Type of *G. rupestris* Kütz. ex Wille, L; isotype, S); à St. Auberc, *de Brébisson* (as *G. rupestris* in Rabenh. Alg. no. 2030, FC, FH, L, NY); sur les rochers inondés, Falaise, *de Brébisson* (as *G. pellucida* in Roumeguère, Alg. de Fr. no. 410, L); Bayeux, de *Brébisson 483*, Sept. 1861 (L, UC); Vire, herb. Lenormand (Type of *G. compacta* Kütz., L [Fig. 49]; isotype, UC); sur un mur humide, Vire, *P. Frémy 663*, Jun. 1923 (NY). Cher: carrières calcaires, Bourges, *C. Ripart 631*, Jun. 1877 (S). Gironde: sur les pièrres d'un fontaine à Cerons, *Durieu de Maisonneuve* (as *G. rupestris* in Roumeguère, Alg. de Fr. no. 956, L). Haute Garonne: sur les rochers de Luchon, *C. Fourcade*, 1886 (as *G. ambigua* f. *pellucida* in Roumeguère, Alg. de Fr. no. 825, L, TA). Hautes Pyrénées: sur rocher suintant, Cirque de Gavarnie, *J. Mabille*, Sept. 1951 (FC); roches inondées au dessus de Cauterets, *Lamy de la Chapelle* (as *G. ianthina* in Roumeguère, Alg. de Fr. no. 826, L, TA). Hérault: jardin botanique de Montpellier, *G. Lagerheim*, 1893 (S). Landes: Dax, muraille au nord du bassin émergé, *A. de Flers*, 16 Nov. 1886 (PC). Loire Inférieure: in fissuris rupium prope mare, Le Croisic, *E. Bornet*, 11 May 1877 (PC). Manche: Lepay, *A. de Brébisson* (FC, NY); rochers du Boule, *G. Thuret*, 5 Jan. 1853 (FH, S), *E. Bornet*, Jul. 1874 (FH, N); Cherbourg, *Thuret*, 5 Jan. 1853 (FC); mûr de l'église de St. Pièrre de Sémilly, *P. Frémy 688*, Oct. 1923 (NY); St. Lo, mûrs, *Frémy 320*, Feb. 1931 (NY), Mar. 1924 (NY); La Meauffe, sur rochers calcaires, *Frémy*, Mar. 1933 (NY); carrières de La Meauffe, *Frémy*, Mar. 1925 (NY). Orne: Domfront, *A. de Brébisson* (Type of *Protococcus haematodes* Kütz., L [Fig. 88]; isotypes, S, and as *Gloeocapsa haematodis* in Rabenh. Alg. no. 2192, L, S). Seine: écorce de *Populus tremula*, Paris, *Brébisson 452* (Type of *G. violacea* Kütz., L [Fig. 42]; isotypes, S, UC, and as *G. lignicola* in Rabenh. Alg. no. 2031, FC, L, NY, S, UC). Seine et Marne: in aquario ad Reutilly, *G. Thuret*, Jan. 1849 (FH, L, NY; as *Aphanocapsa parietina* in Wittr., Nordst., & Lagerh., Alg. Exs. no. 1547, L, NY); Ferrières, sur *Populus alba*, Versailles, *E. Bornet*, 6 Apr. 1850 (Type of *Gloeocapsa Itzigsohnii* Born., FH); La Ferré-Alais, *Thuret*, Jun. 1852 (NY). Seine Inférieure: mûr humide, Le Havre, *Dupray*, Jan. 1887 (as *G. aeruginosa* in Roumeguère, Alg. de Fr. no. 824, K, L). Vosges: étang des Moines, Remiremont, *E. Demangeon* (as *G. magma* in Roumeguère, Alg. de Fr. no. 114, L); Jardin Gremille à Remiremont, *Demangeon*, Jan. 1852 (as *Chroococcus fusco-ater* in Roumeguère, Alg. de Fr. no. 264, L); Vosges, ex herb. Lenormand (DT, FH, L, NY); in locis campestribus etc. (as *Collema scotinum* in Moug. &

Nestl., Stirp. Crypt. Vogeso-Rhen. no. 637, FC); ad saxa granitosa quartzosaque Vogesorum (as *Gloeocapsa magma* in Moug. & Nestl., Stirp. Crypt. Vogeso-Rhen. no. 1291, TA). Yonne: rochers de Pontigny, *Brébisson*, Jan. 1852 (PC). FRENCH SUDAN: Bamako, *M. Serpette SU6*, Aug. 1953 (D). UGANDA: in Lake Albert Nyanza, *R. T. Leiper*, July 1907 (Type of *Microcystis densa* G. S. West in slide collection, BIRM). TANGANYIKA: Amani, an Kalkwänden eines alten Gewächshauses bei den Saatbeeten, *B. Schröder 19b*, Aug. 1910 (S). RUANDA URUNDI: Usumbura, bei Kwambugu an Baumrinden in den Hochwäldern, *Holst 1436*, Sept. 1892 (Type of *Gloeocapsa Holstii* Hieron. in slide collection of N. Wille, O). MOZAMBIQUE: Marramwe lake, *E. L. Stephens 36*, 14 June 1928 (D). ANGOLA: Golungo Alto, ad saxa juxta cataractam riv. Congo prope Sange, *F. Welwitsch 134*, Sept. 1856 (FC, UC). ALGERIA: Amoureh, *F. Debray*, May 1891 (NY); a Calabona, stillicidii d'acqua calcarifera (as *G. atrata* with *Schizosiphon crustiformis* in Marcucci, Un. Itin. Crypt. no. XXIIIa, FC). CANARY ISLANDS: Funchal, *C. Lindman*, 7 Jan. 1885 (D, S). AZORES: ad pumices pr. Furnas ins. San Miguel, *K. Bohlin*, Jul. 1898 (as *Gloeocapsa magma* in Wittr., Nordst., & Lagerh., Alg. Exs. no. 1612, L). BERMUDA: causeway between St. Davids island and Hamilton island, *A. J. Bernatowicz 51-826*, 31 Jan. 1951 (FC); on flat rock near Hungry bay, *F. S. Collins*, 23 Apr. 1912 (as *G. fusco-lutea* in Coll., Hold., & Setch., Phyc. Bor.-Amer. no. 2153, D, FC, L); on inland cliff near Hillcrest, *Collins*, 15 Aug. 1913 (as *G. atrata* in Coll., Hold., & Setch., Phyc. Bor.-Amer. no. 2152, D, FC, L). GREENLAND: Maligiak, *T. M. Fries*, Jul. 1871 (S); Danmarks φ, *N. Hartz*, Mar. 1892 (FH).

NOVA SCOTIA: on littoral rocks, St. Margaret's Bay, Halifax county, *T. A. & A. Stephenson NSM9b*, Jul.—Sept. 1948 (FC). NEW BRUNSWICK: on ledges and rocks, Grand Falls, Victoria county, *H. Habeeb 10057, 10283, 10314, 10456, 10556, 10558, 10580, 10586, 10646, 10652, 10687, 10696, 10704, 10741, 10746, 13341, 13434, 13452*, Jul. 1947—Sept. 1948 (FC, HA); twig in lake 8 miles south of Grand Falls, *Habeeb 11651*, 22 June 1951 (FC, HA); on concrete tub at a spring, mouth of Salmon river, Victoria county, *Habeeb 11673, 11674*, 24 June 1951 (FC, HA); ledge-pool at falls, Pokiok, York county, *Habeeb 11581, 11582*, 6 June 1950 (FC, HA); on ledge, Point Wolf, Albert county, *Habeeb 11683*, 26 July 1951 (FC, HA); Island lake, 40—50 miles due south of Dalhousie (in Northumberland county?), *M. Le Mesurier 4*, Aug. 1953 (FC). MAINE: Canton Point, Oxford county, *Parlin*, May 1935 (NY); Round Pond, Lincoln county, *F. S. Collins 4118*, July 1901 (NY); on rocks above high water-mark, Cape Rosier, *Collins*, July 1900 (FH; as *Entophysalis magnoliae* in Coll., Hold., & Setch., Phyc. Bor.-Amer. no. 1802, FC, L, MIN, TA, UC); in a pond, Hollis, York county, *R. M. Whelden*, 31 Aug. 1929 (D); sink hole, Hollis, *F. C. Seymour*, 11 Sept. 1938 (FC); edge of Basin pond, Mt. Katahdin, *B. M. Davis 7*, 1902 (FC, MICH). NEW HAMPSHIRE: Jackson, *S. Higginson*, 1888 (FC); at the Flume, *W. G. Farlow*, Sept. 1882 (FH), 1901 (MICH); top of Page hill, Chocorua, *Farlow*, Sept. 1909 (FH); Acworth, Sullivan county, *B. M. Davis*, 1903 (MICH). VERMONT: Ripton gorge, Addison county, *W. G. Farlow*, Sept. 1896 (FC, S).

MASSACHUSETTS: Magnolia, *W. G. Farlow*, Aug. 1877 (Type of *Entophysalis magnoliae* Farl., FH; isotype, PC); path, Fells station, Melrose, *F. S. Collins 2503*, Apr. 1893 (NY); on perpendicular wet rock, Middlesex Fells, *Collins*, 9 June 1895 (as *Gloeocapsa magma* in Coll., Hold., & Setch., Phyc. Bor.-Amer. no. 151, FC, L, MICH, TA), *Collins 2565*, May 1893 (NY) Siasconset, Nantucket, *Collins 3394*, 23 Aug. 1896 (FH, NY); Nobska pond, Woods Hole, *W. Trelease*, 1881 (MO); Saugus, *Collins 1279*, Apr. 1890 (NY); Marblehead, *Collins 1296*, Apr. 1889 (NY); Chelmsford, *Russell* (FC); on walls of greenhouse, Botanic Garden, Cambridge, *Collins 3582*, 12 Jan. 1899 (FC, UC; as *Chroococcus varius* in Coll., Hold., & Setch., Phyc. Bor.-Amer. no. 1202, FC, D, L, TA); Juniper point, Salem, *Collins*, 11 Aug. 1889 (FC, UC); South Quincy, *F. C. Seymour*, 3 Jul. 1943 (FC); on dripping masonry under railroad bridge, Melrose, *Collins*, 9 Aug. 1902 (as *Gloeocapsa ianthina* in Coll., Hold., & Setch., Phyc. Bor.-Amer. no. 1205, FC, L, TA). RHODE ISLAND: limerock quarry, Lincoln, *Collins 5437*, 19 Apr. 1906 (FC, UC). CONNECTICUT: abutment, Factory pond dam, Bridgeport, *I. Holden 642*, 19 Jul. 1892 (UC), *Holden 776*, 15 Dec. 1892 (UC); Sage's ravine, Bridgeport, *Holden 1052*, 14 Oct. 1894 (FC, UC); Salisbury, *Holden 699*, 22 Aug. 1892 (FC, UC); moist limestone, Gaylordsville, *Holden*, 30 Oct. 1898 (as *G. violacea* in Coll., Hold., & Setch., Phyc. Bor.-Amer. no. 551, FC, L, TA); on rocks, Round pond, Lantern Hill, North Stonington, *W. A. Setchell 560*, 13 Sept. 1892 (FC, UC); Sperry's falls, Sargents river, Woodbridge, *Setchell 911*, 21 Oct. 1894 (FC, UC), *Setchell*, 7 Nov. 1891 (FC, UC).

ARCTIC ARCHIPELAGO: Cornwallis island, *J. Michéa 1*, 1949 (FC); in rock pools, Koojesse inlet near Sylvia Grinnell river, upper Frobisher bay, Baffin island, *H. A. Senn 3641a*, 29 June 1948 (FC). LABRADOR: sur un rocher de la toundra, sur le ruisseau Nakvak, Monts Torngat, *J. Rousseau 1096*, 1—2 Aug. 1951 (FC). QUEBEC: fond d'un ruisseau, embouchure de la Korok, Baie d'Ungava, *Rousseau 483*, 22 Jul. 1951 (FC); fond de l'étang, au pied de Saglek, fourche de la Korok, *Rousseau 996*, 31 Jul. 1951 (FC); petite source tourbeuse, plateau au voisinage du Mt. Pyramid (57° 30′ lat. N), *Rousseau 823*, 5 Aug. 1947 (FC); fond d'un lac, et sur rocher humide, embouchure de la Kogaluk, baie d'Hudson, *Rousseau 209—211*, 16 Jul. 1948 (FC); dans une tjaele et blocs de quartz, Baie Kayak (dans l'estuaire de la baie Payne),

Rousseau 1513, 1514, 19 Aug. 1948 (FC); sur un "ventre de boeuf" argilleux, portage entre le lac Tashwak et le lac Payne (entre la baie d'Hudson et la baie d'Ungava), *Rousseau 785,* 1 Aug. 1948 (FC); roche moutonée, rivière Payne vers 70° 25′ long. O., *Rousseau 1134,* 11 Aug. 1948 (FC); on rock in a small lake, table-top of Mt. Albert, Gaspé, *H. Habeeb 1792,* 28 Jun. 1951 (FC, HA); on wet rocks inside a cave, Ilet au Massacre, Le Bic, *Rousseau 27015,* 24 Aug. 1927 (FC); sur les rochers, carrière d'Outremont, Mont Royal près Montréal, *C. Lanouette 100, 110, 123,* Sept., Nov., 1940 (FC); carrière abandonée, St. Vincent de Paul, *Lanouette & R. Barabé 119,* 3 Oct. 1940 (FC); Aqueduc lake, Rolland, Montcalm county, *P. Giraudon J179.2,* 15 Jul. 1949 (FC); dans une petite caverne à l'embouchure du rivière Saguenay, *T. L. Doré,* Aug. 1949 (FC).

NEW YORK: Au Sable Chasm, Clinton county, *C. J. Sprague,* Aug. 1880 (FH); Niagara, *F. Wolle,* Oct. 1876 (FH); Kullum pond north of Warrensburg, Warren county, *J. Bader 174,* 4 Aug. 1938 (FC); walls of gorge below fall, Rensselaerville, *M. S. Markle 16,* 28 Jul. 1946 (FC); greenhouse, Barnard College, New York city, *H. C. Bold B20, B21, B24,* Feb. 1941 (FC). NEW JERSEY: Gardner's pond near Andover, Sussex county, *E. T. Moul 7512,* 16 Oct. 1952 (FC); on rocks in spray from a spring, Towaco, Morris county, *F. Drouet & J. Kells 2067,* 17 Jun. 1937 (D). PENNSYLVANIA: dripping rocks, Lehigh valley, *F. Wolle* (D, PENN, S); Glen Onoko, Carbon county, *Wolle,* Oct. 1875 (D, PENN); Bethlehem, an nassen Felswänden, *Wolle* (as *Gloeocapsa Itzigsohnii* in Rabenh. Alg. no. 3529, FC, L, TA, UC); culture, University of Pennsylvania, Philadelphia, *W. A. Cassel 14, 15,* 14 Dec. 1951 (FC); on a wet brick wall, North Warren, *T. Flanagan 70,* 19 Jul. 1943 (FC); wet face of a dam along Cresheim creek, Fairmount park, Philadelphia, *Drouet & C. Hodge 3892,* 20 Jul. 1941 (FC); upper falls, Buck Hill Falls, Monroe county, *B. M. Davis,* 18 Jul. 1915 (FC, MICH). DELAWARE: pond in the dunes one mile south of Point Henlopen, north of Rehoboth Beach, *Drouet & H. B. Louderback 8522, 8565, 8575,* Aug. 1948 (FC). MARYLAND: limestone cliff, Swallow falls of Deep creek, Garrett county, *L. P. McCann, M. W. Woods, D. S. Stoddard, & Drouet 2334,* 29 Aug. 1938 (D); shallow water of the abandoned Chesapeake & Ohio canal, Cabin John, Montgomery county, *Drouet, E. P. Killip, & D. Richards 5539,* 6 Aug. 1944 (FC, US); pool in rocks, south shore of Plummers island in the Potomac river west of Cabin John, *Drouet, Killip, & Richards 5555,* 6 Aug. 1944 (FC, US). VIRGINIA: rock, Bald Knob, Mountain Lake, Giles county, *R. F. Smart 123,* 31 Jul. 1938 (FC, ST); on concrete dam of Farrier's pond near Pembroke, Giles county, *J. C. Strickland 515,* 15 Aug. 1939 (FC, ST); wet cliff, summit of White Top mountain, Grayson county, *Strickland 1072,* 26 Aug. 1941 (FC, ST); limestone west of Dot, Lee county, *Strickland & E. S. Luttrell 1361,* 7 Aug. 1942 (FC, ST); dripping rock about 2 miles north of Pennington Gap, Lee county, *Strickland & Luttrell 1347,* 7 Aug. 1942 (FC, ST); on limestone about 5 miles north of Tazewell, *Strickland & Luttrell 1272,* 5 Aug. 1942 (FC, ST). NORTH CAROLINA: on rocks etc., Highlands, Macon county, *H. C. Bold 3a, H91c, H228, H231, H246,* Jul. 1938, Jun. 1939 (FC); on rocks, Cullasaja Gorge, Macon county, *Bold H159,* 17 June 1939 (FC); stonework, Fort Macon, Beaufort, *C. S. Nielsen & W. L. Culberson 1745, 1746, 1751, 1754, 1755, 1757, 1758, 1770, 1772,* 25 Aug. 1949 (FC, T); Dry falls on U.S. route 64, Pisgah national forest, *Nielsen & Culberson 1909* (FC, T); rocky cliff on state route 209 south of Hot Springs, *Nielsen & Culberson 1820,* 1 Sept. 1949 (FC, T). SOUTH CAROLINA: on wet cliffs, Eastitoe river, Pickens county, *Bold H262,* Jul. 1939 (FC); on damp clay and on *Gleditschia triacanthos,* Aiken, *H. W. Ravenel 548, 554,* Dec. 1884 (FC). GEORGIA: on abrk of *Maclura aurantiaca,* Savannah, *Ravenel 1* (FH); on rocks, Tallulah Gorge, Rabun county, *Bold 295a,* 5 July 1939 (FC).

FLORIDA: on *Taxodium distichum* and other trees and on logs, Gainesville, Alachua county, *Ravenel 9* (FH), *15,* (FH), *18* (FH, UC), *34* (UC); Gainesville, *M. A. Brannon 100, 325,* Aug. 1942, May 1948 (FC, PC); on soil in greenhouse, U. S. Forestry Station, Olustee, Baker county, *Brannon 226,* 1 May 1944 (FC, PC); stones in pine woods about five miles northwest of Beacon Hill, Bay county, *F. Drouet & C. S. Nielsen 11630,* 31 Jan. 1949 (FC, T); on sand above high tide-mark, St. Andrews bay at west end of Hathaway bridge, west of Panama City, *Drouet & Nielsen 10930,* 15 Jan. 1949 (FC, T); on an Australian pine tree south of South lake, Hollywood, Broward county, *Drouet 10287,* 29 Dec. 1948 (FC); Marco island, Collier county, *P. C. Standley 92836,* 14 Mar. 1946 (FC); Little Marco, *O. Tollin,* May 1886 (S); Flag pond about 2 miles east of Little Marco, *Tollin,* May 1886 (S); on cement wall 5 miles west of Goulds, Dade county, *C. Jackson 1,* 15 Jul. 1952 (FC); limestone, Aspalaga area, Gadsden county, *M. H. Voth 6, 9, 3* Nov. 1950 (FC, T), *Nielsen, G. C. Madsen, & D. Crowson 783,* 12 Feb. 1949 (FC, T); on trees beside U. S. highway 90 four miles east of Quincy, *Drouet, Nielsen, Madsen, & Crowson 10427,* 4 Jan. 1949 (FC, T); Highlands county, *P. C. Standley 92700,* 15 Mar. 1946 (FC); wayside park one mile west of Cottondale, Jackson county, *Nielsen & Madsen 851,* 19 Feb. 1949 (FC, T); on limestone etc., Florida Caverns state park north of Marianna, *Nielsen, Madsen, & Crowson 320,* 31 Aug. 1948 (FC, T), *Drouet, Nielsen, Madsen, & Crowson 10343, 10350, 10379, 10380, 10391,* Jan. 1949 (FC, T); 2 miles south of Marianna, *Nielsen, Madsen, & Crowson 336,* Aug. 1948 (FC, T); Leesburg, Lake county, *Brannon 330,* 31 May 1948 (FC, PC); on trees, Hendry Creek, about 10 miles south of Fort Myers, Lee county, *Standley 73196, 73239,* Mar. 1940 (FC); Natural Well, Leon county, *Nielsen 121,* June 1948 (FC, T); Nymphaea lake on Meridian road 14 miles north of Tallahassee, *Nielsen, Madsen, & Crowson 353,* 31 Aug. 1948 (FC, T); on trees on the northwest

59

shore of Lake Iamonia, Leon county, *Drouet, H. Kurz, & Nielsen 11308*, 24 Jan. 1949 (FC, T); edge of river, Rock Bluff, Liberty county, *J. E. Harmon 4*, 4 Oct. 1950 (FC, T); plaster on stairway, Fort Jefferson, Garden key, Dry Tortugas, *W. R. Taylor 318*, Jul. 1924 (TA); on trees and cement walks, Palm Beach, *Drouet & H. B. Louderback 10214, 10216*, 24 Dec. 1948 (FC); on a sand dollar in the dunes, Pensacola Beach, *Drouet, Nielsen, Madsen, Crowson, & A. Pates 10611*, 8 Jan. 1949 (FC, T); on a tree beside St. Marks river, Newport, Wakulla county, *Drouet, Madsen, & Crowson 10824*, 13 Jan. 1949 (FC, T); Log Sulfur spring, Newport, *Nielsen 249, 252*, Aug. 1948 (FC, T); spillway dam at Phillips Pool, St. Marks Wildlife Refuge, Wakulla county, *Nielsen, Madsen, & Crowson 484*, 9 Oct. 1948 (FC, T); on a tree, Wakulla Springs, *Drouet, Madsen, & Crowson 11499*, 27 Jan. 1949 (FC, T); Chipley, Washington county, *Nielsen & Madsen 866*, 19 Feb. 1949 (FC, T).

OHIO: Adams county, *W. A. Daily 294*, 25 May 1940 (DA, FC); road beyond White's Mill, the Plains near Athens, *A. H. Blickle & M. Wright*, 10 Jul. 1941 (DA, FC); cement wall, greenhouse, Ohio State University, Columbus, *Daily & F. K. Daily 260*, 29 Dec. 1939 (DA, FC); wet stone wall under bridge near Addyston, Hamilton county, *L. Lillick & I. Lee 705*, 23 Sept. 1933 (DA, FC); stone wall, Eden Park reservoir, Cincinnati, *Daily 408, 410*, 3 Jul. 1940 (DA, FC); flower pot dish, Peterson's greenhouse, Cincinnati, *Daily 438*, 9 Jul. 1940 (DA, FC); pond in Arlington cemetery, Cincinnati, *Daily 420*, 5 Jul. 1940 (DA, FC); near Seven Caves, Highland county, *W. B. Cooke*, Aug. 1932 (DA, FC); rocks and cliffs, Seven Caves park, Paint, Highland county, *Daily & F. K. Daily 313, 320, 321*, 16 Jun. 1940 (DA, FC); on sandstone, Old Mans Cave state park, Hocking county, *Daily & F. K. Daily 503, 504, 506—508, 517, 518, 520, 525, 526, 528, 529, 532, 540, 541, 545—547, 549, 555—557, 561*, 27 Jul. 1940 (DA, FC); sandstone, Cantwell Cliff state park, *Daily & F. K. Daily 683, 684, 687, 692, 699, 703*, 31 Aug. 1940 (DA, FC); *708—710, 714*, 1 Sept. 1940 (DA, FC); cliff at Ash Cave, Hocking county, *J. Wolfe*, 11 Oct. 1935 (DA, FC); on sandstone cliff, Rock Run, 3 miles northwest of Jackson, *Wolfe*, Nov. 1935 (DA, FC), *Cooke*, 1 May 1937 (DA, FC); cliff, White's Gulch, Jackson county, 2 May 1936, ex coll. Max Britton no. 33 (DA, FC); Kelly Isle quarry near Marblehead, Catawba island, Ottawa county, *Daily & F. K. Daily 578, 584*, 25 Jul. 1940 (DA, FC); limestone cliff, Gibraltar isle, Put-in-Bay, Ottawa county, *Daily & C. E. Taft 665*, 24 Jul. 1940 (DA, FC); stone in damp place in woodland near Apple Creek, Wayne county, *L. J. King 2319*, Aug. 1944 (EAR, FC).

KENTUCKY: on sandstone near Ohio river, Boone county, *Daily & J. A. Tucker 352*, 25 Jun. 1940 (DA, FC); Carter's caves, Carter county, *L. Walp*, Fall 1940 (FC), *D. Parker*, 16 May 1942 (FC, ST); on rock 4 miles east of Maysville, Mason county, *W. L. Culberson 545*, Aug. 1950 (FC); on sandstone rocks, Natural Bridge state park, Powell county, *Daily & F. K. Daily 275*, 14 Apr. 1940 (DA, FC), *365, 368, 369, 373, 374, 379—381, 386, 386a, 388—391, 394, 397—400*, 30 Jun. 1940 (DA, FC), *H. K. Phinney 533, 537, 541*, Apr., May, 1939 (FC, PHI), *J. H. Hoskins*, 1938 (DA, FC), *B. B. McInteer 1026*, 1 Sept. 1939 (DA, FC), *MacFarland 1088*, 10 Jul. 1940 (DA, FC); sandstone, railroad tunnel, Sky Bridge road near Nada, Powell county, *Daily & R. Kosanke 841, 842*, 6 Oct. 1940 (DA, FC); on wet moss, Cumberland Falls, Whitley county, *L. Lillick*, 5 May 1933 (DA, FC); sandstone overhang, Torrent, Wolfe county, *Daily, F. K. Daily, & Kosanke 460, 467, 472, 474, 479, 481, 483, 490*, 14 July 1940 (DA, FC); sandstone, Tight Holler, Pine Ridge, Wolfe county, *Daily, A. T. Cross & J. Tucker 763, 764, 768, 769a—772, 774—776, 778, 779, 783, 784, 787, 788, 790, 798—801, 804—806, 808, 811, 812, 814—820, 822*, 6 Sept. 1940 (DA, FC), *Daily 839, 840, 840b, 844*, 5 Oct. 1940 (DA, FC); *Daily & F. K. Daily 848, 851—853*, 25 May 1941 (DA, FC).

TENNESSEE: water tank, Scout Camp, Anderson county, *H. Silva 2339*, 25 Aug. 1950 (FC, TENN); rocks in spring, Valley Lake, Davidson county, *Silva 1809*, 20 Aug. 1949 (FC, TENN); Glen Echo Lake, Davidson county, *H. C. Bold 3931*, 19 Feb. 1939 (FC); concrete dam, Neuberts Springs, Knox county, *Silva 1526, 1532*, 20 Jul. 1949 (FC, TENN); fish pond at greenhouse, University of Tennessee, Knoxville, *Silva 586*, 5 Apr. 1947 (FC, TENN); limestone roof of Fosters Cave, Montgomery county, *Silva 1402*, 15 Jul. 1949 (FC, TENN); cliff near Rugby, Morgan county, *Silva 1665*, 6 Aug. 1949 (FC TENN); rock in spray of fall, Centennial lake, Nashville, *Bold 161*, 14 Oct. 1938 (FC); Sprout's Spring, Obion county, *Silva 1294*, 12 Jul. 1949 (FC, TENN); wet rock, tunnel near Chimneys, Sevier county, *Silva 454*, 11 Jan. 1947 (FC, TENN); Little River trail, Sevier county, *Silva 66*, 18 Aug. 1941 (FC, TENN); Myrtle Point, Sevier county, *A. J. Sharp 413*, 4 Sept. 1941 (FC, TENN); rock wall, Greenbrier Cove, Sevier county, *H. H. Iltis 2009*, 1 Oct. 1947 (FC, TENN); rocks and cliffs, Ramsey Cascades, Sevier county, *Bold H332a, H349, H356, H359*, 18 Jul. 1939 (FC), *Silva 413, 2350a*, 29 Dec. 1946, 9 Sept. 1950 (FC, TENN); concrete bridge, highway no. 70, Shelby county, *Silva 1086*, 30 June 1949 (FC, TENN); seepage on concrete basin, Union county, *Silva 309*, 1 Nov. 1946 (FC, TENN). ALABAMA: Fort Morgan, Baldwin county, *R. L. Caylor 51-7*, Aug. 1953 (FC); on retaining wall beside Water street between Eslava and New Jersey streets, Mobile, *Drouet & H. B. Louderback 10108*, 19 Dec. 1948 (FC). MICHIGAN: on damp rocks, Wishing spring, West Shore drive, Mackinac island, *H. K. Phinney 472*, 25 Jul. 1943 (FC, PHI); in artesian spring, east shore of Burt lake, Cheboygan county, *Phinney 501*, 29 June 1943 (FC, PHI).

INDIANA: on limestone, Scenic Falls, near Hartsville, Bartholomew county, *F. K. & W. A. Daily*

1042, 1044, 11 Oct. 1942 (DA, FC); cement wall of Lake Shaffer dam, Monticello, Carroll county, *Daily 868,* 29 July 1941 (DA FC); on limestone, Clifty Falls state park near Madison, Jefferson county, *Daily & F. K. Daily 671, 672, 682,* 4 Aug. 1940 (DA, FC); on limestone, Tunnel Mills, Vernon, Jennings county, *F. K. & W. A. Daily 908, 909, 1113, 1753, 1754,* 23 May 1942, 1 June 1947 (DA, FC); limestone, Muscatatuck state park, Vernon, Jennings county, *F. K. & W. A. Daily 912, 1104, 1105,* 23 May 1942 (DA, FC); limestone in quarry, North Vernon, Jennings county, *F. K. & W. A. Daily 1109,* 23 May 1942 (DA, FC); limestone along Sand creek 6 miles west of Zania, Jennings county, *C. M. Palmer, F. K. & W. A. Daily 982,* 24 May 1942 (DA, FC); in a sink hole pool, Lawrence county, *Palmer 1331d,* 2 Sept. 1932 (D); on sandstone along trail no. 1, Shades state park, Waveland, Montgomery county, *F. K. & W. A. Daily 966, 968, 970, 974, 976,* 6 Jun. 1942 (DA, FC); on shale and sandstone, Pine Hills, Montgomery county, *F. K. & W. A. Daily 1443, 1444,* 11 May 1946 (DA, FC); on limestone, cement, etc., McCormicks Creek state park, Owen county, *F. K. & W. A. Daily 1787, 1794, 1800—1806, 1809—1811, 1819, 1821, 1829,* 9—10 Aug. 1947 (DA, FC); on sandstone, Turkey Run state park, Parke county, *F. K. & W. A. Daily 1400, 1413, 1417, 1419, 1420, 1426, 1440,* 10—11 May 1946 (DA, FC), *1841, 1848,* 15 Aug. 1947 (DA, FC), *Drouet & D. Richards 2475, 2478, 2481, 2487, 2498,* 15—16 Jun. 1939 (FC); sandstone cliff, Blue Wells Hollow 5 miles east of Cannelton, Perry county, *W. Welch 8198,* 11 Jul. 1937 (DA, FC); on limestone wall, quarry at St. Paul, Shelby county, *F. K. & W. A. Daily 886,* 31 Aug. 1941 (DA, FC); wall under N. W. Fifth street bridge, Richmond, *L. J. King 2334,* Aug. 1944 (EAR, FC); railroad tunnels, Richmond, *King 61, 64, 76, 77,* and *s. n.,* Aug. 1940, Oct. 1941 (EAR, FC); sediment in edge of Lake Wehi, Wayne county, *M. S. Markle & King 737,* 7 Sept. 1942 (EAR, FC); in a spring near Thistlethwaites falls north of Richmond, *King,* 21 Mar. 1941 (EAR, FC).

WISCONSIN: laboratory cultures A and C, University of Wisconsin, Madison, *R. Evans,* 9 Dec. 1949, 6 June 1952 (FC); in a pool in limestone one mile east of Prairie du Chien, Crawford county, *Drouet 5079* (FC). ILLINOIS: laboratory cultures, University of Chicago, Chicago, *P. D. Voth,* 16 Jan., 20 & 29 Mar., 31 Jul., 1 Aug., 15 Sept. 1941 (FC); on cement walks in greenhouse, University of Chicago, *Voth, with J. M. Beal, & Drouet 2371,* 5 Oct. 1938 (FC); tanks in greenhouse, *J. A. Steyermark, C. Shoop, & Drouet 2444,* 30 Oct. 1938 (FC); on stone, Lake Shore park, Evanston, *H. K. Phinney 387,* 18 Sept. 1943 (FC, PHI); Starved Rock state park, La Salle county, *L. J. King 383,* 5 May 1941 (EAR, FC), *Steyermark & P. C. Standley 40646,* 4 Jul. 1941 (FC); plankton of Goose lake northwest of Quincy, *T. C. Dorris 56,* 18 Jul. 1950 (FC); Lake Glendale, Robbsville recreational area in Shawnee national forest, Pope county, *Phinney 1062,* 1 Sept. 1944 (FC, PHI). MISSISSIPPI: on the trunk of a tree at west limits of Biloxi, Harrison county, *Drouet 9995,* 14 Dec. 1948 (FC). MINNESOTA: Gunflint, Cook county, *B. Fink,* July 1897 (UC); rocks of old dock, north arm of Lake Itasca, Clearwater county, *H. F. Buell 462,* 9 Jul. 1946 (FC); in sluggish stream, Minneapolis, *J. E. Tilden,* 10 Aug. 1894 (as *Palmella uvaeformis* in Tild., Amer. Alg. no. 49, FC); on limestone in a quarry by the greenhouses, University of Minnesota, Minneapolis, *Drouet, E. C. Abbe, & J. W. Moore 4995,* 5 Aug. 1943 (FC). IOWA: on shaded limestone, Fayette, *B. Fink 7,* 17 May 1906 (FC, UC); rocks in fish pond, Guttenberg, Clayton county, *W. Kiener 21533,* 15 Sept. 1946 (FC, KI). MISSOURI: in a salt spring, Chouteau Springs, Cooper county, *Drouet & H. B. Louderback 5641b,* 25 Aug. 1945 (FC); rocks in quarry, Hill park, Independence, *Drouet 321,* 25 Dec. 1928 (D); on limestone ledge by Mill creek north of Bigsprings, Montgomery county, *Drouet 726,* 9 Nov. 1930 (D, FH); top of bluff along Muddy creek 3 miles east of Newland, Pettis county, *C. Shoop 339,* 4 Oct. 1938 (FC); in stream from Elk Lick springs 7 miles southwest of Nelson, Saline county, *Shoop 400,* 6 Oct. 1938 (FC); on walls at entrance to Jam-up Bluff 6 miles northwest of Montier, Shannon county, *J. A. Steyermark 28843,* 8 June 1941 (FC); ledges along White river 11½ miles southeast of Mincy, Taney county, *Steyermark 40107b,* 20 June 1941 (FC); moist shaded cliff, Blackwell, Washington county, *C. Russell 19,* May 1900 (FC, UC).

ARKANSAS: still pool, Warren, Bradley county, *D. Demaree 24813,* 23 Oct. 1943 (FC); in a shaded bayou, Wilmar, Drew county, *Demaree 24755,* 9 Oct. 1943 (FC); on rock, Hot Springs mountain, Garland county, *N. E. Gray 131,* 18 Aug. 1939 (FC). LOUISIANA: on trunk of tree near Loranges, Tangipahoa parish, *L. H. Flint,* 21 Mar. 1953 (FC); on an epiphytic orchid, Varnado, Washington parish, *Flint,* 24 Jan. 1953 (FC); on trees east of Lacombe, St. Tammany parish, *Drouet & Flint 9557,* 2 Dec. 1948 (FC); on a tree beside Bayou du Large just north of Theriot, *Drouet 9308a,* 22 Nov. 1948 (FC); on the bank of the Intracoastal canal, Weeks Island, Iberia parish, *R. P. Ehrhardt & Drouet 68,* 15 Nov. 1948 (FC); on fence-posts, Grand Isle, Jefferson parish, *Drouet & P. Viosca Jr. 9523,* 27 Nov. 1948 (FC). MANITOBA: Fort Churchill, *J. M. Gillett 2005, 2092,* 18 June, 15 July 1948 (FC). SOUTH DAKOTA: on quartz about 20 miles southeast of Rapid City, Custer county, *F. A. Hayes & B. H. Williams 12208,* 8 May 1941 (FC, KI); in a watering tank 2½ miles south of Freeman, Hutchinson county, *P. D. Voth,* 2 Jul. 1945 (FC). NEBRASKA: culture from sandpit north of David City, Butler county, *W. Kiener 22391b,* 10 Nov. 1947 (FC, KI); shallow fishpond, Valentine, Cherry county, *Kiener 23799,* 18 Jun. 1948 (FC, KI); rocks, Chadron state park, Dawes county, *Kiener 11534a, 20275,* 26 Aug. 1941, 15 May 1946 (FC, KI); fishponds, Rock Creek Hatchery, Dundy county, *Kiener 19557, 19558,* 3 Aug. 1945 (FC, KI);

culture from seepage, Blue creek north of Oshkosh, Garden county, *Kiener 16219, 21719a*, 4 Nov. 1945, 28 Oct. 1947 (FC, KI); culture from a spring, Lonergan creek, Lemoyne, Keith county, *Kiener 23068*, 20 Aug. 1948 (FC, KI); on flower pot, University greenhouse, Lincoln, *Kiener 10122, 13671—13674*, 23 May 1941, 13 Mar. 1943 (FC, KI); culture from tank of farm well, Gretna, Sarpy county, *Kiener 12624e*, 16 Jul. 1944 (FC, KI); pebble 2 miles southwest of Orella, Sioux county, *Kiener 22192*, 24 May 1947 (FC, KI); culture from Dismal river near Halsey, Thomas county, *Kiener 16750c*, 7 Oct. 1945 (FC, KI).

OKLAHOMA: under wet bluff, Turner Falls, Murray county, *A. J. Sharp OA-413*, 21 Dec. 1940 (FC, TENN). TEXAS: trail to Boot Springs, Brewster county, Chisos mountains, *E. Whitehouse 24634*, 23 Dec. 1950 (FC); on *Prosopis juliflora* DC. in woodland northeast of Resaca park, Brownsville, *R. Runyon 3722*, 30 May 1944 (FC). WYOMING: bottom of Trail's Divide and Brooklyn lakes, Medicine Bow national forest, *W. G. Solheim 61, 114*, 30 June, 17 July 1933 (FC). COLORADO: in a pond at 12,000 feet, Bald Mt., a few miles south of Pikes Peak, *H. L. Shantz*, 17 Sept. 1904 (Type of *Eucapsis alpina* Clem. & Shantz, US; isotype, D); on siliceous cliff, Chasm gorge, Longs Peak, *W. Kiener 1277*, 18 Sept. 1934 (FC, KI); Middle St. Vrain valley, Boulder county, *Kiener 6245*, 11 Oct. 1937 (FC, KI); South Colony creek, Custer county, *Kiener 10392, 10394*, 9 Jul. 1941 (FC, KI); granite rock, Lily mountain, Larimer county, *Kiener 13171*, 26 Sept. 1942 (FC, KI). NEW MEXICO: Carlsbad Caverns, Eddy county, *Bailey*, Apr. 1924 (NY), *W. B. Lang*, 1944 (FC); in a hot spring, Montezuma (Hot Springs), San Miguel county, *Drouet & D. Richards 2534*, 12 Oct. 1939 (FC); on wet ground near the city park, Hot Springs, Sierra county, *Drouet & Richards 2713*, 26 Oct. 1939 (FC); Little Tesuque, vicinity of Santa Fe, *Bro. G. Arsène 23304*, 6 Apr. 1937 (D). ALBERTA: Middle hot springs, Banff, *W. A. Setchell*, 6 Aug. 1923 (FC, UC). UTAH: Zion canyon, Washington county, *N. E. Gray*, 13 Aug. 1941 (FC); in Quartzite falls, Mt. Timpanogos, Utah county, *A. O. Garrett 56*, 13 Aug. 1925 (FC, UC). ARIZONA: on rocks beside dam of pond in the Coyote mountains, 35 miles southwest of Tucson, Pima county, *Drouet, Richards, & R. & B. Darrow 2757*, 29 Oct. 1939 (FC); above falls in Emory canyon, Lake Mead, *E. U. Clover 77a*, 20 Apr. 1941 (FC, MICH). NEVADA: in water at 132° F., Wabuska, Lyon county, *I. La Rivers 1800*, 9 Jun. 1953 (FC); inside cold tank, Steamboat, Washoe county, *M. J. Groesbeck 190, 375*, 4 Sept. 1940, 3 Apr. 1941 (FC). YUKON: on *Hygrohypnum palustre* in a spring, Dawson, *R. S. Williams*, 30 Jul. 1899 (FC). ALASKA: Juneau, *D. Saunders*, Jun. 1899 (UC); Prince William sound, *Saunders* (FH); Akutan island, *G. N. Jones 9397*, Aug. 1936 (TA); pool in swamp, Glacier valley, Unalaska island, *Setchell 5023a*, 7 Aug. 1899 (FC, UC); Mileposts 72, 83, 170 on Glenallen highway between Anchorage and Fairbanks, *D. E. Hilliard 31, 45, 62*, Sept. 1953 (FC). BRITISH COLUMBIA: Spillamacheen valley lakes near Bald mountain, Purcell range, Kootenay county, *W. R. Taylor*, Sept. 1925 (Type of *Aphanothece uliginosa* W. R. Taylor, TA); dripping rocks, Fish lake south of Beavermouth, *Taylor 14, 16*, Sept. 1921 (TA); Cougar valley, Glacier, Kootenay county, *Taylor*, 1927 (TA). OREGON: in "hot springs" near Madras, Crook county, *L. E. Griffin*, 23 Apr. 1939 (FC).

CALIFORNIA: on a dripping water tank, Berkeley, *N. L. Gardner*, 20 Apr. 1894 (as *Synechocystis aquatilis* in Coll., Hold., & Setch., Phyc. Bor.-Amer. no. 1206, FC, L, TA), 26 Feb. 1906 (as *Aphanothece Richteriana* in Coll., Hold., & Setch. Phyc. Bor.-Amer. no. 1801, FC, L, TA); Berkeley, *Gardner 637*, Mar. 1902 (UC); on floor in the greenhouse, Department of Botany, University of California, Berkeley, *Gardner 6807, 6938*, 10 Sept., 28 Nov. 1931 (FC, UC); on damp boards in the agricultural greenhouse, University of California, Berkeley, *Gardner 7428, 7429*, 24 July 1933 (FC, UC); in an aquarium in the courtyard of the Life Sciences building, University of California, Berkeley, *Gardner 7883*, 3 Nov. 1935 (FC, UC); culture (of algae collected by H. E. Parks) in the University of California, Berkeley, *Gardner 7819* (FC, UC); on trunk of bay tree, Strawberry creek, Berkeley, *Gardner*, 8 Mar. 1903 (as *A. conferta* in Coll., Hold., & Setch., Phyc. Bor.-Amer. no. 1152, FC, L, TA); culture (from Eureka, Humboldt county, collected by H. E. Parks), *Gardner 6569, 7203*, 12 Jan. 1931, 22 Apr. 1933 (FC, UC); on rock in steam and spray, the Geysers, Inyo county, *M. J. Groesbeck 140*, 6 Sept.. 1940 (FC); in cold stream, Keough hot springs, Inyo county, *Groesbeck 230*, 23 Nov. 1940 (FC); about vent, Hot Creek Geysers, Inyo county, *Groesbeck 418a*, 7 Jul. 1941 (FC); first spring south of Triangle spring, Death Valley, Inyo county, *J. & H. W. Grinnell 7625*, 24 Oct. 1933 (FC, UC); fountain, Miller's Patio cafe, Long Beach, *E. C. Jennings*, 28 Feb. 1937 (D); Old El Encino hot springs, Los Angeles county, *B. C. Templeton 1a*, Jul. 1944 (FC); Lagunitas creek, Marin county, *V. Duran 6604*, Jan. 1931 (FC, UC); culture (from Almonte, Marin county, collected by H. E. Parks), *Gardner 6906*, 16 Oct. 1931 (FC, UC); in a cavern above high tide line, Mendocino City, *M. A. Howe*, 7 May 1896 (UC); stone in overflow of hot artesian well on north shore of Mono lake, *Groesbeck 129*, 6 Sept. 1940 (FC); Hot Creek Geysers, Mono county, *Groesbeck 202, 418*, 25 Nov. 1940, 7 Jul. 1940 (FC); Bridgeport, Mono county, *Groesbeck 256, 331*, 24 Nov 1940, 3 Apr. 1941 (FC); pure cultures (one from A. Frenkel, another from Tassajara hot springs, Monterey county) at Hopkins Marine Station, Pacific Grove, *M. B. Allen*, 29 Jan. 1952 (FC); wall of reservoir, Del Monte, Monterey county, *W. J. V. Osterhout*, 13 Sept. 1902 (as *Aphanothece saxicola* in Coll., Hold., & Setch., Phyc. Bor.-Amer. no. 1203, FC, L); Calistoga, Napa county (Gardner no. 7832), *H. E. Parks 3256* (FC, UC); in an urn, Mountain View cemetery, Oakland, *Gardner 8020*, 19 Jan. 1927

(FC, UC); on quartz, Mountain Spring Camp northeast of Essex, San Bernardino county, *Drouet &*
J. F. Macbride 4609, 12 Oct. 1941 (FC); hot water creek in a canyon, Arrowhead hot springs, San
Bernardino county (Gardner no. 7835), *Parks 3250,* Dec. 1929 (FC, UC); Waterman hot springs,
San Bernardino county, *Setchell 1554,* 19 Dec. 1896 (FC, UC); on the trunk of a palm tree,
Needles, *Drouet & Macbride 4598,* 12 Oct. 1941 (FC, UC); Mountain lake, San Francisco,
Gardner 743, Sept. 1902 (UC); Lake Merced, San Francisco, *Gardner 1583,* Nov. 1905 (UC); in
the conservatory, Golden Gate park, San Francisco, *Gardner 7259, 7264,* 3 May 1933 (FC, UC); on
sand in a pond, Golden Gate park, San Francisco, *Gardner 1121,* 15 Jul. 1903 (FC, UC); on stones
in Feldt lake, Stanford farm, San Mateo county, *Gardner 3437,* 2 Aug. 1916 (FC, UC); on rocks
at entrance to ice caves on south side of Mt. Shasta, Siskiyou county, *W. B. Cooke 16673,* 2 Jul. 1942
(FC); in warm water, the Geysers, Sonoma county, *Gardner 7524,* 8 Aug. 1933 (FC, UC); in shallow
water of Tule river, Porterville, Tulare county, *Drouet & Groesbeck 4451,* 4 Oct. 1941 (FC);
culture (from Exeter, Tulare county, collected by H. E. Parks), *Gardner 7786,* 28 July 1935 (FC, UC).

BAHAMA ISLANDS: New Providence, on bark, limestone, pipe of water tank, etc., *L. J. K.*
Brace 197, 9624, 9684a, 9748, 9991, 10036(609), Oct.—Dec. 1918, Feb. 1919 (NY); on damp
walls in a garden, Nassau, *Brace,* Dec. 1921 (NY); brackish marsh, rifle range, New Providence,
E. G. Britton 3282, Feb. 1905 (NY); Moss Hill ponds, Long Cay, *Brace 4235,* Dec. 1905 (Type
of *Chondrocystis Bracei* Howe, NY [Fig. 82]); freshwater lake off Stafford creek, Andros island,
A. D. Peggs, 22 Aug. 1942 (FC). PUERTO RICO: on rocks, Arroyo de los Corchos, Adjuntas to
Jayuya, *N. Wille 1718,* Mar. 1915 (Type of *Aphanocapsa intertexta* Gardn., NY [Fig. 32]); isotype,
FH); bark, Aibonito to Cayey, *Wille 1904,* Mar. 1915 (NY); on rocks, etc., Arecibo to Hatillo, *Wille*
1155a, 1372b, 1387a (Type of *Anacystis nigropurpurea* Gardn., NY [Fig. 75]; isotype, FC),
1393 bis (NY); on rocks on walls in and near Arecibo, *Wille 1337, 1363, 1368, 1457,*
Feb.—Mar. 1915 (NY); on limestone, Arecibo to Utuado, *Wille 1465a* (Type of
Gloeothece endochromatica Gardn., NY [Fig. 53]), *1476, 1476a,* Mar. 1915 (NY);
limestone between Cabo Rojo and San German, *Wille 1194a,* Feb. 1915 (Type of *Anacystis*
minutissima Gardn., NY [Fig. 73]); on trees, walls, etc., Caguas, *Wille 439* (Type of *A. compacta*
Gardn., NY [Fig. 25]), *444* (FC, NY); *460c, 462* (NY), *462a* (Type of *Endospora olivacea*
Gardn., NY [Fig. 41]), Jan. 1915; on walls, rocks, trees, etc., Coamo Springs, *Wille 254a, 273, 277,*
278, 299a (Type of *Anacystis microsphaeria* Gardn., NY), *300* (Type of *A. irregularis* Gardn., NY),
300a, 300d, 301 (Type of *Chroococcus muralis* Gardn., NY [Fig. 26]), *303a, 304a, 307a, 308, 313,*
321, 402b, 405 , 1869e, 1869f (Type of *Anacystis radiata* var. *major* Gardn., NY [Fig. 51, 56]),
1870x, 1901a (Type of *A. radiata* Gardn., NY [Fig. 60]), *1901c, 1901d, 1905c, 1916a* (Type of
Endospora rubra Gardn. NY [Fig. 57]), *1923a,* Jan., Mar. 1915 (NY); Fajardo, *Wille 656* (NY),
679b, 712 (FC, UC), *713, 713a,* (NY), *717a* (D, NY), *719a, 719b,* (FC, UC), *732a* (NY),
732b (Type of *Anacystis anomala* Gardn., NY), *734a, 736a,* Jan. 1915 (NY); on limestone,
Coamo Springs, *N. L. & E. G. Britton 9033,* Mar. 1927 (NY); on bricks and bark, Guanica,
Wille 1838, 1841a, Mar. 1915 (NY); on wood, stones, etc., Hacienda Catalina near Palmer, *Wille*
752, 755, (NY), *758* (FC, UC), *765* (NY), *768b* (FC, UC), *791a* (NY), Jan.—Feb. 1915;
Hato Arriba near Arecibo, *Wille 1397a, 1397b, 1399c, 1399d* (NY), *1406a* (Type of *Gloeocapsa*
livida var. *minor* Gardn., NY [Fig. 52]), *1401b* (Type of *Gloeothece parvula* Gardn., NY [Fig.
35]), *1433* (Type of *Entophysalis violacea* Gardn., NY), Mar. 1915; on Flytteblok, Humacao, *Wille*
572a, 573b, 577a, Jan. 1915 (NY); on stones, Jayuya, Wille *1770a, c, d,* Mar. 1915 (NY); an
Kalkfelsen bei Guanica, *Wille 1843,* 17 Mar. 1915 (FC, UC); Laguna Joyuda, Mayaguez, *Wille*
1204, 1206 (FC, UC), *1207a* (Type of *Anacystis nidulans* Gardn., NY [Fig. 62]), *1209a* (Type of
A. Willei Gardn., NY), *1210* (FC, UC), *1304* (Type of *A. gloeocapsoides* Gardn., NY [Fig. 44]),
1307a, 1316d, Feb. 1915 (NY); Laguna Tortuguero, *Wille 835a* (FC, UC), *866* (Type of
Gloeocapsa calcicola Gardn., NY [Fig. 39]), *871, 872* (NY), Feb. 1915; Maricao, *Wille 1050, 1061*
(NY), *1135* (FC, UC), *1144e* (Type of *Aphanothece opalescens* Gardn., NY [Fig. 28]), *1148b,c*
(NY), *1151a,b* (FC, UC), *1155e* (Type of *Anacystis distans* Gardn., NY), *1155f,* Feb. 1915 (NY);
Mayaguez, *Wille 981a,b, 982a,b, 985, 1000e* (NY), *1012* (FC, UC), *1190,* Feb. 1915 (NY);
road to Monte Montoro, Maricao, *Wille 1060, 1062, 1065d* (NY), *1077e* (Type of *Endospora*
bicoccus Gardn., NY [Fig. 34]), *1073* (FC, UC), *1087* (NY), *1087a* (Type of *Anacystis*
amplivesiculata Gardn., NY [Fig. 20]), *1904* (Type of *A. pulchra* Gardn., NY [Fig. 63]), *1094a*
(NY), *1109* (FC, UC); by "Campo", Maricao, *N. L. Britton 1290c,* Feb. 1915 (Type of
Placoma Willei Gardn., NY [Fig. 66]); on a concrete basin near the experiment station, Rio Piedras,
Wille 177, Dec. 1914 (NY); on logs north of Sabana Grande, *Wille 925,* Feb. 1915 (Type of
Endospora nigra Gardn., NY [Fig. 24]); on a church wall, Sabana Grande, *Wille 953b, 956, 957,*
961b,d, 962b,c, Feb. 1915 (NY); on wall, Governor's Palace, San Juan, *Wille 58b,* Dec. 1914 (NY);
on walls near San Juan, *Wille 66a, 92b,* Dec. 1914 (NY); on a wall at Caleta de San Juan, *Wille*
66g, Dec. 1914 (NY); on walls, Fort San Cristobal, San Juan, *Wille 1989c, 1991c, 1993, 1993a*
(NY), *1995* (FC, UC), *1998c, 1999, 2002,* (NY), *2004* (FC, UC), *2004a,b,c, 2007, 2008, 2011,*
2014c (NY), *2016a* (Type of *Anacystis magnifica* Gardn., NY [Fig. 58]), *2019* (FC, UC), *2021,*
2023a (NY), *2025* (FC, NY); Santurce, *Wille 2* (NY), *2c* (Type of *A. consociata* Gardn., NY), *3*
(Type of *Chroococcus cubicus* Gardn., NY [Fig. 31]), *4a, 5* (NY), *38* (FC, UC), *56, 56j* (NY),
141 (FC, UC), *142, 143a* (NY), *149* (FC, UC), Dec. 1914; on bark at Utuado, *Wille 1506a,* Mar.

1915 (NY); on rocks and soil about 10 km. north of Utuado, *Wille 1533a* (Type of *Endospora mellea* Gardn., NY [Fig. 29]), *1553* (Type of *Anacystis nigroviolacea* Gardn., NY).

GUADELOUPE: Moule, grottes des falaises près le morne Aliard, *Mazé & Schramm* (Type of *Gloeocapsa tropica* Crouan, PC [Fig. 36]; isotype, K). ST. VINCENT: on damp wall of dam, Sharp's River, *W. R. Elliott 477* (Type of *G. gigas* W. & G. S. West, with *Cosmarium pseudopyramidatum*, in slide collection, BM). DOMINICA: on lime trees, Shanford Estate, *Elliott 901*, Nov.—Dec. 1892 (Type of *Chroococcus minor* f. *minimus* W. & G. S. West, with *Hassallia byssoidea* (Berk.) Hass., FC; isotype, UC); on trees, summit of Trois Poitons, *Elliott 903*, 15 Nov. 1892 (Type of *Aphanocapsa elachista* W. & G. S. West, BIRM).

JAMAICA: rocks, Cascade river east of Cinchona, *D. S. Johnson*, Jul. 1919 (NY); Port Antonio, *A. E. Wight* (PC); on rocks, Arntully, *C. R. Orcutt*, 7 Nov. 1928 (FC, US); "decayed Macromitrium", *Orcutt 5097* (BM, D); on wet rocks, *I. F. Lewis*, July 1906 (as *Gloeocapsa versicolor* in Coll., Hold., & Setch., Phyc. Bor.-Amer. no. 2203, FC, D; as *Microcystis marginata* in Coll., Hold., & Setch., Phyc. Bor.-Amer. no. 2204, FC). MEXICO: gorge below dam east of San Miguel de Allende, Guanajuato, *M. C. Carlson 1344b*, 17 Jan. 1949 (FC); caverns of Cacahuamilpa, Guerrero, *G. O. Lee*, Aug. 1938 (TA); rocks, Barranca de Oblatos, Guadalajara, *W. Kiener 18199*, 21 Dec. 1944 (FC, KI); house wall, Rio Primavera, Guadalajara, *Kiener 18293, 18297, 18300*, 27 Dec. 1944 (FC, KI); on a gravel bank 7 miles north of the village, Bahia Kino, Sonora, *F. Drouet & D. Richards 2892*, 9 Nov. 1939 (FC); crevices of cliffs in mountains on east side of Guaymas, Sonora, *Drouet & Richards 3087a*, 3 Dec. 1939 (FC).

GUATEMALA: Dept. Alta Verapaz: caves southwest of Lanquín, *J. A. Steyermark 44084, 44085*, 21 Feb. 1942 (FC); near Cobán, *P. C. Standley 71763*, Mar.—Apr. 1939 (FC); along Rio Carchá between Cobán and San Pedro Carchá, *Standley 90008*, Mar. 1941 (FC); wet cliff near Chirriacté on the Petén highway, *Standley 91880*, 9 Apr. 1941 (FC); on trunk of royal palm near Dolores, 1—2 miles northeast of Cubilguitz, *Steyermark 44861*, 8 Mar. 1942 (FC). Dept. Chiquimula: on cañon slopes on Socorro mountain southeast of Concepción de las Minas, *Steyermark 31119a*, 4 Nov. 1939 (FC). Dept. Huehuetenango: Rio Pucal about 14 km. south of Huehuetenango, *Standley 82335*, 4 Jan. 1941 (FC); on bark near crossing of Rio San Juan Ixtán east of San Rafael Pétzal, *Standley 82952*, 9 Jan. 1941 (FC); entrance to cave, summit of Sierra de los Cuchumatanes, Chémal, *Steyermark 50256*, 8 Aug. 1942 (FC); damp shaded bank north of Chiantla, *Standley 82585*, 6 Jan. 1941 (FC); Cerro Chémal, *Steyermark 50304D*, 8 Aug. 1942 (FC). Dept. Jalapa: rocks about 10 miles south of Jalapa, *Steyermark 32201*, 29 Nov. 1939 (FC). Dept. Sacatepéquez: on old bricks and on bark of Bursera near Antigua, *Standley 58523a, 59832*, Nov. 1938—Feb. 1939 (FC). Dept. San Marcos: on rock around thermal spring, Tajumulco, *Steyermark 36873*, 28 Feb. 1940 (FC); dry boulder just below summit of Volcán Tacaná, *Steyermark 36117*, 19 Feb. 1940 (FC). Dept. Santa Rosa: region of La Sepultura west of Chiquimulilla, *Standley 79340*, 5 Dec. 1940 (FC); on tree truck, Guazacapán, *Standley 78641*, Nov.—Dec. 1940 (FC); on trees near Cuilapilla, *Standley 78111, 78114*, 23 Nov. 1940 (FC).

HONDURAS: Dept. Atlántida: on dead wood near Rio San Alejo south of San Alejo, *Standley 7989*, Apr. 1947 (FC). Dept. Comayagua: on plaster of urinal, Hotel Jerusalém, Comayagua, *Standley 5780*, Mar. 1947 (FC); on cliffs, Comayagua, *Standley 5689, 5726*, Mar. 1947 (FC); soil, sandstone, and fence rails, vicinity of Siguatepeque, *Standley 5518, 6576, 6592, 6671*, Mar.—Apr. 1947 (FC). Dept. Cortés: on *Citrus aurantiifolia*, vicinity of La Lima, *Standley & J. Chacón P. 7112*, Apr. 1947 (FC). Dept. El Paraíso: open sandy bank, Las Casitas, *Standley, L. Williams, & P. Allen 581*, 4 Dec. 1946 (FC). Dept. Morazán: on rocks and trees, El Zamorano, *Standley 189, 215, 2171, 2365, 2396*, Nov. 1946—Jan. 1947 (FC), *11708*, Aug. 1947 (FC); on rocks, Quebrada de Santa Clara near Rio Yeguare, *Standley & L. O. Williams 1614*, 17 Dec. 1946 (FC); on log, Agua Amarilla above El Zamorano, *Standley & Willams 21*, Nov. 1946 (FC); on old wood along Rio Yeguare east of El Zamorano, *Standley 1073*, Dec. 1946 (FC). EL SALVADOR: on bark, San Vicente, *Standley & E. Padilla V. 3519*, Feb. 1947 (FC). NICARAGUA: on masonry and bark, Chichigalpa, dept. Chinangeda, *Standley 11461, 11536*, July 1947 (FC); on tree trunks, Jinotepe, dept. Carazo, *Standley 8453, 8494*, 20 May 1947 (FC); on tree trunk, La Libertad, dept. Chontales, *Standley 8854*, May—Jun. 1947 (FC). COSTA RICA: Alrededores de San Ramón, *A. M. Brenes 14937*, 4 Jan. 1932 (FC); Cerro de la Muerte, 5 km. above Millsville, Cordillera de Talamanca, prov. Cartago, *R. W. Holm & H. H. Iltis 1005a*, July 1949 (FC); U. S. Department of Agriculture Rubber Experiment Station, Los Diamantes on Rio Santa Clara, 1.6 km. east of Guapiles, prov. Limon, *Holm & Iltis 1021*, 11 July 1949 (FC). PANAMA: in ponte ligneo oppidi Colon, *G. Lagerheim*, Dec. 1892 (as *Gloeocapsa magma* in Wittr., Nordst., & Lagerh., Alg. Exs. no. 1541a, L); in ancient grinding stone, Panama City, *G. W. Prescott CZ19*, Aug. 1939 (FC); Juan Mina, Chagres river (Canal Zone), *H. H. Bartlett CZ198*, Sept. 1944 (FC); moist lime bluff near Salamanca Hydrographic Station in the gorge of Rio Pequení, *C. W. Dodge, Steyermark, & P. H. Allen 16972, 16975*, 14 Dec. 1934 (D, MO).

VENEZUELA: Amazonas: on igneous rock, Sanariapo, *Steyermark 58509*, Sept. 1944 (FC). Anzoátegui: limey bluff along Rio Zumbador northeast of Bergantín, *Steyermark 61282*, Mar. 1945 (FC); on shaded bluff, Fila Grande between Bergantín and Cerro Peonia, *Steyermark 61550*, 19 Mar. 1945 (FC); bluff between Cerro San José and Cerro Peonia, *Steyermark 61567*, 20 Mar.

64

1945 (FC). Bolívar: on face of bluff, Ptari-tepuí, *Steyermark 59881*, 6 Nov. 1944 (FC); on hematite outcrops, Sororopán-tepuí, *Steyermark 60130*, 15 Nov. 1944 (FC); on rocks, summit of Mt. Roraima, *Steyermark 58820*, 27 Sept. 1944 (FC). Dist. Federal: on concrete, house at Los Venados, *Steyermark 55120*, 2 Jan. 1944 (FC). Sucre: hot springs near La Toma southeast of Cumaná, *Steyermark 62863*, 21 May 1945 (FC). BRAZIL: Ceará: in seepage water at base of dunes, Urubú, Fortaleza, *Drouet 1381*, 24 Aug. 1935 (D); pool below dam, Açude Bôa Aqua, Quixadá, *Drouet 1404*, 2 Sept. 1935 (D); pools in Rio Pacoty 6 km. south of Aquiraz, *Drouet 1453*, 15 Oct. 1935 (D); Rio Pacoty 20 km. southeast of Mecejana, Pacatuba, *Drouet 1456*, 18 Oct. 1935 (D); roadside pool near Primavera, Soure, *S. Wright*, 28 Oct. 1937 (D). Parahyba: Açude Bodocongó near Campina Grande, *Wright 1583*, 6 Aug. 1934 (D). Pernambuco: wall along street below monastery, Olinda, *Drouet 1267*, 17 Jun. 1935 (D). Rio Grande do Sul: *H. Kleerekoper 28*, 1941 (FC). URUGUAY: in saxis, Punta Carretas, dept. Montevideo, *G. Herter 840*, Sept. 1947 (MO, UC). ECUADOR: Quito, *G. Lagerheim*, May 1892 (Isotypes of *Pleurocapsa muralis* Lagerh. in Wittr. & Nordst., Alg. Exs. no. 1097, L, MIN, NY); Puente de Chimbo, *Lagerheim*, Aug. 1891 (S; as *Gloeocapsa magma* in Wittr., Nordst., & Lagerh., Alg. Exs. no. 1541b, L); in rupibus madidis ad San Nicolas prov. Pichincha, *Lagerheim*, Oct. 1891 (with *Scytonema mirabile* in Wittr., Nordst., & Lagerh., Alg. Exs. no. 1320, FC). PERU: on rocks and "sobre tierra", *A. Maldonado*, Jul. 1950, 1952 (FC); agua mineral, Arequipa, *Maldonado 6, 10, 12*, Jul. 1950 (FC). CHILE: fuktige bergvaggar vid Eberhard, Magellanes, *O. Borge*, Mar. 1899 (S).

HAWAIIAN ISLANDS: damp cliff at west end of beach, Kalalau valley, Kauai, *W. J. Newhouse 1953-5*, 2 Jan. 1953 (FC); mud, Kaali spring, Niihau, *H. St. John 22837*, 16 Aug. 1947 (FC); rocks, Rainbow falls near Hilo, Hawaii, *W. A. Setchell 3210*, 14 July 1900 (FC, UC); culture, University of Hawaii, Honolulu, *M. Lohman 8134*, Oct. 1951 (FC). MARSHALL ISLANDS: "mesa" island on ocean side of Jam-Wa, Arno atoll, *L. Horwitz 9617*, 19 Aug. 1951 (FC); palm on Jiland island, Arno atoll, *Horwitz 9104*, 17 Jul. 1951 (FC); Airukiraru island, Bikini atoll *W. R. Taylor 46-244a*, Apr. 1946 (FC, MICH); water hole behind the site of the village, Bikini island, Bikini atoll, *Taylor 46-567*, 3 Jul. 1946 (FC, MICH); crust on coral sand, Lae islet, Lae atoll, *F. R. Fosberg 34007*, 6 Jan. 1952 (FC, US); on Pandanus, Majuro island, Majuro atoll, *A. Spoehr*, May 1947 (FC); on coral rocks in woods, Burok island, Rongelap atoll, *Taylor 46-599*, 18 Jul. 1946 (FC, MICH); on coral boulders in scrub, Ujae islet, Ujae atoll, *Fosberg 34420*, 13 Mar. 1952 (FC, US); open sand in scrub, Kabben islet, Wotho atoll, *Fosberg 34432*, 19 Mar. 1952 (FC, US). TUAMOTU ISLANDS: Ngarumova, Raroia atoll, *M. S. Doty & J. Newhouse 11257*, 2 Jul. 1952 (FC). SOCIETY ISLANDS: with *Coccochloris stagnina* Spreng., Tahiti, *J. E. Tilden*, Oct. 1909 (as *Aphanothece microspora* in So. Pacific Alg. no. 3, MIN). SAMOA: on dead Tridacne shells, Rose atoll, *A. G. Mayor 1055*, 6 June 1920 (Type of *Pleurocapsa Mayorii* Setch., UC; isotype, FC [Fig. 40]); an Felsen, Savaii, *K. H. Rechinger 2834* (Type of *Gloeothece samoensis* Wille in the slide collection, O); auf Baumrinde, Upolu, *Rechinger 3621* (Type of *Gloeocapsa aeruginosa* f. *lignicola* Wille in the slide collection, O). GILBERT ISLANDS: fish-ponds on Nikunau, *Mrs. R. Catala 5, 10*, July. 1951 (FC).

NEW ZEALAND: terrestrial, Rangitoto island, Waitemata, *V. J. Chapman*, Sept. 1947 (FC); Rotorua, *L. A. Doore & A. Nash*, Feb. 1935 (as *Aphanocapsa thermalis* in Tild., So. Pacific Pl., 2nd Ser., no. 545, FC); hotsprings, Tokano, *S. Berggren 195, 199*, 1878 (S); Rotorua, Ohinemuru, *Capra A, B*, Mar. 1909 (DT); cooking pool, native block, Ohinemutu, Rotorua, *L. M. & M. H. Jones*, 10 Feb. 1935 (Tild., So. Pacific Pl., 2nd Ser., no. 574, FC); in cold water, Great Geyser, and in steam, edge of Crystal Pool, Wairakei geyser valley, Taupo county, *W. A. Setchell 5962 (53), 5965 (56)*, May 1904 (FC, UC); on rocks in hot stream, Whakarewarewa, Rotorua, *Setchell*, 1904 (FC, UC). JAPAN: *Y. Tanaka* in herb. A. Grunow (W); brook near Rogashima, Miyanoshita, Hakone, *H. M. Richards*, May 1900 (FH). PALAU ISLANDS: on coralline limestone on the Peleliu airfield, *A. C. Mason*, 10 Feb. 1949 (FC). PHILIPPINES: Manila and Rizal province: *J. D. Soriano 783, 831, 874*, Jan. & Apr. 1951 (FC, PUH); on walls of Malate church, *Soriano 57-2*, 17 Jun. 1948 (FC, PUH); on decaying organic matter kept moist by rain, third floor of Rizal building, University of the Philippines, *Soriano 76*, 29 June 1948 (FC, PUH); aquaria, University of the Philippines, Manila, *G. T. Velasquez 444, 591*, 21 Feb. 1940, 11 Jan. 1941 (FC); on cement walls, botany garden, University of the Philippines, Manila, *Velasquez 445*, 21 Feb. 1940 (FC); on walls, Reina Regente street, Manila (Binondo), *Soriano 309*, 17 Jul. 1949 (FC, PUH); on wall, Dapitan and Trabajo streets, Sampaloc, *Soriano 405*, 9 Oct. 1949 (FC, PUH); on adobe wall, Ermita, *Soriano 200-3*, 9 Nov. 1948 (FC, PUH); walls of ice plant, Marikina, *Soriano 898*, 27 Apr. 1952 (FC, PUH); cultivated at Fisheries experiment station, Dagat-Dagatan, Navotas, *Velasquez 628*, 19 Jan. 1941 (FC). Mindoro: bark of coconut, Tabinay Maliit, Puerto Galera, *Velasquez 737*, 7 Apr. 1941 (FC); ʰhot spring at Bisayaan, Puerto Galera, *Velasquez 1039*, 26 Apr. 1941 (FC); Big Balatero, Puerto Galera, *Velasquez 9*, 15 Apr. 1953 (FC, PUH). Palawan: Araceli, *Velasquez 2922, 2932*, 15 Jun. 1951 (FC, PUH); Coron, *Velasquez 2793*, 3 June 1951 (FC, PUH); Cuyo, *Velasquez 2801*, 10 Jun. 1951 (FC, PUH). Panay island: on walls and fences, Iloilo City, *Soriano 1009, 1020, 1056*, Jun. Jul. 1952 (FC, PUH); dikes in Griños fishponds, Loganes, Iloilo prov., *Soriano 1590*, 11 Jan. 1953

(FC, PUH); walls of cemetery, Pototan, *Soriano 1836*, 23 Feb. 1953 (FC, PUH); on adobe walls, Passi, Iloilo province, *Soriano 1066, 1073*, 20 Jul. 1952(FC, PUH); wall along road to the beach, San Jose, Antique province, *Soriano 1525*, 1 Dec. 1952 (FC, PUH). Sulu archipelago: Busbos, Jolo, *Velasquez 3246, 3247*, 22 May 1952 (FC, PUH). NEW CALEDONIA: lily pool, Ile Ouin, *J. T. Buchholz 1667B*, 8-9 Feb. 1948 (FC, UC).

CHINA: Fukien: on the wall of a hot spring, Huangchun, Foochow, *H. H. Chung A433*, Sept. 1926 (Type of *Gloeocapsa minutula* Gardn., FH; isotype, UC), *A436*, Sept. 1926 (D, FH); on wet rock, Kushan near Foochow, *Chung A363c*, Aug. 1926 (Type of *G. multispherica* Gardn., FH; isotype, D [Fig. 21]). Hopeh: on stone by stream, Lying Buddha Temple, Peiping, *Y. C. H. Wang 195*, 1930—31 (FC, UC); on rock, Jade Spring Hill, Peiping, *Wang 287a*, 1930—31 (FC, UC). Kiangsu: Suchau, *N. G. Gu* (NY); on rock at a spring pool, Tshishiasan, *C. C. Wang 335*, 20 Apr. 1930 (FC, UC); on rocks and walls, Nanking, *Y. C. H. Wang 53, 60, 75, 95*, 1930 (FC, UC); on rocks and walls, Nanking, *C. C. Wang 302, 303, 332*, Apr. 1930 (FC, UC); on tree bark, National Central University, Nanking, *C. C. Wang 353, 357, 409*, Apr. 1930, May 1931 (FC, UC). Szechwan: Chin-Chen-Shan, *H. Chu 1343*, 7 Aug. 1941 (Types of *G. incrustata* H. Chu, *Entophysalis sinensis* H. Chu, and *Anacystis? hyalina* H. Chu, FC); on wet rocks, Omei, *Chu 1015, 1353, 1363, 1389*, 23—25 Aug. 1942 (FC); wet rock, Ko-Fu-Tung, Mt. King-Fu, Nan-chwan, *Chu 849*, 15 Apr. 1945 (FC); wet rock, Lung-chi bridge, Nanzechwan, *Chu 716*, 7 Aug. 1945 (FC). Yünnan: an den Sinterrassen der Quelle von Bödö(Peti)se von Dschungdien, *H. Handel-Mazzetti 4478*, 4 Aug. 1914 (Type of *Gloeothece sophacea* Skuja, W; isotype, D); an Kalkfelsen bei der heissen Schwefelquelle unter Baoschi bei Dschungdien, *Handel-Mazzetti 7729*, 17 Aug. 1915 (Type of *Synechococcus ambiguus* Skuja, W [Fig. 67]; isotype, D).

NEW GUINEA: on rocks along a stream, Finschhafen, Papua, *L. B. Martin*, Jun. 1945 (FC); Cycloopgebergte, *Lorenz*, 11—15 Apr. (L). INDONESIA: groten rivier bij Pankadjene, *A. Weber-van Bosse 938, 939*, 1899—1900 (L). Bira: grot, *Weber-van Bosse 879*, 1899—1900 (FC, L). West Borneo: warme bron bij Moeka-Moeka, *Weber-van Bosse 674*, 1888 (FC, L). Celebes: grot bij Maros, *Weber-van Bosse 789*, Sept. 1888 (FC, L). Java: bij Kampong Nglijep, zuidkust, *P. Groenhart d*, Oct. 1936 (L). Kangean: grot, *Weber-van Bosse*, 1899—1900 (L). Sumatra: Ngalan bij Pajakombah, *Weber-van Bosse 704, 706*, 1888 (FC, L); an Steinen, Frauenbadeteich, Kadjaj, *A. Thienemann*, 7 Mar. 1929 (Type of *Chroococcidiopsis thermalis* Geitl., slide no. SKW2∝ in collection of L. Geitler). Talaud (Talauer) island: à l'embouchure de la rivière, Liroeng, Salebaboe island, *Weber-van Bosse*, 1899—1900 (L). AUSTRALIA: Queensland: on tree trunk, Myora, *A. B. Cribb G*, 26 Feb. 1946 (FC); bottom of Day's lagoon, Moreton island, *Cribb 143-1*, 26 May 1951 (FC). BURMA: Cantonment Gardens, Rangoon, *L. P. Khanna 335*, 9 May 1935 (FC); Kyauktan, Rangoon, *Khanna 451*, 22 Aug. 1935 (Type of *Gloeothece rhodochlamys* Skuja, FC [Fig. 72]); Maymyo, Northern Shan States, *Khanna 639*, 12 May 1936 (FC). INDIA: Calcutta, *H. B. C.* (L); auf Steinen im Victoria-Garden in Bombay, *A. Hansgirg*, 20 Nov. 1895 (FC, W); Felsen vor Poona, *Hansgirg*, 20 Oct. 1895 (FC, W). MAURITIUS: sur les arbres, *J. McGregor*, Apr. 1819 (BM, D).

4b. ANACYSTIS MONTANA f. MINOR Drouet & Daily, Butler Univ. Bot. Stud. 10: 221. 1952. *Aphanothece saxicola β aquatica* f. *minor* Wille, Öfvers. K. Sv. Vet.-Akad. Förh. 36(5): 22. 1879. *A. saxicola* f. *minor* Wille ex Forti, Syll. Myxophyc., p. 81. 1907. —Type from Matotchkin, Nova Zembla (S).

Microhaloa firma Kützing, Tab. Phyc. 1: 6. 1846. *Palmella firma* Brébisson & Lenormand *pro synon.* in Kützing, loc. cit. 1846. *Micraloa firma* Kützing ex Trevisan, Sagg. Monogr. Alg. Coccot., p. 38. 1848. *Polycystis firma* Rabenhorst, Fl. Eur. Algar. 2: 53. 1865. *Coccochloris firma* Richter in Hauck & Richter, Phyk. Univ. 13: 649A. 1894. *Microcystis firma* Schmidle, Engler Bot. Jahrb. 23: 57. 1902. *Anacystis firma* Drouet & Daily, Lloydia 11: 77. 1948. —Type from Falaise, France (L). Kützing's notation of the original specimen in Tab. Phyc. 1:6 (1846) as "Heisse Bäder der Euganeen: Lenormand!" is obviously an editorial error. We can surmise that he intended to correct this mistake in Sp. Algar., p. 207 (1849), where he says simply, "Prope Falaise legit amic. Brebisson (Lenormand)".

Palmella hyalina γ seriata Kützing, Sp. Algar., p. 215. 1849. —Type from Falaise, France (L). FIG. 95.

Palmella sudetica Rabenhorst, Alg. Sachs. 11 & 12: 105. 1851. —Type from the Riesengebirge, Silesia (L). FIG. 94.

Aphanothece Trentepohlii Grunow in Rabenhorst, Fl. Eur. Alg. 2: 65. 1865. *Nostoc Trentepohlii* Mohr *pro synon.* ex Grunow in Rabenhorst, loc. cit. 1865. *Coccochloris Trentepohlii* Richter in Hauck & Richter, Phyk. Univ. 4: 194. 1888. —Type from Kiel, Germany (W).

Anacystis glauca Wolle, Bull. Torr. Bot. Club 6: 182. 1877. *Microcystis glauca* Drouet & Daily, Field Mus. Bot. Ser. 20: 72. 1939. *Polycystis glauca* Drouet & Daily, Field Mus. Bot. Ser. 20: 125. 1942. —Type from Bethlehem, Pennsylvania (FC). FIG. 93.

Nostoc microscopicum var. *linguiforme* Hansgirg, Prodr. Algenfl. Böhmen 2: 65. 1892. —Type from Krummau, Bohemia (W). FIG. 96.

66

Chroococcus parallelepipedon Schmidle, Engler Bot. Jahrb. 30: 242. 1901. —Type from Lake Nyasa, Tanganyika (ZT). FIG. 91.

Aphanocapsa elachista var. *irregularis* Boye Petersen, Bot. Icel. 2(2, 7): 265. 1923. *Microcystis pulverea* f. *irregularis* Elenkin, Monogr. Algar. Cyanophyc., Pars Spec. 1: 124. 1938. —Type from Grimsa, Iceland (in the collection of J. Boye Petersen). FIG. 97.

Aphanocapsa Farlowiana Drouet & Daily, Field Mus. Bot. Ser. 20: 125. 1942. —Type from Falmouth, Massachusetts (D). FIG. 92.

Radiocystis geminata Skuja, Symbolae Bot. Upsal. 9(3): 43. 1948. —Type from Uppland, Sweden (in the collection of H. Skuja).

Original specimens have not been available to us for the following names; their original descriptions are here designated as the Types until the specimens can be found:

Pleurococcus pulvereus Wood, Smiths. Contrib. Knowl. 241: 79. 1872. *Anacystis pulverea* Wolle, Fresh-w. Alg. U. S., p. 329. 1887. *Polycystis pulverea* Wolle ex Hansgirg, Prodr. Algenfl. Böhmen 2: 145. 1892. *Microcystis pulverea* Forti, Syll. Myxophyc., p. 92. 1907; Migula ex Lemmermann, Krypt. Fl. Mark Brandenb. 2: 77. 1907.

Aphanocapsa Grevillei var. *microgranula* West, Journ. Linn. Soc. Bot. 29: 199. 1892.

Aplococcus natans Roze, Journ. de Bot. 10: 321. 1896.

Polycystis pallida Lemmermann, Bot. Centralbl. 76: 154. 1898. *P. stagnalis* Lemmermann, Ber. Deutsch. Bot. Ges. 18: 24. 1900. *Microcystis stagnalis* Lemmermann, Forschungsber. Biol. Sta. Plön 10: 150. 1903. *M. pulverea* f. *stagnalis* Elenkin, Monogr. Algar. Cyanophyc., Pars Spec. 1: 124. 1938.

Microcystis stagnalis var. *pulchra* Lemmermann, Arch. f. Hydrobiol. & Planktonk. 5: 303. 1910. *M. pulverea* f. *pulchra* Elenkin, Monogr. Algar. Cyanophyc., Pars Spec. 1: 124: 1938.

Microcystis minima Bernard, Protococc. & Desmid. d'Eau Douce, p. 49. 1908.

Chroococcus dispersus var. *minor* G. M. Smith, Bull. Wisconsin Geol. & Nat. Hist. Surv. 57: 28. 1920. *Gloeocapsa minima* f. *Smithii* Hollerbach in Elenkin, Monogr. Algar. Cyanophyc., Pars Spec. 1: 242. 1938.

Aphanocapsa elachista var. *planctonica* G. M. Smith, Bull. Wisconsin Geol. & Nat. Hist. Surv. 57: 42. 1920. *Microcystis pulverea* f. *planctonica* Elenkin, Monogr. Algar. Cyanophyc., Pars Spec. 1: 124. 1938.

Alphanocapsa endophytica G. M. Smith, Bull. Wisconsin Geol. & Nat. Hist. Surv. 57: 42. 1920. *Microcystis endophytica* Elenkin, Monogr. Algar. Cyanophyc., Pars Spec. 1: 139. 1938.

Aphanocapsa siderosphaera Naumann, K. Sv. Vet.-Akad. Handl. 62(4): 18. 1921.

Aphanocapsa sideroderma Naumann, K. Sv. Vet.-Akad. Handl. 62(4): 18. 1921.

Aphanocapsa Koordersii Strøm, Nyt Mag. Naturvid. 61: 128. 1923. *Microcystis Koordersii* Elenkin, Monogr. Algar. Cyanophyc., Pars Spec. 1: 140. 1938.

Microcystis exigua Zalessky, Rév. Gén. de Bot. 38: 34. 1926.

Microcystis pulverea var. *racemiformis* Nygaard, K. Danske Vidensk. Selsk. Biol. Skr. 7(1): 183. 1949.

Plantae aerugineae, aquaticae, cellulis in divisione binis depresso—sub-sphaericis, aetate provecta globosis diametro 2—3µ crassis, sine ordine (saepe conferte) per gelatinum vaginale distributis; gelatino vaginale homogeneo, hyalino, firmo vel (saepe omnino) diffluente. FIGS. 91—97.

In shallow water of freshwater ponds, lakes, and streams, often in the plankton, seldom in brackish water.

Specimens examined:

NOVA ZEMBLA: Matotchkin, *F. Kjellman 24b*, 1875 (Type of *Aphanothece saxicola* β *aquatica* f. *minor* Wille, S). NORWAY: Valders, in rupibus ad Skogstad, *N. Wille*, 31 Jul. 1879 (D, S); Aas pr. Kristiania, *Wille*, Aug. 1890 (S). SWEDEN: plankton, Uppland, *H. Skuja* (Type of *Radiocystis geminata* Skuja in plankton collection of H. Skuja); Grimstorp, Vestergötland, *O. Nordstedt*, Sept. 1867 (S); Vassitjokko, Torne Lappmark, *O. Borge 3*, Jul. 1909 (S); Möja, *Borge*, Jun. 1931 (S); Lassby backar, Uppsala, *A. Grevillius*, May 1883 (S); in turfosis ad Stromsberg prope Jonkoping, *Nordstedt*, Aug. 1879 (as *Cylindrospermum macrospermum* in Wittr. & Nordst., Alg. Exs. no. 396a, FC); Dalsland, Hallan, *V. Wittrock*, 10 Jun. 1865 (D, S), 12 Jul. 1866 (D, S; as *Aphanocapsa Castagnei* in Wittr. & Nordst., Alg. Exs. no. 299, FC, L, NY, S); in stagno ad Kvikkjokk Lapponiae Lulensis, *G. Lagerheim*, Jul. 1883 (D, S). DENMARK: Markerup Moor, *L. Hansen* (MO, NY), *W. F. R. Suringar 127*, 3 Sept. 1849 (L). POLAND: Silesia: in stagno ad Falkenberg, *B. Schröder*, Jul. 1893 (D, S), *Schmula*, July 1893 (S); im Aquarium des pflanzenphysiol.

Instituts der Universität, Breslau, *Schröder,* Sept. 1892 (as *Coccochloris firma* in Hauck & Richt., Phyk. Univ. no. 649B, L, MIN); in verlassenen Steinbrüchen am Galgenberge bei Strehlen, *Hilse,* Aug. 1860 (as *Palmella botryoides* in Rabenh. Alg. no. 1037, L); Felswände am Zackenfall im Riesengebirge, *Peck* (Type of *Palmella sudetica* Rabenh., in Rabenh. Alg. no. 105, L; isotypes, D, FC [Fig. 94]). GERMANY: Kiliae, *Mohr* (Type of *Aphanothece Trentepohlii* Grun., W; isotypes, BM, L, MO); bei Angeln, Schleswig-Holstein (L); in Torfgruben bei Flensburg, *R. Häcker* (as *Coccochloris stagnina* in Breutel, Alg. Exs. no. 415, L, UC); im Rehwinkel (Brandenburg), *Itzigsohn & Rothe,* Jul. 1852 (as *Microhaloa firma* in Rabenh. Alg. no. 203, FC, L, NY); Gr. Plöner See, *W. Rohde,* 15 Sept. 1950 (FC); im Teiche der Parthenquelle bei Schönbach um Colditz (Sachsen), *P. Richter,* Jul. 1892 (as *Coccochloris firma* in Hauck & Richt., Phyk. Univ. no. 649A, L, MIN, NY); Parkteiche in Erdmannsdorf bei Schmiedeberg (Brandenburg), *G. Hieronymus 710,* July 1887 (MICH), Aug. 1887 (as *C. Trentepohlii* in Hauck & Richt., Phyk. Univ. no. 194, L); Münster (Westfalen), *Sendtner,* Apr. 1850 (B); schwimmend in Titisee (Baden), *A. Braun 127,* May 1848 (L); im Moor bei Freiburg-im-Baden, *Braun 128,* Aug. 1848 (L).

CZECHOSLOVAKIA: Moravia: rybník, *F. Novácek,* 16 Jul. 1934 (D, FC). Bohemia: Dessendorf, *A. Hansgirg,* Jul. 1885 (FC, W); am Urkalk bei Krummau, *Hansgirg,* Aug. 1884 (Type of *Nostoc microscopicum* var. *linguiforme* Hansg., W; isotype, FC [Fig. 96]); Plas prope Pilsen, *Hansgirg,* Aug. 1887 (FC, W); Pocátky, *Hansgirg,* Aug. 1888 (FC, W); Riesengrund—Petzer im Riesengebirge, *Hansgirg,* Jul. 1887 (FC, W). SWITZERLAND: in einem Sumpfe des Torfmoores zwischen Einsiedeln und dem Etzel, Schwyz, *C. O. Harz* (as *Polycystis firma* in Wartm. & Wint., Schweiz. Krypt. no. 845, MIN). SCOTLAND: in a hill loch near Faskally House, Pitlochry, Perthshire, *A. J. Brook,* Autumn 1953 (FC). NETHERLANDS: in vijvers bij Beek, Noord Brabant, *A. Weber-van Bosse 146,* Apr. 1886 (L); plassen van Brenkelveen en Tienhoven, *W. Vervoort,* May 1942 (L); Nieuwkoopsche plassen, *S. J. van Ooststroom 7208,* May 1937 (L). FRANCE: eaux tranquilles (as *Palmella hyalina* in Desmaz., Pl. Crypt. Fr., ed. 2, no. 215, L). Calvados: Caen, *A de Brébisson,* ex herb. Lenormand (FC, FH, L); Caen etc., eaux douces (as *P. hyalina* in Chauvin, Alg. de Normandie no. 130, L, UC); marais de Plainville, *de Brébisson* (NY); Vire, herb. Lenormand no. 111 (L), *de Brébisson 812* (L); Falaise, ex herb. Chauvin no. 74 (UC); Falaise, ex herb. Lenormand (LD, UC); Falaise, *de Brébisson 182* (L), *439* (Type of *Palmella hyalina* γ *seriata* Kütz., L [Fig. 95]; isotype, UC), *533* (Type of *Microhaloa firma* Kütz., L; isotypes, B, D, FH, S, UC), *963* (FC). Manche: Mortain, *de Brébisson* (S), *782* (L). Seine: Lac du Bourget, *Joly,* 10 Mar. 1951 (FC, PC). Vosges: Bruyères, *J. B. Mougeot* (as *Palmella hyalina* in Roumeguère, Alg. de Fr. no. 111, L). TANGANYIKA: am Seeufer, Langenburg am Nyassasee, *W. Goetze 866,* Apr. 1899 (Type of *Chroococcus parallelepipedon* Schmidle, ZT [Fig. 91]). ICELAND: among Myriophyllum, stagnant water in a branch to Grimsá, Vallanes, *J. Boye Petersen,* 26 Jun. 1914 (Type of *Aphanocapsa elachista* var. *irregularis* Boye Petersen in collection of J. Boye Petersen; isotype, D [Fig. 97]). BERMUDA: pool near Harrington sound, *F. S. Collins 7118,* Apr. 1912 (NY).

MASSACHUSETTS: Bateman's pond, Concord, *R. H. Howe Jr.,* Dec. 1908 (FH); Horn pond, Woburn, *F. S. Collins 5376,* 9 Sept. 1905 (FH; as *Aphanothece saxicola* in Coll., Hold., & Setch., Phyc. Bor.-Amer. no. 1301, FC, D, L, TA); Hammond pond, Brookline, *B. M. Davis,* Oct. 1893 (MICH); Hammond pond, Newton, *W. G. Farlow,* Oct. 1890 (FC, TA, UC); in shallow water, Oyster pond, Falmouth, *R. N. Webster & Drouet 2156,* 16 July 1938 (Type of *Aphanocapsa Farlowiana* Dr. & Daily, D [Fig. 92]; isotypes, DA, FH, L, NY, S, UC), *J. Bader 119,* 30 July 1938 (D), *V. Trombetta & Drouet 2184,* 8 Sept. 1937 (D). RHODE ISLAND: Hope reservoir, Providence, *Appleton,* Dec. 1883 (FC, NY). QUEBEC: in a lake, Maniwaki, Hull county, *J. Brunel 390,* 7 Jul. 1933 (FC); Yamaska mountain, on cliffs, *Brunel 234,* 2 Jun. 1931 (FC); près du 17e portage de la Kogaluk, à environ 74 milles de la baie d'Hudson, boue d'un étang, *J. Rousseau 517,* 24 Jul. 1948 (FC); Lac Brennen, Rawdon, comté de Montcalm, *S. M. J. Eudes 292-28,* 29 Aug. 1941 (FC). NEW YORK: in Pond brook at Buffalo creek, Elma, Erie county, *J. Blum 324,* 23 Jul. 1949 (FC); in (middle) Lehn springs, Williamsville, Erie county, *Blum 173,* 5 Feb. 1949 (FC). PENNSYLVANIA: in fonte rupium calcareorum ad Bethlehem, *F. Wolle,* 1884 (Type of *Anacystis glauca* Wolle in Wittr. & Nordst., Alg. Exs. no. 796, FC; isotypes, FI, L [Fig. 93], NY, UC); lime stone springs, *Wolle* (L). VIRGINIA: culture of plankton from York river at Yorktown, *J. C. Strickland 1413,* 24 Aug. 1942 (FC, ST). NORTH CAROLINA: rocks below Dry falls, Highlands, Macon county, *H. C. Bold H215, H247,* 29 Jun. 1939 (FC); in Lake Anthony, Pineola, Forest county, *Bold H377,* 25 Jul. 1939 (FC); on wet rocks, Cullasaja gorge, Macon county, *Bold H157, H158, H161b,* 17 Jun. 1939 (FC). SOUTH CAROLINA: in Alligator lake near Myrtle Beach, Horry county, *P. J. Philson SC80, SC82,* 21 Sept. 1938 (FC); Eastitoe river, Pickens county, *H. C. Bold,* Jul. 1939 (FC). FLORIDA: Leesburg, *M. A. Brannon 333,* 31 May 1948 (FC, PC); Gainesville, *Brannon 70, 268, 326,* 1942—46 (FC, PC); sink[hole], Gainesville, *R. K. Strawn 308,* 10 May 1948 (FC, PC); Lake Mise, 16 miles northwest of Gainesville, *Brannon 615,* 15 May 1948 (FC, PC). OHIO: wet sandstone, Old Mans Cave state park, Hocking county, *W. A. Daily & F. K. Daily 519, 536, 554,* 27 Jul. 1940 (DA, FC); wet sandstone, Cantwell Cliff state park, Hocking county, *Daily 713,* 1 Sept. 1940 (DA, FC); Biddles pond, Athens county, *A. H. Blickle,* 6 May 1941 (DA, FC).

KENTUCKY: on rock in Mammoth Cave park, *B. B. McInteer 1074*, 11 Jun. 1940 (DA, FC). TENNESSEE: sluiceway just above last homestead from Elkmont to Blanket mountain, Sevier county, *H. Silva 194*, 28 Sept. 1941 (FC, TENN); in pond near Mason Hall, Obion county, *Silva 1248*, 8 Jul. 1949 (FC, TENN). INDIANA: outlet of Lost lake, Lake Maxinkuckee, Marshall county, *H. W. Clark & B. W. Evermann 265*, 27 Jul. 1906 (FC, US). WISCONSIN: Clear lake, Bradley, Lincoln county, *F. C. Seymour*, 17 Sept. 1951 (FC); Bolton lake, Vilas county, *G. W. Prescott 2W243*, 8 Jul. 1937 (FC). ILLINOIS: north shore of Lake Glendale, Robbsville recreational area in Shawnee national forest, Pope county, *H. K. Phinney 1067*, 1 Sept. 1944 (FC, PHI). MINNESOTA: in and near Itasca state park, *K. C. Fan 10267, F. Drouet 12199*, Jun., Jul. 1954 (D, MIN). ARKANSAS: from a large spring-slough, Imboden, Lawrence county, *B. C. Marshall P*, May 1931 (NY). NEBRASKA: culture from a roadside ditch 4 miles east of Maxwell, Lincoln county, *W. Kiener 16463c*, 1 Oct. 1944 (FC, KI). MONTANA: in Bitterroot slough below country club, Missoula, *F. H. Rose 4117*, 3 May 1941 (FC). COLORADO: pool, Peaceful valley slope, Boulder county, *Kiener 1268*, 5 Aug. 1934 (FC, KI). UTAH: Columbine falls, Mt. Timpanogos, Utah county, *A. O. Garrett 32*, 3 Aug. 1925 (FC, UC). ALASKA: Point Barrow, 1883 (FH); 124 and 229 miles south of Fairbanks on Richardson highway, *D. Hilliard 3, 8*, Sept. 1953 (FC). CALIFORNIA: in ditch by railroad, Emery, Alameda county, *N. L. Gardner 1100*, 20 Jun. 1903 (FC, UC); laboratory cultures, Berkeley, *Gardner 6973, 7147* (FC, UC); plankton of Lily lake near Fallen Leaf lake, Eldorado county, *W. A. Setchell 13*, 12 Aug. 1908 (FC, UC); Benton hot springs, Mono county, *V. Duran 7797*, 13 Apr. 1935 (FC, UC); culture from an irrigation ditch southwest of Folsom, Sacramento county, *Gardner 7394*, 30 June 1933 (FC, UC).

ECUADOR: in rupibus humidis, Baños, *G. Lagerheim*, Dec. 1891 (S). PERU: Lago de Pachacamac, *A. Maldonado 73a*, Jan. 1942 (FC). PHILIPPINES: aquarium, San Augustin College, Iloilo City, *J. D. Soriano 1006*, 20 Jun. 1952 (FC, PUH). CHINA: on a rope in a stream, Wuhsien (Soochow), Kiangsu, *C. C. Wang 367*, Apr. 1926 (FC, UC).

4c. ANACYSTIS MONTANA f. GELATINOSA Drouet & Daily, Butler Univ. Bot. Stud. 10: 221. 1952. *Coccochloris stagnina* f. *gelatinosa* Hennings, Phyk. March. 1: 43. 1893. *Aphanothece gelatinosa* Lemmermann, Krypt.-Fl. Mark Brandenb. 3: 69. 1907. *A. stagnina* f. *gelatinosa* Hennings ex Forti, Syll. Myxophyc., p. 77. 1907. —Type from Berlin, Germany (S). FIG. 98.

Aphanocapsa Roeseana de Bary in Rabenhorst, Alg. Eur. 215—217: 2156. 1870; Hedwigia 1870: 74. 1870. *Microcystis Roeseana* Elenkin, Monogr. Algar. Cyanophyc., Pars Spec. 1: 129. 1938. *Anacystis Roeseana* Drouet & Daily, Lloydia 11: 77. 1948. —Type from Reinhardsbrunn, Thuringia, Germany (L).

Synechocystis Willei Gardner, Mem. New York Bot. Gard. 7: 2. 1927. —Type from Mayaguez, Puerto Rico (NY). FIG. 99.

Original specimens have not been available to us for the following names; their original descriptions are here designated as the Types until the specimens can be found:

Aphanocapsa arctica Whelden in Polunin, Bot. Canadian Eastern Arctic 2: 21. 1947. *Synechocystis planctonica* Proschkina-Lavrenko, Akad. Nauk S. S. R. Bot. Inst. Otd. Sporovykh. Rost. Bot. Mater. 6: 69. 1950.

Plantae aerugineae, aquaticae, cellulis in divisione binis depresso—sub-sphaericis, aetate provecta globosis diametro 5—6μ crassis, sine ordine per gelatinum vaginale distributis; gelatino vaginale homogeneo, hyalino, saepe omnino diffluente. FIGS. 98, 99.

In shallow water of permanent freshwater ponds, lakes, etc., seldom appearing in brackish water. That the type specimens concerned here are large forms of *Anacystis montana* is unquestionable. Some other specimens cited (*Humm & Nielsen 1790, 1791, 1794*, for example) may in the future prove to be actively dividing growth-forms of *Coccochloris stagnina*, in which the cells do not become elongate before division begins again.

Specimens examined:

SWEDEN: Stockholm, Danviks krokar, *G. Lagerheim*, Jun. 1881 (D, S). GERMANY: in den Teichen bei Reinhardsbrunn, Thüringen, *A. de Bary* (Type of *Aphanocapsa Roeseana* de Bary in Rabenh. Alg. no. 2156, L; isotypes, D, FC, UC); Teich in Berliner botanischen Garten, *P. Hennings*,

Jul. 1891, Apr. 1892 (Type of *Coccochloris stagnina* f. *gelatinosa* Henn. in Henn., Phyk. March. no. 43, S; isotypes, FH, L, NY [Fig. 98]); Berlin, *Hennings*, Jan. 1892 (as *C. stagnina* f. *gelatinosa* in Hauck & Richt., Phyk. Univ. no. 747, L, MIN, NY, S); an einem Wassertroge in Königstein (Sachsen), *W. Krieger*, Sept. 1884 (TA). NORTH CAROLINA: Mullet pond, Shackleford Banks near Beaufort, *H. J. Humm & C. S. Nielsen 1790, 1791, 1794*, 23 Aug. 1949 (FC, T). TEN-NESSEE: seepage, Scottish Pike at L. & N. railroad overpass, Knoxville, *H. Silva 477*, 17 Jan. 1947 (FC, TENN). COLORADO: Howard lake, Boulder county, *R. W. Pennak*, 13 Jun. 1948 (FC). PUERTO RICO: Mayaguez, *N. Wille 1318*, Feb. 1915 (NY); 4 km. north of Mayaguez, *Wille 1329* (NY), *1329b* (Type of *Synechocystis Willei* Gardn., NY [Fig. 99]), Feb. 1915. TRINIDAD: Pitch lake, La Brea, *H. Field*, 25 Mar. 1942 (FC). MEXICO: plankton, Lago Patzcuaro, Michoacan, *B. F. Osorio Tafall*, 1941 (FC), *E. S. Deevey*, 10 Jul. 1941 (FC). GUATEMALA: in hot pool and in edge of lake, Lago Amatitlán, *P. C. Standley 89443, 89472*, 15 Mar. 1941 (FC). PHILIP-PINES: submerged on quiet sides of Okoy river, Palinpinon, Oriental Negros, *J. D. Soriano 1897*, 6 Apr. 1953 (FC, PUH).

5. ANACYSTIS DIMIDIATA Drouet & Daily, Butler Univ. Bot. Stud. 10: 221. 1952. *Trochiscia dimidiata* Kützing, Linnaea 8: 593. 1833. *Protococcus dimidiatus* Kützing, Phyc. Gener., p. 168. 1843. *Pleurococcus dimidiatus* Trevisan, Sagg. Monogr. Alg. Coccot., p. 33. 1848. *Chroococcus dimidiatus* Nägeli, Gatt. Einzell. Alg., p. 46. 1849. *C. turgidus* var. *dimidiatus* Brébisson in Rabenhorst, Alg. Eur. 201—204: 2033. 1867. *Gloeocapsa dimidiata* Drouet & Daily, Lloydia 11: 78. 1948. —Type from Nordhausen, Germany (L). FIG. 105.

Pleurococcus thermalis Meneghini, Consp. Algol. Eugan., p. 337. 1837. *Protococcus thermalis* Kützing, Tab. Phyc. 1: 5. 1846. *Chroococcus thermalis* Nägeli, Gatt. Einzell. Alg., p. 46. 1849. *C. turgidus* f. *thermalis* Rabenhorst, Fl. Eur. Algar. 2: 33. 1865. *Pleurococcus vulgaris* f. *thermalis* Brébisson in Mougeot & Roumeguère in Louis, La Departement des Vosges, 1887: Algues, p. 18 of reprint. 1887. *Chroococcus turgidus* var. *thermalis* Rabenhorst ex Hansgirg, Prodr. Algenfl. Böhmen 2: 161. 1892. *C. turgidus* γ *thermalis* Rabenhorst ex Wille, Nyt Mag. Naturvid. 62: 180. 1925. —Type from Abano, Italy (FI).

Protococcus turgidus Kützing, Tab. Phyc. 1: 5. 1846. *P. turgidus* ∝ *chalybeus* Kützing, loc. cit. 1846. *Pleurococcus turgidus* Trevisan, Sagg. Monogr. Alg. Coccot., p. 34. 1848. *Chroococcus turgidus* Nägeli, Gatt. Einzell. Alg., p. 46. 1849. *C. chalybeus* Rabenhorst, Alg. Eur. 115—116: 1144. 1861. *C. aeruginosus* Rabenhorst *pro synon.*, loc. cit. 1861. *Protococcus chalybeus* Kützing *pro synon.* ex Rabenhorst, Krypt.-Fl. Sachs. 1: 69. 1863. *Chroococcus turgidus* a. *chalybeus* Rabenh. ex Kirchner, Krypt.-Fl. Schlesien 2(1): 262. 1878. *Gloeocapsa turgida* Hollerbach in Elenkin, Monogr. Algar. Cyanophyc., Pars Spec. 1: 211. 1938. —Type from Freiburg im Breisgau, Germany (L). FIG. 100.

Protococcus turgidus β *fuscescens* Kützing, Tab. Phyc. 1: 5. 1846. *Chroococcus fuscescens* Richter in Bot. Ergebn. Drygalski's Grönlandexped., A. Kryptog. 1: 3. 1897. *C. turgidus* var. *fuscescens* Forti, Syll. Myxophyc., p. 13. 1907. —Type from Freiburg im Breisgau, Germany (L). FIG. 100.

Chroococcus thermophilus Wood, Amer. Journ. Sci. & Arts, ser. 2, 46: 34. 1868. —Type from Benton hot springs, Mono county, California (PH).

Chroococcus turgidus var. *Hookeri* Lagerheim, Ofvers. K. Sv. Vet.-Akad. Förh. 40(2): 38. 1883. —Type from Kristineberg, Sweden (S). FIG. 102.

Chroococcus turgidus var. *tenax* Kirchner, Krypt.-Fl. Schlesien 2(1): 262. 1878. *C. tenax* Hieronymus, Beitr. Biol. Pfl. 5: 483. 1892. *C. turgidus* β *tenax* Kirchner ex Wille, Nyt. Mag. Naturvid. 62: 180. 1925. *Gloeocapsa tenax* Hollerbach in Elenkin, Monogr. Algar. Cyanophyc., Pars Spec. 1: 210. 1938. —Material presumably seen by the author, designated as the Type (W).

Chroococcus turgidus var. *subnudus* Hansgirg, Prodromus Algenfl. Böhmen 2: 161. 1892. *C. turgidus* f. *subnudus* Hansgirg in Kerner, Fl. Exs. Austro-Hung. 2398. *C. turgidus* δ *subnudus* Hansgirg ex Wille, Nyt Mag. Naturvid. 62: 181. 1925. *Gloeocapsa turgida* f. *subnuda* Hollerbach in Elenkin, Monogr. Algar. Cyanophyc., Pars Spec. 1: 212. 1938. —Type from Stechovic, Bohemia (W).

Chroococcus turgidus var. *submarinus* Hansgirg, Sitzungsber. K. Böhm. Ges. Wiss., Math.-Nat. Cl., 1892: 230. 1892. —Type from Zaule, Trieste (W). FIG. 104.

Chroococcus giganteus West, Trans. Roy. Microsc. Soc. 1892: 741. 1892. *C. turgidus* var. *giganteus* Cedergren, Bot. Not. 1926: 299. 1926. *Gloeocapsa gigantea* Hollerbach in Elenkin, Monogr. Algar. Cyanophyc., Pars Spec. 1: 220. 1938. —Type from Westmorland, England (BIRM).

Chroococcus schizodermaticus var. *badio-purpureus* W. & G. S. West, Journ. of Bot. 35: 302. 1897. —Type from Golungo Alto, Angola (BM). FIG. 103.

Chroococcus (Rhodococcus) insignis Schmidle, Allgem. Bot. Zeitschr. 1897: 108. 1897. —Type from Freiburg im Breisgau, Germany (ZT). FIG. 101.

Chroococcus Westii Boye Petersen, Bot. Icel. 2(2): 263. 1923. —Cotype from Bielatal in Bohemia (FC, L, UC).

Gloeocapsa palustris Nägeli ex Wille, Nyt Mag. Naturvid. 62: 195. 1925. —Type from Naumattweier, Switzerland (O).

Chroococcus turgidus var. *maximus* Nygaard, Vidensk. Medd. Dansk Naturh. Foren. Kjøbenh. 82: 201. 1926. —Type from the Kai islands, Indonesia (in the private collection of G. Nygaard).

Chroococcus giganteus var. *occidentalis* Gardner, Mem. New York Bot. Gard. 7: 8. 1927. *C. giganteus occidentalis* Gardner, New York Acad. Sci., Sci. Surv. Porto Rico 8: 255. 1932. —Type from Laguna Tortuguero, Puerto Rico (NY). FIG. 106.

Original specimens have not been available to us for the following names; their original descriptions are here designated as the Types until the specimens can be found:

Chroococcus multicoloratus Wood, Proc. Amer. Philos. Soc. 11: 122. 1869.

Chroococcus turgidus var. *violaceus* West, Trans. Roy. Microsc. Soc. 1892: 741. 1892. *C. turgidus violaceus* West ex Gardner, New York Acad. Sci., Sci. Surv. Porto Rico 8: 256. 1932. *Gloeocapsa turgida* f. *violacea* Hollerbach in Elenkin, Monogr. Algar. Cyanophyc., Pars Spec. 1: 212. 1938. —G. S. West's sketch and notes in the British Museum (Natural History) are reproduced here: FIG. 107.

Chroococcus solitarius Eichler, Pamietn. Fizyograf. 14(3): 134. 1896.

Chroococcus indicus Bernard, Protococc. & Desmid. d'Eau Douce, p. 47. 1908. *C. Bernardii* van Oye, Hedwigia 63: 177. 1921.

Chroococcus turgidus var. *japonicus* Bernard, Alg. Unicell. Récolt. dans Dom. Malais, p. 16. 1909.

Chroococcus turgidus var. *Pullei* Bernard, Nova Guinea. Résult. de l'Exped. Sci. Néerl. Nouv.-Guinée en 1907 & 1909, 8(Bot. 2): 259. 1910.

Chroococcus turgidus var. *mipitanensis* Woloszynskia, Bull. Int. Acad. Sci. Cracovie, Cl. Sci Math. & Nat., ser. B, 1912: 692. 1913. *C. mipitanensis* Geitler in Pascher, Süsswasserfl. 12: 79. 1925. *Gloeocapsa turgida* f. *mipitanensis* Hollerbach in Elenkin, Monogr. Algar. Cyanophyc., Pars Spec. 1: 212. 1938.

Chroococcus turgidus var. *violaceus* f. *minor* Wille, Nyt Mag. Naturvid. 61: 76. 1924.

Chroococcus spelaeus Ercegovic, Acta Bot. Inst. Bot. Univ. Zagreb. 1: 76. 1925.

Chroococcus spelaeus var. *aerugineus* Ercegovic, Acta Bot. Inst. Bot. Univ. Zagreb. 1: 77. 1925.

Chroococcus spelaeus var. *violascens* Ercegovic, Acta Bot. Inst. Bot. Univ. Zagreb. 1: 77. 1925.

Chroococcus indicus var. *epiphyticus* Ghose, Journ. Burma Res. Soc. 16: 220. 1926.

Chroococcus quaternarius Zalessky, Rév. Gén. de Bot. 38: 33. 1926. *Gloeocapsa turgida* f. *quaternaria* Hollerbach in Elenkin, Monogr. Algar. Cyanophyc., Pars Spec. 1: 212. 1938.

Chroococcus turgidus var. *solitarius* Ghose, Journ. Burma Res. Soc. 17(3): 245. 1927.

Synechococcus euryphyes Beck-Mannagetta, Vestn. Král. Ces. Spolec. Nauk 1926: 17. 1927.

Chroococcus turgidus f. *minor* Wille, Deutsche Südpolar-Exped. 1901—03, 8: 415. 1928.

Chroococcus turgidus f. *lamellosus* Beck-Mannagetta, Lotos 77: 99. 1929.

Chroococcus turgidus f. *pallidus* Beck-Mannagetta, Lotos 77: 99. 1929.

Synechocystis sallensis Skuja, Acta Horti Bot. Univ. Latv. 4: 12. 1929.

Chroococcus luteolus Woronichin, Trav. Mus. Bot. Acad. Sci. U. R. S. S. 25: 443. 1932.

Chroococcus tenax var. *boeticus* Gonzalez Guerrero, Anal. Jard. Bot. Madrid 6: 242. 1946.

Chroococcus Mutisii Gonzalez Guerrero, Anal. Jard. Bot. Madrid 7: 436. 1947.

Plantae aerugineae, violaceae, roseae, vel olivaceae, praecipuius microscopicae 2—16-cellulares, cellulis in divisione binis truncato-hemisphaericis, aetate provecta truncato-globosis vel raro exacte sphaericis, diametro (8—)12—50μ crassis, sine ordine (aut in seriebus eucapsoideis) per gelatinum vaginale distributis; gelatino vaginale hyalino (rarissime pallidi-luteo), tenue, homogeneo vel plus minusve lamelloso; protoplasmate aerugineo, violaceo, roseo, vel olivaceo, tenui- vel sparse grossi-granuloso. FIGS. 100—107.

In shallow fresh, brackish, and marine waters and in seepage, usually mixed with other algae; often encountered in the plankton. Macroscopic growth-forms of *Anacystis dimidiata* are rarely found; as a rule, a plant breaks in two before it reaches the 8-celled stage. Certain subaerial growths are reminiscent of and may be confused with enlarged cells of *Palmogloea protuberans* (Sm. & Sow.) Kütz. In marine collections, some cells may be suspected of originating from the super-

71

ficial layers of *Entophysalis deusta*. Where the gelatinous matrix hydrolyzes completely, the cells may resemble somewhat certain short growth-forms of *Coccochloris aeruginosa*. The plants may become parasitized by chytridiaceous fungi.

Specimens examined:
NORWAY: ad Reiersdal in convalle Sätersdalen, *O. Nordstedt,* Jul. 1872 (as *Chroococcus turgidus* in Wittr. & Nordst., Alg. Exs. no. 100, FC, L). SWEDEN: in Lapponia Lulensi in alpe Njammats prope Qvikkjokk, *G. Lagerheim,* 5 Aug. 1883 (as *C. turgidus* in Wittr. & Nordst., Alg. Exs. no. 699, FC, L, S); Djurö, Uplandia, *O. Borge,* Jun. 1909 (S, TA); Möja, Uplandia, *Borge,* Aug. 1934 (S); Tjörnays, Scania, *Borge,* May 1893 (S); Naturvetenskap. Station, Torne Lappmark, *Borge,* Jul. 1909 (S); Vaddö, *O. F. Anderson,* Jul. 1889 (S); in arena regionis littoralis ad Kristineberg Bahusiae, *Lagerheim,* Aug. 1882 (Type of *C. turgidus* var. *Hookeri* Lagerh., with *Holopedium sabulicola* in Wittr., Nordst., & Lagerh., Alg. Exs. no. 1549, S; isotype, L [Fig. 102]). POLAND: Silesia: zwischen Sphagnum im "Moosebruch" bei Reihwiesen, *J. Nave,* Jul. 1861 (as *Chroococcus chalybeus* in Rabenh. Alg. no. 1144b, FC, FH, L, S); Ost-Schlesien, *Nave* (NY); Kochelschlucht unterhalb des Kochelfalles im Riesengebirge, *G. Hiernoymus,* Aug. 1889 (designated as Type of *C. turgidus* var. *tenax* Kirchn., as *C. tenax* in Hauck & Richt., Phyk. Univ. no. 549, W; isotypes, FH, L, NY, UC); Wasserlöchen auf der weissen Wiese bei der Wiesenbaude im Riesengebirge, *Hieronymus,* Aug. 1887 (as *C. turgidus* var. *chalybeus* in Hauck & Richt., Phyk. Univ. no. 145, L), *Hieronymus 751,* Sept. 1887 (MICH); Wittgenauer Berge bei Grünberg, Grätzbergwiese auf Ochelhermsdorf zu, *B. Schröder,* Aug. 1890 (as *C. turgidus* var. *dimidiatus* in Hauck & Richt., Phyk. Univ. no. 482, L, UC). GERMANY: Baden: im Höllenthal bei Freiburg im Breisgau, *A. Braun,* Sept. 1847 (Type of *Protococcus turgidus* Kütz. and β *fuscescens* Kütz., L [Fig. 100]); an feuchten, schattigen Felsen des Hirschsprunges im Höllenthal bei Freiburg im Breisgau, *G. Schmidle,* Aug. 1896 (Type of *Chroococcus insignis* Schmidle, ZT [Fig. 101]). Bavaria: prope Nürnberg, *J. Kaulfuss,* Jun. 1902 (S). Brandenburg: Torfsümpfe des Grunewaldes, *P. Hennings,* Jul. 1892 (as *C. turgidus* in Henn., Phyk. March. no. 49, FH, L), Jun. 1899 (S). Hesse-Nassau: unter Cladophora, Frankfurt am Main, *G. Fresenius* (as *C. turgidus* in Rabenh. Alg. Suppl. no. 104, FC, FH, L, UC). Saxony: am Schöneck, *P. Richter* (UC); in einem Sumpfe zwischen den Rauensteinen, *Gerstenberger,* Oct. 1860 (as *C. thermalis* in Rabenh. Alg. no. 1413, FC, L, NY); bei Wurzen, 1858, in herb. *O. Kuntze* (NY); Schönfeld bei Leipzig, *Auerswald* (FH); bei Dresden, *Biene,* Apr. 1862 (as *C. turgidus* in Rabenh. Alg. no. 1333, L, NY, UC); Eschdorf, Dresden, *Richter,* Aug. 1879 (S), Jul. 1880 (as *C. turgidus* in Wittr. & Nordst., Alg. Exs. no. 472, S); in einem Sumpf bei Leipzig, *L. Rabenhorst* (as *C. turgidus* var. *chalybeus* in Rabenh. Alg. no. 65, L, UC); Leipzig, *O. Bulnheim,* Dec. 1860 (as *Protococcus chalybeus* in Roumeguère, Alg. de Fr. no. 335, L). Thuringia: an feuchten Felsen der Ludwigsklamm, Eisenach, *W. Migula,* 23 Jul. 1931 (S; as *Chroococcus turgidus* in Mig., Krypt. Germ., Austr., & Helv. Exs., Alg. no. 254, FC, TA); in dem Moor der Teufelskreise am Schneekopf, *A. Röse* (as *Pleurococcus angulosus* in Rabenh. Alg. no. 327, FC); Bruchberg, *Römer* (NY); Nordhausen, in einer Wasserflasche, *F. T. Kützing* (Type of *Trochiscia dimidiata* Kütz., L [Fig. 105]).

CZECHOSLOVAKIA: Moravia: in stagnirenden Buchten der Tess bei Winkelsdorf, *Kolenati,* Sept. 1861 (as *Chroococcus chalybeus* in Rabenh. Alg. no. 1144, FC, L, UC). Slovakia: in fossa turfosa ad lacum Csorber-See in Tatra alta, *V. Wittrock,* 25 Jul. 1885 (as *C. turgidus* in Wittr. & Nordst., Alg. Exs. no. 799, FC, D, L, S). Bohemia: im Bielathale an nassen Felswänden, *L. Rabenhorst* (cotype of *C. Westii* Boye Petersen, as *Pleurococcus turgidus* in Rabenh. Alg. no. 104, FC, L, UC); Davle a údol. Libric, *A. Hansgirg,* Jul. 1887 (FC, W); Ouzice, *Hansgirg,* Oct. 1883 (FC, W); in rupibus madidis prope Stechowic, *Hansgirg* (Type of *Chroococcus turgidus* var. *subnudus* Hansg. in Fl. Exs. Austro-Hung. no. 2398I., W; isotypes, FC, FH, MIN, MO, S), Apr. 1887 (FC, W). AUSTRIA: Ötscher, Lower Austria, *G. de Beck* (FC, W); lacus prope Ossiach et St. Leonhard ad Villach, *A. Hansgirg,* Sept. 1889 (FC, UC); an der Oberfläche eines Moores des Leiterberges (Tyrol) im hohen Gesenke, *Kolenati,* Sept. 1861 (as *C. chalybeus* in Rabenh. Alg. no. 1144, FC, L, UC); Matrei, *Hansgirg,* 1891 (FC, W). JUGOSLAVIA: ex caverna (Jelenska Jarna), Franzdorf (Slovenia), *Hansgirg,* Aug. 1889 (FC, W); Laak prope Steinbrück, Slovenia, *Hansgirg,* Aug. 1890 (FC, W); Fasano prope Pisino, Istria, *Hansgirg,* Jul. 1889 (FC, W); lacus ad Boca-Niazzo prope Zaram, *Hansgirg,* Aug. 1888 (FC, W). TRIESTE: in Salinen zwischen Capodistria und Zaule, *Hansgirg,* Aug. 1889 (Type of *C. turgidus* var. *submarinus* Hansg., W; isotype, FC [Fig. 104]). SWITZERLAND: [Naumattweier, *K. von Nägeli,* Aug. 1948] (Type of *Gloeocapsa palustris* Näg. in the slide collection of N. Wille, O); am nördlichen Ufer des Katzensees, Zürich, *C. Cramer & C. von Brügger,* Jun. 1862 (as *Chroococcus turgidus* in Rabenh. Alg. no. 1444, UC). ITALY: Monfalcone, in aqua subsalsa, *Hansgirg,* Aug. 1889 (FC, W); S. Margarita prope Ala, *Hansgirg,* 1891 (FC, W); Abano, herb. Meneghini (Type of *Pleurococcus thermalis* Menegh., FI); therm. Eugan., *Meneghini* (L, S, UC, W); S. Pietro, herb. Meneghini (FI); dans les grottes de St. Avendrace, Cagliari, Sardinia, *Marcucci,* 1866 (as *Gloeocapsa polydermatica* in Marcucci, Un. Itin. Crypt. no. XXVIII, FC).

72

BRITISH ISLES: on mosses, Scotland, herb. W. Joshua (FC); Bowness, Westmorland, *W. West*, Sept. 14 (Type of *Chroococcus giganteus* West, BIRM); Penzance, Cornwall, *T. E. Hazen 3*, 1 Aug. 1926 (FC, DA). NETHERLANDS: Friesland, *W. F. R. Suringar 55*, Jul. 1854 (L). FRANCE: flaques de marais tourbeux des environs de Falaise, Calvados, *A. de Brébisson* (as *C. turgidus* var. *dimidiatus* in Rabenh. Alg. no. 2033, FC); Falaise, *de Brébisson* (L, UC); dans les flaques des rochers élevés remplies d'eau salée, Le Croisic, *G. Thuret*, 1873 (FC, FH, NY); flaques d'eau pluviale à Bercugnas (Pyrénées Centrales), *C. Fourcade* (as *C. turgidus* in Roumeguère, Alg. de Fr. no. 724, L). BRITISH TOGOLAND: Keta, *G. W. Lawson A516*, 30 Dec. 1952 (D). ANGOLA: ad rupes juxta rivum Coango, Golungo Alto, *F. Welwitsch 5* (FC, UC), *139* (Type of *C. schizodermaticus* var. *badio-purpureus* W. & G. S. West, BM; isotype, BIRM), *140* (FC [Fig. 103], UC), May-Jun. 1856. BERMUDA: *W. G. Farlow*, Jan. 1900 (FH); Lovers lake, western St. Georges island, *A. J. Bernatowicz 49-111*, 1 Mar. 1949 (FC). LABRADOR: flaque d'eau sur un rocher de la toundra, Monts Torngat, col de Saglek, *J. Rousseau 1097*, Aug. 1951 (FC). NEW BRUNSWICK: brooklet, lower basin at falls, Grand Falls, *H. Habeeb 10208, 10852*, 2 Aug. 1947, 30 Sept. 1948 (FC, HA); pond, St. Anne, Madawaska county, *Habeeb 11620*, 22 Aug. 1950 (FC, HA); Island lake, 40—50 miles due south of Dalhousie, *M. Le Mesurier 2*, Aug. 1953 (FC). MAINE: Ragged island, *F. S. Collins 4858*, Jul. 1903 (FH, NY). NEW HAMPSHIRE: on Sphagnum, bog pond, Lakewood, *W. G. Farlow*, Aug. 1917 (FH). MASSACHUSETTS: Salisbury Beach, *Collins 1700, 1702*, May 1890 (NY); rocks, Essex street, Saugus, *Collins 1696*, Jun. 1890 (NY); high ledge, Saugus, *Collins 1560*, Apr. 1890 (NY); Medford, *Collins*, May 1887 (NY); Cedar Swamp, Woods Hole, *F. Drouet 1941*, 13 Sept. 1936 (D); in tide pools near town landing, Cuttyhunk island, Gosnold, *Drouet 2122*, 3 Aug. 1937 (D, DT, FH, NY, S); in Sheep pond, Nantucket island, *B. F. D. Runk, R. N. Webster, & F. Drouet 2236*, 24 Jul. 1938 (D, FH, NY); in a brackish swamp (Chara pond) near Oyster pond, Falmouth, *Drouet, R. Patrick, & C. Hodge 3564*, 16 Jul. 1940 (FC); in Sphagnum, Lost pond, Brookline, *C. Bullard*, 14 Jul. 1910 (as *C. turgidus* var. *chalybeus* in Coll., Hold., & Setch., Phyc. Bor.-Amer. no. 1701, FC, L, TA, UC). RHODE ISLAND: tide-pool near Eastons point, Newport, *Collins*, Apr. 1898 (NY). CONNECTICUT: brackish ditches, Eastern point, Groton, *W. A. Setchell*, Jul. 1888 (UC).

QUEBEC: mousses d'un étang, embouchure de la Kogaluk, baie d'Hudson, *J. Rousseau 208*, 16 Jul. 1948 (FC); ruisseau de la toundra et fond humide d'une vallée suspendue, pic Rousseau dans les monts Otish, *Rousseau 223, 398*, Aug. 1949 (FC); Lac Tashwak, à la source de la rivière Kogaluk, 95 milles de la baie d'Hudson, *Rousseau 591*, 27 Jul. 1948 (FC); petit source tourbeuse au voisinage du Mt. Pyramid, *Rousseau 823*, 5 Aug. 1947 (FC); Lac Brennen, Rawdon, Montcalm county, *S. M. J. Eudes 291*, 29 Aug. 1941 (FC). NEW YORK: Bergen swamp, Genessee county, *A. T. Hotchkiss 222*, 1949 (D); gorge below falls, Rensselaerville, *M. S. Markle 1*, 1 Jul. 1946 (FC). MARYLAND: Somerset county: marsh pool at Chance, *P. W. Wolle & Drouet 2268*, 22 Aug. 1938 (D, UC); marsh pool west of Ewell, Smiths island, *Wolle & Drouet 2306*, 26 Aug. 1938 (FC, D, FH); Tylerton, Smiths island, *P. W. Wolle*, 12 Aug. 1938 (FC); pools in the salt marsh south of Jenkins creek, Crisfield, *Drouet & Wolle 3633, 3639, 3640*, 23 Jul. 1940 (FC). VIRGINIA: wet rocks beside Cascades, Giles county, *H. Silva 2787*, 12 Jul. 1953 (FC, TENN); wet cliff by Moormans river in "Sugar Hollow", Albemarle county, *J. C. Strickland 662*, 6 Oct. 1940 (FC, ST); in ditch, Chincoteague island, Accomac county, *L. C. Goldstein, E. S. Luttrell, & Strickland 1178*, 7 Jun. 1942 (FC, ST). NORTH CAROLINA: Ocracoke, *W. D. Hoyt*, Aug. 1906 (NY); on rocks below Dry Falls, Highlands, Macon county, *H. C. Bold H235*, 29 Jun. 1939 (FC); outlet draining a salt marsh, Hatteras, Dare county, *L. G. Williams 2a*, 12 Jul. 1949 (FC); tidal flats, Cape Lookout near Beaufort, *H. J. Humm & C. S. Neilsen 1797*, 26 Aug. 1949 (FC, T). SOUTH CAROLINA: Walhalla tunnel, Walhalla, Oconee county, *Bold H126a, H128*, 12 June 1939 (FC). FLORIDA: Bivins Arm, Gainesville, *M. A. Brannon 126*, 7 Nov. 1942 (FC, PC); Marco island, Collier county, *P. C. Standley 92809*, 14 Mar. 1946 (FC); Station 5 along Flamingo road, Everglades national park, Dade county, *D. Blake & R. Hauke*, 3 Feb. 1953 (FC, T); in a depression in sand dunes, shore of Apalachicola bay at the east end of John Gorrie causeway, west of Eastpoint, Franklin county, *Drouet & C. S. Nielsen 10960*, 16 Jan. 1949 (FC, T); on Ruppia, Hernando county, *Brannon 559*, 23 Oct. 1948 (FC, PC); region of Hendry Creek, about 10 miles south of Fort Myers, Lee county, *Standley 73254*, Mar. 1940 (FC); laboratory culture, Florida State University, Tallahassee, *Nielsen 59*, May 1948 (FC, T); with *Anacystis thermalis* in a brackish pond, Key West, *M. A. Howe 1736*, Nov. 1902 (FH, NY, UC); pools, Long key, Dry Tortugas. *W. R. Taylor 21*, 77, Jun. 1924 (TA); saline flats and drying pools, Big Pine key, Monroe county, *E. P. Killip 41806, 43205*, 22 Jan. 1952, 5 Apr. 1953 (FC, US).

OHIO: on moss and rock, Adams county, *W. A. Daily 298, 300*, 25 May 1940 (FC, DA); quarry pools, Kellys island, Erie county, *C. E. Taft*, 6 July 1938 (FC, DA); Eden park, Cincinnati, *Daily 407*, 3 July 1940 (FC, DA), *F. K. Daily*, 24 Aug. 1940 (FC, DA); "Seven Caves", Paint, Highland county, *W. A. & F. K. Daily 322, 332, 334, 336*, 16 June 1940 (FC, DA); sandstone, Old Mans Cave state park, Hocking county, *W. A. & F. K. Daily 511, 527, 534, 535, 544, 548, 553*, 27 July 1940 (DA, FC); Cantwell Cliff state park, Hocking county, *W. A. Daily 688, 693, 727, 731*, Aug.-Sept. 1940 (DA, FC); in quarry, Junction City, Perry county, *J. H. Hoskins & A. H. Blickle*, 14 Oct. 1939 (DA, FC). KENTUCKY: sandstone, Natural Bridge state park, Powell county, *W. A.*

73

& F. K. Daily 372, 30 Jun. 1940 (DA, FC); sandstone, "Tight Holler" near Pine Ridge, Wolfe county, W. A. Daily, A. T. Cross, & J. Tucker 785, 6 Sept. 1940 (DA, FC); sandstone, Torrent, Wolfe county, W. A. & F. K. Daily & R. Kosanke 468, 470, 471, 475, 482, 484, 486, 493, 494, 14 Jul. 1940 (DA, FC). TENNESSEE: in lake, Hidden Lake, Davidson county, H. C. Bold 138b, 1 Oct. 1938 (FC); lily pool, Vanderbilt University, Nashville, Bold 10, 1937 (FC); wet limestone at entrance of Bellamy cave, Montgomery county, H. Silva 1405, 15 Jul. 1949 (FC, TENN); wet rocks etc., Ramsey Cascades, Sevier county, Silva 136, 2350b, Sept. 1941, 1950 (FC, TENN), Bold H352a, H353, 18 Jul. 1939 (FC). MICHIGAN: Livingston's Bog, Cheboygan county, A. H. Gustafson, 1937 (FC), H. K. Phinney 2M40/2, 25 Jun. 1940 (FC, PHI); Mud lake, Douglas Lake region, Phinney 515, 8 Aug. 1943 (FC, PHI); pool, North Fishtail bay, Douglas Lake, Phinney 151, 31 Jul. 1942 (FC, PHI); stream near Twin Lakes on highway M28, Luce county, Phinney 20M40/51, 27 Jul. 1940 (FC, PHI); Echo lake, Bois Blanc island, Mackinac county, Phinney 35, 8 Jul. 1942 (FC, PHI). INDIANA: limestone, Clifty Falls state park, Madison, Jefferson county, W. A. & F. K. Daily 675, 676, 4 Aug. 1940 (DA, FC); Daniels lake 5 miles southeast of North Webster, Kosciusko county, F. K. & W. A. Daily 2265, 26 Jul. 1950 (DA, FC); Hawk's marsh, Lake Maxinkuckee, Marshall county, H. W. Clark & B. W. Evermann 14, 41, 13 Aug. 1906 (FC, US); sandstone, Shades state park, Waveland, Montgomery county, F. K. & W. A. Daily 969, 6 Jun. 1942 (DA, FC); Gentian lake 6 miles north of Angola, Steuben county, F. K. & W. A. Daily 2672, 24 Aug. 1953 (DA, FC); pond, Glenn Home, Vigo county, B. H. Smith 101, 3 Jul. 1929 (DA, FC); pond, Spring Hill, Vigo county, Smith 136, 18 Jul. 1929 (DA, FC); Richmond, L. J. King 131, 201, 371, Sept., Nov. 1940 (EAR, FC).

WISCONSIN: laboratory culture, University of Wisconsin, Madison, R. Evans, 6 Jun. 1949 (FC); bogs near Woodruff, Oneida county, and Sand Lake, Vilas county, G. W. Prescott 2W173, 3W279, 4 Jul. 1937, 2 Sept. 1938 (FC). ILLINOIS: Honey lake northeast of Barrington, in Lake county, P. C. Standley 92448, 29 Jul. 1945 (FC); salt marsh along creek, Illinois river bottoms 4 miles east of Starved Rock, La Salle county, J. A. Steyermark & Standley 40632, 4 Jul. 1941 (FC); in swamp in the abandoned quarry at Stony Island avenue and 94th street, Chicago, Drouet & H. & J. Rubinstein 5798, 23 Aug. 1947 (FC). MISSISSIPPI: in seepage, south shore of Back bay between Keesler Field and d'Iberville bridge, Biloxi, Drouet 10012, 15 Dec. 1948 (FC). MINNESOTA: Itasca state park, K. C. Fan 10395, 24 Jun. 1954 (D). IOWA: culture from the alkaline fen at Silver Lake, J. D. Dodd, 3 Jul. 1953 (FC). MISSOURI: in a jar of water in the greenhouse, University of Missouri, Columbia, Drouet 292, 24 Nov. 1928 (D); in salt water about the springs, Chouteau Springs, Cooper county, Drouet & H. B. Louderback 5648a, 25 Aug. 1945 (FC). LOUISIANA: in pools and shallow water, north shore of Lake Charles, Lake Charles, Drouet 8754, 8766, 28 Oct. 1948 (FC). NEBRASKA: shallow water, sandpit lakes, Fremont, Dodge county, W. Kiener 14657, 23 Jul. 1943 (FC, KI); cultures from floodplain pond southeast of Grand Island, Kiener 1636a,b, Apr., Jul. 1944 (FC, KI); culture from Lake McConaughey, Ogallala, Keith county, Kiener 23020a, 20 Aug. 1948 (FC, KI); drainage ditch, Kingsley dam, Ogallala, Kiener 24079, 31 Jul. 1948 (FC, KI); swampy bank of Whitehorse creek 5 miles north of North Platte, Kiener 17504, 1 Sept. 1944 (FC, KI). TEXAS: limestone grotto near Blowout, Blanco county, F. B. Plummer & F. A. Barkley 200, 20 Jun. 1943 (FC, TEX); Tornillo creek about 1 mile north of Hot Springs, Brewster county, E. Whitehouse 24640, 24 Dec. 1950 (FC); in ditch, municipal water plant, Brownsville, R. Runyon 3917a, 17 Jan. 1945 (FC); in seepage below sewage-disposal lake at Camp Barkeley, Taylor county, W. L. Tolstead 7751, 29 Jan. 1944 (FC); pool, Austin, F. A. Barkley 44a1g22, 22 Jul. 1944 (FC, TEX). MONTANA: spring, Stillwater county, F. H. Rose 4029, 4056, Sept. 1940 (FC). COLORADO: siliceous ledge, South Colony creek, Custer county, W. Kiener 10393, 9 Jul. 1941 (FC, KI). UTAH: Steam Mill lake, Cache county, G. Piranian, 13 Aug. 1933 (FC); Zion canyon, Washington county, N. E. Gray 68, 13 Aug. 1941 (FC). ARIZONA: Cliff spring, north rim, Grand Canyon national park, Mohave county, Gray 53, Aug. 1941 (FC); in an irrigation pool on the campus, University of Arizona, Tucson, Drouet, D. Richards, D. M. Crooks, L. Benson, & R. Darrow 2783, 30 Oct. 1939 (FC). ALBERTA: in a warm pool below the main spring, Banff, H. M. Richards 6, 23 Aug. 1900 (FC, UC). NEVADA: in water, Sodaville, I. La Rivers 1792, 9 Jun. 1953 (FC); swamp 9 miles north of Glendale, Clark county, G. Piranian, 6 Apr. 1934 (FC). BRITISH COLUMBIA: with Coccochloris aeruginosa (Näg.) Dr. & Daily in wet moss near Glacier, W. R. Taylor 10, Aug. 1921 (FH). WASHINGTON: Whidbey island, W. A. Setchell & N. L. Gardner 472 (UC).

CALIFORNIA: culture from an aerating tank, West Berkeley, Alameda county, N. L. Gardner 6745, 6 Oct. 1931 (FC, UC); campus, University of California, Berkeley, Gardner 3219, 3220, Jan. 1916 (UC); culture from soil on the University campus, Berkeley, Gardner 6758, 7 Jan. 1933 (FC, UC); water culture, botanical laboratory, University of California, Berkeley, N. L. Gardner, Jan. 1916 (as Chroococcus turgidus in Coll., Hold., & Setch., Phyc. Bor.-Amer. no. 2202a, FC, L, MICH, TA), Mar. 1916 (as C. turgidus in Coll., Hold., & Setch., Phyc. Bor.-Amer. no. 2202b, FC, D, L, MICH, TA); banks of Bear creek, Colusa county, J. F. Macbride & H. L. Mason 8018, 29 Sept. 1946 (FC); in stagnant water, Keough hot springs, Inyo county, M. J. Groesbeck 399, 7 Jul. 1941 (FC); stream in Gnome's Workshop, Death Valley, Inyo county, Groesbeck 215, 12 Oct. 1940 (FC); in tepid spring, Furnace creek wash, Death Valley national monument, Groesbeck 298,

1 Jan. 1941 (FC); first spring south of Triangle spring, Death Valley, *J. & H. W. Grinnell 7626,* 24 Oct. 1933 (FC, UC); Lagunitas Creek, Marin county, *V. Duran 6604c,* Jan. 1931 (FC, UC); laboratory culture from Marin county collected by H. E. Parks, *Gardner 6586, 6587,* 15 Jan. 1931 (FC, UC); in Benton's hot spring near Partzwick, in the northern part of Owen's valley, Mono county, *Mrs. Partz,* Aug. 1866 (Type of *Chroococcus thermophilus* Wood, PH; isotype, FC); hot springs on side of mountain, Bridgeport, Mono county, *Groesbeck 46,* 5 Apr. 1940 (FC); large warm pool in the travertine quarry, Bridgeport, *Groesbeck 101,* 16 Jun. 1940 (FC); culture (from R. Hecker), Hopkins Marine Station, Pacific Grove, *M. B. Allen,* 29 Jan. 1952 (FC); culture from the geyser, Calistoga, Napa county, *Gardner 7247,* 20 Apr. 1933 (FC, UC); in rock pools, Laguna, Orange county, *Gardner,* 27 Nov. 1908 (as *C. turgidus* var. *submarinus* in Coll., Hold., & Setch., Phyc. Bor.-Amer. no. 1551, FC, L, TA); salt marsh and brackish water west of Newport Beach, Orange county, *G. J. Hollenberg 1561, 1625,* 16 Aug. 1934 (FC); Waterman hot springs, San Bernardino county, *W. A. Setchell 1554a,* 19 Dec. 1896 (FC, UC); culture from Plath's greenhouse, San Francisco, *Gardner 6964,* 20 Apr. 1933 (FC, UC); in seepage on a cliff in the canyon of the Sacramento river, Dunsmuir, Siskiyou county, *Drouet & D. Richards 4215,* 15 Sept. 1941 (FC); in shallow water of the Tule river, Porterville, Tulare county, *Drouet & Groesbeck 4449,* 4 Oct. 1941 (FC).

BAHAMA ISLANDS: aquarium in Lerner Laboratory, Bimini, *H. J. Humm 3a,* 4 Apr. 1948 (FC); freshwater lake off Stafford creek, Andros, *A. D. Peggs,* 22 Aug. 1942 (FC). PUERTO RICO: in Laguna Tortuguero, *N. Wille 829e* (NY), *849b* (Type of *Chroococcus giganteus* var. *occidentalis* Gardn., NY [Fig. 106]), Feb. 1915; on damp earth, Coamo Springs, *Wille 288, 304,* Jan. 1915 (NY). BRITISH WEST INDIES: Cockburn Harbor, South Caicos, *M. A. Howe 5584,* Dec. 1907 (D, FH); with *Johannesbaptistia pellucida* (Dick.) Taylor & Dr. in basins of salt works, Cockburn Harbor, South Caicos, *Howe 5584b,* Dec. 1907 (NY); Bath, Jamaica, *C. E. Pease* (NY); among various algae, Jamaica, *C. E. Pease & E. Butler,* Jul. 1900 (as *Chroococcus turgidus* in Coll., Hold., & Setch., Phyc. Bor.-Amer. no. 571, FC, NY, TA). MEXICO: Sonora: on exposed soil on mountains south of Hermosillo, *D. Richards, Drouet, & L. D. Alvarado 584,* 2 Nov. 1939 (FC); backwaters of Rio de Sonora, Unión, Hermosillo, *Drouet, Richards, & Alvarado 2840,* 4 Nov. 1939 (FC), *Drouet & Richards 3032a,* 24 Nov. 1939 (FC); on tidal flat, Playa Miramar, Guaymas, *Drouet & Richards 3123,* 5 Dec. 1939 (FC); in a slightly brackish pool by railroad near the station, Empalme, *Drouet & Richards 3413,* 23 Dec. 1939 (FC). Yucatan: with *Chara Hornemannii,* Progreso, *G. F. Gaumer* (FC). GUATEMALA: in warm water of hot spring, Baños de San Lorenzo near Tejar, Dept. Sacatepéquez, *P. C. Standley 59839a,* 14 Dec. 1938 (FC); in springs near Tajumulco, Dept. San Marcos, *J. A. Steyermark 36890, 36923,* 28 Feb. 1940 (FC). COSTA RICA: in pool, Ojo de Agua, along Pan-American Highway near Cerro Las Vueltas, *R. W. Holm & H. H. Iltis 1008,* 26 Jul. 1949 (FC, MO). PANAMA: in north arm of Gigante bay, Gatun lake, Canal Zone, *C. W. Dodge,* 9 Aug. 1925 (D, MO).

VENEZUELA: edge of pond, summit of Mount Roraima, Bolívar, *J. A. Steyermark 58915,* 28 Sept. 1944 (FC); sides of bluff between Cerro San José and Cerro Peonia, Anzoátegui, *Steyermark 61567,* 20 Mar. 1945 (FC); hot springs near La Toma, southeast of Cumaná, Sucre, *Steyermark 62858a,* 21 May 1945 (FC). BRAZIL: Pará: on mud along creek in Bosque Rodriguez Alves, Belém, *Drouet 1304,* 8 Jul. 1935 (D); on water plants, Lagôa Agua Preta, Belém, *Drouet 1531,* 29 Jun. 1935 (D); in reservoir 1 km. northeast of Santa Izabel, east of Belém, *S. Wright 1612,* 10 Jul. 1935 (D). Ceará: in lake at Maracanahú, Maranguape, *Drouet 1416,* 13 Sept. 1935 (D); in a pool in Rio Pacoty at Fortaleza—Recife road, Pacatuba, *Drouet 1503,* 24 Nov. 1935 (D); in pool below dam, Açude Bôa Agua, Quixadá, *Drouet 1404,* 2 Sept. 1935 (D). ECUADOR: in an inland salt pool near Albemarle point, north end of Isabela island, Galapagos islands, *W. R. Taylor 123,* 12 Jan. 1934 (D, TA). PERU: Laguna Saraja, Dept. Ica, *A. Maldonado 93a,* Feb. 1942 (FC); Cuistalima, Lago Chilca, *Maldonado 65a,* Jan. 1942 (FC); Las Delicias, Trujillo, *N. Ibañez H. "I",* 18 May 1952 (FC).

HAWAIIAN ISLANDS: in paludibus montis Mauna Koa, Hawaii, *S. Berggren,* 1875 (as *C. turgidus* in Wittr. & Nordst., Alg. Exs. no. 250, S). MARSHALL ISLANDS: on outer reef and in channel, Amen island, Bikini atoll, *R. F. Palumbo Y28,* 27 Jul. 1949 (FC); water in bottom of bomb-crater, Utirik island, Utirik atoll, *D. P. Rogers 1780,* 2 Sept. 1946 (FC, NY); Igurin island, Eniwetok atoll, *W. R. Taylor 46-316,* 22 May 1946 (FC, MICH). SOCIETY ISLANDS: moist rocks, cliffs near Huau, Tahiti, *W. A. Setchell & H. E. Parks 5119,* 5 Jun. 1922 (FC, UC). GILBERT ISLANDS: freshwater pit and brackish water pool, North island, Onotoa, *E. T. Moul 8166, 8346,* Jul., Aug. 1951 (FC); in taro pit, southern fish-ponds of Nikunau, *Mrs. R. Catala 18a,* Jul. 1951 (FC). PHILIPPINES: embankment of stream at San Isidro, Puerto Galera, Mindoro, *G. T. Velasquez 952,* 23 Apr. 1941 (FC); Big Balatero, Puerto Galera, Mindoro, *Velasquez 10,* 6 May 1953 (FC, PUH); from Mr. Vincent's yard, Jolo, Sulu archipelago, *Velasquez 3252,* 22 May 1952 (FC, PUH). INDONESIA: kleine poel ten noorden van Kampong Gersik Poetih, Madoera island, *L. G. M. Baas-Becking,* Aug. 1936 (L); plankton of Lake Ohoitiel, Kai islands, *H. Jensen & O. Hagerup,* 1922 (Type of *Chroococcus turgidus* var. *maximus* Nygaard, in the private slide collection of G. Nygaard). CHINA: Lungchi bridges, Nanzechwan, Szechwan, *H. Chu 716,* 7 Aug. 1945 (FC).

75

6. ANACYSTIS AERUGINOSA Drouet & Daily, Lloydia 11: 77. 1948. *Palmogloea aeruginosa* Zanardini, Mem. R. Ist. Veneto 17: 162. 1872. *Gloeocapsa Zanardinii* Hauck, Oesterr. Bot. Zeitschr. 27: 230. 1877. *Aphanocapsa Zanardinii* Hansgirg, Oesterr. Bot. Zeitschr. 39: 6. 1889. —Type from Sarawak (FI). FIG. 110.

Chroococcus Raspaigellae Hauck, Hedwigia 27: 16. 1888. *Aphanocapsa Raspaigellae* Frémy in Feldmann, Arch. de Zool. Exp. & Gén. 75(Jubelaire, Fasc. 25): 389. 1933. —Type from Zaule, Trieste (L). FIG. 112.

Chroococcus smaragdinus Hauck, Hedwigia 27: 15. 1888. *Aphanocapsa smaragdina* Hansgirg, Oesterr. Bot. Zeitschr. 39: 6. 1889. —Type from Isola, Trieste (L). FIG. 108.

Aphanocapsa Howei Collins in Britton & Millspaugh, Bahama Fl., p. 618. 1920. —Type from Watling island, Bahama islands (NY). FIG. 109.

Aphanocapsa Feldmannii Frémy in Feldmann, Arch. de Zool. Exp. & Gén. 75(Jubelaire, Fasc. 25): 395. 1933. —Type from Banyuls sur Mer, France (NY). FIG. 113.

Aphanocapsa sesciacensis Frémy, Bull. Mus. d'Hist. Nat. Paris 34: 383. 1928. —Type from Chausey islands, France (NY). FIG. 111.

Chroococcus sonorensis Drouet & Daily, Field Mus. Bot. Ser. 20: 127. 1942. —Type from Guaymas, Mexico (FC).

Original specimens have not been available to us for the following names; their original descriptions are here designated as the Types until the specimens can be found:

Coelosphaeriopsis halophila Lemmermann, Abh. Nat. Ver. Bremen 16: 353. 1899. *Coelosphaerium halophilum* Geitler in Pascher, Süsswasserfl. 12: 102. 1925.

Gloeocapsa endocodia Vouk, Rad Jugoslav. Akad. Znan. i Umjetn. Zagreb, Mat.-Prirod. Razr., 254(79): 6. 1936.

Plantae pallide aeruginosae, roseae, violaceae, vel decoloratae, primum microscopicae demum bullosae ad plura centimetra crassae, cellulis in divisione binis depresso—sub-sphaericis, aetate provecta globosis diametro 6—12 (aut minoria)μ crassis, sine ordine aut in seriebus eucapsoideis per gelatinum vaginale distributis; gelatino vaginale hyalino, homogeneo nonnumquam indistincte lamelloso, evidenter limitato vel (saepe omnino) diffluente; protoplasmate pallide aeruginoso, roseo, vel violaceo, homogeneo vel sparse tenui-granuloso. FIGS. 108—113.

In brackish and more or less quiet marine waters, often forming gelatinous masses on rocks, wood, larger plants, and attached animals. Plants from deeper waters are usually more roseate in color than those developing near the surface. Young microscopic plants are often very similar to growth forms of *Anacystis thermalis*. In some cases where the plants grow as strata on rocks or other algae, the plants look as if they might have been produced by the hydrolyzation of the superficial layers of *Entophysalis deusta* or *E. conferta*.

Specimens examined:

JUGOSLAVIA: in Meere an Steinen, Parenzo, Istrien, *F. Hauck*, Sept. 1875 (L). TRIESTE: Zaule, [*C von Marchesetti*] (Type of *Chroococcus? Raspaigellae* Hauck, L [Fig. 112]); in der Rindenschichte von *Raspaigella brunnea* O. Schm. bei Zaule, *A. Valle*, Aug. 1887 (as *C. Raspaigellae* in Hauck & Richt., Phyk. Univ. no. 242, D, L, MIN); Isola, *Hauck*, Sept. 1887 (Type of *C.? smaragdinus* Hauck, L). NETHERLANDS: along the Krabbekreek northwest of Oud-Vossemer, isle of Tholen, Zeeland, *W. G. Beeftink 99*, 11 Jul. 1952 (FC, L). FRANCE: Iles Chausey, *P. Frémy*, 25 Aug. 1926 (Type of *Aphanocapsa sesciacensis* Frémy, NY [Fig. 111]); dans une mare littorale, Ile aux Oiseaux, Iles Chausey, *Frémy*, Aug. 1926 (isotypes of *A. sesciacensis* Frémy in Hamel, Alg. de Fr. no. 56, FC, MICH); dans l'éponge *Hircinia variablis* O. Schm., Banyuls-sur-Mer, Pyrénées Orientales, *J. Feldmann* (Type of *A. Feldmannii* Frémy, NY [Fig. 113]; isotype, S); Le Croisic, *G. Thuret*, Aug. 1873 (FC, FH); St. Malo, *E. Bornet*, 7 Aug. 1872 (NY); pointe de Querqueville, Manche, *Thuret*, 2 Aug. 1856 (FH, L); sur les Cladophorae â haute mer, Cherbourg, *Bornet*, 2 Aug. 1856 (FH, NY); ad Cladophoras marinas parasitica, Cherbourg, *Thuret*, 2 Aug. 1856 (D, L, S, YU). BELGIAN CONGO: gelatinous crusts in salt water, *J. J. Symoens 440* (FC). BERMUDA: on Penicillus in shallow water, Mangrove lake, Hamilton island, *A. J. Bernatowicz 50-168*, 17 Nov. 1950 (FC). MAINE: on Rhodymenia, Cape Rosier, Hancock county, *F. S. Collins 3* (FH); Kennebunkport, *Collins*, Jul. 1883 (NY); on Rhodymenia, Spectacle island, Penobscot bay, *Collins*

2990, 2 Sept. 1895 (NY, UC). MASSACHUSETTS: on Zostera, East Falmouth, *Collins 4,* Jul. 1882 (FH); on Zostera, Little Harbor, Woods Hole, *W. Trelease,* 1 Aug. 1881 (FC, NY); rocks, entrance of Hadley Harbor, Naushon island, *W. R. Taylor,* 11 Jul. 1931 (D, TA); Menauhant, *Collins,* Jul. 1882 (NY), *G. W. Perry,* Aug. 1882 (NY); Eastham, *Collins 6118,* Sept. 1909 (NY); Gloucester, *W. G. Farlow* (MICH). RHODE ISLAND: Newport, *Farlow* (as *Polycystis pallida* in Farl., Anders., & Eaton, Alg. Am. Bor. Exs. no. 179, FC, FH, MICH, NY, UC). FLORIDA: Piney point, Tampa bay, Pinellas county, *H. J. Humm 1,* 2 Apr. 1946 (FC); Biscayne bay, Cutler, Dade county, *Humm 11,* 13 Jan. 1946 (FC); saline pool, southeast hammock, Big Pine key, Monroe county, *E. P. Killip 42479,* 7 Jan. 1953 (FC, US); 0.5 mile off Shell Point, Wakulla county, *J. Morrill,* 11 Sept. 1954 (D, T).

CALIFORNIA: San Diego, *D. Cleveland 149* (FC); Pacific Beach, *M. S. Snyder* (UC); on Chaetomorpha, La Jolla, *M. S. Snyder,* 5 Nov. 1898 (as *Anacystis elabens* in Coll., Hold., & Setch., Phyc. Bor.-Amer. no. 2251, FC, NY, TA). BAHAMA ISLANDS: in 3 dm. of water in salt lake and the lagoon, Watling island, *M. A. Howe 5091* (Type of *Aphanocapsa Howei* Coll., NY [Fig. 109]; isotypes, D, UC), *5092, 5100,* 25 Nov. 1907 (NY); Waterloo lake near Nassau, *Howe 3068,* Apr. 1904 (NY); bay bottom opposite Lyon's estate, Bimini island, *H. J. Humm 26,* 1 Apr. 1948 (FC); pilings, Lerner Laboratory pier, Bimini, *Humm 20,* 27 Mar. 1948 (FC). PUERTO RICO: on Halodule, Condado lagoon, San Juan, *H. L. Blomquist 12109,* 31 Oct. 1941 (FC). GUADE-LOUPE: Les Saintes, dragué entre Terre de Haut et l'ilet â Cabrit, *J. Feldmann 3391 bis,* Mar. 1936 (TA). MEXICO: in brackish water near the beach 4 km. east of Guaymas, Sonora, *Drouet & D. Richards 3296* (FC), *3298* (Type of *Chroococcus sonorensis* Dr. & Daily, FC), 16 Dec. 1939. ECUADOR: in mangrove swamps on southeast side of Narborough island, Galapagos islands, *J. T. Howell 575,* 1 Jun. 1932 (FC, UC). SOCIETY ISLANDS: rocks in 1 foot of water, Tufaehau, at southernmost tip of Mehetia, *M. L. Grant 4667, 4773,* 15 Dec. 1930 (FC, UC). PHILIPPINES: on sponge, Puerto Princesa, Palawan, *R. C. McGregor,* Oct. 1925 (FC, UC); Tangalan and Small Balatero coves, Puerto Galera, Mindoro, *G. T. Velasquez 1671, 1880,* 5 May 1948, 11 May 1949 (FC, PUH). SARAWAK: [Tangion Datu] ad Sargassa, *E. Beccari* (Type of *Palmogloea aeruginosa* Zanard., FI; isotype, L [Fig. 110]). INDONESIA: Saleyer, *A. Weber-van Bosse 1149,* Jan. 1889 (FC, L). AUSTRALIA: on rock near beach, Curl Curl near Sydney, *V. May 1158,* 1 Dec. 1945 (FC); culture, Marion bay, South Australia, *L. G. M. Baas-Becking,* Aug. 1937 (L).

7. ANACYSTIS THERMALIS (Menegh.) Dr. & Daily.

Plantae aerugineae, olivaceae, luteolae, violaceae, vel roseae, praecipuius micro-scopicae, cellulis in divisione binis depresso—sub-sphaericis (raro truncato-hem-isphaericis), aetate provecta globosis, diametro (4—)6—12μ crassis, sine ordine vel in seriebus eucapsoideis per gelatinum vaginale distributis; gelatino vaginale hyalino, homogeneo vel lamelloso, firmo vel (saepe omnino) diffluente; proto-plasmate aerugineo, olivaceo, luteolo, violaceo, vel roseo, homogeneo vel granuloso. FIGS. 114—132.

Key to forms of *Anacystis thermalis:*

Plants 1—8-celled, sheaths thin and often lamellose, protoplasm usually
granular .... 7a. f. THERMALIS

Plants 8—128-celled, the cells distributed in more or less cubical (eucapsoid)
arrangement within the ample gelatinous matrix, protoplasm usually homogeneous .... 7b. f. MAJOR

7a. ANACYSTIS THERMALIS f. THERMALIS. *A. thermalis* Drouet & Daily, Butler Univ. Bot. Stud. 10: 221. 1952. *Trochiscia thermalis* Meneghini, Consp. Algol. Eugan., p. 334. 1837. *Protococcus nudus* Kützing, Phyc. Germ., p. 145. 1845. *Pleurococcus nudus* Rabenhorst, Fl. Eur. Algar. 3: 26. 1868. —Type from the Euganean springs, Italy (PC).

Pleurococcus membraninus Meneghini, Mem. R. Accad. Torino, ser. 2, 5 (Sci. Fis. & Mat.): 34. 1843. *Protococcus membraninus* Meneghini in Kützing, Tab. Phyc. 1: 5. 1846. *Chroococcus membraninus* Nägeli, Gatt. Einzell. Alg., p. 46. 1849. *Gloeocapsa membranina* Drouet & Daily in Habeeb & Drouet, Rhodora 50: 68. 1948. —Type from the Euganean springs, Italy (FI). FIG. 123.

77

Pleurococcus cohaerens Brébisson in Meneghini, Mem. R. Accad. Torino, ser. 2, 5(Sci. Fis. & Mat.): 35. 1843. *Protococcus cohaerens* Kützing, Tab. Phyc. 1: 5. 1846. *Chroococcus cohaerens* Nägeli, Gatt. Einzell. Alg., p. 46. 1849. *Gloeocapsa cohaerens* Hollerbach in Elenkin, Monogr. Algar. Cyanophyc., Pars Spec. 1: 231. 1938. —Type from Falaise, France (PC); cotypes from Euganean springs, Italy (L, UC). FIG. 125.

Protococcus rufescens Kützing, Tab. Phyc. 1: 9. 1846. *Pleurococcus rufescens* Brébisson *pro synon.* in Kützing, loc. cit. 1846. *Chroococcus rufescens* Nägeli, Gatt. Einzell. Alg., p. 45. 1849. —Type from Arromanches, Calvados, France (PC).

Protococcus pallidus Nägeli in Kützing, Sp. Algar., p. 201. 1849. *Chroococcus pallidus* Nägeli, Gatt. Einzell. Alg., p. 46. 1849. —Type from Zürich, Switzerland (L). FIG. 129.

Palmella testacea A. Braun in Kützing, Sp. Algar., p. 211. 1849. *Aphanocapsa testacea* Nägeli, Gatt. Einzell. Alg., p. 52. 1849. *Microcystis testacea* Elenkin, Monogr. Algar. Cyanophyc., Pars Spec. 1: 138. 1938. —Type from Freiburg im Breisgau, Germany (L). FIG. 127.

Chroococcus helveticus Nägeli, Gatt. Einzell. Alg., p. 46. 1849. —Type from Lucerne, Switzerland (FC); paratype from Zürich named by the author (FC). FIG. 114.

Chroococcus decorticans A. Braun, Betracht. ü. d. Erscheinung d. Verjüngung i. d. Natur, p. 194. 1849—50. *Gloeocapsa decorticans* Richter ex Wille, Nyt Mag. Naturvid. 62: 186. 1925. —Type from St. Aubin, Switzerland (L). FIG. 128.

Chroococcus turgidus var. *rufescens* Wartmann in Rabenhorst, Alg. Sachs. 63 & 64: 631. 1857. —Type from Zürich, Switzerland (FC).

Chroococcus virescens Hantzsch in Rabenhorst, Alg. Sachs. 133 & 134: 1332. 1862. *C. minutus* var. *virescens* Hansgirg, Prodr. Algenfl. Böhmen 2: 162. 1892. —Type from Dresden, Germany (L). FIG. 116.

Chroococcus obliteratus Richter in Hauck & Richter, Phyk. Univ. 1: 41. 1886; Hedwigia 25: 215. 1886. *C. minutus* var. *obliteratus* Hansgirg, Prodr. Algenfl. Böhmen 2: 162. 1892. —Type from Leipzig, Germany (L).

Chroococcus minutus var. *salinus* Hansgirg, Oesterr. Bot. Zeitschr. 36: 333. 1886. —Type from Ouzic, Bohemia (W). FIG. 118.

Aphanocapsa salinarum Hansgirg, Physiol. & Algol. Stud., p. 153. 1887. —Type from Ouzic, Bohemia (W). FIG. 118.

Chroococcus membraninus var. *crassior* Hansgirg, Sitzungsber. K. Böhm. Ges. Wiss., Math.-Nat. Cl., 1891(1): 360. 1891. —Type from Tüffer, Slovenia, Jugoslavia (W). FIG. 126.

Chroococcus helveticus var. *aureo-fuscus* Hansgirg, Prodr. Algenfl. Böhmen 2: 162. 1892. —Type from Slichov, Bohemia (W). FIG. 119.

Chroococcus helveticus var. *aurantiofuscescens* Hansgirg, Prodr. Algenfl. Böhmen 2: 162. 1892. —Type from Podmoran, Bohemia (W). FIG. 122.

Chroococcus turgidus var. *glomeratus* Hansgirg, Sitzungsber. K. Böhm. Ges. Wiss., Math.-Nat. Cl., 1892: 154. 1893. —Type from Kufstein, Tyrol, Austria (W). FIG. 120.

Gloeocapsa calcarea Tilden, Minnesota Bot. Stud., ser. 2, 1: 29. 1898; Amer. Alg. 3: 299. 1898. —Type from Osceola, Wisconsin (MIN). FIG. 124.

Chroococcus polyedriformis Schmidle, Engler Bot. Jahrb. 30: 241. 1901. —Type from Lake Nyasa, Tanganyika (ZT).

Chroococcus mediocris Gardner, Mem. New York Bot. Gard. 7: 6. 1927. —Type from Laguna Tortuguero, Puerto Rico (NY). FIG. 121.

Chroococcus subsphaericus Gardner, Mem. New York Bot. Gard. 7: 6. 1927. —Type from Coamo Springs, Puerto Rico (NY). FIG. 117.

Chroococcus aeruginosus Gardner, Mem. New York Bot. Gard. 7: 7. 1927. —Type from Coamo Springs, Puerto Rico (NY).

Chroococcus turgidus var. *uniformis* Gardner, Mem. New York Bot. Gard. 7: 7. 1927. *C. turgidus uniformis* Gardner, New York Acad. Sci., Sci. Surv. Puerto Rico 8: 256. 1932. —Type from Coamo Springs, Puerto Rico (NY).

Chroococcus beangloios Gardner, Mem. New York Bot. Gard. 7: 9. 1927. —Type from Utuado, Puerto Rico (NY).

Gloeocapsa acervata Gardner, Mem. New York Bot. Gard. 7: 10. 1927. —Type from near Arecibo, Puerto Rico (NY).

Gloeocapsa ovalis Gardner, Mem. New York Bot. Gard. 7: 11. 1927. —Type from near Adjuntas, Puerto Rico (NY). FIG. 115.

Synechocystis crassa var. *major* Geitler, Arch. f. Hydrobiol., Suppl. XII, Trop. Binnengewässer 4: 623. 1933. —Type from Tjurup, Sumatra, Indonesia (in the collection of L. Geitler).

Chroococcus schizodermaticus var. *incoloratus* f. *paucistratosus* Geitler ex J. de Toni, Diagn. Alg. Nov., I. Myxophyceae 9: 833. 1946. —Type from Kuripan, Java (in the collection of L. Geitler).

Original specimens have not been available to us for the following names; their original descriptions are here designated as the Types until the specimens can be found:

Chroococcus rufescens var. *turicensis* Nägeli, Gatt. Einzell. Alg., p. 46. 1849. *C. rufescens* f.

78

turicensis Nägeli ex Rabenhorst, Fl. Eur. Algar. 2: 33. 1865. *C. turicensis* Hansgirg, Physiol. & Algol. Stud., p. 165. 1887. —Notes and sketches of this were found in the notebooks of Carl Nägeli in the Pflanzenphysiologisches Institut, Eidgenossische Technische Hochschule, Zürich.

Aphanothece globosa Elenkin, Kamchatskaya Eksped. Fiod. Pavlov. Ryakushinskavo, Bot. 2: 151. 1914.

    Chroococcus bataviae van Oye, Hedwigia 63: 177. 1921.

    Chroococcus cumulatus Bachmann, Mitt. Naturf. Ges. Luzern 8: 11. 1921.

    Chroococcus minutus var. *amethystaceus* Wille, Deutsche Südpolar-Exped. 1901—03, 8: 383. 1924. *C. minutus* var. *amethysteus* Wille ex Geitler, Rév. Algol. 2: 345. 1925.

    Chroococcus helveticus var. *consociato-dispersus* Elenkin, Not. Syst. Inst. Crypt. Hort. Bot. Petropol. 2: 23. 1923. *Gloeocapsa minuta* f. *consociato-dispersa* Hollerbach in Elenkin, Monogr. Algar. Cyanophyc., Pars Spec. 1: 233. 1938.

    Chroococcus Scherffelianus Kol, Folia Crypt. Szeged 1: 614. 1928.

    Myxosarcina chroococcoides Geitler, Arch. f. Protistenk. 60: 443. 1928.

    Chroococcus irregularis Huber-Pest., Verh. Int. Ver. Theor. & Ang. Limnol. 4: 366. 1929.

    Synechocystis crassa Woronichin, Bull. Jard. Bot. Princip. U. R. S. S. 28: 31. 1929.

    Chroococcus lithophilus var. *rotae* Gonzalez Guerrero, Anal. Jard. Bot. Madrid 8: 266. 1948.

    Myxosarcina sphaerica Proschkina-Lavrenko, Akad. Nauk S. S. S. R. Otd. Sporovykh. Rost. Bot. Mater. 6: 72. 1950.

Plantae aerugineae, olivaceae, luteolae, violaceae, vel roseae, microscopicae vulgo 1—8-cellulares, cellulis in divisione binis depresso—sub-sphaericis (raro truncato-hemisphaericis), aetate provecta globosis diametro (4—)6—12μ crassis, sine ordine per gelatinum vaginale distributis; gelatino vaginale hyalino, tenue, homogeneo vel lamelloso, firmo vel (etiam omnino) diffluente; protoplasmate aerugineo, olivaceo, luteolo, violaceo, vel roseo, plerumque (saepe grossi-) granuloso. FIGS. 114—129.

Among other algae in seepage and in shallow fresh water, rarely in brackish water. Macroscopic plants of *Anacystis thermalis* f. *thermalis* rarely develop. In brackish water, plants interpreted as of this form may appear to be derived from the superficial cells of *Entophysalis crustacea*.

Specimens examined:

    NORWAY: Valders, in rupibus ad Skogstad, *N. Wille,* 31 Jul. 1879 (D). SWEDEN: Möja, Uppland, *O. Borge,* Aug. 1933 (S); in fossa turfosa ad Källstorp par. Holm in Dalia, *V. Wittrock,* 22 Aug. 1878 (as *Merismopoedia glauca* in Wittr. & Nordst., Alg. Exs. no. 300, FC); in aquaeductu ad Nacka prope Stockholmiam, *Wille,* 1884 (D). POLAND: Silesia: an einem eisernen Brunnentroge in Wüstewaldersdorf, *B. Schröder 9,* 4 Aug. 1893 (S, UC). GERMANY: Baden: Freiburg, *A. Braun,* Oct. 1847 (Type of *Palmella testacea* A. Br., L [Fig. 127]; isotypes, L, NY), May 1850 (NY); Karlsruhe, in einem mit Bachwasser angesetzten Kulturtopf, *W. Migula,* Oct. 1904 (as *Chroococcus minutus* in Mig., Crypt. Germ., Austr., & Helv. Exs., Alg. no. 84, MICH, TA); Badensweiler, *Braun 26,* Aug. 1848 (L). Saxony: auf Torfmoorschlamm, cultivirt (as *C. minutus* in Rabenh. Alg. no. 1214, FC, L, TA); Dresden, gesammelt im Aug. 1860, cultivirt bis jetzt, *C. A. Hantzsch,* May 1862 (Type of *C. virescens* Hantzsch in Rabenh. Alg. no. 1332, L; isotypes, FH, NY [Fig. 116], UC); Schönfeld bei Dresden, *P. Richter,* Aug. 1876 (FC), Aug. 1877 (L), Aug. 1875 (as *C. virescens* in Rabenh. Alg. no. 2533, FC, L, NY, S, TA, UC); étangs, *Richter* (as *C. virescens* in Roumeguère, Alg. de Fr. no. 506, L); Leipzig, in einem flachen Teller, *C. Werner,* Sept. 1885 (Type of *C. obliteratus* Richt. in Hauck & Richt., Phyk. Univ. no. 41, L; isotypes, MICH, MIN). CZECHOSLOVAKIA: Bohemia: Arnau, *A. Hansgirg,* Jul. 1885 (FC, W); Eichwald, *Hansgirg,* Jul. 1883 (FC, W); Mühle am Olafsgrund im Riesengebirge, *Hansgirg,* Jul. 1884 (FC, W); in stagnis subsalsis prope Ouzic ad Kralup, *Hansgirg* (with *Gomphosphaeria aponina* Kütz. in Kerner, Fl. Exs. Austro-Hung. no. 3600, MO); in seichtem Wasser und im Schlamme der Salzwassersümpfe bei Ouzic nächst Kralup, *Hansgirg,* Oct. 1883 (Types of *Aphanocapsa salinarum* Hansg. and *Chroococcus minutus* var. *salinus* Hansg., W; isotype, FC [Fig. 118]); auf feuchten Felsen bei Podmoran, *Hansgirg,* Jun. 1885 (Type of *C. helveticus* var. *aurantiofuscescens* Hansg., W; isotype, FC [Fig. 122]); in parietibus subhumidis caldariorum horti botanici Pragae, *Hansgirg,* 1883 (as *C. aurantio-fuscus* in Wittr. & Nordst., Alg. Exs. no. 700, FC, L, S); Sázava, *Hansgirg,* Sept. 1885 (FC, W); Selc u Roztok, *Hansgirg,* Nov. 1886 (FC, W); in rupibus madidis prope Selc in agro Pragensi, *Hansgirg* (as *C. helveticus* in Kerner, Fl. Exs. Austro-Hung. no. 2397 I., FC, FH, L, MO, S); auf feuchten Felsen bei Slichow, *Hansgirg,* May 1885 (Type of *C. helveticus* var. *aureo-fuscus* Hansg., W; isotype, FC [Fig. 119]); St. Prokop ad Tinonicum, *Hansgirg,* May 1885

(FC, W); skály pod Tetínem nad Berounkou, *Hansgirg*, Sept. 1888 (FC, W).
AUSTRIA: Lower Austria: in oppido Baden in receptaculo aquae salientis, *Stockmayer* (as
C. helveticus in Kerner, Fl. Exs. Austro-Hung. no. 2397 II., FC, FH, L, MO, S); in lacu horti Bar.
Doblhoff in oppido Baden, *Stockmayer* (as *C. turgidus* f. *subnudus* in Kerner, Fl. Exs. Austro-Hung.
no. 2398 II., FC, FH, L, MIN, S). Tyrol: im Längsee bei Kufstein, *Hansgirg*, 1891 (Type of *C.
turgidus* var. *glomeratus* Hansg., W; isotype, FC [Fig. 120]). JUGOSLAVIA: Slovenia: Laak prope
Steinbrück, *Hansgirg*, Aug. 1890 (FC, W); in Tüffer im Abflusse der warmen Quelle (im kleinen
Wasserbasin), *Hansgirg*, Aug. 1890 (Type of *C. membraninus* var. *crassior* Hansg., W; isotype, FC
[Fig. 126]); Veldes, *Hansgirg*, Sept. 1889 (FC, W); Podnart, *Hansgirg*, Aug. 1889 (FC, W).
Istria: Pisino, *Hansgirg*, Aug. 1889 (FC, W). Dalmatia: Spalato, *Hansgirg*, Aug. 1889 (FC, W);
ad cataractam fluminis Kerkae ad Scardonam, *Hansgirg*, Aug. 1888 (FC, W). SWITZERLAND:
Luzern, an nassen Felsen, *K. von Nägeli*, Dec. 1847 (Type of *C. helveticus* Näg., FC; isotypes in
Hauck & Richt., Phyk. Univ. no. 483, FH, L); in saxis irrigatis prope Rehalp ad Zürich, *G. Winter*,
Nov. 1878 (as *C. pallidus* in Wittr. & Nordst., Alg. Exs. no. 400, FC, L, NY, S); auf feuchten
Tufffelsen, St. Aubin am Neuchateller See, *A. Braun 114*, 4 Sept. 1848 (Type of *C. decorticans* A.
Br., L [Fig. 128]); an einem Brunnen bei Zürich, *Winter*, Aug. 1878 (S); an Felsen, Erlenbacher
Tobel, Zürich, *Nägeli*, 10 Sept. 1848 (paratypes of *C. helveticus* Näg., FC[Fig. 114], FH); Zürich,
Nägeli 304 (Type of *Protococcus pallidus* Näg. [= *Chroococcus pallidus* Näg.], L[Fig. 129];
isotype, UC); an feuchten Felsen bei Zürich, *Wartmann*, Jul. 1856 (Type of *C. turgidus* var. *rufescens*
Wartm., FC; isotypes, L, NY). ITALY: Görz, *A. Hansgirg*, 1891 (FC, W); in aquis thermalibus
Thermae Euganeae, *Meneghini* (Type of *Trochiscia thermalis* Menegh., PC; isotype, L), *idem* (cotype
of *Pleurococcus cohaerens* Bréb., L[Fig. 125]); in fossa del molino [therm. Eugan.], *Meneghini*, Jun.
1838 (Type of *Pleurococcus membraninus* Menegh., FI[Fig. 123]; isotypes, L, S, UC).
NETHERLANDS: Drammensvei, Bergen, *A. Weber-van Bosse 102*, Aug. 1885 (L). FRANCE:
Calvados: Arromanches, *Brébisson* ex herb. Lenormand (Type of *Protococcus rufescens* Kütz., FC;
isotype, UC); dans une cascade dans les Falaises près d'Arromanches, *A. de Brébisson* (as *P. rufescens* in
Rabenh. Alg. no. 2034, FC); Bayeux, *de Brébisson* (S); Falaise, *de Brébisson* (Type of *Pleurococcus
cohaerens* Bréb., PC; isotypes, FI, UC). Seine Inférieure: Falaises d'Etretat, *M. Gomont*, Jul. 1884
(FC). Basses Pyrénées: sur la terrasse d'une maison, Guéthary, *C. Sauvageau*, May 1898 (PC).
PORTUGAL: Oporto, *I. Newton*, 1887 (as *Aphanocapsa testacea* in Hauck & Richt., Phyk. Univ. no.
331, L). TUNISIA: Gafsa, *M. Serpette TL45, TL64*, 15 Mar. 1948 (D). ALGERIA: Djendel,
F. Debray, May 1891 (NY). AZORES: in aqua fluente, *K. Bohlin*, Aug. 1898 (as *Chroococcus
membraninus* in Wittr., Nordst., & Lagerh., Alg. Exs. no. 1538, L, NY). TANGANYIKA: an
Gneissblöcken in der Nähe der Brandung, Langenburg am Nyassasee, *W. Goetze 867*, 23 Apr. 1899
(Type of *C. polyedriformis* Schmidle, ZT).
ICELAND: in a hot spring, Uxahver, North Amt, *W. F. Palsson 2*, Jul. 1941 (FC). NEW
BRUNSWICK: pools and ledges about the falls, Grand Falls, Victoria county, *H. Habeeb 10264,
10380, 10671, 13455*, Jul.—Aug., 1947—1948 (FC, HA); in lake 8 miles south of Grand Falls,
Habeeb 11653, 22 June 1951 (FC, HA); Island lake, 40—50 miles due south of Dalhousie, *M. Le
Mesurier 1*, Aug. 1953 (FC). MASSACHUSETTS: on moss, Magnolia, *W. G. Farlow*, Sept. 1903
(FH). QUEBEC: sur roches, carrière d'Outremont, Mont-Royal, près Montréal, *C. Lanouette 104,
105, 109, 111*, Sept. 1940 (FC). NEW YORK: incrustation on inside of spring barrel near the
road culvert of Reed's Creek, Ontario county, *D. Haskins 49*, 1944 (FC). VIRGINIA: on mud
in drain on campus of Virginia Polytechnic Institute, Blacksburg, *A. B. Massey, E. S. Luttrell, & J. C.
Strickland 981*, 15 Aug. 1941 (FC, ST). NORTH CAROLINA: in drippings from a wooden
tank, Beaufort Chemical Corporation, Lennoxville road, Beaufort, *H. J. Humm*, 25 May 1948 (FC);
on wet rocks, Cullasaja gorge, Macon county, *H. C. Bold H157*, 17 Jun. 1939 (FC); on wet rocks,
Jackson county, *Bold H33b*, 7 May 1939 (FC). FLORIDA: Sand pond, Alachua county, *J. E.
Davis 19*, 8 Oct. 1941 (FC, PC); on soil, Gainesville, *F. B. Smith 37*, 21 Jan. 1942 (FC, PC);
Gainesville, *M. A. Brannon 290, 305*, Feb., Aug. 1945 (FC, PC); pond 4 miles west of Olustee,
Baker county, *A. M. Scott 102*, 22 Oct. 1947 (FC); Hernando, Citrus county, *Brannon 383*, 20
Oct. 1946 (FC, PC); Marco island, Collier county, *P. C. Standley 73391, 73529*, 19 Mar. 1940
(FC), *92811*, 14 Mar. 1946 (FC); salt meadow 20 miles southeast of Naples, Collier county, *J. A.
Steyermark 63165a*, 10 Mar. 1946 (FC); wet sand in the greenhouse, Florida State University,
Tallahassee, *F. Drouet & D. Crowson 10453*, 5 Jan. 1949 (FC, T); limestone canyon, Aspalaga on
Apalachicola river, Liberty county, *C. S. Nielsen, G. C. Madsen, & Crowson 764*, 12 Feb. 1949
(FC, T); floating in brackish pond, Key West, *M. A. Howe 1736*, Nov. 1902 (FH, NY); saline flats,
east side of bay, Big Pine key, Monroe county, *E. P. Killip 41807*, 22 Jan. 1952 (FC, US); on
flower pots, Orlando, *Brannon 542, 606*, May, Oct. 1948 (FC, PC).
OHIO: dripping cliff, Steam Furnace, Adams county, *M. Britton*, 11 Jun. 1937 (DA, FC);
Sugar Creek Pond, Athens county, *A. H. Blickle & L. Mann 50E*, 22 May 1941 (DA, FC); water
trough, the Plains, Athens county, *Blickle & M. Wright*, 10 Jul. 1941 (DA, FC); state hospital
ponds, Athens, *Blickle*, 3 Oct. 1940 (DA, FC), and laboratory culture of the same, 1941 (DA, FC);
on limestone, Clifton Gorge, Greene county, ex coll. M. Britton, 3 Apr. 1937 (DA, FC); Sharon
woods, Sharonville, Hamilton county, *W. A. Daily 307*, 12 Jun. 1940 (DA, FC); Addyston pond,
Hamilton county, *Daily 163*, 19 Oct. 1939 (DA, FC); reservoirs, Eden park, Cincinnati, *Daily 412*,

415, 3 Jul. 1940 (DA, FC); drain on university campus, Cincinnati, *Daily & R. Kosanke 454*, 11 Jul. 1940 (DA, FC); in quarry pool on Kellys island, Ottawa county, *C. E. Taft 13*, 6 Jul. 1938 (DA, FC); laboratory culture, Marietta College, Marietta, *L. Walp C41117*, Nov. 1941 (FC); on sandstone in quarry near Constitution, Washington county, *H. Noland & F. K. & W. A. Daily 897, 899—902*, 11 Oct. 1941 (DA, FC). KENTUCKY: near Slade, Powell county, *A. T. Cross, J. Tucker, & W. A. Daily 758*, 5 Sept. 1940 (DA, FC); wet rock on U. S. highway 68 near Brooklyn Bridge, Woodford county, *B. B. McInteer 1137*, 1941 (DA, FC). TENNESSEE: ditch west of Clarksville, Montgomery county, *H. Silva 2080*, 18 Mar. 1950 (FC, TENN). MICHIGAN: Little Lake Sixteen, Cheboygan county, *H. K. Phinney 10M40/2*, 13 Jul. 1941 (FC, PHI); bog on U. S. highway 2 south of Rudyard, Chippewa county, *Phinney 30M41/5*, 16 Aug. 1941 (FC, PHI); Sodon lake west of Bloomfield Hills, Oakland county, *S. A. Cain C-1*, 9 Dec. 1947 (FC). INDIANA: on limestone, Scenic Falls, Bartholomew county, *F. K. & W. A. Daily 1043, 1045, 1047, 1048, 1052, 1058*, 11 Oct. 1942 (DA, FC); Lake Shaffer dam, Monticello, Carroll county, *W. A. Daily 872, 875*, 30 Jul. 1941 (DA, FC); in seepage, Lake Freeman, Monticello, *F. K. & W. A. Daily 889*, 1 Jul. 1941 (DA, FC); floating in Lake Cicott near Logansport, Cass county, *F. K. & W. A. Daily 1022*, 10 Sept. 1942 (DA, FC); on wall of Lake McCoy dam near Greensburg, Decatur county, *F. K. & W. A. Daily 1075*, 11 Oct. 1942 (DA, FC); on limestone, North Vernon, Jennings county, *F. K. & W. A. Daily 910, 1107*, 23 May 1942 (DA, FC); on dam in Blue river at Edinburgh, Johnson county, *F. K. & W. A. Daily 1049*, 11 Oct. 1942 (DA, FC); on floor in Pahuds greenhouse, Indianapolis, *F. K & W. A. Daily 1179*, 20 May 1944 (DA, FC); plankton of White river near the sewage disposal plant, Indianapolis, *C. M. Palmer*, 16 Sept. 1931 (DA, FC); in a jar, Butler University, Indianapolis, *Palmer 86*, 17 Aug. 1926 (DA, FC); on limestone in a quarry south of Bloomington, Monroe county, *F. K. & W. A. Daily 1009*, 7 Sept. 1942 (DA, FC); gravel pit 3 miles north of Ade, Newton county, *F. K. & W. A. Daily 2559*, 28 Aug., 1951 (DA, FC); on limestone above falls in Echo canyon, McCormicks Creek state park, Owen county, *F. K. & W. A. Daily 1817*, 10 Aug. 1947 (DA, FC); Turkey Run state park, Parke county, *Palmer 15*, 20 May 1932 (DA, FC), *W. A. Daily 1433*, 11 May 1946 (DA, FC); on limestone in quarry near St. Paul, Shelby county, *F. K. & W. A. Daily 1149, 1156*, 3 Oct. 1943 (DA, FC); plankton of Lake Wehi, south of Germantown, Wayne county, *F. K. & W. A. Daily 1461, 1468*, 15 Jun. 1946 (DA, FC), *M. S. Markle & L. J. King 739, 757*, 7 Sept. 1942 (EAR, FC); on rocks and walls, Richmond, *L. J. King 207, 1114*, 1940—1943 (EAR, FC).

WISCONSIN: culture, University of Wisconsin, Madison, *R. Evans B*, 9 Dec. 1949 (FC); on boards where spring water drips, Osceola, *J. E. Tilden*, 15 Sept. 1897 (Type of *Gloeocapsa calcarea* Tild., Amer. Alg. no. 299, MIN; isotypes, FC [Fig. 124], L). ILLINOIS: cultures, University of Chicago, Chicago, *P. D. Voth*, Dec. 1944, 27 Mar. 1945 (FC); Palos Park, Cook county, *L. J. King & J. O. Young*, Spring 1941 (FC); in seepage on spillway, Lake Glendale, Robbsville recreational area in Shawnee national forest, Pope county, *H. K. Phinney 973*, 14 Jul. 1944 (FC, PHI). MINNESOTA: on moist wall in stone quarry near campus, University of Minnesota, Minneapolis, *G. Lilley*, 21 Nov. 1901 (as *Gloeocapsa rupestris* in Tild., Amer. Alg. no. 599, FC); Elk spring, Itasca state park, *Drouet 12185*, 6 Jul. 1954 (D, MIN). MISSOURI: in a muriatic spring, Kimmswick, Jefferson county, *J. A. Steyermark*, 1 May 1934 (D); in an aquarium, William Jewell College, Liberty, Clay county, *C. J. Elmore*, 20 Feb. 1938 (D); in pond and pool, Thayer, Oregon county, *N. L. Gardner 1406, 1410*, Jul. 1904 (FC, UC); wet concrete about a spring, Chouteau Springs, Cooper county, *F. Drouet & H. B. Louderback 5645*, 25 Aug. 1945 (FC); sandstone cliffs, Mine La Motte, Madison county, *C. Russell 36*, Oct. 1899 (FC, UC). ARKANSAS: artificial lake, Crossett, Ashley county, *M. Thomason 234*, 30 Jun. 1945 (FC). NEBRASKA: culture from pond in state park, Ravenna, Buffalo county, *W. Kiener 20172b*, 1 May 1946 (FC, KI); culture from sandpit north of David City, Butler county, *Kiener 22391a*, 10 Nov. 1947 (FC, KI); culture from rainwater basin, Schickley, Fillmore county, *Kiener 22350a*, 28 Oct. 1947 (FC, KI); culture from floodplain pond southeast of Grand Island, *Kiener 16396b*, 11 Jul. 1944 (FC, KI); cultures from spring, Lonergan creek, Lemoyne, Keith county, *Kiener 23066—23068*, 20 Aug. 1948 (FC, KI); culture from seepage, Blue creek north of Oshkosh, Garden county, *Kiener 16218c*, 4 Nov. 1945 (FC, KI); culture from Gimlet lake, Garden county, *Kiener 16199c*, 4 Nov. 1945 (FC, KI); culture from sandpit southwest of Phillips, Hamilton county, *Kiener 17200c*, 17 Sept. 1945 (FC, KI); culture from roadside ditch 9 miles east of North Platte, *Kiener 16871*, 13 Jul. 1944 (FC, KI); old gravel pit pond south of North Platte, *Kiener 17957*, 4 Nov. 1944 (FC, KI); culture from ice in cut-off pond, Platte river near Cedar Bluffs, Saunders county, *Kiener 16373b,c*, 16 Jul. 1944 (FC, KI). TEXAS: permanent pool, Austin, *F. A. Barkley 44alg22*, 25 Jul. 1944 (FC, TEX); creek, Austin, *B. C. Tharp*, 7 Oct. 1946 (FC, TEX); creek, Bastrop county, *Tharp*, 13 Oct. 1946 (FC, TEX); in water around bathhouse, Hot Springs, Big Bend national park, Brewster county, *E. Whitehouse 19692*, 19 Apr. 1948 (FC); in creek at sulphur well, Palmetto state park near Gonzales, *B. F. Plummer & Barkley I*, 22 May 1943 (FC, TEX).

MONTANA: culture from flower pot saucer in greenhouse, Montana State University, Missoula, *Barkley 4848*, 14 Oct. 1940 (FC); in hot water, hot springs, Lo Lo, Missoula county, *D. Griffiths*, 17 Sept. 1898 (as *Chroococcus varius* in Tild., Amer. Alg. no. 600, FC). WYOMING: on rocks, White Elephant Caves, and in lowermost overflow of Cleopatra Terrace, Mammoth hot springs,

Yellowstone national park, *W. A. Setchell 1978, 2008,* 31 Aug. 1898 (FC, UC). NEW MEXICO: in hot springs, Montezuma (Hot Springs), San Miguel county, *Drouet & D. Richards 2545, 2653,* Oct. 1939 (FC); warm mineral spring, Jermez Springs, Sandoval county, *A. A. Lindsey,* 8 Jun. 1947 (FC). UTAH: in edge of pond from artesian well, Old Resort, Utah lake, Utah county, *E. Snow D26,* 17 Aug. 1930 (FC, UC). ARIZONA: in a lily pool on campus of the University of Arizona, Tucson, *Drouet, Richards, D. M. Crooks, L. Benson, & R. Darrow 2774,* Oct. 1939 (FC). NEVADA: in cold pool, hot springs, Steamboat, Washoe county, *M. J. Groesbeck 293,* 25 Nov. 1940 (FC). WASHINGTON: scum on the ground in overflow from a warm spring, Olympic national park, *W. B. Cooke,* Aug. 1951 (FC); abandoned Epsom salt mine near Seattle, *G. Anderson,* 2 Jun. 1955 (D). OREGON: in "hot springs" near Madras, Crook county, *L. E. Griffin,* 23 Apr. 1939 (FC).

CALIFORNIA: edge of "the Lagoon", Niles, Alameda county, *W. A. Setchell,* 5 Nov. 1898 (as *Chroococcus membraninus* in Coll., Hold., & Setch., Phyc. Bor.-Amer. no. 1201, FC, L, TA); cultivated in the botanical laboratory, University of California, Berkeley, *N. L. Gardner,* May 1916 (as *C. helveticus* in Coll., Hold., & Setch., Phyc. Bor.-Amer. no. 2201, FC, D), *Gardner 7192,* 23 Mar. 1933 (FC, UC), *Gardner 8041,* 24 Mar. 1937 (FC, UC); in a flowerpot in the Agricultural greenhouse, University of California, Berkeley, *Gardner 6977,* 21 Jan. 1932 (FC, UC), and culture, *Gardner 7148,* 7 Jan. 1933 (FC, UC); culture from soil in the northwestern part of the campus, University of California, Berkeley, *Gardner 6757,* 7 Jan. 1933 (FC, UC); Byron hot springs, Contra Costa county, *Gardner 7614, 7718,* 11 Apr. 1933, 21 Oct. 1933 (FC, UC); culture from Bad Water, Death Valley, *R. M. Holman & L. Bonar 7799,* 2 Apr. 1933 (FC, UC); cold alkaline stream at Gnome's Workshop, Death Valley, *M. J. Groesbeck 295,* 1 Jan. 1941 (FC); Nevares Springs near Cow Creek, Death Valley, *P. A. Munz 2280.5,* 7 Apr. 1937 (FC); in a saline pond about 2 miles north of Buckingham Park, Clear Lake, Lake county, *Gardner 7561,* 7 Aug. 1933 (FC, UC), and culture, *Gardner 7700,* 16 Apr. 1934 (FC, UC); culture from seepage in Live Oak canyon, La Verne, Los Angeles county, *G. J. Hollenberg 1648,* 12 Nov. 1934 (FC); dam in San Dimas canyon near La Verne, *Hollenberg 2077a,* 20 Apr. 1934 (D); Old El Encino hot springs, Los Angeles county, *B. C. Templeton 1a,* Jul. 1944 (FC); boiling pools, hot springs formation 2 miles southeast of Bridgeport, Mono county, *A. M. Alexander & L. Kellogg,* 30 Jun. 1944 (FC); in the travertine quarry, Bridgeport, *Groesbeck 42, 101, 257, 342, 469,* Apr. 1940—Jul. 1941 (FC); Hot Creek and Geysers, Mono county, *Groesbeck 206,* 25 Nov. 1940 (FC); culture from north end of beach at Carmel, Monterey county, *Gardner 8013,* 1 Dec. 1936 (FC, UC); near St. Helena, Napa county, *Setchell 1091,* Nov. 1895 (UC); at the geyser, Calistoga, Napa county, *Gardner 6824, 6976,* 22 Aug. 1931 (FC, UC), and cultures, *Gardner 6968, 7012,* 18 Jan. 1932, 10 Aug. 1932 (FC, UC); culture from an urn in Mountain View cemetery, Oakland, *G. J. Hollenberg 8020a,* 23 Nov. 1937 (FC, UC); on damp soil on bluff east of Fort point, San Francisco, *Gardner 7104a,* 25 Nov. 1932 (FC, UC), and culture, *Gardner 7104,* 26 Mar. 1933 (FC, UC); in a pond by the Art building of the 1915 World's Fair, San Francisco, *Gardner 7193,* 26 Mar. 1933 (FC, UC); culture from Laguna Salada, San Mateo county (collected by H. E. Parks), *Gardner 7445,* 4 Aug. 1933 (FC, UC); in shallow water of Tule river, Porterville, Tulare county, *Drouet & Groesbeck 4443,* 4 Oct. 1941 (FC).

PUERTO RICO: on rocks, Arroyo de los Corchos, Adjuntas to Jayuya, *N. Wille 1695,* Mar. 1915 (NY); on rocks, wood, walls, etc. in and about the hot spring, Coamo Springs, *Wille 287a* (Type of *Chroococcus turgidus* var. *uniformis* Gardn., NY), *382a* (Type of *C. aeruginosus* Gardn., NY), *384a* (NY), *402a* (Type of *C. subsphaericus* Gardn., NY [Fig. 117]), *403a* (NY), *408* (FC, NY); on limestone, Hatillo to Arecibo, *Wille 1390,* 27 Feb. 1915 (Type of *Gloeocapsa acervata* Gardn., NY); ditch in a wet field, Ponce, *Wille 1670c,* Mar. 1915 (NY); on rocks about 20 km. north of Ponce, *Wille 1753,* 15 Mar. 1915 (D, NY); in Laguna Tortuguero, *Wille 831b,* Feb. 1915 (Type of *Chroococcus mediocris* Gardn., NY [Fig. 121]); on rocks about 12 km. north of Utuado, *Wille 1512* (NY), *1537c* (Type of *C. heangloios* Gardn., NY), Mar. 1915; on limestone, Utuado to Adjuntas, *Wille 1640* (Type of *Gloeocapsa ovalis* Gardn., NY [Fig. 115]), *1640c,g,* Mar. 1915 (NY). MEXICO: in pool by Rio de Sonora, Unión, Hermosillo, Sonora, *Drouet & D. Richards 3039,* 24 Nov. 1939 (FC). HONDURAS: wet rock, El Quebracho above El Zamorano, Dept. Morazán, *P. C. Standley 310,* 29 Nov. 1946 (FC); wet bank, Quebrada del Ingenio de los Angeles near Yuscarán, Dept. El Paraíso, *Standley, E. D. Merrill, L. O. Williams, & A. Molina R. 4735,* 28 Feb. 1947 (FC). NICARAGUA: on rock under trickling water, Sierra de Managua, region of Las Nubes, *Standley 8705,* 24 May 1947 (FC). PANAMA: rain water in ancient grinding stone, Panama City, *G. W. Prescott CZ19,* Aug. 1939 (FC); pit at Miraflores Locks, Canal Zone, *Prescott CZ67,* Aug. 1939 (FC). VENEZUELA: hot springs near La Toma, southeast of Cumaná, Sucre, *Steyermark 62858a,* 21 May 1945 (FC). BRAZIL: Ceará: in seepage from pond along the beach, Urubú, Fortaleza, *Drouet 1327,* 27 Jul. 1935 (D, S, UC); in outlet of Açude Choró, Quixadá, *Drouet 1399,* 1 Sept. 1935 (D). São Paulo: represa, Estação Experimental de Caça e Pesca, Pirassununga, *H. Kleerekoper 1,* 17 Apr. 1940 (FC); margem do Mogy-Guassú perto de Cachoeira, *Kleerekoper 5,* 17 Apr. 1940 (FC). PERU: in a reservoir, Malabrigo, Dept. Libertad, *A. Maldonado 87,* Feb. 1942 (FC, FH); in fresh water, Lago Villa,

Maldonado 40, Jan. 1942 (FC); agua alcalina, Las Delicias, Trujillo, *N. Ibañez H. "H"*, 18 May 1952 (FC). HAWAIIAN ISLANDS: high up in littoral, Kawela bay, Oahu, *C. J. Engard*, 23 Nov. 1940 (FC); rocks on cliff, Nuuanu Pali, Honolulu, *G. T. Shigeura*, 4 Mar. 1939 (FC). SOCIETY ISLANDS: on dripping rocks, cliffs near Huau, Tahiti, *W. A. Setchell & H. E. Parks 5123*, Jun. 1922 (UC); cliffs near mouth of Papenu river, Tahiti, *Setchell & Parks 5350*, Jun. 1922 (UC). GILBERT ISLANDS: in southern fish-pond of Nikunau, *Mrs. R. Catala 11a*, Jul. 1951 (FC). NEW ZEALAND: Wardell stream, Bay of Islands, *A. Nash*, 27 Jan. 1935 (in Tild., So. Pacific Pl., 2nd Ser., no. 488, FC). PHILIPPINES: in standing water, Cavite street, Manila, *G. T. Velasquez 2437*, 23 Jul. 1950 (FC, PUH); on dripping wall of race tracks, Santa Ana, Manila, *J. D. Soriano 565*, 11 Dec. 1949 (FC, PUH); on moist stone wall, Geliños, Sampaloc, Manila, *Velasquez 2441*, 24 Jul. 1950 (FC, PUH); on concrete pillar, University of the Philippines, Ermita, Manila, *Soriano 166*, 22 Oct. 1948 (FC, PUH); on exterior walk of the ice plant, Marikina, Rizal prov., *Soriano 897*, 27 Apr. 1952 (FC, PUH); on wall, San Nicolas, Rizal prov., *Soriano 545*, 10 Dec. 1949 (FC, PUH); in water at airstrip at Tacloban, Leyte, *M. E. Britton 80, 83*, 4 Jul. 1945 (FC); on moist sides of public Kiosko in the plaza, Cabatuan, Iloilo prov., *Soriano 1112*, 16 Nov. 1952 (FC, PUH); on cement floor, Bungao, Tawi-Tawi, Sulu archipelago, *Velasquez 3206*, 10 May 1952 (FC, PUH). INDONESIA: auf der Oberflache vom Bach A. Djermih durchflossenes Thermalgebiet östlich von Tjurup, Suban Ajer Panas, Stromgebiet des Musi, Südsumatra, *Deutsche Sunda-Expedition*, May 1929 (Type of *Synechocystis crassa* var. *major* Geitl., no. M4d in the collection of L. Geitler); Travertin-Hügel von Kuripan, Java, *Deutsche Sunda-Expedition*, 1928—29 (Type of *Chroococcus schizodermaticus* var. *incoloratus* f. *paucistratosus* Geitl., slide no. BK5c in the collection of L. Geitler). CHINA: on rocks, Omei-shan, Shan-Shan-Kow (Omei), and Thousand Gods Cliff at Nanzechwan, Szechwan, *H. Chu 774, 1329, 1329a*, Aug. 1941—45 (FC); on rock under dripping water, Hsiang-Shan, Peiping, Hopeh, *Y. C. H. Wang 186*, 1930—31 (FC, UC); in Kyam spring in Chang-chenmo valley north of Pang-gong Tso, western Tibet, *G. E. Hutchinson L58*, 20 Jul. 1932 (D). AUSTRALIA: in a dish in glasshouse, Brisbane, Queensland, *A. B. Cribb 101*, 14 Jun. 1949 (FC); in high littoral pools, Middle River, Kangaroo island, South Australia, *H. B. S. Womersley A12664a, K13488a*, 13 Jan. 1950 (FC).

7b. ANACYSTIS THERMALIS f. MAJOR (Lagerheim) Drouet & Daily, COMB. NOV. *Chroococcus helveticus* f. *major* Lagerheim ex Forti, Syll. Myxophyc., p. 17. 1907. *Gloeocapsa limnetica* f. *major* Drouet & Daily, Lloydia 11: 78. 1948. *A. limnetica* f. *major* Drouet and Daily, Butler Univ. Bot. Stud. 10: 221. 1952. —Type from Praestvandet near Tromsö, Norway (NY). FIGS. 131, 132.

Chroococcus limneticus Lemmermann, Bot. Centralbl. 76: 153. 1898. *C. minutus* var. *limneticus* Hansgirg, Beih. z. Bot. Centralbl. 8(2): 520. 1905. *Gloeocapsa limnetica* Hollerbach in Elenkin, Monogr. Algar. Cyanophyc., Pars Spec. 1: 236. 1938. *Anacystis limnetica* Drouet & Daily, Butler Univ. Bot. Stud. 10: 221. 1952. —Paratype, annotated by W. A. Setchell: "fide Lemmermann", from Lough Neagh, Ireland, designated as the Type (FC).

Chroococcus limneticus var. *subsalsus* Lemmermann, Forschungsber. Biol. Sta. Plön 8: 84. 1901. —Paratype, annotated by W. A. Setchell: "fide Lemmermann", from Lough Corrib, Ireland, designated as Type (FC).

Chroococcus minor var. *dispersus* Keissler, Verh. Zool.-bot. Ges. Wien 52: 311. 1902. *C. dispersus* Lemmermann, Ark. f. Bot. 2(2): 102. 1904. *Gloeocapsa minor* f. *dispersa* Hollerbach in Elenkin, Monogr. Algar. Cyanophyc., Pars Spec. 1: 239. 1938. —Type from Wolfgang-See, Salzburg, Austria (in the collection of C. Keissler, W).

Chroococcus Prescottii Drouet & Daily, Field Mus. Bot. Ser. 20: 127. 1942. —Type from West Falmouth, Massachusetts (D). FIG. 130.

Original specimens have not been available to us for the following names; their original descriptions are here designated as the Types until the specimens can be found:

Chroococcus minutus var. *lacustris* Chodat, Bull. Herb. Boissier 6:59. 1898.

Chroococcus minutus var. *carneus* Chodat, Bull. Herb. Boissier 6: 180. 1898. *C. limneticus* var. *carneus* Lemmermann, Ark. f. Bot. 2(2): 101. 1904.

Chroococcus purpureus Snow, Bull. U. S. Fish Comm. 1902: 388, 390. 1903. *C. limneticus* var. *purpureus* Tiffany & Ahlstrom, Ohio Journ. Sci. 31: 456. 1931.

Chroococcus limneticus var. *fuscus* Lemmermann, Ark. f. Bot. 2(2): 102. 1904.

Chroococcus limneticus var. *distans* G. M. Smith, Bull. Torr. Bot. Club 43: 481. 1916. *Gloeocapsa limnetica* f. *distans* Hollerbach in Elenkin, Monogr. Algar. Cyanophyc., Pars Spec. 1: 236. 1938.

Chroococcus limneticus var. *elegans* G. M. Smith, Trans. Wisconsin Acad. Sci. 19: 619. 1918.

Chroococcus Gomontii Nygaard, Vidensk. Medd. Dansk Naturh. Foren. Kjøbenh. 82: 202. 1926.

Plantae aerugineae, olivaceae, violaceae, vel roseae, microscopicae 8—128-cellulares, libere natantes, cellulae in divisione binis depresso—sub-sphaericis, aetate provecta globosis diametro (4—)6—12μ crassis, obscure vel crebre in seriebus eucapsoideis per gelatinum vaginale distributis; gelatino vaginale hyalino, homogeneo, amplo; protoplasmate aerugineo, olivaceo, violaceo, vel roseo, homogeneo vel tenui-granuloso. FIGS. 130—132.

In the plankton of lakes and in shallow fresh (rarely brackish) water along their margins. The cubical (eucapsoid) arrangement of the cells is usually evident in plants growing in the limnoplankton, but it is conspicuous in the more compact plants which develop in shallow water.

Specimens examined:

NORWAY: in fossa turfosa ad Praestvandet prope Tromsö, *G. Lagerheim,* May 1894 (Type of *Chroococcus helveticus* f. *major* Lagerh. in Wittr., Nordst., & Lagerh., Alg. Exs. no. 1537, NY; isotype, L [Figs. 131, 132]). SWEDEN: with *Gomphosphaeria Wichurae* (Hilse) Dr. & Daily, Frisksjön, Loftahammar, Smaland, *O. Borge,* Jun. 1914 (S); plankton of Björkesakrasjön, Malmöhus, *B. Carlin-Nilsson 521,* 17 Aug. 1937 (D). GERMANY: in alten Wassertümpeln, Freiburg im Breisgau, *A. Braun 28,* May 1850 (L). AUSTRIA: Plankton von Aber- oder Wolfgang-See, Salzburg, *C. Keissler,* 1901 (Type of *Chroococcus minor* var. *dispersus* Keissl., slide in the collection of C. Keissler, W). IRELAND: Lough Neagh, *W. West 23,* May 1900 (designated as Type of *C. limneticus* Lemm., FC; isotype, UC); plankton, Lough Corrib, *W. West,* with *Gomphosphaeria lacustris* Chod. (designated as Type of *Chroococcus limneticus* var. *subsalsus* Lemm., FC; isotype, UC). SCOTLAND: Lake Shurrery and nearby Loch Chaluim, *A. J. Brook 3,* 4, Jul., Sept. 1950 (D); Loch Kinardochy, Perthshire, *Brook 5,* 7 Oct. 1953 (D). NEW BRUNSWICK: in lake 8 miles south of Grand Falls, Victoria county, *H. Habeeb 11657,* 22 Jun. 1951 (FC, HA). MAINE: sink hole, Hollis, *F. C. Seymour,* 11 Sept. 1938 (FC). MASSACHUSETTS: Desmid Haven pond, West Falmouth, *C. M. Palmer 192,* 27 Jul. 1936 (DA, FC), Aug. 1937 (Type of *C. Prescottii* Dr. & Daily, D [Fig. 130]; isotypes, DT, FH, L, NY, S, TA, UC), *H. Croasdale,* Jul. 1935 (D); Woods Hole, 12 Jul. 1935, comm. J. H. Hoskins (DA, FC); Tarpaulin pond, Naushon island, *J. M. Furber,* Aug. 1917 (as *C. limneticus* in Coll., Hold., & Setch., Phyc. Bor.-Amer. no. 2252, FC, TA). VIRGINIA: in Hargrove's pond, New Kent county, *M. H. Wood,* 25 Jun. 1939 (FC, ST). FLORIDA: in shallow water, Lake Harris in the municipal park, Leesburg, Lake county, *F. Drouet & M. A. Brannon 11064,* 19 Jan. 1949 (FC). OHIO: quarry pool near Columbus, *C. E. Taft,* 7 Aug. 1940 (DA, FC).

MICHIGAN: Livingston's bog, Douglas lake region, Cheboygan county, *H. K. Phinney 28M39-2,* 28 Jul. 1939 (FC, PHI); Bryant's bog near Douglas lake, Cheboygan county, *H. A. Gleason Jr.,* Jul. 1935 (TA); plankton of Manistique lake at Emory's resort, Luce county, *Phinney 9M41/14,* 19 Jul. 1941 (FC, PHI). INDIANA: Lake Cicott, Cass county, *F. K. & W. A. Daily 1024,* 10 Sept. 1942 (DA, FC); Indiana lake, 5 miles north of Bristol, Elkhart county, *F. K. & W. A. Daily 2630,* 27 Aug. 1952 (DA, FC); in south end of Lake Wawasee at Johnson's Hotel, Syracuse, Kosciusko county, *F. K. & W. A. Daily 2190, 2238,* Jul. 1950 (DA, FC); in Dewart lake, Kosciusko county, *F. K. & W. A. Daily 2119,* 9 Jul. 1949 (DA, FC); plankton at Maxinkuckee boat landing, Lake Maxinkuckee, Marshall county, *F. K. & W. A. Daily 1544,* 5 Jul. 1946 (DA, FC); Bass lake, 5 miles south of Knox, Starke county, *F. K. & W. A. Daily 1560, 1566,* 6 Jul. 1946 (DA, FC); Lake George, 6 miles north of Angola, Steuben county, *F. K. & W. A. Daily 1934A,* 13 Sept. 1947 (DA, FC); Clear lake, 6 miles east of Fremont, Steuben county, *F. K. & W. A. Daily 1933A, 2452,* 13 Sept. 1947, 17 June 1951 (DA, FC); plankton of Hamilton lake at pier in Circle park, Hamilton, Steuben county, *F. K. & W. A. Daily 2443,* 17 Jun. 1951 (DA, FC); Fox lake, 1 mile southwest of Angola, Steuben county, *F. K. & W. A. Daily 2487,* 19 Jun. 1951 (DA, FC). WISCONSIN: in Pell lake near Lake Geneva, Walworth county, *W. E. Lake,* 29-30 Aug. 1942 (FC). WYOMING: inlet, Half Moon lake, Fremont county, *Bond F264,* 21 Aug. 1934 (FC); Fremont lake, Fremont county, *Bond F259, F260,* Aug. 1934 (FC). PUERTO RICO: in a pool, Mayaguez, *N. Wille 1318a,* Feb. 1915 (NY). BRAZIL: from an open cistern, Campina Grande, Paraíba, *S. Wright 1593,* 28 Aug. 1934 (D). HAWAIIAN ISLANDS: in pools in the swamp of an old crater in the mountains, Kauai, *M. Reed 11,* Aug. 1909 (FC, UC).

GENUS 3. JOHANNESBAPTISTIA

J. de Toni, Noter. Nomencl. Algol. 1: 6. 1934. *Cyanothrix* Gardner, Mem. New York Bot. Gard. 7: 30. 1927; non Schmidle, 1897. —Type species: *Cyanothrix primaria* Gardn.

84

Plantae microscopicae cylindricae elongatae rectae vel curvatae, cellulis discoideis (ad apices depresso-sphaericis) uniseriatim in gelatino vaginale dispositis; divisione cellulae in uno plano ad axem fili perpendiculare procedente; gelatino vaginale hyalino.

In this genus each discoid cell divides only in a plane along the diameter of the disc (perpendicular to the axis of the plant); such cell divisions result in a uniseriate arrangement of cells within a long cylindrical gelatinous matrix.

One species:

1. JOHANNESBAPTISTIA PELLUCIDA W. R. Taylor & Drouet in Drouet, Bull. Torr. Bot. Club 65: 285. 1938. *Hormospora pellucida* Dickie, Journ. Linn. Soc. Bot. 14: 365. 1874. —Type from Fernando de Noronha, Brazil (BM). FIG. 182.

   Geminella scalariformis f. *marina* G. S. West in Collins & Hervey, Proc. Amer. Acad. Arts & Sci. 53: 31. 1917. —Isotype from Bermuda (FC).
   Cyanothrix primaria Gardner, Mem. New York Bot. Gard. 7: 30. 1927. *Johannesbaptistia primaria* J. de Toni, Noter. Nomencl. Algol. 1: 6. 1934. *J. Gardneri* Frémy, Bull. Soc. Hist. Nat. Afr. du Nord 26: 95, 99. 1935. —Type from Laguna Tortuguero, Puerto Rico (NY). FIG. 184.
   Cyanothrix Willei Gardner, Mem. New York Bot. Gard. 7: 31. 1927. *Johannesbaptistia Willei* J. de Toni, Noter. Nomencl. Algol. 1: 6. 1934. —Type from Laguna Tortuguero, Puerto Rico (NY). FIG. 184.
   Nodularia? fusca W. R. Taylor, Carnegie Inst. Wash. Papers Tortugas Lab. 25: 48. 1928. —Type from Dry Tortugas, Florida (TA). FIG. 183.

   Original specimens have not been available to us for the following names; their original descriptions are here designated as the Types until the specimens can be found:

   Hormospora scalariformis G. S. West, Journ. of Bot. 42: 282. 1904. *Geminella scalariformis* G. S. West in Collins, Tufts Coll. Stud. 4(7): 55. 1918.
   Cyanothrix Cavanillesii Gonzalez Guerrero, Anal. Jard. Bot. Madrid 6: 250. 1946.

Fila aeruginea vel olivacea, recta vel curvata, ad 1 mm. longa, cellulis discoideis vel sphaerico-discoideis, ad apices filorum rotundis, diametro 3—20μ crassis, in gelatino vaginale uniseriatim distributis; gelatino vaginale hyalino, homogeneo obscure lamelloso, firmo vel (saepe omnino) diffluente; protoplasmate aeruginео, olivaceo, vel luteolo, homogeneo. FIGS. 182—184.

In shallow brackish, salt, or fresh water; often found among other algae in old marsh pools. Seurat & Frémy in Bull. Soc. d'Hist. Nat. de l'Afrique du Nord 28: 294 (1937) and Frémy & J. de Toni in Atti R. Inst. Veneto Sci., Lett. & Arti 99(2): 401—406 (1940) have contended that plants of *Johannesbaptistia pellucida* are growth-forms of *Lyngbya semiplena* (Ag.) J. Ag. We have been unable to observe any transitional stages between filaments of this species and those of any of the hormogonial Myxophyceae. The absence of *Lyngbya semiplena* from the several inland habitats occupied by *Johannesbaptistia pellucida* seems to give further support to our position that the latter is an autonomous species.

Specimens examined:

   JUGOSLAVIA: Salona prope Spalato, Dalmatia, *A. Hansgirg*, Aug. 1889 (FC, W). BERMUDA: spray pool, south shore of Nonsuch island, *A. J. Bernatowicz 49-307*, 15 Mar. 1949 (FC, MICH); tide pool at entrance to Hungry bay, *F. S. Collins*, 23 Apr. 1912 (isotype of *Geminella scalariformis* f. *marina* G. S. West in Coll., Hold., & Setch., Phyc. Bor.-Amer. 2002, FC). MASSACHUSETTS: in a brackish swamp (Chara pond) near Oyster pond, Falmouth, *F. Drouet, R. Patrick & C. Hodge 3569*, 16 Jul. 1940 (FC). CONNECTICUT: Fresh pond, Stratford, *I. Holden 1035*, Sept. 1894 (D, FH). MARYLAND: in a marsh pool west of Ewell, Smiths island, *P. W. Wolle & Drouet 2307*, 26 Aug. 1938 (FC, D, FH); in the salt marsh south of Jenkins creek, Crisfield, *Drouet & Wolle 3630, 3641*, 23 Jul. 1940 (FC); marsh pools between Dames Quarter and Chance, Somerset county, *P. W. Wolle*, 30 Sept. 1951 (FC); in pools of brackish water, Jericho marshes west of Fairmount, Somerset county, *Drouet & Wolle 3649*, 23 Jul. 1940 (FC). FLORIDA: Long Key, Dry Tortugas, *W. R. Taylor*, June 1925 (Type of *Nodularia? fusca* W. R. Taylor, TA [Fig. 183]; isotypes, D, N); on mud of sinkhole and marsh near the Inn, Big Pine key, Monroe county, *E. P. Killip 41729, 41943*, 11 Jan., 18 Feb. 1952 (FC, US); in brackish pool, Marco island, Collier county, *P. C. Standley 73407*, 19 Mar. 1940 (FC). INDIANA: bottom of Lake Wehi, south of Germantown, Wayne county, *F. K. & W. A. Daily 1185*, 16 Jul. 1944 (DA, FC). IOWA: pond near shore of Silver lake, *J. D. Dodd*, 2 Jul. 1953 (D, FC). TEXAS: on sand at Blind bayou, Galveston island, *H. K. Phinney 2T41/2*, 30 Aug. 1941 (FC, PHI); permanent pool, Austin, *F. A. Barkley 44alg27*, 25 Jul. 1944 (FC, TEX); southwest tip of lake west of Padre island, 65 miles north of Brownsville, in Kenedy county, *J. O. Barry*, 27 Sept. 1948 (FC); in a pool of fresh water (seepage from a tank) on a roof, Brownsville, *R. Runyon 4207a*, 19 Aug. 1946 (FC). CALIFORNIA: in brackish water west of Newport Beach, Orange county, *G. J. Hollenberg 1533e*, 26 Sept. 1934 (D).

BAHAMA ISLANDS: Lake Cunningham, New Providence, *A. D. Peggs*, 10 May 1941 (FC). BRITISH WEST INDIES: in basins of salt works, Cockburn Harbor, South Caicos, *M. A. Howe 5584b*, Dec. 1907 (NY). VIRGIN ISLANDS: lagoon, Lindbergh Field, Charlotte Amalie, St. Thomas, *W. A. Hoffman 146/593* (D, TA). PUERTO RICO: Laguna Tortuguero, *N. Wille 830b* (Type of *Cyanothrix primaria* Gardn., NY [Fig. 184]), *830e* (Type of *C. Willei* Gardn., NY [Fig. 184]), 5 Feb. 1915. MEXICO: in brackish water of a sand pit on beach 4 km. east of Guaymas, Sonora, *Drouet & D. Richards 3297*, 16 Dec. 1939 (FC). BRAZIL: in rock pools, Fernando de Noronha, *H. N. Moseley* on Challenger Expedition (Type of *Hormospora pellucida* Dickie, BM [Fig. 182]). ECUADOR: inland salt pool near Albemarle point, north end of Albemarle island, Galapagos islands, *W. R. Taylor 123*, 12 Jan. 1934 (D, TA). INDONESIA: binnenmeer van Roesa Linguette, *van der Sande*, 28 Mar., 21 Jul. 1908 (FC, L).

Genus 4. Agmenellum

de Brébisson, Mém. Soc. Acad. Sci., Art. & Belles-Lettres Falaise 1839: p. 2 of reprint. 1839. —Type species: *Trochiscia quadruplicata* Menegh.

Original specimens of the Type species for the following generic and subgeneric names have not been available to us for study:

Gonidium Ehrenberg ex Meneghini, Linnaea 14: 213. 1849. —Type species: *Gonium glaucum* Ehrenb.

Merismopoedia Kützing, Phyc. Gener., p. 163. 1843. *Merismopoedia* Subgenus *Eumerismopedia* Ryppowa, Acta Soc. Bot. Polon. 3(1): 45. 1925. —Type species: *M. punctata* Meyen.

Holopedium Subgenus *Euholopedium* Forti, Syll. Myxophyc., p. 110. 1907. —Type species: *Merismopoedia irregularis* Lagerh.

Coccopedia Troitzkaja, Not. Syst. Inst. Crypt. Hort. Bot. Petropol. 1: 131. 1922. *Holopedium* Subgenus *Coccopedia* Elenkin, Not. Syst. Inst. Crypt. Hort. Bot. Petropol. 2: 66. 1923. —Type species: *Coccopedia limnetica* Troitzk.

Pseudoholopedia Elenkin, Monogr. Algar. Cyanophyc., Pars. Spec. 1: 86. 1938. *Merismopoedia* Subgen. *Pseudoholopedia* Ryppowa, Acta Soc. Bot. Polon. 3(1):45. 1925. —Type species: *Merismopoedia gigas* Ryppowa.

Plantae (laminae) microscopicae vel macroscopicae, cellulis sphaericis vel ovoideis vel cylindraceis, in lamina in seriebus regularibus aliis ad alias perpendicularibus dispositis; divisione cellularum seriatim in doubus planis unoquodque ad alia et ad superficiem laminae perpendicularibus procedente; gelatino vaginali hyalino.

In this genus, the plant is a flat or foliose plate in which the cells are arranged in series of rows in two directions at right angles to each other. Cell division proceeds successively in two planes perpendicular to each other and to the surface of the plate.

Key to species of Agmenellum:

Cells 1—3.5μ in diameter, plants 1—256-celled .... 1. A. QUADRUPLICATUM
Cells 4—10μ in diameter, plants larger and often foliose .... 2. A. THERMALE

1. AGMENELLUM QUADRUPLICATUM Brébisson, Mém. Soc. Acad. Sci., Art. & Belles-Lettres de Falaise 1839: 2 (of reprint). 1839. *Trochiscia quadruplicata*

86

Meneghini, Consp. Algol. Eugan., p. 334. 1837. *Gonidium quadruplicatum* Meneghini, Linnaea 14: 213. 1840. *Merismopoedia quadruplicata* Trevisan, Nomencl. Algar. 1: 28. 1845. —Paratype from Falaise, France, designated as Type (L).

Trochiscia Prasiola Meneghini, Consp. Algol. Eugan., p. 334. 1837. *Agmenellum Prasiola* Meneghini ex Brébisson, Mém. Soc. Acad. Sci., Art. & Belles-Lettres de Falaise 1839: 2 (of reprint). 1839. *Gonidium Prasiola* Meneghini, Linnaea 14: 214: 1840. *Merismopoedia Prasiola* Trevisan, Nomencl. Algar. 1: 28. 1845. —Isotype from Euganean springs, Italy (W).

Merismopoedia hyalina subsp. *Warmingiana* Lagerheim, Öfvers. K. Sv. Vet.-Akad. Förh. 40(2): 41. 1883. *M. Warmingiana* Lagerheim ex Forti, Syll. Myxophyc., p. 109. 1907. —Type from Kristineberg, Sweden (S).

Merismopoedia punctata f. *minor* Lagerheim, Öfvers. K. Sv. Vet.-Akad. Förh. 40(2): 41. 1883. —Type from Kristineberg, Sweden (S).

Merismopoedia glauca var. *fontinalis* Hansgirg, Sitzungsber. k. Böhm. Ges. Wiss., Math.-Nat. Cl., 1890(2): 98. 1891. —Type from Sct. Procop near Prague, Czechoslovakia (W). FIG. 133.

Merismopoedia affixa Richter, Stizungsber. Naturf. Ges. Leipzig 19—21: 152. 1895; Hedwigia 34: 25. 1895; in Hauck & Richt., Phyk. Univ. 13: 648. 1895. —Type from Kiel, Germany (L).

Merismopoedia minima Beck, Ann. K. K. Nat.-Hist. Hofmus. Wien 12: 83. 1897. —Type from Bombay, India (W).

Merismopoedia elegans var. *remota* G. S. West, Journ. Linn. Soc. Bot. 38: 185. 1907. —Type from Baraka, Tanganyika (FC).

Merismopoedia convoluta f. *minor* Wille in S. Hedin, Southern Tibet 6(3, Bot.): 166. 1922. *M. convoluta* var. *minor* Tiffany & Ahlstrom, Ohio Journ. Sci. 31: 457. 1931. —Type from Pamir (O).

Merismopoedia hyalina f. *salina* Wille in S. Hedin, Southern Tibet 6(3, Bot.): 183. 1922. —Material named by the author from Atjik-bulaks, Tibet, designated as the Type (O).

Holopedium pulchellum Buell, Bull. Torr. Bot. Club 65: 385. 1938. *Microcrocis pulchella* Geitler in Engler & Prantl, Natürl. Pflanzenfam., ed. 2, 1b: 56. 1942. —The original specimens are reported by Dr. Helen Foot Buell and Dr. Josephine E. Tilden to have been lost. A topotype from the original locality in Minneapolis, Minnesota, has been designated as the Type (FC).

Original specimens have not been available to us for the following names; their original descriptions are here designated as the Types until the specimens can be found:

Gonium tranquillum Ehrenberg, Abh. K. Akad. Wiss. Berlin 1833: 251. 1834. *Gonidium tranquillum* Ehrenberg ex Meneghini, Linnaea 14: 213. 1840. *Agmenellum tranquillum* Trevisan, Prosp. Fl. Eugan., p. 58. 1842. *Merismopoedia tranquilla* Trevisan, Nomencl. Algar. 1: 28. 1845.

Gonium glaucum Ehrenberg, Infusionsth., p. 58. 1838. *Merismopoedia glauca* Kützing, Phyc. Germ., p. 142. 1845.

Merismopoedia punctata Meyen, Neues Syst. Pfl.-Physiol. 3: 440. 1839. *M. Kuetzingii* Nägeli, Gatt. Einzell. Alg., p. 55. 1849. *Sarcina punctata* Meyen ex Bizzozero, Fl. Veneta Crittog. 1: 22. 1885. *Merismopoedia glauca* var. *punctata* Hansgirg, Prodr. Algenfl. Böhmen 2: 142. 1892.

Merismopoedia nova Wood, Proc. Amer. Philos. Soc. 11: 123. 1869.

Merismopoedia paludosa Bennett, Journ. Roy. Microsc. Soc. London, ser. 2, 6: 4. 1886.

Merismopoedia tenuissima Lemmermann, Bot. Centralbl. 76: 154. 1898.

Merismopoedia glauca var. *fontinalis* f. *irregularis* Selk, Jahrb. Hamburg. Wiss. Anstalt 25 (Beih. 3): 49. 1907.

Merismopoedia punctata var. *oblonga* Playfair, Proc. Linn. Soc. New South Wales 19: 135. 1914.

Merismopoedia cyanea Playfair, Proc. Linn. Soc. New South Wales 43: 501. 1918.

Merismopoedia duplex Playfair, Proc. Linn. Soc. New South Wales 43: 501. 1918.

Merismopoedia glauca f. *minor* Häyrén, Bidrag t. Kännedom af Finlands Nat. & Folk 80(3): 92. 1921.

Coccopedia limnetica Troitzkaja, Not. Syst. Inst. Crypt. Horti Bot. Petropol. 1: 131. 1922.

Merismopoedia glauca f. *rosea* Geitler in Pascher, Süsswasserfl. 12: 106. 1925.

Merismopoedia insignis Schkorbatow, Not. Syst. Inst. Crypt. Horti Bot. Petropol. 2: 89. 1923; Arch. Russes de Protistol. 6: 127. 1927. *M. glauca* f. *insignis* Geitler in Pascher, Süsswasserfl. 12: 106. 1925.

Merismopoedia punctata f. *arctica* Kossinskaja, Acta Inst. Bot. Acad. Sci. U. R. P. S. S., ser. 2, Pl. Crypt. 1: 47. 1933.

Coccopedia turkestanica Kisseliova ex Elenkin, Monogr. Algar. Cyanophyc., Pars Spec. 1: 68. 1938. —We have been unable to find the original reference reported by Elenkin: Trav. de l'Inst. d'Ousbekistan de la Medicine Tropicale 1(3): 375. 1931.

Merismopoedia Haumanii Kufferath, Bull. Soc. Roy. Bot. Belg. 74: 95. 1942.

Holopedium Mutisianum Gonzalez Guerrero, Anal. Jard. Bot. Madrid 6: 243. 1946.

Holopedium Clementei Gonzalez Guerrero, Anal. Jard. Bot. Madrid 6: 243. 1946.

Plantae (laminae) aerugineae, olivaceae, violaceae, vel roseae, rectangulares, 1—256-cellulares, cellulis post divisionem globosis diametro 1—3.5μ crassis, per gelatinum vaginale laxe vel conferte distributis; gelatino vaginale firmo vel (saepe omnino) diffluente; protoplasmate aerugineo, olivaceo, violaceo, vel roseo, homogeneo vel tenui-granuloso. FIG. 133.

In shallow fresh and brackish water and on mud and wet sand; often seen in the plankton of rivers and lakes. Among the specimens enumerated, there seems to be no good reason to separate a small-celled (ca. 1μ in diameter) form prevalent especially in the plankton of rivers. It is a common phenomenon to find the sheath material hydrolyzed and the cells completely dissociated. Here the cells are apt to be confused with those of *Anacystis montana*. An occasional pseudovacuole may be observed in the cells of *Agmenellum quadruplicatum*, but all collections seen which are predominantly pseudovacuolate prove to be of the bacterium, *Erythroconis littoralis* Oerst.

Specimens examined:

SWEDEN: in arena regionis littoralis ad Kristineberg Bahusiae, *G. Lagerheim*, Aug. 1882 (Types of *Merismopoedia punctata* f. *minor* Lagerh. and *M. hyalina* subsp. *Warmingianum* Lagerh. with *Holopedium sabulicola* in Wittr., Nordst., & Lagerh., Alg. Exs. no. 1549, S; isotypes, L, MIN); Möja, Uppland, *O. Borge*, Aug. 1933 (S); Vaddö, Uppland, *O. F. Andersson*, Aug. 1890 (S); Lassby backar, Uppsala, *Borge*, Sept. 1884 (S). GERMANY: Kieler Föhrde, *T. Reinbold*, Aug. 1889 (NY); Kiel, im Salzwasser, *Reinbold* (Type of *Merismopoedia affixa* Richt. in Hauck & Richt., Phyk. Univ. no. 648, L; isotype, MIN); in einem Gartenbassin in Niederlössnitz bei Kotzschenbroda, Dresden, *R. Wollny*, Nov. 1886 (as *M. irregularis* in Hauck & Richt., Phyk. Univ. no. 146, MIN, MICH); Leipzig, *P. Richter* (L), um Grimma, *Richter* (UC); Lössnitzgrund bei Dresden, *C. A. Hantzsch* (NY); Oberlössnitz bei Dresden in einem Bache, *Hantzsch* (as *M. violacea* in Rabenh. Alg. no. 857, FC, NY, UC); aus einem Moortümpel bei Bernbruch um Lausigk, *Richter*, Aug. 1867 (as *M. glauca* in Rabenh. Alg. 2056, UC); Schönfeld bei Dresden, *Richter*, Aug. 1876 (FC); in Torftümpeln bei Wurzen, *Bulnheim*, 1857 (as *M. violacea* in Rabenh. Alg. no. 650, FC, L, NY); am Strande auf Wangerooge, Oldenburg, *F. T. Kützing*, Jul. 1839 (L). CZECHOSLOVAKIA: am schlammige Grunde in Felsenquellen bei Sct. Prokop nächst Prag, *A. Hansgirg*, May 1890 (Type of *M. glauca* var. *fontinalis* Hansg., W; isotype, FC [Fig. 133]). SWITZERLAND: Bieler See, Kt. Bern, *Kützing* (L). ITALY: ne'vasi delle piante acquatiche all'Orto Botanico di Pisa, *Savi*, 1867 (as *Microcystis olivacea* in Erb. Critt. Ital., ser. II, no. 30, L); aq. therm. Euganaei, *Meneghini* (isotype of *Trochiscia Prasiola* Menegh., W). NETHERLANDS: groene Strand, West Terschelling, *J. T. Koster 177, 178*, Jul. 1937 (L). FRANCE: *Agmenellum quadruplicatum* Bréb., Falaise, *A. de Brébisson* (designated here as Type of *Trochiscia quadruplicata* Menegh., L); ruisseau, Falaise, *Brébisson 414* (L); Falaise, sur la vase d'un étang desséché, *Brébisson* (as *Protococcus minutus* var. *caerulescens* in Rabenh. Alg. no. 2035, FC, L, TA); flottant dans les fossés et les flaques des marais spongieux, Falaise, *Brébisson* (as *Merismopoedia aeruginea* in Rabenh. Alg. no. 1356, MIN, L). TANGANYIKA: with *Anabaena tanganyikae* G. S. West in plankton near Baraka, *W. A. Cunnington 240*, 24 Feb. 1905 (Type of *Merismopoedia elegans* var. *remota* G. S. West, FC; isotype, UC). UNION OF SOUTH AFRICA: Zeekoe Vlei near Cape Town, *Mrs. G. E. Hutchinson 35, 5* Feb. 1928 (D). MAINE: Long pond, Mt. Desert island, *F. S. Collins*, July 1900 (NY). MASSACHUSETTS: pond on Naushon island, *W. R. Taylor*, Jul. 1921 (TA), *H. T. Croasdale*, 17 Jun. 1934 (D). RHODE ISLAND: Portsmouth, *Collins*, 14 Jun. 1883 (FC). FLORIDA: Robinson lake, Leesburg, *M. A. Brannon 244*, 28 Jul. 1944 (FC, PC); 7 miles south of La Belle, Lee county, *J. E. Davis 20*, 8 Oct. 1941 (FC, PC). OHIO: pond in Lakeview fish hatchery, Cincinnati, *W. A. Daily 434, 8* Jul. 1940 (DA, FC); culture in University of Cincinnati greenhouse, Cincinnati, *L. Lillick & I. Lee 603*, 21 Feb. 1934 (DA, FC); pond at fish hatchery, Newton, Hamilton county, *F. Brinley, 29* Oct. 1941 (DA, FC); Big Miami river, North Bend, Hamilton county, *R. H. Fink 600*, Aug. 1932 (DA, FC); Beaver creek near Celina, Mercer county, *J. B. Lackey*, 22 Aug. 1940 (DA, FC); above falls, Bear creek east of Foster, Warren county, *J. H. Hoskins & Daily 230*, 4 Nov. 1939 (DA, FC). KENTUCKY: pond on U. S. highway no. 31 south of Glasgow, Barren county, *B. B. McInteer 612*, 19 Aug. 1929 (DA, FC); new reservoir at Earlington, Hopkins county, *McInteer 592*, 16 Aug. 1929 (DA, FC). TENNESSEE: in spring near Cherokee dam, Jefferson county, *H. Silva 908, 908a*, 25 Jun. 1949 (FC, TENN). MICHIGAN: Bryant's bog and Lancaster lake, near Douglas lake, Cheboygan county, *C. E. Taft 42, 80*, Jul. 1936 (DA, FC). INDIANA: in lake one mile north of Columbus, Bartholomew county, *F. K. & W. A. Daily 1041*, 11 Oct. 1942 (DA, FC), 2 miles north of Columbus, *W. A. Daily, F. K. Daily, & C. Kenoyer 1162*, 19 Sept. 1943 (DA, FC); Lake Papakeechee, Kosciusko county, *B. H. Smith 146*, 15 Sept. 1929 (DA, FC); Aubendaubee bay,

Lake Maxinkuckee, Marshall county, *F. K. & W. A. Daily 1485*, 30 Jun. 1946 (DA, FC); in a quarry pond at St. Paul, Shelby county, *F. K. & W. A. Daily 1154*, 19 Sept. 1946 (DA, FC); Lake James, lower basin, Pokagon state park, Steuben county, *F. K. & W. A. Daily 1918A*, 12 Sept. 1947 (DA, FC); stream east of Snow Lake, Steuben county, *C. M. Palmer B34*, Sept. 1933 (DA, FC); Fox Lake, Steuben county, *Palmer 1019*, Jul. 1933 (DA, FC); pond one mile north of Liggett creek, Vigo county, *B. H. Smith 430*, 4 Aug. 1930 (DA, FC); pond, Rose Polytechnic Institute, Terre Haute, *Smith 120*, 3 Jul. 1929 (DA, FC); Wabash river, Vigo county, *Smith 406*, 15 Aug. 1930 (DA, FC); in muck, north end of Lake Wehi, south of Germantown, Wayne county, *F. K. & W. A. Daily 1457*, 15 Jun. 1946 (DA, FC); culture from spray pond, Crystal Ice Company, Richmond, *L. J. King 34*, 22 Aug. 1940 (EAR, FC). WISCONSIN: in Pell lake near Lake Geneva, Walworth county, *W. E. Lake*, Aug. 1942 (FC). MINNESOTA: in Lake of the Isles lagoon, Minneapolis, *M. F. Buell, R. L. Cain, & M. F. Moore C*, 26 Jul. 1938 (Type of *Holopedium pulchellum* Buell, FC); Mud Lake Wildlife Refuge, Marshall county, *Drouet 12092*, 28 June 1954 (D, MIN). IOWA: bottom ooze, Des Moines river, Boone county, *W. Starrett 142B*, 6 Sept. 1946 (D, FC).

ARKANSAS: in a pond in the fish hatchery, Mammoth Spring, Fulton county, *F. Drouet 171*, 19 Aug. 1928 (D). NEBRASKA: old gravel pit pond south of North Platte, *W. Kiener 17955*, 4 Nov. 1944 (FC, KI); plankton in pool, Lincoln, *Kiener 16593*, 15 May 1944 (FC, KI); in farm tank, Fremont, Dodge county, *Kiener 14668*, 23 Jul. 1943 (FC, KI). TEXAS: creek below spring near swimming pool, Lampasas, *E. Whitehouse 24900*, 18 Feb. 1951 (FC); permanent pool, Austin, *F. A. Barkley 44alg20*, 25 Jul. 1944 (FC, TEX). MONTANA: Logan pass, Glacier national park, 23 Aug. 1934 (FC). CALIFORNIA: Berkeley, *N. L. Gardner 1361* (UC), *1698* (FC, UC), Apr. 1904, May 1906; on sandy ground, Lake Merced, San Francisco, *Gardner 1107*, Jun. 1903 (S; as *Merismopoedia glauca* var. *fontinalis* Hansg. in Coll., Hold., & Setch., Phyc. Bor.-Amer. no. 1156, FC, L, TA). BAHAMA ISLANDS: freshwater lake off Stafford creek, Andros, *A. D. Peggs*, 22 Aug. 1942 (FC). PUERTO RICO: in a pool 4 km. north of Mayaguez, *N. Wille 1325, 1326a, 1327a, 1329d*, Feb. 1915 (NY). COSTA RICA: Laguna del Misterio, al norte de San José, *J. M. Orozco C. 53*, 1953 (FC). VENEZUELA: moist bases of sandstone bluffs, Ptari-tepui, Bolívar, *J. A. Steyermark 59871*, 6 Nov. 1944 (FC). BRAZIL: Ceará: seepage from pond along beach, Urubú, Fortaleza, *Drouet 1327*, 27 Jul. 1935 (D); pond at Aroeiras, Quixadá, *Drouet 1401, 1402*, 2 Sept. 1935 (D); pool below the dam, Açude Cedro, Quixadá, *Drouet 1393*, 31 Aug. 1935 (D); pool in Rio Pacoty at Fortaleza-Recife road, Pacatuba, *Drouet 1503*, 24 Nov. 1935 (D). PERU: Cocha Zapote, Rio Pacaya, Bajo Ucayali, dept. Loreto, *F. Anceita*, Jul. 1947 (FC); Laguna Saraja, dept. Ica, *A. Maldonado 93a*, Feb. 1942 (FC). FORMOSA: fish pond, Tainan, *C. C. Chou*, 11 May 1951 (D). TIBET: Atjik-bulaks, *S. Hedin*, 1 Aug. 1899 (designated as Type of *M. hyalina* f. *salina* Wille in the slide collection of N. Wille, O). PAMIR: *S. Hedin 13* (Type of *M. convoluta* f. *minor* Wille in the slide collection of N. Wille, O). INDIA: in Victoria Garten, Bombay, *A. Hansgirg* (Type of *M. minima* Beck with *Polycystis insignis* Beck in Mus. Vindob. Krypt. Exs. no. 227, W; isotypes, DT, FC, L, S).

2. AGMENELLUM THERMALE (Kütz.) Dr. & Daily, COMB. NOV.

Merismopoedia thermalis Kützing, Phyc. Gener., p. 163. 1843. *Aphanocapsa thermalis* Brügger, Jahresber. Naturf. Ges. Graubündens, N. F., 8: 244. 1863. *Sarcina thermalis* Kützing ex Bizzozero, Fl. Veneta Crittog. 1: 23. 1885. —Type from Battaglia, Italy (L). FIG. 140.

Merismopoedia elegans A. Braun in Kützing, Sp. Algar., p. 472. 1849. —Since the original material from Freiburg-in-Breisgau (L) contains no plants as described, we have selected as the Type a specimen (L) from Leipzig, Germany, collected and annotated as of this species by the author. FIG. 134.

Merismopoedia aeruginea Brébisson in Kützing, Sp. Algar. p. 472. 1849. *M. convoluta* var. *aeruginea* Rabenhorst, Alg. Sachs. 71—72: 719. 1858. *M. convoluta* b. *aeruginea* Rabenhorst, Fl. Eur. Algar. 2: 58. 1865. —Type from Falaise, France (L). FIG. 136.

Merismopoedia major Kützing, Sp. Algar., p. 472. 1849. *M. thermalis* b. *major* Rabenhorst, Fl. Eur. Algar. 2: 58. 1865. —Type from St. Hilaire, France (L). FIG. 135.

Merismopoedia convoluta Brébisson in Kützing, Sp. Algar., p. 472. 1849. *Pseudoholopedia convoluta* Elenkin, Monogr. Algar. Cyanophyc., Pars Spec., 1: 86. 1938. —Type from Falaise, France (L). FIG. 139.

Merismopoedia elegans var. *marina* Lagerheim, Öfvers. K. Sv. Vet.-Akad. Förh. 40(2): 40. 1883. —Type from Kristineberg, Sweden (S).

Merismopoedia glauca subsp. *amethystina* Lagerheim, Öfvers. K. Sv. Vet.-Akad. Förh. 40(2): 41. 1883. *M. glauca* var. *amethystina* Lagerheim ex Forti, Syll. Myxophyc., p. 107. 1907. —Type from Kristineberg, Sweden (S). FIG 138.

Merismopoedia sabulicola Lagerheim, Öfvers. K. Sv. Vet.-Akad. Förh. 40(2): 43. 1883. *Holopedium sabulicola* Lagerheim, Nuova Notarisia 4: 210. 1893. *Microcrocis sabulicola* Geitler

89

in Engler & Prantl, Natürl. Pflanzenfam., ed. 2, lb: 56. 1942. —Type from Kristineberg, Sweden (S).

Prasiola Gardneri Collins, Phyc. Bor.-Amer. 24: 1185. 1904. *Merismopoedia Gardneri* Setchell in Gardner, Univ. Calif. Publ. Bot. 2: 238, 239. 1906. —Type from Alameda, California (NY). FIG. 141.

Merismopoedia aeruginea f. *ovata* Wille, Nyt Mag. f. Naturvidenskab. 61: 76. 1924. —Type from Husøy, Norway (O).

Merismopoedia Willei Gardner, Mem. New York Bot. Gard. 7: 3. 1927. —Type from Mayaguez, Puerto Rico (NY). FIG. 137.

Original specimens have not been available to us for the following names; their original descriptions are here designated as the Types until the specimens can be found:

Merismopoedia mediterranea Nägeli, Gatt. Einzell Alg., p. 56. 1849. *M. glauca* f. *mediterranea* Collins in Collins, Holden & Setchell, Phyc. Bor.-Amer. 31: 1651. 1910. —Nägeli's sketches and notes, from his notebooks in the Pflanzenphysiologisches Institut, Eidgenossische Technische Hochschule, Zürich, are reproduced here: FIG. 142.

Merismopoedia irregularis Lagerheim, Öfvers. K. Sv. Vet.-Akad. Förh. 40(2): 43. 1883. *Holopedium irregulare* Lagerheim, Nuova Notarisia 4: 209. 1893. *Microcrocis irregularis* Geitler in Engler & Prantl, Natürl. Pflanzenfam., ed. 2, lb: 56. 1942.

Merismopoedia revolutiva Askenasy, Flora 78: 1. 1894; Askenasy in Bailey, Bot. Bull. Queensland Dept. Agric. 11: 49. 1895.

Merismopoedia elegans var. *ulvacea* Bernard, Protococc. & Desmid. d'Eau Douce, p. 51. 1908.

Merismopoedia Terekii Henckel, Scripta Bot. Horti Univ. Imp. Petropol. 26: 55. 1908; ibid. 27: 120. 1909. *M. solitaria* Henckel, ibid. 27: 227. 1909.

Merismopoedia boreocaspica Henckel, Scripta Bot. Horti Univ. Imp. Petropol. 26: 55. 1908; ibid. 27: 120. 1909.

Merismopoedia caspica Henckel, Scripta Bot. Horti Univ. Imp. Petropol. 26: 55. 1908; ibid. 27: 120. 1909.

Merismopoedia elegans var. *constricta* Playfair, Proc. Linn. Soc. New South Wales 43: 501. 1918.

Merismopoedia elegans var. *major* G. M. Smith, Bull. Wisconsin Geol. & Nat. Hist. Surv. 57: 32. 1920. *M. major* Geitler in Pascher, Süsswasserfl. 12: 107. 1925. *M Smithii* J. de Toni, Archivio Bot. 15: 290. 1939.

Merismopoedia gigas Ryppowa, Acta Soc. Bot. Polon. 3: 46. 1925. *Holopedium dednense* Pevalek, Nuova Notarisia 1925: 293. 1925.

Merismopoedia elongata Kossinskaja, Bull. Jard. Bot. Princip. U. R. S. S. 29: 111, 128. 1930.

Merismopoedia Jeaneli Bachmann, Zeitschr. f. Hydrol. 8: 135. 1938.

Eucapsis salina Gonzalez Guerrero, Anal. Jard. Bot. Madrid 8: 267. 1948.

Pseudoholopedia convoluta var. *subsalina* Proschkina-Lavrenko, Akad. Nauk S. S. S. R. Otd. Sporovykh. Rost. Bot. Mater. 6: 69. 1950.

Plantae (laminae) aerugineae, olivaceae, vel violaceae, primum rectangulares et microscopicae, deinde ad 4 cm. longae ambitu irregulares increscentes, cellulis post divisionem globosis, ovoideis, vel cylindraceis diametro $4—10\mu$ crassis, $4—20\mu$ longis, per gelatinum vaginale plerumque conferte distributis; gelatino vaginale hyalino, firmo vel (saepe omnino) diffluente; protoplasmate aerugineo, homogeneo vel tenui-granuloso. FIGS. 134—142.

In shallow fresh, brackish, or salt water or on mud. After considerable restudy of the material at hand, we have come to the conclusion that, if there are two species of Agmenellum, the separation must be made on the basis of size of cells and size and configuration of the plate. In *A. thermale* there appears to be a gradual elongation of the cells in a direction at right angles to the surface of the plate as the plant ages and enlarges. Well developed plants with the cells somewhat disorganized can be mistaken for those of *Microcrocis geminata*. Where the gelatinous matrix has hydrolyzed completely, the single cells can be confused with those of *Anacystis* and *Coccochloris* spp.

Specimens examined:

NORWAY: [i ferskvand paa strandklipperne paa] Husøy, *N. Wille*, 23 Apr. 1919 (Type of

Merismopoedia aeruginea f. *ovata* Wille in slide collection of N. Wille, O). SWEDEN: in arena regionis littoralis ad Kristineberg Bahusiae, *G. Lagerheim,* Aug. 1882 (Types of *M. sabulicola* Lagerh., *M. glauca* subsp. *amethystina* Lagerh. [Fig. 138], and *M. elegans* var. *marina* Lagerh., as *Holopedium sabulicola* in Wittr., Nordst., & Lagerh., Alg. Exs. no. 1549, S; isotypes, MIN, L); in fossa turfosa ad Källstorp par. Holm in Dalia, V. Wittrock, 22 Aug. 1878 (as *Merismopoedia glauca* in Wittr. & Nordst., Alg. Exs. no. 300, FC, L, S; in Aresch., Alg. Scand. Exs., Ser. Nov., no. 430, FC, L, S). GERMANY: auf Elbschlamm kultivirt, *Reinicke* (as *M. convoluta* var. *aeruginosa* in Rabenh. Alg. no. 719, FC, L, MIN, UC); Eisenach, Teich am Weg nach der Elfengrotte, *W. Migula,* Aug. 1930 (as *M. glauca* in Mig., Krypt. Germ., Austr., & Helv. Exs., Alg. no. 263, FC, TA); Leipzig, in dem grossen Lautzscher Teiche, *O. Bulnheim,* Mar. 1856 (as *M. elegans* in Rabenh. Alg. no. 515, FC, FH, L, NY, UC); Leipzig, *A. Braun,* May 1856 (designated as Type of *M. elegans* A. Br., L [Fig. 134]; isotype, S), *P. Richter,* Jun. 1877 (S), Jul. 1877 (L); Leipzig, um Lausigk, *Richter,* 1867 (as *Chroococcus cohaerens* in Rabenh. Alg. no. 1852, L, NY, UC); Rassnitz bei Leipzig, *Richter* (as *Merismopoedia convoluta* in Hauck & Richt., Phyk. Univ. no. 682, L, S). AUSTRIA: in einer Quelle in der Längepiesting, Lower Austria, *G. de Beck 367,* May 1883 (FC, W). ITALY: auf dem Badenwasser von Battaglia, *Meneghini* (Type of *M. thermalis* Kütz., L [Fig. 140]). SCOTLAND: Loch Shurrery, Caithness, *A. J. Brook 3,* 12 Sept. 1950 (D). ENGLAND: edge of River Coln, Fairford, Gloucestershire, *W. L. Tolstead 8451,* 13 Sept. 1944 (FC). NETHERLANDS: Goes, *van den Bosch 362* (L). FRANCE: Jardin des Plantes, Paris, *M. Gomont,* May 1888 (NY, DT); Falaise, herb. Lebel (FC), *A. de Brébisson* (S), *Brébisson 415* (Type of *M. convoluta* Bréb., L [Fig. 139]; isotype, UC); dans les flaques d'un étang abandonné, Falaise, *de Brébisson* (as *M. convoluta* in Rabenh. Alg. no. 1355, L, MIN; as *M. convoluta* var. *aeruginosa* in Roumeguère, Alg. de Fr. no. 617, L); St. Hilaire, herb. Lenormand (Type of *M. major* Kütz., L [Fig. 135]); eaux saumâtres, parcs d'huitres du Calvados, *de Brébisson 448* (Type of *M. aeruginea* Bréb., L [Fig. 136]); eaux saumâtres des environs de Brest, MM. *Crouan* (as *M. glauca* in Desmaz., Pl. Crypt. Fr., ed. 2, no. 301, FC, UC). BRITISH TOGOLAND: Keta, *G. W. Lawson A515,* 30 Dec. 1952 (D).

MAINE: Long pond, Mt. Desert island, *F. S. Collins 3964,* Jul. 1900 (NY). MASSA-CHUSETTS: near Linden station, Malden, *Collins 2074* (NY, UC), *2727* (NY), Jun. 1891, Oct. 1895; fresh water and mud, Woods Hole, *W. Trelease,* 1881 (MO). QUEBEC: in the St. Lawrence river, Longueuil near Montreal, *J. Brunel 33,* 30 Aug. 1930 (FC). NEW YORK: Van Cortlandt park, New York City, *J. Cohn 108,* 1933 (D). MARYLAND: in the Potomac river, Plummers island west of Cabin John, Montgomery county, *F. Drouet, E. P. Killip, & D. Richards 5533,* 6 Aug. 1944 (FC, US). FLORIDA: intertidal in St. Andrews bay, Cove Hotel, Panama City, *Drouet & C. S. Nielsen 11625,* 30 Jan. 1949 (FC, T); Dunnellon, Marion county, *M. A. Brannon 380,* 20 Oct. 1946 (FC, PC); in a very shallow stream near Tallahassee, *Nielsen 48a,* 2 May 1948 (FC, T); Cocoplum beach, Biscayne bay, Miami, *H. J. Humm 2,* 14 Jan. 1946 (FC); swamp 8 miles north of Sanderson, Baker county, *A. M. Scott 149,* 2 Jan. 1948 (FC). KENTUCKY: pond at St. Vincent, Union county, *B. B. McInteer 970,* 26 Jun. 1938 (DA, FC); pond on U. S. highway 60 one mile east of Winchester, Clark county, *McInteer 673,* 23 Aug. 1929 (DA, FC). TENNESSEE: Knoxville, *H. C. Bold 6a,* Jul. 1938 (FC). MICHIGAN: Livingston bog, Douglas lake, Cheboygan county, *H. K. Phinney,* 28 Jul. 1939 (DA, FC, PHI). INDIANA: in muck from North Mud lake 3 miles northeast of Fulton, *F. K. & W. A. Daily 1473,* 29 Jun. 1946 (DA, FC); artificial lake north of Marion, Grant county, *F. K. & W. A. Daily 2657,* 23 Aug. 1953 (DA, FC); in Tippecanoe lake, 2½ miles from North Webster, Kosciusko county, *F. K. & W. A. Daily 2229,* 25 Jul. 1950 (DA, FC); debris on bottom of Messick lake, 1½ miles north of Eddy, Lagrange county, *F. K. & W. A. Daily 2690,* 26 Aug. 1953 (DA, FC); Lake Sullivan in Riverside park, Indianapolis, *F. K. & W. A. Daily 1035,* 27 Sept. 1942 (DA, FC); Morgan county, *C. M. Palmer M2,* 5 Oct. 1925 (DA, FC); pond on state road no. 41, 2 miles north of Rockville, Parke county, *F. K. & W. A. Daily 1879,* 16 Aug. 1947 (DA, FC). MISSOURI: culture from mud in the lake north of Chouteau Springs, Cooper county, *F. Drouet 562,* 27 Apr. 1930 (D). LOUISIANA: roadside pool near Bastrop, *G. W. Prescott La25,* 26 Jun. 1938 (D). NEBRASKA: old gravel-pit pond south of North Platte, *W. Kiener 17956, 17957,* 4 Nov. 1944 (FC, KI); Dewey lake, Cherry county, *E. Palmatier & T. R. Porter 13775,* 28 Jun. 1936 (FC, KI). TEXAS: on sand at Blind Bayou, Galveston island, *H. K. Phinney 2T41/2,* 30 Aug. 1941 (FC, PHI); in a stream, Travis county, *B. C. Tharp 46A500,* 24 Oct. 1946 (FC, TEX); in a resaca pond south of Seventh street, Brownsville, *R. Runyon 3891a,* 28 Nov. 1944 (FC).

CALIFORNIA: floating in a pool of very salt water, Alameda, *W. J. V. Osterhout & N. L. Gardner,* 16 Sept. 1903 (Type of *Prasiola Gardneri* Coll. in Coll., Hold., & Setch., Phyc. Bor.-Amer. no. 1185, NY; isotypes, FC [Fig. 141], TA); Richmond, *Gardner 810,* Sept. 1902 (TA); in a small spring near San Pablo, *Gardner,* 27 Sept. 1902 (as *Merismopoedia glauca* in Coll., Hold., & Setch., Phyc. Bor.-Amer. no. 1155, FC, L, TA); on sand wet by spring water, Lake Merced, San Francisco, *Gardner,* 15 Nov. 1905 (as *M. glauca* f. *mediterranea* in Coll., Hold., & Setch., Phyc. Bor.-Amer. no. 1651, FC, L, TA), *Gardner 1578,* Nov. 1905 (FC, S). BAHAMA ISLANDS: aquarium, Lerner Laboratory, Bimini, *H. J. Humm 25a,* 26 Mar. 1948 (FC). PUERTO RICO: in a pool about 4 km. north of Mayaguez, *N. Wille 1310a,* Feb. 1915 (Type of *M. Willei* Gardn.,

91

NY [Fig. 137]); with *Anacystis thermalis* (Menegh.) Dr. & Daily in a warm stream, Coamo Springs, *Wille 384a*, Jan. 1915 (NY). GUADELOUPE: shore, Vx. Bourg, *A. Questel 255*, 22 Feb. 1944 (FC, TA). BRAZIL: Estação Biologica do Alto da Serra, entre Santos e São Paulo, *A. B. Joly 3*, 4 May 1953 (D). NEW ZEALAND: Long Beach, Bay of Islands, *M. A. Pocock & V. F. Lindauer*, 1950 (D, FC). PHILIPPINES: mud at Dagat-Dagatan, Navotas, Rizal prov., *G. T. Velasquez 441, 442*, 12 Feb. 1940 (FC). NEW GUINEA: with *Najas flexilis* Delile in brackish lagoon, *H. J. Rogers NG9*, 20 Sept. 1944 (FC).

GENUS 5. MICROCROCIS

Richter in Hauck & Richter, Phyk. Univ. 11: 548. 1892. —Type species: *M. Dieteli* Richt.

Holopedium Lagerheim, Nuova Notarisia 4: 209. 1893. *Merismopoedia* Subgenus *Holopedium* Lagerheim, Öfvers. K. Sv. Vet.-Akad. Förh. 40(2): 43. 1883. *Holopedium* Subgenus *Microcrocis* Forti, Syll. Myxophyc., p. 111. 1907. —Type species: *Merismopoedia geminata* Lagerh.

Plantae microscopicae vel sub-macroscopicae, cellulis cylindraceis vel ovoideo-cylindraceis, in lamina irregulariter dispositis; divisione cellulae seriatim in duobus planis unoquodque ad alium et ad superficiem laminae perpendicularibus procedente; gelatino vaginale hyalino.

It is highly probable that plants of this genus are growth-forms of *Agmenellum thermale*, as indicated by Rabenhorst in Alg. Sachs. 71 & 72: 719 (1878): "Unter B finden sich Exemplare, wo die Gonidien nicht mehr der Gattung [Merismopoedia] entsprechend goerdnet sind." Earlier, O. Bulnheim in Rabenh. Alg. Sachs. 51 & 52: 515 (1856) said: "Was als grösseres grünes Häutchen der Merismopoedia sich findet, hält Hr. A. Braun nicht für eine Prasiola, sondern für einen höhern Alterszustand bei Merismopoedia, in welchem die Theilung minder regelmässig geworden . . ." The few specimens available for study are inadequate to allow us to satisfy ourselves on this point.

One species:

MICROCROCIS GEMINATA Geitler in Engler & Prantl, Natürl. Pflanzenfam., ed. 2, lb: 56. 1942. *Merismopoedia geminata* Lagerheim, Öfvers. K. Sv. Vet.-Akad. Förh. 40(2): 43. 1883. *Holopedium geminatum* Lagerheim, Nuova Notarisia 4: 209. 1893. —Paratype studied by the author, the type of *Microcrocis Dieteli* Richt. (L). See Lagerheim in Nuova Notarisia 1892: 207 and 1894: 655, and Richter, *ibid*. 1892: 292. The personal herbarium of G. Lagerheim is to be sought at the University of Quito. FIG. 181.

Microcrocis Dieteli Richter in Hauck & Richter, Phyk. Univ. 11: 548. 1892. *Holopedium Dieteli* Migula, Krypt.-Fl. Deutschl. 2(Alg. la): 41. 1905. —Type from Leipzig, Germany (L). FIG. 181.
Holopedium obvolutum Tiffany, Contrib. F. T. Stone Lab. Ohio State Univ. 6: 19. 1934. —Type from Put-in-Bay, Ohio (in the collection of L. H. Tiffany).

Original specimens have not been available to us for the following names: their original descriptions are here designated as the types until the specimens can be found:

Holopedium bellum Beck-Mannagetta, Arch. f. Protistenk. 66: 10. 1929. *Beckia bella* Elenkin, Monogr. Algar. Cyanophyc., Pars Spec. 1: 71. 1938.
Merismopoedia angularis Thompson, Univ. Kansas Sci. Bull. 25: 10. 1938.
Holopedium granulatum Skuja, Symbolae Bot. Upsal. 9(3): 41. 1948.

Plantae (laminae) aerugineae vel olivaceae, primum microscopicae deinde ad

1 mm. latae ambitu irregulares, cellulis post divisionem cylindraceis vel ovoideo-cylindraceis, ad apices rotundis, diametro 4—7μ crassis, ad 15μ longis, per gela-tinum vaginale conferte distributis; gelatino vaginale hyalino, firmo vel diffluente; protoplasmate aerugineo vel olivaceo, homogeneo vel tenui-granuloso. FIG. 181.

In long-standing shallow pools, in old cultures, and seldom in the plankton.

Specimens examined:
GERMANY: Leipzig, in einem Wassergraben, *P. Dietel,* Apr. 1891 (Type of *Microcrocis Dieteli* Richt. in Hauck & Richt., Phyk. Univ. no. 548, L; isotypes, FH, MIN, N [Fig. 181]; paratypes of *Merismopoedia geminata* Lagerh.); Leipzig in dem grossen Lautzschen Teiche, *O. Bulnheim,* Mar. 1856 (with *M. elegans* in Rabenh. Alg. no. 515, FC, FH, L, NY, UC); auf Ebschlamm kultivirt, *Reinicke* (with *M. convoluta* var. *aeruginosa* in Rabenh. Alg. no. 719, FC, L). FRANCE: Calvados? ex herb. Lebel (FC). OHIO: Terwilliger's pond, Put-in-Bay, *E. H. Ahlstrom,* Aug. 1932 (Type of *Holopedium obovatum* Tiff. in the private collection of L. H. Tiffany).

GENUS 6. GOMPHOSPHAERIA

Kützing, Alg. Aq. Dulc. Germ. Dec. 16: 151. 1836. *Gomphosphaeria* Subgenus *Eugomphosphaeria* Elenkin, Not. Syst. Inst. Crypt. Horti Bot. Petropol. 2: 67. 1923. —Type species: *Gomphosphaeria aponina* Kütz.

Original specimens of the type species for the following names have not been available to us for study:
Coelocystis Nägeli in Kützing, Sp. Algar., p. 209. 1849. *Coelosphaerium* Nägeli, Gatt. Einzell. Alg., p. 54. 1849. *Gomphosphaeria* Subgenus *Coelosphaerium* Elenkin, Not. Syst. Inst. Crypt. Horti Bot. Petropol. 2; 67. 1923. —Type species: *Coelocystis Kuetzingianum* Näg. (= *Coelosphaerium Kuetzingianum* Näg.)
Thamniastrum Reinsch, Notarisia 3: 513. 1888. —Type species: ..*T. cruciatum* Reinsch.
Marssoniella Lemmermann, Ber. Deutsch. Bot. Ges. 18: 275. 1900. —Type species: *M. elegans* Lemm.
Woronichinia Elenkin, Acta Inst. Bot. Acad. Sci. U. R. P. S. S., ser. II, 1: 28. 1933. —Type species: *Coelosphaerium Naegelianum* Ung.
Snowella Elenkin, Monogr. Algar. Cyanophyc., Pars Spec. 1: 278. 1938. —Type species: *Coelosphaerium roseum* Snow.

Plantae microscopicae, multicellulares, sphaericae vel ovoideae, varie tuberculosae et in divisione constrictae, cellulis sphaericis, ovoideis, cylindraceis, obovoideis, vel pyriformibus, radiatim in superficie thalli irregulariter aut in seriebus aliis ad alias perpendicularibus dispositis; divisione cellulae seriatim in duobus planis unoquodque ad alium et ad superficiem thalli perpendicularibus procedente; gelatino vaginale hyalino, homogeneo vel circum cellulas laminoso, saepe conspicua filiola dichotome ramosa a cellulis ad centrum thalli exhibente.

Plants of Gomphosphaeria are typically sphaerical, with the cells arranged radially in a single layer in the periphery of the gelatinous matrix. Cell division proceeds successively in two planes perpendicular to each other and radial within the plant. Individual sheaths are evident about the cells in many plants of *G. aponina* and *G. lacustris;* the remains of the old parts of these sheaths form a dichotomously branched structure extending radially toward the center of the gelatinous matrix. In some plants, the rest of the gelatinous matrix hydrolyzes, and the dichotomous structure remains intact.

Key to species of Gomphosphaeria:

Cells before division 3—5μ in diameter, containing pseudovacuoles; plants de-veloping as water blooms in fresh water ....1. G. WICHURAE

93

Cells before division 2—4µ in diameter, without pseudovacuoles; plants not developing as water blooms ....2. G. LACUSTRIS

Cells before division 4—15µ in diameter, without pseudovacuoles; plants not developing as water blooms ....3. G. APONINA

1. GOMPHOSPHAERIA WICHURAE Drouet & Daily, Butler Univ. Bot. Stud. 10: 222. 1952. *Coelosphaerium Wichurae* Hilse in Rabenhorst, Alg. Eur. 153—156: 1523. 1863; Hedwigia 1863: 151. 1863. —Type from Reichenbach, Silesia (FC). FIG. 173.

Hydroepicoccum genuense de Notaris, Erb. Critt. Ital., ser. 2, 4: 178. 1869; Hedwigia 1869: 86. 1869. *Coelosphaerium genuense* Ardissone & Strafforello, Enum. Alg. Liguria, p. 61. 1877. —Isotypes from Genoa, Italy (FC, FH, L, NY, UC). FIG. 174.

Coelosphaerium Naegelianum var. *Lemmermannii* Elenkin & Hollerbach, Not. Syst. Inst. Crypt. Hort. Bot. Petropol. 2: 156. 1923. *Woronichinia Naegeliana* var. *Lemmermannii* Elenkin & Hollerbach, Acta Inst. Bot. Acad. Sci. U. R. P. S. S., ser. 2, Pl. Crypt. 1: 32. 1933. *W. Naegeliana* f. *Lemmermannii* Elenkin in Elenkin & Hollerbach, loc. cit. 1933. —Specimens from Oswitz near Breslau, Silesia, designated as isotypes (L, S, TA).

Gomphosphaeria fusca Skuja, Symbolae Bot. Upsal. 9(3): 39. 1948. —Topotype from Erken, Uppland, Sweden, received from the author (D). FIG. 177.

Original specimens have not been available to us for the following names; their original descriptions are here designated as the Types until the specimens can be found:

Gomphosphaeria lilacea Virieux, Ann. Biol. Lac. 8: 69. 1916.

Microcystis elabens var. *major* Bachmann, Mitt. Naturf. Ges. Luzern 8: 11. 1921.

Synechococcus endobioticus Elenkin & Hollerbach, Not. Syst. Inst. Crypt. Hort. Bot. Petropol. 2: 160. 1923. *Synechocystis endobiotica* Elenkin & Hollerbach in Elenkin, Monogr. Algar. Cyanophyc., Pars Spec. 1: 23. 1938.

Plantae smaragdinae vel laete aerugineae, sphaericae vel ovoideae varie tuberculosae et constrictae, cellulis ovoideis vel ovoideo-cylindraceis, diametro ante divisionem 3—5µ crassis, in superficie thalli dense et irregulariter vel in seriebus aliis ad alias perpendicularibus ordinatis; gelatino vaginale hyalino, firmo vel (interdum omnino) diffluente, homogeneo vel filiola dichotoma ad centrum thalli inconspicue exhibente; protoplasmate aeruginoso, pseudovaculois farcto. FIGS. 173, 174, 177.

In the plankton of bodies of fresh water; often found as conspicuous and heavy water blooms during the warmer months of the year. In heavy water blooms, the cells of many plants of *Gomphosphaeria Wichurae* dissociate as the gelatinous matrix hydrolyzes. The firmer matrices of other plants may harbor bacteria, *Phormidium mucicola* Naum. & Hub., and other organisms. Gomphosphaerioid growth-forms of *Lamprocystis rosea* (Kütz.) Dr. & Daily, the cells of which contain a red pigment, are sometimes confused with plants of *Gomphosphaeria Wichurae*. Also a growth-form of *Anacystis cyanea* in which most of the cells are distributed in the periphery of the gelatinous matrix may simulate plants of this species.

Specimens examined:

SWEDEN: Järlasjön, Stockholm, *G. Lagerheim*, Sept. 1898 (S); Frisksjön, Loftahammar, Smolandia, *O. Borge*, Jun. 1914 (S); in rivulo, Ljusterö, Stockholm, *K. Bohlin*, Aug. 1896 (S); Valloxen, Uplandia, *Borge*, Oct. 1897 (S); Erken, Uppland, *H. Skuja*, 1 Oct. 1950 (topotype of *Gomphosphaeria fusca* Skuja, FC [Fig. 177]); lacus ad Tenhult Smolandiae, *O. Nordstedt*, 13 Aug. 1900 (NY, S), 18 Jun. (as *Coelosphaerium Naegelianum* in Mus. Vindob. Krypt. Exs. no. 1631, FC, L, NY, S); Sövdesjön, Malmöhus, *B. Carlin-Nilsson 520*, 17 Aug. 1937 (D). POLAND: Silesia: auf einem Teiche am Schlosse von Habendorf, Kr. Reichenbach, *Hilse*, Sept. 1862 (Type of *C. Wichurae* Hilse in Rabenh. Alg. no. 1523, FC [Fig. 173]; isotypes, L, NY, UC); in einem

94

Wasserloche hinter Oswitz bei Breslau, *O. Kirchner*, Aug. 1874 (designated as isotypes of *C. Naegelianum* var. *Lemmermannii* Elenk. & Hollerb. in Rabenh. Alg. no. 2423, L, S, TA). GERMANY: Boxheim im Plankton, *Lauterborn*, Aug. 1902 (as *C. Naegelianum* in Mig., Krypt. Germ., Austr., & Helv. Exs., Alg. no. 14, NY); Leipzig, Bernbruch bei Grimma, *P. Richter* (NY); auf einem Teiche in Bernbruch bei Lausigk, *Richter*, Jul. 1864 (as *C. Kuetzingianum* in Rabenh. Alg. no. 1791, NY), Jul. 1874 (TA); Lausigk, *Richter* (N); Dumetsweiher bei Kosback bei Erlangen, *H. Glück*, 9 Sept. 1865 (as *C. Kuetzingianum* in Hauck & Richt., Phyk. Univ. no. 683, L, NY, S). CZECHO-SLOVAKIA: Bohemia: Jordan u Tábora, *A. Hansgirg*, Aug. 1883 (FC, W); piscina Radov prope Blatná, *B. Fott*, 20 Sept. 1948 (FC); great pond of Doksy, *B. Fott*, 15 Sept. 1948 (FC). ITALY: in aquariis horti botanici Genuensis, *de Notaris*, 1868—69 (isotypes of *Hydroepicoccum genuense* de Not. in Rabenh. Alg. no. 2127, FC [Fig. 174], FH, L, NY, UC; Erb. Critt. Ital., ser. 2, no. 178, FH); in una vasca dell'Orto botanico di Pisa, *A. Mori & A. Lodi*, Aug. 1882 (as *Coelosphaerium Kuetzingianum* in Erb. Critt. Ital., ser. 2, no. 1252, FC, UC). SCOTLAND: Loch Chaluim near Loch Shurrery, Caithness, *A. J. Brook 4*, 4 Jul. 1950 (D). MASSACHUSETTS: Spot pond, Medford, *F. S. Collins*, 2 Oct. 1881 (NY); Hammond pond, Brookline, *B. M. Davis*, 20 Oct. 1893 (D, MICH, MO, YU); Fresh pond, Cambridge, *W. G. Farlow*, Oct. 1886 (MO); South Framingham, *Farlow*, Oct. 1879 (MO); Framingham, *A. K. Stone* (N); ad Framingham, *Farlow*, 1883 (as *C. Kuetzingianum* in Wittr. & Nordst., Alg. Exs. no. 692, FC, L, NY, S); on Winchester reservoir, Winchester, *Collins*, 2 Oct. 1898 (as *C. Kuetzingianum* in Coll. Hold. & Setch., Phyc. Bor.-Amer. no. 553, FC, D, L, MICH, NY, TA), *Collins 3554*, 14 Aug. 1899 (NY), Aug. 1899 (as *C. Kuetzingianum* in Coll., N. Amer. Alg. no. 2, D, NY); Arlington, *E. Dewart 2160*, 27 Jul. 1891 (FC, D, L, NY, S, TA). CONNECTICUT: water-bloom, Housatonic river, *H. C. Bold B-3*, 30 Sept. 1939 (DA, FC); in Avery's brook, Ledyard, *W. A. Setchell 549*, Sept. 1892 (UC). NEW YORK: Basic reservoir, Rensselaerville, *M. S. Markle 25*, Jul. 1947 (EAR, FC); plankton of Lake Mohansic, Westchester county, *Bold B121*, 4 Oct. 1941 (FC).

NEW JERSEY: bloom on protected bay, Lake Farrington, New Brunswick, *E. T. Moul 6108*, 19 Sept. 1948 (FC). PENNSYLVANIA: Mercer county: Shenango river, New Hamburg, *L. Walp & W. Murchie 1*, 26 Jun. 1941 (FC), Greenville, Sharon, and Wheatland, *Walp & Murchie 3—5*, 6 Jul. 1941 (FC). ONTARIO: Experiment farm, Ottawa, *J. Fletcher*, Aug. 1888 (FH). OHIO: south reservoir, Portage Lake near Akron, *E. H. Ahlstrom*, 28 Dec. 1932 (DA, FC); Chippewa lake, *Ahlstrom*, 30 Aug. 1931 (DA, FC); Lake Erie, *L. H. Tiffany*, 1929 (DA, FC); Decker lake near Piqua, Miami county, *C. F. Clark*, 23 Oct. 1942 (DA, FC); plankton in Arlington cemetery lake, Cincinnati, *W. A. Daily & J. H. Hoskins 108*, 4 Oct. 1939 (DA, FC); in State Hospital pond, Athens, *A. H. Blickle*, 8 Oct. 1941 (DA, FC). MICHIGAN: Benton Harbor, *M. E. Britton*, Aug. 1939 (DA, FC); Sodon lake west of Bloomfield Hills, Oakland county, *S. A. Cain*, 9 Dec. 1947 (FC); in Marble lake near Quincy, *G. W. Prescott M411a*, 21 Sept. 1937 (FC); "Three Lakes", Ann Arbor, *L. N. Johnson 5*, 27 Sept. 1892 (FC). INDIANA: Anderson's pond 3 miles south of Hartsville, Bartholomew county, *F. K. & W. A. Daily 1040*, 11 Oct. 1942 (DA, FC); Jimmy Strahl lake, Brown County state park, *R., E. A., & M. Fritsche 1—3*, 2 Nov. 1947 (DA, FC); Lake Cicott, Cass county, *F. K. & W. A. Daily 1023*, 9 Sept. 1942 (DA, FC); Lake McCoy, *3—5* miles east of Greensburg, Decatur county, *F. K. & W. A. Daily 1038, 1182*, 11 Oct. 1942, 19 Sept. 1943 (DA, FC); Cups pond, 4 miles east of Greensburg, *F. K. & W. A. Daily 1039*, 11 Oct. 1942 (DA, FC); southeast shore of Heaton lake, 6 miles northeast of Elkhart, *F. K. & W. A. Daily 2619, 2621*, 26 Aug. 1952 (DA, FC); South Mud lake, 4 miles east of Fulton, *F. K. & W. A. Daily 1482*, 29 Jun. 1946 (DA, FC); in quarry on Washington street, Kokomo, *F. K. & W. A. Daily 1026*, 10 Sept. 1942 (DA, FC); west side of Dewart lake, Kosciusko county, *F. K. & W. A. Daily 2115a*, 9 Jul. 1949 (DA, FC); Center lake, Warsaw, *F. K. & W. A. Daily 2180*, 22 Jul. 1950 (DA, FC); Durham (Daniel) lake, Kosciusko county, *F. K. & W. A. Daily 2262, 26* Jul. 1950 (DA, FC); Lake Wawasee, Syracuse, *F. K. & W. A. Daily 221, 2238, 2245, 24—26* Jul. 1950 (DA, FC); Carr's lake northeast of Claypool, Kosciusko county, *F. K. & W. A. Daily 2174*, 22 Jul. 1950 (DA, FC); Tippecanoe lake near North Webster, *H. B. Metcalf 2*, 12 Oct. 1947 (DA, FC), *F. K. & W. A. Daily 2228, 2229, 25* Jul. 1950 (DA, FC); Little Tippecanoe (James) lake, Kosciusko county, *F. K. & W. A. Daily 2227*, 25 Jul. 1950 (DA, FC); Wabee lake southeast of Milford, *F. K. & W. A. Daily 2636*, 28 Aug. 1952 (DA, FC); Webster lake at North Webster, *F. K. & W. A. Daily 2222, 2226, 25* Jun. 1950 (DA, FC); Syracuse lake at city park, Syracuse, *F. K. & W. A. Daily 2195, 23* Jul. 1950 (DA, FC); Atwood lake west of Wolcottville, Lagrange county, *F. K. & W. A. Daily 2687, 26* Aug. 1953 (DA, FC); Indian lake near Fort Benjamin Harrison, Marion county, *F. K. & W. A. Daily 1050, 1051*, 18 Oct. 1942 (DA, FC); Lake Maxinkuckee, Marshall county, *G. T. Moore*, 28 Aug. 1899 (FC), *F. K. & W. A. Daily 1533, 1542*, 4-5 Jul. 1946 (DA, FC); small lake at north edge of Rome City, *F. K. & W. A. Daily 2278, 29* Jul. 1950 (DA, FC); outlet of Sylvan lake, Rome City, *F. K. & W. A. Daily 2277, 29* Jul. 1950 (DA, FC); Cree lake, 4 miles north of Kendallville, Noble county, *F. K. & W. A. Daily 1914*, 12 Sept. 1947 (DA, FC); Skinner lake, east of Albion, Noble county, *F. K. & W. A. Daily 1912*, 12 Sept. 1947 (DA, FC); Thomas lake east of Lena, Parke county, *F. K. & W. A. Daily 2531*, 25 Aug. 1951 (DA, FC); Eliza lake, Moss lake, and Spectacle lake near Valparaiso, Porter county, *F. K. & W. A. Daily 2567, 2574, 2576, 30* Aug. 1951 (DA, FC); Bass lake, 6 miles west of Angola, *F. K. & W. A.*

Daily 1917, 12 Sept. 1947 (DA, FC); Hamilton lake, Hamilton, Steuben county, *F. K. & W. A. Daily 2444, 2449*, 17 Jun. 1951 (DA, FC); Hogback lake, Steuben county, *C. M. Palmer 1015*, Jul. 1933 (DA, FC); Lake James, Pokagon state park, Steuben county, *F. K. & W. A. Daily & I. Spangler 1931*, 13 Sept. 1947 (DA, FC); Lake Lonidaw, Pokagon state park, *S. A. Joyner*, 31 Aug. 1949 (DA, FC); Otter lake, Steuben county, 13 Sept. 1947 (DA, FC); Pleasant lake, Steuben county, *F. K. & W. A. Daily 2458*, 18 Jun. 1951 (DA, FC); Snow lake, 7 miles north of Angola, *F. K. & W. A. Daily 1943, 1945*, 13—14 Sept. 1947 (DA, FC); Shakamak lake, Shakamak state park, Sullivan county, *Palmer B18*, 8 Sept. 1933 (DA, FC), *F. K. & W. A. Daily, F. Geisler, & F. Hueber 2378*, 23 Sept. 1950 (DA, FC); pond at Rose Polytechnic Institute, Terre Haute, *B. H. Smith 134*, 30 Jul. 1929 (DA, FC); pond on state road no. 9 near Columbia City, Whitley county, *F. K. & W. A. Daily*, 14 Sept. 1947 (DA, FC).

WISCONSIN: Lake Mendota, Madison, *W. Trelease*, Oct. 1886 (D, FH), *R. E. Blount*, 2 Oct. 1886 (FC); small lake near Clear lake, Rock county, *J. B. Lackey*, Aug. 1943 (FC); Hog creek, Chippewa flowage, Sawyer county, *G. W. Prescott 3W185*, 30 Aug. 1938 (FC); Little Star lake, *Prescott 3W95*, 21 Aug. 1938 (FC). ILLINOIS: Cook county, *M. E. Britton 987*, 8 Sept. 1939 (DA, FC); Maple lake, Cook county, *Britton 993*, 8 Sept. 1939 (DA, FC); Lake Zurich, Fish lake, Petit lake, and Lake Marie, Lake county, *Britton 916, 940, 956, 959, 1030, 1032, 1051*, Aug.—Sept. 1939 (DA, FC); Channel lake 3 miles west of Antioch, Lake county, *Britton 1038, 1043*, 14 Sept. 1939 (DA, FC). MINNESOTA: Cass lake, Cass county, *C. B. Reif*, 11 Sept. 1936 (DA, FC); Lake Itasca, Clearwater county, *S. Eddy*, 20 Aug. 1905 (DA, FC); Lake of the Isles, Minneapolis, *F. Drouet 5593*, 19 Aug. 1944 (FC); Cross lake, just north of Minneapolis, *C. B. Reif*, 22 Sept. 1936 (DA, FC); Lake Minnetonka at Spring Park, Hennepin county, *Drouet, A. A. Cohen, & E. Cohen 5007, 5016*, 8 Aug. 1943 (FC); Waterville, Le Sueur county, *J. C. Arthur* (FH); Little Pelican and Sear lakes, Ottertail county, *S. Eddy*, Jul.—Aug. 1940 (DA, FC); Keller lake north of St. Paul, *Drouet & E. Cohen 5045, 5050*, 9 Aug. 1943 (FC); Lake Gervais and Lake Owasso, Ramsey county, *S. Eddy*, 25 Jun., 15 Aug. 1931 (DA, FC); Poplar lake, Superior national forest, *C. B. Reif*, 28 Jun. 1935 (DA, FC); Osakis lake, Todd county, *Reif*, 19 Jul. 1938 (DA, FC); Long lake northeast of North St. Paul, in Washington county, *Drouet & E. Cohen 5062*, 9 Aug. 1943 (FC). MISSOURI: St. Charles, *B. Ennis*, 12 Nov. 1930 (D); Reed's lake, Callaway county, *J. R. Hurt R-1, R-8*, 14 Oct. 1939, 12 Jun. 1940 (FC); New and Old Fayette lakes near Fayette, Howard county, *Hurt NF-2, OF-1, NF-11a*, 24 Nov. 1939, 4 Jun. 1940 (FC). NEBRASKA: sandpit pond north of David City, Butler county, *W. Kiener 21570a*, 8 Nov. 1946 (FC, KI); sandpit lakes, Fremont, Dodge county, *Kiener 19700, 23934*, 1 Jul. 1948, 15 Sept. 1945 (FC, KI); lake 3 miles east of Stratton, Hitchcock county, *Kiener 10470*, 24 Jul. 1941 (FC, KI); pond in city park, Norfolk, Madison county, *Kiener 22916*, 15 Oct. 1947 (FC, KI). BRITISH COLUMBIA: Deer lake near Burnaby Lake, New Westminster county, *J. E. Nielsen 81, 91*, 14 Jun. 1924, 22 Nov. 1925 (FC). WASHINGTON: Phantom Lake, *W. T. Edmondson*, 30 Jun. 1953 (FC); Fidalgo island, *L. E. Griffin*, summer 1938 (FC). CALIFORNIA: Big Bear lake, San Bernardino county, *G. J. Hollenberg 469, 1624*, 26 Aug. 1934 (FC, D, UC).

2. GOMPHOSPHAERIA LACUSTRIS Chodat, Bull. Herb. Boiss. 6: 180. 1898. *Coelosphaerium lacustre* Ostenfeld, Hedwigia 46: 396. 1907. —Topotypic material from Lac de Joux, Switzerland (in the collection of Otto Jaag).

Chroococcus minutus var. *minimus* Keissler, Verh. K. K. Zool.-bot. Ges. Wien 51: 394. 1901. *C. minimus* Lemmermann, Ark. f. Bot. 2(2): 102. 1904. *Gloeocapsa minima* Hollerbach in Elenkin, Monogr. Algar. Cyanophyc., Pars Spec., 1: 242. 1938. —Type from Attersee, Upper Austria (W).

Gomphosphaeria lacustris var. *compacta* Lemmermann, Abh. Nat. Ver. Bremen 16: 341. 1899. *G. compacta* Strøm, Naturw. Unters. Sarekgebirges in Schwed.-Lappl. 3(Bot., Lief. 5): 446. 1923. —Material named by the author from Lake Wombsjön, Sweden, designated as the Type (S).

Coelosphaerium Collinsii Drouet & Daily in Daily, Amer. Midl. Nat. 27: 653. 1942. —Type from Naushon island, Massachusetts (D). FIG. 176.

Original specimens have not been available to us for the following names; their original descriptions are here designated as the Types until the specimens can be found:

Coelocystis Kuetzingiana Nägeli in Kützing, Sp. Algar., p. 209. 1849. *Coelosphaerium Kuetzingianum* Nägeli, Gatt. Einzell. Alg., p. 54. 1849. —Thorough search through the F. T. Kützing herbarium at Leiden and the Nägeli herbaria at Zürich and Munich has failed to discover that the original specimens were preserved. Nägeli's sketches and notes, from the notebooks in the Pflanzenphysiologisches Institut, Eidgenossische Technische Hochschule, Zürich, are reproduced here: FIG. 175.

Coelosphaerium Naegelianum Unger, Denkschr. K. Akad. Wiss. Wien, Math.-Nat. Kl., 7: 195. 1854. *Gomphosphaeria Naegeliana* Lemmermann, Krypt.-Fl. Mark Brandenb. 3: 80. 1907.

Woronichinia Naegeliana Elenkin, Acta Inst. Bot. Acad. Sci. U. R. P. S. S., ser. 2, Pl. Crypt. 1: 28. 1933.
   Thamniastrum cruciatum Reinsch, Notarisia 3: 513. 1888. —See Taft in Bull. Torr. Bot. Club 72: 246. 1945.
   Coelosphaerium aerugineum Lemmermann, Bot. Centralbl. 76: 154. 1898. *C. anomalum* var. *aerugineum* Hansgirg, Beih. z. Bot. Centralbl. 8(2): 516. 1905.
   Marssoniella elegans Lemmermann, Ber. Deutsch. Bot. Ges. 18: 275. 1900.
   Coelosphaerium roseum Snow, Bull. U. S. Fish Comm. 1902: 387, 390. 1903. *Gomphosphaeria rosea* Lemmermann, Krypt.-Fl. Mark Brandenb. 3: 80. 1907. *Snowella rosea* Elenkin, Monogr. Algar. Cyanophyc., Pars Spec. 1: 278. 1938.
   Gomphosphaeria litoralis Häyrén, Bidrag t. Kännedom Finlands Nat. och Folk 20(3): 91. 1921.
   Coelosphaerium radiatum G. M. Smith, Roosevelt Wild Life Bull. 2: 136. 1924.
   Coelosphaerium pusillum Van Goor, Rec. Trav. Bot. Neerl. 21: 318. 1924.
   Coelosphaerium Geitleri Schiller, Sitzungsber. Oesterr. Akad. Wiss., Math.-Nat. Kl., I, 163: 131. 1954.

Plantae aerugineae, olivaceae, violaceae, vel roseae, sphaericae vel ovoideae varie tuberculosae vel constrictae, cellulis globosis vel ovoideis, etiam obovoideis, diametro ante divisionem 2—4µ crassis, in superficie thalli irregulariter aut in seriebus aliis ad alias perpendicularibus ordinatis; gelatino vaginale hyalino, firmo vel (interdum omnino) diffluente, homogeneo vel filiola dichotoma ad centrum thalli vulgo exhibente; protoplasmate aerugineo, olivaceo, vel roseo, homogeneo. FIGS. 175, 176.

In the plankton of lakes, ponds, and rivers, and in shallow water and pools along their margins.   Plants of *Gomphosphaeria lacustris* developing in shallow water are usually firmer and more compact than those in the plankton, and their cells are often more ovoid or obovoid than are those of the latter.   Growth-forms of the bacterial *Lamprocystis rosea* (Kütz.) Dr. and Daily, especially those in which the pseudovacuoles have been lost, are often confused with this species. Likewise, certain growth-forms of *Anacystis montana* may superficially resemble this species.

Specimens examined:
   SWEDEN: Saxorp, Vestergötland, *O. Nordstedt*, Jun. 1883 (S); in lacu Wombsjön Scaniae, *Nordstedt*, Jun. 1901 (designated as Type of *Gomphosphaeria lacustris* var. *compacta* Lemm., S). AUSTRIA: Attersee, Upper Austria, *C. von Keissler*, 13 Aug. 1900 (Type of *Chroococcus minutus* var. *minimus* Keissl. in the slide collection of C. Keissler, W). SWITZERLAND: Lac de Joux, Vaud, *O. Jaag* (topotype of *Gomphosphaeria lacustris* in the collection of O. Jaag). IRELAND: Lough Corrib, *G. S. West 53* (FC, UC). NEW BRUNSWICK: pool, lower basin, Grand Falls, Victoria county, *H. Habeeb 10849*, 30 Sept. 1948 (FC, HA). MAINE: Long Pond, Mt. Desert island, *F. S. Collins 3964*, July 1900 (NY, UC). MASSACHUSETTS: French Watering Place, Naushon island, *F. Drouet 2125*, 12 Aug. 1937 (Type of *Coelosphaerium Collinsii* Dr. & Daily, D [Fig. 176]); Fresh pond, Nobska point, Falmouth, *Drouet 1948*, 16 Sept. 1936 (D). QUEBEC: étang envahi par les grandes marais, Baie Kopaluk (estuaire de rivière George), Ungava, *J. Rousseau 1099a*, 10 Aug. 1947 (FC). NORTH CAROLINA: Mullet pond, Shackleford Banks, Beaufort, *H. J. Humm & C. S. Nielsen 1792*, 23 Aug. 1949 (FC, T). FLORIDA: Leesburg, *M. A. Brannon 332*, 31 May 1948 (FC, PC); on Chara, Battery Point, Hernando county, *Brannon 554*, 23 Oct. 1948 (FC, PC). OHIO: Terwilliger's pond, South Bass island, Ottawa county, *E. H. Ahlstrom*, 4 Sept. 1931 (DA, FC); ponds, Newtown fish hatchery, Hamilton county, *J. B. Lackey*, Jul. 1944 (FC). MICHIGAN: Lake La Grange northeast of Dowagiac, Cass county, *J. P. Woods, L. L. Pray, & F. H. Letl*, Aug. 1942 (FC); Interlochen, Grand Traverse county, *J. H. Hoskins*, 29 Aug. 1934 (DA, FC).
   INDIANA: Elkhart county: Heaton lake 6 miles northeast of Elkhart, *F. K. & W. A. Daily 2621*, 26 Aug. 1952 (DA, FC). Jackson county: sluiceway of dam, Driftwood state fish hatchery 2 miles south of Vallonia, *F. K. & W. A. Daily 2343*, 26 Aug. 1950 (DA, FC). Kosciusko county: Dewart lake, *F. K. & W. A. Daily 2119*, 9 Jul. 1949 (DA, FC); Laka Wawasee, *F. K. & W. A. Daily 2183, 2203, 2207, 2214, 2238, 2250, 2272*, 23—26 Jul. 1950 (DA, FC); Wabee lake southeast of Milford, *F. K. & W. A. Daily 2636*, 28 Aug. 1952 (DA, FC). Marion county: Geist reservoir 8 miles northeast of Indianapolis, *F. K. & W. A. Daily 2485*, 16 Sept. 1951 (DA, FC). Porter county: Flint lake 3 miles north of Valparaiso, *F. K. & W. A. Daily 2569*, 30 Aug. 1951 (DA,

FC). Starke county: pond, Bass Lake state fish hatchery 5 miles south of Knox, *E. Warkentien & F. K. & W. A. Daily 1553*, 6 Jul. 1946 (DA, FC). Steuben county: Bass lake 6 miles west of Angola, *F. K. & W. A. Daily 1917A*, 12 Sept. 1947 (DA, FC); Clear lake 6 miles east of Fremont, *F. K. & W. A. Daily 2452*, 17 Jun. 1951 (DA, FC); Crooked lake, *C. M. Palmer B49b*, Sept. 1933 (DA, FC); Fox lake, *Palmer 1019*, Jul. 1933 (DA, FC); Gentian lake 6 miles north of Angola, *F. K. & W. A. Daily 2659, 2666, 2671, 2674*, 24 Aug. 1953 (DA, FC); Hamilton lake, Hamilton, *F. K. & W. A. Daily 2449*, 17 Jun. 1951 (DA, FC); Lake James, Pokagon state park, *S. A. Joyner*, 6 Jul. 1949 (DA, FC), *F. K. & W. A. Daily 2474, 2677*, 19 Jun. 1951, 25 Aug. 1953 (DA, FC); Jimerson lake 7 miles northwest of Angola, *F. K. & W. A. Daily 2479A*, 18 Jun. 1951 (DA, FC).

WISCONSIN: Pell lake near Lake Geneva, Walworth county, *W. E. Lake*, Aug. 1942 (FC); Round lake and Connors lake, *G. W. Prescott 3W181, 3W213* (DA, FC); small lake near Clear lake, Rock county, *J. B. Lackey*, Aug. 1945 (FC). MINNESOTA: Lida lake, Ottertail county, *S. Eddy*, 5 Aug. 1940 (DA, FC); Lake Itasca, Clearwater county, *Eddy*, 18 Aug. 1933 (DA, FC), *F. O. Fortich 18, 22*, 1 Jul. 1954 (D). NEBRASKA: Jackson lake 9 miles north of North Platte, Lincoln county, *W. Kiener 17516*, 1 Sept. 1944 (FC, KI). SASKATCHEWAN: Last Mountain lake, Humboldt county, *P. E. Kuehne*, 7 Sept. 1940 (FC). MONTANA: Middle Quartz lake, Glacier national park, *Long F175*, 1 Sept. 1934 (FC). WYOMING: Fayette lake, Fremont county, *Bond F268*, 23 Aug. 1934 (FC). NEW MEXICO: pond in the lava-beds 6—8 miles east of Grant, Valencia county, *A. A. Lindsey 8*, 12 Aug. 1944 (FC). UTAH: Steam Mill and Tony Grove lakes, Cache county, *G. Piranian*, 13—14 Aug. 1933 (FC).

3. GOMPHOSPHAERIA APONINA Kützing, Alg. Aq. Dulc. Dec. 16: 151. 1836. *Sphaerastrum cuneatum* Kützing in Meneghini, Consp. Algol. Eugan., p. 337. 1837. —Type from Abano, Italy (L). FIG. 179.

Gomphosphaeria aponina var. *cordiformis* Wolle, Bull. Torr. Bot. Club 9: 25. 1882. *G. cordiformis* Hansgirg, Oesterr. Bot. Zeitschr. 36: 333. 1886. *G. aponina* b. *cordiformis* Wolle ex Hansg., Prodr. Algenfl. Böhmen 2: 144. 1892. *G. aponina* f. *cordiformis* Elenkin, Monogr. Algar. Cyanophyc., Pars Spec., 1: 286. 1938. —Type from Bethlehem, Pennsylvania (FC). FIG. 178.

Gomphosphaeria cordiformis var. *olivacea* Hansgirg, Oesterr. Bot. Zeitschr. 36: 333. 1886. *G. aponina* var. *olivacea* Hansgirg, Prodr. Algenfl. Böhmen 2: 144. 1892. *G. aponina* f. *olivacea* Forti, Syll. Myxophyc., p. 98. 1907. —Type from Ouzic, Bohemia (W). FIG. 180.

Gomphosphaeria aponina var. *multiplex* Nygaard, Vidensk. Medd. Dansk Naturh. Foren. Kjøbenh. 82: 204. 1926. *G. aponina* f. *multiplex* Elenkin, Monogr. Algar. Cyanophyc., Pars Spec. 1: 286. 1938. —Type from Kai islands, Indonesia (in the collection of G. Nygaard).

Original specimens have not been available to us for the following names; their original descriptions are here designated as the Types until the specimens can be found:

Gomphosphaeria aponina var. *limnetica* Virieux, Ann. Biol. Lac. 8: 68. 1916. *G. aponina* f. *limnetica* Elenkin, Monogr. Algar. Cyanophyc., Pars Spec. 1: 286. 1938.

Gomphosphaeria aponina var. *delicatula* Virieux, Ann. Biol. Lac. 8: 69. 1916. *G. aponina* f. *delicatula* Elenkin, Monogr. Algar. Cyanophyc., Pars. Spec. 1: 286. 1938.

Gomphosphaeria aponina f. *major* Pevalek, Geobot. & Algol. Istraz. Cretova u Hrvalskoj & Sloven. Rad 230: 86. 1924.

Gomphosphaeria aponina var. *gelatinosa* Prescott, Hydrobiologia 2(1): 93. 1950.

Plantae aerugineae, olivaceae, luteolae, violaceae, vel roseae, sphaericae vel ovoideae aetate provecta varie constrictae, cellulis ovoideis, cylindraceis, obovoideis, vel pyriformibus, in divisione saepe cordiformibus, diametro ante divisionem 4—15μ crassis, in superficie thalli irregulariter vel in seriebus aliis ad alias perpendicularibus ordinatis; gelatino vaginale hyalino, vulgo firmo nonnumquam (etiam omnino) diffluentes, homogeneo vel lamellosa, filiola dichotoma ad centrum thalli plerumque exhibente; protoplasmate aerugineo, olivaceo, luteolo, violaceo, vel roseo, homogeneo vel tenui-granuloso. FIGS. 178—180.

In shallow fresh, brackish, and marine waters, in seepage, and in the plankton. Under certain conditions, the protoplasts shrink and evidently secrete additional material within their individual sheaths; later they may continue growth again as a strain with narrower cells. Where the plants of *Gomphosphaeria aponina* dis-

98

integrate, groups of cells sometimes remotely resemble those of *Anacystis thermalis*.

Specimens examined:
POLAND: Mergelsgruben bei Warkotsch bei Strehlen in Schlesien, *Bleisch* (as *Gomphosphaeria aponina* in Rabenh. Alg. no. 1497, FC, UC). GERMANY: Leipzig, um Altmanndorf, *P. Richter* (L, UC); Pötzschau bei Leipzig, *Auerswald*, Aug. 1959 (UC); in einem Graben am Eisenbahndamme oberhalb Beiersdorf in Franken, *P. Reinsch*, 9 Oct. 1864 (as *G. aponina* in Rabenh. Alg. no. 1911, FC, L, UC). CZECHOSLOVAKIA: in seichtem Wasser und in Schlamme am Rande der Salzwassersümpfe bei Ouzic nachst Kralup in Böhmen, *A. Hansgirg*, Apr. 1884 (Type of *G. cordiformis* var. *olivacea* Hansg., W; isotype, FC [Fig. 180]; isotypes in Fl. Exs. Austro-Hung. no. 3600, L, MO, S, W). JUGOSLAVIA: Salona prope Spalato,, *Hansgirg*, Aug. 1889 (FC, W). ITALY: Abano, *F. T. Kützing* (DT, FC, L, MO, UC), May 1835, inter *Confervam aponinam* (Type of *Gomphosphaeria aponina* Kütz., L [Fig. 179]; isotypes in Kütz., Alg. Aq. Dulc. Dec. no. 151, L, UC); in thermis Euganeis, *G. Meneghini* (L, S); Monfalcone, in aqua subsalsa, *Hansgirg*, Aug. 1889 (FC, W). TRIESTE: in aqua subsalsa, Capo d'Istria, *Hansgirg*, Aug. 1889 (FC, W). SCOTLAND: Loch Shurrery, Caithness, *A. J. Brook 2*, 3 Aug. 1950 (D). BERMUDA: Lovers lake (brackish), western St. Georges island, *A. J. Bernatowicz 49-111*, 1 Mar. 1949 (FC, MICH). NEW BRUNSWICK: in a pond, St. Anne, Madawaska county, *H. Habeeb 11617, 11620*, 22 Aug. 1950 (FC, HA). MASSACHUSETTS: in Gardiner's ditch, Woods Hole, *F. Drouet 1908*, 10 Aug. 1936 (D, FH, NY); in tide-pools of harbor, Cuttyhunk island, Gosnold, *Drouet 2122*, 3 Aug. 1937 (D, DT, FH). CONNECTICUT: basin, Norwich, *W. A. Setchell*, Jul. 1888 (UC). PENNSYLVANIA: in stagnis ad Bethlehem, *F. Wolle*, 30 Aug. 1881 (Type of *G. aponina* var. *cordiformis* Wolle in Wittr. & Nordst., Alg. Exs. no. 498, FC [Fig. 178]; isotypes, L. S). MARYLAND: marsh pool between Chance and Dames Quarter, Somerset county, *P. W. Wolle*, 12 Jun. 1938 (D), *P. W. Wolle & Drouet 2260*, 22 Aug. 1938 (FC, D, FH). FLORIDA: pools, Long key, Dry Tortugas, *W. R. Taylor 22, 501*, Jun. 1924 (TA); with *Chara Hornemannii* Wallm., in deep water near bridge, Big Pine key to Torch key, Monroe county, *E. P. Killip 41970b*, 4 Mar. 1951 (FC, US); freshwater ditch, Hendry creek, about 10 miles south of Fort Myers, Lee county, *P. C. Standley 73254*, Mar. 1940 (FC); on a tidal flat along East river one mile northeast of St. Marks lighthouse, Wakulla county, *Drouet, G.C. Madsen, & D. Crowson 11771*, 1 Feb. 1949 (FC, T).

OHIO: reservoirs in Eden park, Cincinnati, *F. K. Daily*, 24 Aug. 1940 (DA, FC), *W. A. Daily 177, 404, 406*, 24 Oct. 1939, 3 Jul. 1940 (DA, FC), *J. B. Lackey*, Sept. 1942 (FC). KENTUCKY: wet sandstone and soil, Torrent, Wolfe county, *W. A. & F. K. Daily & R. Kosanke 457, 473, 478, 485, 489*, 14 Jul. 1940 (DA, FC). MICHIGAN: pond on north side of Mackinac island, *F. K. & W. K. Daily 1727*, 24 May 1947 (DA, FC); slough on Northport road opposite Cedar lake, Leelanau county, *J. H. Hoskins*, 30 Aug. 1934 (DA, FC). INDIANA: Lake Cicott 8 miles west of Logansport, Cass county, *F. K. & W. A. Daily 1099*, 9 Sept. 1942 (DA, FC); Wabee lake southeast of Milford, *F. K. & W. A. Daily 2636*, 28 Aug. 1952 (DA, FC); in fish-hatchery pond 2 miles northwest of Culver, Marshall county, *F. K. & W. A. Daily 1495*, 1 Jul. 1946 (DA, FC); Flint lake 3 miles north of Valparaiso, Porter county, *F. K. & W. A. Daily 2569*, 30 Aug. 1951 (DA, FC); Crooked lake, Steuben county, *C. M. Palmer B54*, Sept. 1933 (DA, FC); Gentian lake 6 miles north of Angola, *F. K. & W. A. Daily 2669, 2675*, 24 Aug. 1953 (DA, FC). MINNESOTA: in covered tank, Zoological Laboratory, University of Minnesota, Minneapolis, *J. E. Tilden*, 20 Apr. 1898 (as *Gomphosphaeria aponina* in Tild., Amer. Alg. no. 300, FC). NEBRASKA: old meander pond of Platte river, Scottsbluff, *W. Kiener 22044*, 20 May 1947 (FC, KI). TEXAS: on *Digenia simplex* (Wulf.) Ag., Rockport, *G. L. Fisher 41090*, 14 Aug. 1941 (FC). MONTANA: *F. A. Barkley*, 1940 (FC); just below forest boundary on Lolo creek, Missoula county, *F. H. Rose 4211*, 12 Jun. 1941 (FC). NEVADA: Springdale hot springs 10 miles north of Beatty, Nye county, *G. Waltenspiel 1740*, 8 Feb. 1953 (FC). WASHINGTON: Whidbey island, *W. A. Setchell & N. L. Gardner 301, 472* (UC).

CALIFORNIA: Bad Water, Death Valley, Inyo county, *R. M. Holman & L. Bonar 7213*, 2 Apr. 1933 (FC, UC), *M. J. Groesbeck 1, 217, 303*, Feb., Oct. 1940, 1 Jan. 1941 (FC); in fresh water, Los Angeles, *S. P. Monks*, Nov. 1895 (UC); in the travertine quarry, Bridgeport, Mono county, *Groesbeck 101, 161*, 13 Jun. 1940, 3 Sept. 1940 (FC); Monterey, *T. E. Hazen* (DA, FC); brackish water one mile west of Newport Beach, Orange county, *G. J. Hollenberg 447b, 1618*, 16 Aug. 1934 (FC, UC); pond north of Torrey Pines state park, San Diego county, *W. E. Allen 41a*, 8 Dec. 1935 (FC). JAMAICA: Morant cays, *V. J. Chapman*, 1939 (FC). GUATEMALA: *S. E. Meek 84*, 1906 (FC, US). ECUADOR: in stagno in Valle de Guano prope Riobamba, *G. Lagerheim*, Jul. 1891 (as *Gomphosphaeria aponina* in Wittr., Nordst., & Lagerh., Alg. Exs. no. 1545, L); salt pool near Albermarle point, north end of Isabela island, Galapagos islands, *W. R. Taylor 123*, 12 Jan. 1934 (D, TA). ARGENTINA: laguna frente al puerto, Mar del Plata, Buenos Aires, *S. A. Guarrera 2729*, 6 Apr. 1939 (FC). HAWAIIAN ISLANDS: Niihau island, *A. & S. Robinson 33*, Dec. 1907 (FC, UC); Hamauma bay, Oahu, *W. J. Newhouse 59*, 3 May 1953 (FC). MARSHALL ISLANDS: rocks on reef flats, seaward side of Igurin island, Eniwetok atoll, *W. R. Taylor 46-316*, 22 May 1946 (FC). RYUKYU ISLANDS: Unten, Nakijin-son, *Y. K. Okada*, 3 Apr.

1937 (as *G. aponina* var. *multiplex* in Okada, Alg. Aq. Japon. no. 2, FC). INDONESIA: Lake Ohoitiel, Kai islands, *H. Jensen & O. Hagerup*, 1922 (Type of *G. aponina* var. *multiplex* Nyg. in the slide collection of G. Nygaard). WESTERN TIBET: in Kyam Spring, in Chang-chenmo valley north of Pang-gong Tso, *G. E. Hutchinson L58*, 19 Jul. 1932 (D, DT, FH, L, NY).

FAMILY II. CHAMAESIPHONACEAE

Borzi, N. Giorn. Bot. Ital. 10(3): 298. 1878. *Chamaesiphonaceae* Subfam. *Euchamaesiphonaceae* Hansgirg, Notarisia 3: 588. 1888. —Type genus: *Chamaesiphon* A. Br. & Grun.

Pleurocapsaceae Geitler, Beih. z. Bot. Centralbl., II, 41: 238. 1925. *Chamaesiphonaceae* Subfam. *Cystogoneae* Hansgirg, Notarisia 3: 588. 1888. —Type genus: *Pleurocapsa* Thur.

Entophysalidaceae Geitler, Beih. z. Bot. Centralbl., II, 41: 235. 1925. —Type genus: *Entophysalis* Kütz.

Chlorogloeaceae Geitler, Beih. z. Bot. Centralbl., II, 41: 236. 1925. —Type genus: *Chlorogloea* Wille.

Dermocarpaceae Geitler, Beih. z. Bot. Centralbl., II, 41: 247. 1925. —Type genus: *Dermocarpa* Crouan fr.

Siphononemataceae Geitler, Beih. z. Bot. Centralbl., II, 41: 251. 1925. —Type genus: *Siphononema* Geitl.

Xenococcaceae Ercegovic, Bull. Int. Acad. Yougosl. Sci. & Arts, Cl. Sci. & Math., 26: 38. 1932. —Type genus: *Xenococcus* Thur.

Hyellaceae Ercegovic, Bull. Int. Acad. Yougosl. Sci. & Arts, Cl. Sci. Math. & Nat., 26: 38. 1932. —Type genus: *Hyella* Born. & Flah.

Original specimens of the type species of the type genus of the following family have been unavailable for our study:

Scopulonemataceae Ercegovic, Bull. Int. Acad. Yougosl. Sci. & Arts, Cl. Sci. Math. & Nat., 26: 38. 1932. —Type genus: *Scopulonema* Erceg.

Plantae uni—multi-cellulares, aquaticae, microscopicae vel macrascopicae, cellulis primum solitariis basim ad substratum affixis, unaquidque seriatim in cellulas-filias primo inaequales deinde aequales dividente et mox ab aliis cum gelatino vaginale se separante, deinde sursum supra superficiem substrati radiatim in stratum vel pulvinum solidum et deorsum intus substratum in fila ramosa increscentibus; cellulis quibuspiam se amplificare atque interne in paucas vel multas endosporas se dividere potentibus; reproductione a fragmentatione vel ab endosporis.

In this family, the plants, originally unicellular, grow eventually into strata or cushions from which filaments of cells penetrate the substratum. The solitary cells are basally attached to the substratum by a sheath of gelatinous material; cell division proceeds, at right angles to the axis of the cell, in an unequal fashion: the apical daughter cell is as a rule much smaller than the basal daughter cell. The upper part of the sheath is burst open, and the small daughter cell passes out of the mother cell sheath or develops *in situ* within the open sheath. By aggregate growth of these solitary cells and/or by successive equal divisions of the daughter cells in three planes perpendicular to each other, first a stratum, then a cushion (sometimes large and bullose) of vaguely radial structure is formed. The basal cells of these cushions and strata elongate and grow as uni- or pluriseriate, often branched, filaments downward into the substratum. Any of the solitary cells and any of the cells in the superficial layers of the stratum or cushion may enlarge and divide internally, wholly or in part, into numerous small cells (endospores), each of which is capable of growing as a solitary cell or as an endosporangium. Reproduction takes place also by fragmentation of the cushions. The gelatinous matrix of the cushion often becomes hydrolyzed, and the cells are thus dissociated.

One genus:

Kützing, Phyc. Gener., p. 177. 1843. *Gloeocapsa* Subgenus *Entophysalis* Elenkin, Not. Syst. Inst. Crypt. Horti Bot. Petropol. 2: 69. 1923. —Type species: *Entophysalis granulosa* Kütz.

Hydrococcus Kützing, Linnaea 8: 380. 1833; non Link, 1833. *Chroococcus* Subgenus *Hydrococcus* Elenkin, Not. Syst. Inst. Crypt. Horti Bot. Petropol. 2: 68. 1923. —Type species: *Hydrococcus rivularis* Kütz.

Exococcus Nägeli, Die Neuern Algensyst., p. 170. 1847. —Type species: *E. ovatus* Näg.

Thaumaleocystis Trevisan, Sagg. Monogr. Alg. Coccot., p. 79. 1848. —Type species: *Coccochloris deusta* Menegh.

Dermocarpa Crouan fr., Ann. Sci. Nat. IV. Bot 9: 70. 1858. —Type species: *D. violacea* Crouan fr.

Chamaesiphon A. Braun & Grunow in Rabenhorst, Fl. Eur. Algar. 2: 148. 1865. *Brachythrix* A. Braun *pro synon.* in Rabenhorst, loc. cit. 1865. *Chamaesiphon* Sectio *Brachythrix* A. Braun ex Hansgirg, Prodr. Algenfl. Böhmen 2: 124. 1892. *Chamaesiphon* Subgenus *Brachythrix* A. Braun ex Forti, Syll. Myxophyc., p. 138. 1907. *Chamaesiphon* Sectio *Euchamaesiphon* Geitler, Beih. z. Bot. Centralbl., II, 41: 250. 1925. —Type species: *Chamaesiphon confervicola* A. Br.

Sphaenosiphon Reinsch, Contrib. Algol. & Fungol. 1: 15. 1874. —Type species: *S. cuspidatus* Reinsch.

Placoma Schousboe & Thuret in Bornet & Thuret, Notes Algol. 1: 2. 1876. *Gloeocapsa* Subgenus *Placoma* Elenkin, Not. Syst. Inst. Crypt. Horti Bot. Petropol. 2: 69. 1923. —Type species: *Placoma vesiculosa* Schousboe & Thur.

Xenococcus Thuret in Bornet & Thuret, Notes Algol. 2: 73, 75. 1880. —Type species: *X. Schousboei* Thur.

Sphaerogonium Rostafinski, Rozpr. Akad. Umiej. Krakow., Wydz. Mat.-Przyr., 10: 304. 1883. *Chamaesiphon* Sectio *Sphaerogonium* Rostafinski ex Hansgirg, Prodr. Algenfl. Böhmen 2: 123. 1892. *Chamaesiphon* Subgenus *Sphaerogonium* Hansgirg ex Forti, Syll. Myxophyc., p. 136. 1907. —Type species: *Chamaesiphon incrustans* Grun.

Godlewskia Janczewski, Ann. Sci. Nat. VI. Bot. 16: 227. 1883. *Chamaesiphon* Sectio *Godlewskia* Geitler, Beih. z. Bot. Centralbl., II, 41: 250. 1925. —Type species: *Godlewskia aggregata* Jancz.

Pleurocapsa Thuret in Hauck, Meeresalg., p. 515. 1885. —Type species: *P. fuliginosa* Hauck.

Hyella Bornet & Flahault, Journ. de Bot. 2: 163. 1888. —Type species: *H. caespitosa* Born. & Flah.

Cyanoderma Subgenus *Myxoderma* Hansgirg, Notarisia 3: 588. 1888. —Type species: *Cyanoderma rivulare* Hansg.

Radaisia Sauvageau, Journ. de Bot. 9: 374. 1895. —Type species: *R. Gomontiana* Sauvag.

Chlorogloea Wille, Nyt Mag. Naturvid. 38(1): 5. 1900. *Chroococcus* Subgenus *Chlorogloea* Elenkin, Not. Syst. Inst. Crypt. Horti Bot. Petropol. 2: 68. 1923. —Type species: *Palmella tuberculosa* Hansg.

Hyellococcus Schmidle, Allgem. Bot. Zeitschr. 1905: 64. 1906. —Type species: *H. niger* Schmidle.

Guyotia Schmidle, Allgem. Bot. Zeitschr. 1905: 64. 1906. —Type species: *G. singularis* Schmidle.

Siphononema Geitler, Beih. z. Bot. Centralbl. II, 41: 251. 1925; Arch. f. Protistenk. 51: 332. 1925. —Type species: *Pleurocapsa polonica* Racib.

Gloeocapsopsis Geitler, Beih. z. Bot. Centralbl. II, 41: 229. 1925. —Type species: *Protococcus crepidinum* Thur.

Myxohyella Geitler, Beih. z. Bot. Centralbl. II, 41: 246. 1925. —Type species: *Hyella socialis* Setch. & Gardn.

Nematoradaisia Geitler, Beih. z. Bot. Centralbl., II, 41: 242. 1925. —Type species: *Radaisia Laminariae* Setch. & Gardn.

Radaisiella Geitler, Beih. z. Bot. Centralbl., II, 41: 242. 1925. *Geitleriella* J. de Toni, Noter. Nomencl. Algol. 8. 1936. —Type species: *Radaisia subimmersa* Setch. & Gardn.

Chroococcopsis Geitler, Arch. f. Protistenk. 51: 342. 1925; Beih. z. Bot. Centralbl., II, 41: 241. 1925; in Pascher, Süsswasserfl. 12: 125. 1925. —Type species: *C. gigantea* Geitl.

Chamaesiphon Sectio *Brachythrix* Geitler, Beih. z. Bot. Centralbl., II, 41: 250. 1925. —Type species: *Sphaerogonium fuscum* Rostaf.

Chamaesiphonopsis Fritsch, New Phytol. 28: 173. 1929. —Type species: *Xenococcus britannicus* Fritsch.

Cyanodermatium Geitler, Arch. f. Hydrobiol., Suppl. XII, 4: 627. 1933. —Type species: *C. gelatinosum* Geitl.

101

The following generic names have type species for which no original specimens have been available to us:

Cyanocystis Borzi, N. Giorn. Bot. Ital. 14: 314. 1882. —Type species: *C. versicolor* Borzi.

Askenasya Möbius, Ber. Deutsch. Bot. Ges. 5: 1xii. 1887. —Type species: *A. polymorpha* Möb.

Chondrocystis Lemmermann, Abh. Nat. Ver. Bremen 16: 353. 1899. *Gloeocapsa* Subgenus *Chondrocystis* Elenkin, Not. Syst. Inst. Crypt. Horti Bot. Petropol. 2: 69. 1923. —Type species: *Chondrocystis Schauinslandii* Lemm.

Dermocarpella Lemmermann, Engler Bot. Jahrb. 38: 349. 1907. —Type species: *Chamaesiphon hemisphaericus* Lemm.

Lithocapsa Ercogovic, Acta Bot. Inst. Bot. Univ. Zagreb. 1: 82. 1925. —Type species: *L. fasciculata* Erceg.

Lithococcus Ercegovic, Acta Bot. Inst. Bot. Univ. Zagreb. 1: 83. 1925. —Type species: *L. ramosa* Erceg.

Pseudocapsa Ercegovic, Acta Bot. Inst. Bot. Univ. Zagreb. 1: 95. 1925. —Type species: *P. dubia* Erceg.

Solentia Ercegovic, Acta Bot. Inst. Bot. Univ. Zagreb. 2: 78. 1927. —Type species: *S. stratosa* Erceg.

Aspalatia Ercegovic, Acta Bot. Inst. Bot. Univ. Zagreb. 2: 81. 1927. —Type species: *A. crassior* Erceg.

Hormathonema Ercegovic, Arch. f. Protistenk. 66: 165. 1929. —Type species: *H. paulocellulare* Erceg.

Dalmatella Ercegovic, Acta Bot. Inst. Bot. Univ. Zagreb. 4: 39. 1929. —Type species: *D. buaensis* Erceg.

Tryponema Ercegovic, Arch. f. Protistenk. 66: 168. 1929. —Type species: *T. endolithicum* Erceg.

Scopulonema Ercegovic, Arch. f. Protistenk. 71: 365. 1930. —Type species: *S. Hansgirgianum* Erceg.

Podocapsa Ercegovic, Acta Bot. Inst. Bot. Univ. Zagreb. 6: 33. 1931. —Type species: *P. pedicellatum* Erceg.

Brachynema Ercegovic, Acta Bot. Inst. Bot. Univ. Zagreb. 6: 35. 1931. *Ercegovicia* J. de Toni, Noter. Nomencl. Algol. 8. 1936. —Type species: *Brachynema litorale* Erceg.

Epilithia Ercegovic, Rad. Jugoslov. Acad. 244 (Razr. Mat.-Prirod. 75): 141. 1932. —Type species: *E. adriatica* Erceg.

Plantae primo microscopicae aetate provecta macroscopicae, supra et intus substratum increscentes, cellulis solitariis forma diversis; cellulis pulvini sphaericis, ellipticis, cylindraceis, vel polyhedroideis, in seriebus indistincte radialibus vel erectis ordinatis; cellulis filorum intus substratum penetrantium cylindricis, hemisphaericis, sphaericis, ovoideis, vel polyhedroideis; endosporangiis diversiformibus; membrana endosporangii tenue.

All the species of Entophysalis are aquatic in at least relatively permanent bodies of water; none appears to survive long in habitats which dry out periodically. The growth-forms commonly encountered differ with the species: in *E. deusta* and *E. rivularis* the strata and cushions are most often seen; in *E. conferta* the endosporangia are the most generally collected growth-forms; in *E. Lemaniae* and *E. rivularis* the solitary vegetative cells are familiar; in *E. endophytica, E. deusta, E. conferta,* and *E. rivularis* the penetrating filaments are often found.

Key to species of Entophysalis:

1. Marine .... 2.

1. Freshwater .... 4.

2. On rocks, wood, and shells .... 1. E. DEUSTA

2. On larger algae and living animals .... 3.

3. Cells 1—2μ in diameter; plants yellowish in color ...... 2. E. ENDOPHYTICA

3. Cells larger; plants blue-green, violet, or red .... 3. E. CONFERTA

4. On rocks, wood, and shells .... 4. E. RIVULARIS

4. On larger plants .... 5. E. LEMANIAE

1. ENTOPHYSALIS DEUSTA Drouet & Daily, Lloydia 11: 79. 1948. *Coccoch-loris deusta* Meneghini, Atti 2. Riun. Sci. Ital. Torino 1840: 173. 1841. *Micro-cystis deusta* Meneghini, Mem. R. Accad. Sci. Torino, ser. 2, 5 (Sci. Fis. & Mat.): 81. 1843. *Bichatia deusta* Trevisan, Nomencl Algar. 1: 60. 1845. *Thaumaleo-cystis deusta* Trevisan, Sagg. Monogr. Alg. Coccot., p. 79. 1848. *Gloeocapsa deusta* Kützing, Sp. Algar., p. 224. 1849. —Type from Genoa, Italy (FI). FIG. 191.

Myrionema crustaceum J. Agardh, Alg. Mar. Medit. & Adriat., p. 32. 1842. *Entophysalis crustacea* Drouet & Daily, Butler Univ. Bot. Stud. 10: 222. 1952. —Type from Brioni island, Istria, Jugoslavia (LD). FIG. 250.

Palmella mediterranea Kützing, Phyc. Gener., p. 171. 1843. *Brachtia mediterranea* Trevisan, Sagg. Monogr. Alg. Coccot., p. 58. 1848. *Palmella submarina* Crouan fr., Fl. Finistère, p. 109. 1867. —Type from Naples, Italy (L). FIG. 193.

Entophysalis granulosa Kützing, Phyc. Gener., p. 177. 1843. *Corynephora granulosa* Kützing pro synon., loc. cit. 1843. —Type from Split, Jugoslavia (L). FIG. 187.

Protococcus crepidinum Thuret, Mém. Soc. Nat. Sci. Nat. Cherbourg 2: 388. 1854. *Pleurococcus crepidinum* Rabenhorst, Fl. Eur. Algar. 3: 25. 1868. *Gloeocapsa crepidinum* Thuret in Bornet & Thuret, Notes Algol. 1: 1. 1876. *Chroococcus crepidinum* Hansgirg, Physiol. & Algol. Stud., p. 97, 152. 1887. *Pleurocapsa crepidinum* Ercegovic, Arch. f. Protistenk. 71: 364. 1930. *P. Ercegovicii* J. de Toni, Noter. Nomencl. Algol. 1: 7. 1934. —Type from Cherbourg, France (PC). FIG. 186.

Dermocarpa violacea Crouan fr., Ann. Sci. Nat. IV. Bot. 9: 70. 1858. —Type from Brest, France (PC).

Palmella oceanica Crouan fr., Fl. Finistère, p. 110. 1867; Crouan fr. in Desmazières, Pl. Cryptog. France, ed. 2, no. 535. —Type from near Brest, France (PC). FIG. 190.

Placoma vesiculosa Schousboe & Thuret in Bornet & Thuret, Notes Algol. 1: 2. 1876. —Type from Tangier (PC).

Pleurocapsa fuliginosa Hauck, Meeresalg. Deutschl. & Oesterr., p. 515. 1885. —Type from Trieste (L). FIG. 189.

Hyella caespitosa Bornet & Flahault, Journ. de Bot. 2: 163. 1888. —Type from Le Croisic, France (PC).

Xenococcus concharum Hansgirg, Oesterr. Bot. Zeitschr. 39: 5. 1899. —Type from Zadar, Jugoslavia (W). FIG. 192.

Pleurocapsa fluviatilis var. *subsalsa* Hansgirg, Sitzungsber. K. Böhm. Ges. Wiss., Math.-Nat. Cl., 1890(1): 18. 1890. *Oncobyrsa rivularis* var. *subsalsa* Hansgirg ex Forti, Syll. Myxophyc., p. 115. 1907. —Type from near Parenzo, Istria, Jugoslavia (W). FIG. 188.

Aphanocapsa concharum Hansgirg, Sitzungsber. K. Böhm. Ges. Wiss., Math.-Nat. Cl., 1890(1): 19. 1890. —Type from Pola, Jugoslavia (W). FIG. 247.

Hyella caespitosa var. *spirorbicola* Hansgirg, Sitzungsber. K. Böhm. Ges. Wiss., Math.-Nat. Cl., 1892: 226. 1892. —Type from Ragusa, Jugoslavia (W).

Aphanocapsa litoralis Hansgirg, Sitzungsber. K. Böhm. Ges. Wiss., Math.-Nat. Cl., 1892: 229. 1892. —Type from Pola, Jugoslavia (W). FIG. 248.

Aphanocapsa litoralis var. *macrococca* Hansgirg, Sitzungsber. K. Böhm. Ges. Wiss., Math.-Nat. Cl., 1892: 229. 1892. —Type from Pirano, Trieste (W).

Chroococcus atrochalybeus Hansgirg, Sitzungsber. K. Böhm. Ges. Wiss., Math.-Nat. Cl., 1892: 230. 1892. —Type from Orsera, Istria, Jugoslavia (W). FIG. 194.

Hyella caespitosa var. *nitida* Batters, Journ. of Bot. 34: 385. 1896. —Type from Plymouth, England (BM).

Pleurocapsa crepidinum Collins, Rhodora 3: 136. 1901. —Type from Magnolia, Massachusetts (FH).

Hyella Balanii Lehmann, Nyt Mag. Naturvid. 41: 85. 1903. —Topotypes and paratypes from near Aalesund, Norway (D, FC, L, NY, WU). FIG. 185.

Entophysalis violacea Weber-van Bosse, Siboga Exped., Liste des Algues 1: 7. 1913. —Type from Solor island, Indonesia (L).

Pleurocapsa magna Weber-van Bosse, Siboga Exped., Liste des Algues 1: 9. 1913. —Type from Solor, Indonesia (L).

Hyella Littorinae Setchell & Gardner, Univ. Calif. Publ. Bot. 6: 441. 1918. —Type from Carmel, California (UC).

Placoma violacea Setchell & Gardner in Gardner, Univ. Calif. Publ. Bot. 6: 456. 1918. —Type from Neah bay, Washington (UC).

Pleurocapsa entophysaloides Setchell & Gardner in Gardner, Univ. Calif. Publ. Bot. 6: 463. 1918. —Type from Carmel bay, California (UC).

Pleurocapsa gloeocapsoides Setchell & Gardner in Gardner, Univ. Calif. Publ. Bot. 6: 465. 1918. *Gloeocapsa gloeocapsoides* Geitler ex J. de Toni, Diagn. Alg. Nov., I. Myxophyc 2: 135. 1937. —Type from Alameda, California (UC).

Gloeocapsa bahamensis Collins in Britton & Millspaugh, Bahama Fl., p. 619. 1920. —Type from Mariguana, Bahama islands (NY).

Entophysalis violacea Collins, in Britton & Millspagh, Bahama Fl., p. 619. 1920. *E. Collinsii* J. de Toni, Noter. Nomencl. Algol. 1: 6. 1934. —Type from Atwood cay, Bahama islands (NY).

Polycystis clarionensis Setchell & Gardner, Proc. Calif. Acad. Sci., ser. 4, 22: 66. 1937. *Microcystis clarionensis* J. de Toni, Diagn. Algar. Nov., I. Myxophyc. 5: 494. 1938. —Type from Clarion island, Revilla Gigedo islands, Mexico (CAS).

Dermocarpa sphaerica var. *galapagensis* Setchell & Gardner, Proc. Calif. Acad. Sci., ser. 4, 22: 67. 1937. —Type from Narborough island, Galapagos islands (CAS).

Chroococcus calcicola Anand, Journ. of Bot. 1937 (Suppl. 2): 37. 1937. —Type from Westgate, England (in collection of F. E. Fritsch). FIG. 249.

Myxohyella seriata Hollenberg, Bull. Torr. Bot. Club 66: 489. 1939. —Type from San Pedro, California (in the herbarium of G. J. Hollenberg).

Entophysalis marginalis Hollenberg, Bull. Torr. Bot. Club 66: 490. 1939. —Type from San Pedro, California (in the herbarium of G. J. Hollenberg).

Microcystis splendens Hollenberg, Bull. Torr. Bot. Club 66: 493. 1939. —Type from culture at La Verne, California (in the herbarium of G. J. Hollenberg).

Microcystis ovalis Hollenberg, Bull. Torr. Bot. Club 66: 493. 1939. —Type from culture in La Verne, California (in the herbarium of G. J. Hollenberg).

Pleurocapsa Deeveyi Drouet, Field Mus. Bot. Ser. 20: 126. 1942. —Type from Willacy county, Texas (FC).

Original specimens have not been available to us for the following names; their original descriptions are here designated as the Types until the specimens can be found:

Haematococcus dalmaticus Zanardini, Sagg. di Class. d. Ficee, p. 64. 1843. *Protosphaeria dalmatica* Trevisan, Sagg. Monogr. Alg. Coccot., p. 28. 1848.

Protococcus glaucus Crouan fr., Fl. Finistère, p. 109. 1867. *Pleurococcus glaucus* Crouan fr. ex de Toni, Syll. Algar. 2: 692. 1889.

Palmella crustacea Zanardini, N. Giorn. Bot. Ital. 10: 40. 1878.

Hyella voluticola Chodat, Bull. Herb. Boissier, ser. 1, 5: 716. 1897.

Chondrocystis Schauinslandii Lemmermann, Abh. Nat. Ver. Bremen 16: 353. 1899.

Aphanocapsa litoralis var. *natans* Wille in Hansen, Ergebn. Plankton-Exped. Humboldt-Stift. 4Mf, Schizophyceen, p. 47. 1904.

Solentia stratosa Ercegovic, Acta Bot. Inst. Bot. Univ. Zagreb. 2: 78. 1927.

Solentia intricata Ercegovic, Acta Bot. Inst. Bot. Univ. Zagreb. 2: 80. 1927.

Aspalatia crassior Ercegovic, Acta Bot. Inst. Bot. Univ. Zagreb. 2: 81. 1927. *A. crassa* Ercegovic ex J. de Toni, Diagn. Algar. Nov. I. Myxophyc. 4: 351. 1938.

Aspalatia tenuior Ercegovic, Acta Bot. Inst. Bot. Univ. Zagreb. 2: 82. 1927.

Dalmatella buaensis Ercegovic, Acta Bot. Inst. Bot. Univ. Zagreb. 4: 39 . 1929.

Hormathonema paulocellulare Ercegovic, Arch. f. Protistenk. 66: 165. 1929.

Tryponema endolithicum Ercegovic, Arch. f. Protistenk. 66: 168. 1929.

Scopulonema Hansgirgianum Ercegovic, Arch. f. Protistenk. 71: 365. 1930.

Hormathonema luteo-brunneum Ercegovic, Arch. f. Protistenk. 71: 372. 1930.

Hormathonema violaceo-nigrum Ercegovic, Arch. f. Protistenk. 71: 372. 1930.

Solentia foveolarum Ercegovic, Arch. f. Protistenk. 71: 374. 1930.

Podocapsa pedicellatum Ercegovic, Acta Bot. Inst. Bot. Univ. Zagreb. 6: 33. 1931.

Brachynema litorale Ercegovic, Acta Bot. Inst. Bot. Univ. Zagreb. 6: 35. 1931.

Synechococcus marinus Ercegovic, Rad Jugoslov. Akad. 244(Razr. Mat.-Prirod. 75): 138. 1932.

Chroococcus membraninus var. *salinus* Ercegovic, Rad Jugoslov. Akad. 244(Razr. Mat. Prirod. 75): 138. 1932.

Entophysalis major Ercegovic, Rad Jugoslov. Akad. 244(Razr. Mat.-Prirod. 75): 139. 1932.

Pleurocapsa fissurarum Ercegovic, Rad Jugoslov. Akad. 244(Razr. Mat.-Prirod. 75): 140. 1932.

Epilithia adriatica Ercegovic, Rad Jugoslov. Akad. 244(Razr. Mat.-Prirod. 75): 141. 1932.

Scopulonema mucosum Ercegovic, Rad Jugoslov. Akad. 244(Razr. Mat.-Prirod. 75): 144. 1943.

Scopulonema Hansgirgianum f. *roseum* Ercegovic, Rad Jugoslov. Akad. 244(Razr. Mat.-Prirod. 75): 144. 1932.

Scopulonema brevissimum Ercegovic, Rad Jugoslov. Akad. 244(Razr. Mat.-Prirod. 75): 145. 1932.

Hyella dalmatica Ercegovic, Rad Jugoslov. Akad. 244(Razr. Mat.-Prirod. 75): 147. 1932.

Hyella tenuior Ercegovic, Rad Jugoslov. Akad. 244(Razr. Mat.-Prirod. 75): 147. 1932.

Dalmatella polyformis Ercegovic, Rad Jugoslov. Akad. 244(Razr. Mat.-Prirod. 75): 149. 1932.

Dalmatella violacea Ercegovic, Rad Jugoslov. Akad 244(Razr. Mat.-Prirod. 75): 150. 1932.
Dalmatella anomala Ercegovic, Rad Jugoslov. Akad. 244(Razr. Mat.-Prirod. 75): 151. 1932.
Dalmatella litoralis Ercegovic, Rad Jugoslov. Akad. 244(Razr. Mat.-Prirod. 75): 151. 1932.
Solentia achromatica Ercegovic, Rad Jugoslov. Akad. 244(Razr. Mat.-Prirod. 75): 153. 1932.
Hormathonema sphaericum Ercegovic, Rad Jugoslov. Akad. 244(Razr. Mat.-Prirod. 75): 155. 1932.
Hormathonema longicellulare Ercegovic, Rad Jugoslov. Akad. 244(Razr. Mat.-Prirod. 75): 155. 1932.
Hormathonema epilithicum Ercegovic, Rad Jugoslov. Akad. 244(Razr. Mat.-Prirod. 75): 155. 1932.
Pleurocapsa minuta Geitler, Rabenh. Krypt.-Fl. 14: 355. 1932. *Scopulonema minutum* Geitler in Engler & Prantl, Natürl. Pflanzenfam., ed. 2, 1b: 93. 1942.
Aphanocapsa Roberti-Lamii Frémy, Mém. Soc. Nat. Sci. Nat. & Math. Cherbourg 41: 16. 1934.
Aphanocapsa endolithica var. *marina* Frémy, Mém. Soc. Nat. Sci. Nat. & Math. Cherbourg 41: 17. 1914.
Myxosarcina salina Frémy in J. de Toni, Diagn. Algar. Nov. I, Myxophyc. 8: 702. 1946.
Nematoradaisia gasteropodum Gonzalez Guerrero, Anal. Jard. Bot. Madrid 7: 438. 1947.
Aspalatia andalousica Gonzalez Guerrero, Anal. Jard. Bot. Madrid 7: 438. 1947.
Xenococcus gaditanus Gonzalez Guerrero, Anal. Jard. Bot. Madrid 7: 438. 1947.
Dermocarpa gaditana Gonzalez Guerrero, Anal. Jard. Bot. Madrid 7: 442. 1947.

Plantae in stratum aut pulvinum microscopicum vel macroscopicum aerugineum, olivaceum, violaceum, roseum, brunneum, vel nigrum, firmum vel molle vel dissolvens increscentes, cellulis solitariis sphaericis, ovoideis, vel pyriformibus, diametro ad 10μ crassis; cellulis pulvini sphaeroideis, post divisiones praecipue polyhedroideis, diametro (1—)3—6μ crassis; cellulis filorum intus substratum penetrantium ovoideis, sphaericis, cylindraceis, vel polyhedroideis, vulgo uniseriatis interdum multiseriatis, diametro (1—)2—15μ crassis; endosporangiis sphaericis vel ovoideis, diametro ad 30μ crassis; gelatino vaginale primum hyalino demum lutescente vel raro rubrescente vel coerulescente, homogeneo vel lamelloso; protoplasmate aerugineo, olivaceo, luteolo, violaceo, vel roseo, vulgo homogeneo. FIGS. 185—194, 247—250.

Intertidal or below low tide mark on marine rocks, shells, wood, and soil, and on similar substrata in brackish water. If the substratum has been transported to a freshwater habitat, apparently this alga persists for a considerable period. The stratum or cushion type of growth-form, often with well developed filaments penetrating the substratum, is most usually encountered in *Entophysalis deusta*. On limestone and shells, a profusion of basal filaments is formed; whereas on igneous rocks and wood, these structures are either not produced or are restricted to elongated cells at the base of the stratum or cushion. Where the species persists after being introduced into freshwater habitats, the plants often simulate *Anacystis montana*. Strata on intertidal soil may sometimes appear very similar to those of *Coccochloris stagnina*. Below low tide level and in tide pools and brackish ponds, the gelatinous matrix of the upper layers of the stratum or cushion may hydrolyze, and the cells may appear very much like those of *Anacystis aeruginosa*. Fragments of the stratum or cushion may be mistaken as plants of *Anacystis dimidiata, A. thermalis, A. montana*, or *Palmogloea protuberans* (Sm. & Sow.) Kütz.

Fungi are often found parasitizing the cells; and according to Bornet & Thuret, Notes Algol. 1: 2 (1876), species of Verrucaria result from such parasitization.

Specimens examined:
NORWAY: Tromsö, *M. Foslie*, 10 Jun. 1887 (PC); in conchis Balanorum, Sliningen prope Aalesund, *N. Wille* (topotype and paratype of *Hyella Balanii* Lehm. in Mus. Vindob. Krypt. Exs. no. 2050, D[Fig. 185], FC, L, NY, WU). SWEDEN: Graen, Landskrona, Scania, *O. Nordstedt*, 1879

(D, S). DENMARK: in herb. Lyngbye (C). ROMANIA: apud Constanta, ad scopulos Ponti Euxini, *E. C. Teodorescu 485*, Apr. 1897 (W). JUGOSLAVIA: Istria: insula Brioni, *Biasoletto* (Type of *Myrionema crustaceum* J. Ag., LD [Fig. 250]); Fasana bei Pola, *A. Hansgirg*, Apr. 1889 (FC, W); Lussin, *Hansgirg*, Aug. 1888 (FC, W); auf Patella u. ä. Schalen an der Flutgrenze selten bei Orsera, *Hansgirg*, Apr. 1889 (Type of *Chroococcus atrochalybeus* Hansg., W; isotype, FC [Fig. 194]); auf Steinen in einem Bache, welcher in adriatisches Meer fliesst und angesalztes Wasser führt, zwischen Parenzo und Orsera, *Hansgirg*, Apr. 1889 (Type of *Pleurocapsa fluviatilis* var. *subsalsa* Hansg., W; isotype, FC [Fig. 188]); im adriatischen Meere, Umgebung von Pola, *Hansgirg*, Apr. 1889 (Type of *Aphanocapsa concharum* Hansg., W; isotype, FC [Fig. 247]); auf unreinen Molosteinen zwischen Flut- und Ebberspiegel bei Pola, *Hansgirg*, Aug. 1888 (Type of *A. litoralis* Hansg., W; isotype, FC [Fig. 248]); Pola, *Hansgirg*, Apr. 1889 (FC, W); Rovigno, *Hansgirg*, Apr. 1889 (FC, W); Volosca, *Hansgirg*, Aug. 1885 (FC, W). Croatia: Fiume, *Hansgirg*, Aug. 1885 (FC, W). Dalmatia: Castelnuovo, Cannosa prope Ragusa, Gravosa prope Ragusa, *Hansgirg*, 1891 (FC, W); op kleine kalkrotsen, Promaina, noord van Makarska, *J. J. ter Pelkwijk*, 21 Jul. 1939 (FC, L); an Spirorbis-Schalen bei Ragusa, *Hansgirg*, 1891 (Type of *Hyella caespitosa* var. *spirorbicola* Hansg., W; isotype, FC); ad rupes maritimas prope Spalatam, *F. T. Kützing*, 1835 (Type of *Entophysalis granulosa* Kütz., L [Fig. 187]; isotype, PC); Spalato, *Hansgirg*, Aug. 1889 (FC, W); Stagno Piccolo, *Hansgirg*, 1891 (FC, W); Zara, *Hansgirg* (PC); am Meeresufer bei Zara, *Hansgirg*, Aug. 1888 (Type of *Xenococus concharum* Hansg., W; isotype, FC [Fig. 192]). TRIESTE: ad Miramar prope Triest, *F. Hauck*, Sept. 1878 (as *Entophysalis granulosa* in Wittr. & Nordst., Alg. Exs. no. 294, FC, L, MIN); Muggia, *Hauck*, 19 Mar. 1879 (L, PC), 2 Oct. 1887 (DT, FH; as *E. granulosa* in Hauck & Richt., Phyk. Univ. no. 241b, L, MIN, PC); auf unreinen Molosteinen bei Pirano, *Hansgirg*, Apr. 1889 (Type of *Aphanocapsa litoralis* var. *macrococca* Hansg., W; isotype, FC); Servola, *Hauck*, Mar. 1883 (L; as *Entophysalis granulosa* in Hauck & Richt., Phyk. Univ. no. 241a, L, MIN, PC); St. Andrea, *Hansgirg*, 1891 (FC, W); salinae ad Strogniano, *Hansgirg*, Apr. 1889 (FC, W); Triest, *Hauck* (Type of *Pleurocapsa fuliginosa* Hauck, L [Fig. 189]; isotype, UC).

ITALY: Genova, *Meneghini* (Type of *Coccochloris deusta* Menegh., FI; isotypes, L [Fig. 191], PC, UC); Mergellina bij Napels, *J. T. Koster 205*, Apr. 1939 (L); Principia di Posilippo, *Koster 281*, May 1939 (L); Neapel, *F. T. Kützing* (Type of *Palmella mediterranea* Kütz., L [Fig. 193]; isotype, UC). NETHERLANDS: zeedijk van Morsele, Zuid Beveland, *C. Brakman 64*, Jan. 1941 (L); Ymuiden, *A. Weber-van Bosse 1228*, Sept. 1893 (L); Sloedam, *Brakman 63*, Sept. 1940 (L); zeedijk, Walcheren, *Brakman 62*, Feb. 1941 (L); Zuid-Sloe, Walcheren, *Brakman 345a*, May 1942 (L); Vlieland, *Weber-van Bosse*, May 1891 (L). BRITISH ISLES: Plymouth, *E. A. Batters*, Feb. 1896 (Type of *Hyella caespitosa* var. *nitida* Batt., in the slide collection, BM); chalk cliffs at Westgate, *Anand*, Dec. 1935 (Type of *Chroococcus calcicola* Anand in the collection of F. E. Fritsch; isotype, D [Fig. 249]); Point of Ayr, Flintshire, *Batters*, 20 Jul. 1886 (PC). FRANCE: Alpes Maritimes: Antibes, baie de l'Olivette, *E. Bornet*, 1 Mar. 1872, 23 Dec. 1875 (PC), Plage des Mielles, *Bornet*, 17 Jan. 1872 (PC), ad rupes maritimas, *Bornet*, 2 & 5 Mar. 1872 (FH, L, UC), ad lapides submersas in limite maris, *Bornet*, Dec. 1875 (as *Entophysalis granulosa* in Wittr., Nordst., & Lagerh., Alg. Exs. no. 1200, L, MIN, S); in testis Balanorum ad Trayas, *G. Lagerheim*, Feb. 1893 (as *Hyella caespitosa* in Wittr., Nordst., & Lagerh., Alg. Exs. no. 1199, L, WU). Basses Pyrenées: Biarritz, *Bornet & G. Thuret*, 19 Jun. 1868 (FC, L, PC), 19 Jul. 1870 (FC, FH, NY, PC), Jul. 1870 (as *Placoma vesiculosa* in Lloyd, Alg. de l'Ouest de Fr. no. 461, FC), 5 Jun. 1870 (FC); à Guethary, *C. Sauvageau*, Jul.-Aug. 1896 (FH, NY); Saint Jean de Luz, Jetée Sainte Barbe, *J. Feldmann* (as *P. vesiculosa* in Hamel, Alg. de Fr. no. 1, MICH). Calvados: Arromanches (DT), *Lebel 431* (L). Finistère: sur des fragments de faience dans le rade de Brest, *Crouan* (Type of *Dermocarpa violacea* Crouan fr., PC); sur les bords de la rivière marine de Penfield près Brest, *Crouan* (Type of *Palmella oceanica* Crouan fr., PC [Fig. 190]; isotype in Desmaz., Pl. Cryptog. Fr., ed. 2, no. 535, UC), 3 Nov. 1855 (FC); Brest, *Le Dantec*, 1888 (PC); Roscoff, *G. Thuret*, Oct. 1887 (PC). Hérault: Port de Cette, *C. Flahault*, Jun. 1882 (FH, S), 25 Dec. 1887 (PC), *Flahault 165*, 13 Jan. 1882 (PC). Ille et Vilaine: au quai de la plage des bains, St. Malo, *Bornet*, 19 Jun. 1972 (FC). Loire Inférieure: Le Croisic, *Bornet*, Sept. 1887 (Type of *Hyella caespitosa* Born. & Flah., PC; isotype, LD). Manche: Cherbourg, herb. Lebel no. 636 (FC); Cherbourg, sur les murs des quais, *G. Thuret*, 2 May 1854 (Type of *Protococcus crepidinum* Thur., PC; isotypes, FC, FH, L [Fig. 186], and as *Gloeocapsa crepidinum* in Wittr. & Nordst., Alg. Exs. no. 1100, L, S; as *Protococcus crepidinum* in Rabenh. Alg. no. 2032, FC, FH, L, PC, UC; as *P. crepidinum* in Le Jolis, Alg. Mar. de Cherbourg no. 16, L, PC); Cherbourg, *A. Le Jolis 449* (L, NY), *589* (L), *1561* (L), *1562* (DT, FH), *1563* (FH, L), May 1850—59.

SPAIN: San Vicente de la Barquera, *C. Sauvageau*, Sept. 1896 (PC). PORTUGAL: cultures from Setubal, *L. G. M. Baas Becking*, Sept. 1937 (L). TUNISIA: cote à Monastir, *M. Serpette T6*, Mar. 1947 (D). TANGIER: Tingi, *P. K. A. Schousboe*, Mar. 1827 (Type of *Placoma vesiculosa* Schousb. & Thur., PC; isotype, C); Algah et Dar Hamra prope Tanger, *Schousboe*, Mar., Dec. 1827 (PC, S). SIERRA LEONE: marine, Goderich, Freetown, *G. W. Lawson A628*, 6 Jul. 1953 (FC). GOLD COAST: Teshie, *Lawson A359*, 29 Jan. 1952 (FC); Sekondi, *Lawson A447*, 15 Dec. 1952 (D); Accra Beach, *Lawson A553*, 13 Mar. 1953 (FC). BELGIAN CONGO: crusts in salt water, *J. J. Symoens 430* (FC). CAPE OF GOOD HOPE: Knysna, *A. Weber-van Bosse*, 1894-95 (FC, L);

Mossel bay, *W. A. Setchell,* 12 May 1927 (FC, UC); on rocks, Muizenberg, *Setchell,* 19 May 1927 (FC, UC). BERMUDA: intertidal, Castle island, *T. A. & A. Stephenson BRCA6, BRCA7,* 22 Jul. 1952 (FC); on limestone and concrete, west end of the Causeway between St. Davids and Hamilton islands, *A. J. Bernatowicz 50-528,* 27 Dec. 1950 (FC); Cobbler's island, *Stephenson BRC13,* 4 Jul. 1952 (FC); Gibbet island, *Stephenson BRG3,* 6 Jul. 1952 (FC); crusts in pools, Gravelly bay, Hamilton island, *Bernatowicz 49-1584,* 5 May 1949 (FC); North Rock, *Stephenson BRN7,* 7-8 Jul. 1952 (FC); Smiths island, *Stephenson BRS2,* 7 Aug. 1952 (FC); Whalebone bay, St. Georges island, *Bernatowicz 49-91, W. R. Taylor & Bernatowicz 49-95,* 26 Feb. 1949 (FC), *Stephenson BRW5a,c,d,* 12 Jul. 1952 (FC). NOVA SCOTIA: on littoral rocks, Peggy's Cove, Halifax county, *T. A. & A. Stephenson NSP1a,* Jul.—Sept. 1948 (FC); in tidepool and on shells, *R. A. Lewin,* 3 Oct. 1952, Jul. 1953 (FC). MAINE: Eastport, *W. G. Farlow* (FC, FH, L, NY, PC); on shell, South Harpswell, *F. S. Collins 4704,* Jul. 1903 (NY); Harpswell, *Collins 4813* (NY); in tidepool, Center, *Collins,* Jul. 1892 (FH); Cutler, *Collins 4350,* Jul. 1902 (UC); Portland, *Collins 3,* Jul. 1881 (FH); Seal cove, Mount Desert island, *W. R. Taylor,* Sept. 1920 (TA); on Spirorbis, Spectacle island, *Collins* (FH); in dead shells, Spectacle island, *Collins,* Jul. 1894 (as *Hyella caespitosa* in Coll., Hold., & Setch., Phyc. Bor.-Amer. no. 302, L, TA, WU); Eagle island, Penobscot bay, *Collins,* Jul. 1892 (as *Chroococcus cohaerens* in Coll., Hold., & Setch., Phyc. Bor.-Amer. no. 701, FC, FH, L, NY, TA), Jul. 1896 (as *Gloeocapsa crepidinum* in Coll., Hold., & Setch., Phyc. Bor.-Amer. no. 351, FC, D, L, NY, TA), *Collins 701* (FH), *2137, 2138* (NY), *2139* (NY, UC), Jul. 1953 (FC); Cape Rosier, *Collins 1132, 2442* (FH, NY), *2695* (NY), *Collins,* Jul. 1894 (as *Entophysalis granulosa* in Coll., N. Amer. Alg. no. 1, NY), Jul. 1895 (as *E. granulosa* in Coll., Hold., & Setch., Phyc. Bor.-Amer. no. 152a, FC, FH, L, NY, TA), Jul. 1896 (as *Polycystis elabens* in Coll., Hold., & Setch., Phyc. Bor.-Amer. no. 1101, FC, FH, NY, TA), 14 Jul. 1898 (as *Chroococcus minutus* in Coll., Hold., & Setch., Phyc. Bor.-Amer. no. 951, FC, D, L, MICH).

MASSACHUSETTS: Lackey's bay, Elizabeth islands, *W. R. Taylor,* Jul. 1938 (TA); in shells, Black Rock, Fairhaven, *Taylor,* 27 Jul. 1938 (FC, TA); Sconticut Neck, Fairhaven, *Taylor,* Jul. 1941 (TA); on rocks, Quisset Harbor, Falmouth, *F. Drouet 1919,* 17 Aug. 1936 (D); in pools between Chara pond and Vineyard sound, Falmouth, *Drouet 2187,* 8 Sept. 1937 (D, FH, NY, S, UC); Gloucester, *W. G. Farlow* (FH, L, NY, S; as *Gloeocapsa crepidinum* in Farl., Anders., & Eat., Alg. Exs. Amer. Bor. no. 180, FC, NY, PC, UC); Kettle island, Magnolia, *Farlow,* Sept. 1903 (FH, WU); on Balanus and rocks near exit of drain pipe, Magnolia Point, *Farlow,* Sept. 1903 (Type of *Pleurocapsa crepidinum* Coll., FH; isotypes, FC, PC, and in Coll., Hold., & Setch., Phyc. Bor.-Amer. no. 1157, FC, FH, L, TA, WU); Marblehead, *Collins,* 17 Jun. 1902 (FC, UC); on rock in marsh pool, *Collins,* Sept. 1902 (NY, UC); Medford, *Collins,* May 1887 (NY); Kettle cove, Naushon island, *I. F. Lewis,* Aug. 1921 (TA); in shells, Meganset Beach, North Falmouth, *Taylor,* 21 Jul. 1931 (FC, L, TA); Penikese island, Gosnold, *M. S. Doty et al. 7018, 7283—7285, 7295,* Jul.—Sept. 1947, *Taylor et al. 7273, 7279, 7300,* 3 Aug. 1947 (FC); in marsh pool, Point of Pines, Revere Beach, *Collins 4661,* 8 May 1904 (FC, UC); Rockport, *Collins 1525,* Apr. 1890 (NY); Bakers island, Salem, *Collins 1865,* Jul. 1890 (NY); on bridge over ditch, Wellington, *Collins* (NY); Woods Hole, *Farlow,* Aug. 1876 (FH); in shells on beach, *W. A. Setchell 837,* 25 Aug. 1894 (FC, UC). RHODE ISLAND: on rocks, Grove point, Block island, *Drouet 3658,* 13 Jul. 1941 (FC); bottom of pool, Bateman's beach, Newport, *Farlow,* Jul. 1878 (FH); Silver Spring, *Collins 5404,* Apr. 1906 (NY); rocks, Newport, *Collins,* Jun. 1883 (NY, UC), *Collins 4495* (FC, NY, UC), *4497, 4498* (FC, UC), May 1904.

CONNECTICUT: on stones and woodwork, Bridgeport, *I. Holden 906, 907,* 25 Dec. 1893 (C; as *Pleurocapsa fuliginosa* in Coll., Hold., & Setch., Phyc. Bor.-Amer. no. 101a,b, FC, L, TA, WU), *Holden 596,* May 1892 (FH, NY), *Holden 968,* Aug. 1894 (UC); Seaside park, Bridgeport, *Holden 766* (NY, UC), *770* (UC), *773* (NY, UC), *778* (NY), Nov., Dec. 1892, *Setchell 733,* 22 Oct. 1893 (FC, UC); Middle Ground light, Bridgeport, *Holden,* Sept. 1890 (FH); Fairfield, *L. N. Johnson 1022,* 30 Aug. 1893 (FC); on shells, Ram island, *Setchell 711,* 13 Sept. 1893 (FC, UC); Stratford, *Johnson,* Aug. 1893 (FH); on stones, Fresh pond, Stratford, *Holden 709,* Sept. 1892 (FH), *Holden 870* (FH), *Holden 1132,* 11 Aug. 1895 (FH; as *Entophysalis granulosa* in Coll., Hold., & Setch., Phyc. Bor.-Amer. no. 152b, FC, D, L, TA). NEW JERSEY: Atlantic City, *S. R. Morse,* Apr. 1888 (NY, PC), Jun. 1888 (L), *Collins,* Spring 1891 (FC, UC), *Collins 2088— 2091,* May 1891 (NY); beach on Delaware bay, Cape May point, *W. R. Taylor,* Apr. 1922 (TA). DELAWARE: on wooden jetty near Hotel Henlopen, Rehoboth Beach, *Drouet & H. B. Louderback 8562,* 24 Aug. 1948 (FC). MARYLAND: on hardened soil, shore of Tangier sound near Prickly point, Fairmount, *Drouet & P. W. Wolle 3642,* 23 Jul. 1940 (FC); on mud, Wenona, Somerset county, *Wolle & Drouet 2275,* 22 Aug. 1938 (D); in Patuxent river, Solomons Island, *Drouet, E. P. Killip, & F. R. Fosberg 3995, 3995a,* 26 Jul. 1941 (FC, US); salt marsh flat, Chesapeake Beach, Calvert county, *E. C. Leonard 18741,* 17 Aug. 1939 (FC, US). VIRGINIA: York river at Yorktown, *J. C. Strickland 1218,* 22 Jul. 1942 (FC, ST), *H. J. Humm,* 2 Mar. 1948 (FC). NORTH CAROLINA: on wooden breakwater, Wrightsville Beach, New Hanover county, *H. J. Humm C,* 24 May 1946 (FC); near Beaufort, *T. A. & A. Stephenson BM(S)11, BSB8,* Apr.—May 1947 (FC, UC); Shackleford breakwater, Beaufort, *C. S. Nielsen & Humm 1645,* 6 Aug. 1949 (FC, T); beach near Duke University Laboratory pier, Beaufort, *Humm,* 15 Jan. 1949 (FC); on shells, Pivers island,

Beaufort, *H. L. Blomquist 13697,* 26 Jun. 1945 (FC); breakwater, Bogue sound north of Fort Macon, Beaufort, *Humm,* 28 Apr. 1947 (FC); Harkers Island, Carteret county, *Blomquist 13702,* 25 Aug. 1935 (FC), *Humm & Nielsen 1728,* 10 Aug. 1949 (FC, T). GEORGIA: salt creeks, Darien, *H. W. Ravenel 308* (PC).

FLORIDA: Bay county: St. Andrews bay, Cove Hotel, Panama City, *Drouet & Nielsen 11626,* 30 Jan. 1949 (FC, T); St. Andrews bay at Hathaway bridge west of Panama City, *Drouet & Nielsen 10927,* 15 Jan. 1949 (FC, T), *G. C. Madsen, A. L. Pates, M. N. Hood, & L. Elias 1941,* 11 Sept. 1949 (FC, T). Broward county: in the Intracoastal waterway, Dania Beach, *Drouet & H. B. Louderback 10265,* 28 Dec. 1948 (FC); in mangrove swamp south of South lake, Hollywood, *Drouet 10306, 10311,* 29 Dec. 1948 (FC). Collier county: Marco Island, *P. C. Standley 92803, 92818, 92820, 92822, 92823, 92840,* 14 Mar. 1946 (FC). Dade county: Cocoanut Grove, Soldier's key, and Cape Florida, *R. Thaxter,* 1897 (FH); Biscayne bay, Cutler, *H. J. Humm,* 10 & 13 Jan. 1946 (FC); on coral among mangroves, Miami, *M. A. Howe 2794,* Mar. 1904 (D, NY). Duval county: jetty of St. Johns river mouth and pier 2 miles south, *Humm,* 19 Mar. 1948 (FC). Escambia county: on shells in Pensacola bay north of the bridge on U. S. highway 98, Pensacola, *Drouet, Nielsen, Madsen, D. Crowson, & Pates 10619a,* 8 Jan. 1949 (FC, T). Franklin county: on the west side of East point, southwest of Eastpoint, *Drouet & Nielsen 11668,* 31 Jan. 1949 (FC, T), *Madsen, Pates, Hood, & Elias 1951,* 11 Sept. 1949 (FC, T); shore west of Apalachicola, *Drouet & Nielsen 11655,* 31 Jan. 1949 (FC, T); docks in New river, Carrabelle, *Drouet & Nielsen 10968,* 16 Jan. 1949 (FC, T); at the mouth of New river, Carrabelle, *Drouet & Nielsen 11674,* 31 Jan. 1949 (FC, T); shore in southeast part of Apalachicola, *Drouet & Nielsen 11001b,* 16 Jan. 1949 (FC, T); north shore of St. Vincent sound about 10 miles west of Apalachicola, *Drouet & Nielsen 10973,* 16 Jan. 1949 (FC, T). Gulf county: shore in the municipal park, Port St. Joe, *Drouet & Nielsen 10946a,* 15 Jan. 1949 (FC, T). Lee county: Bonita Beach, *Standley 92797,* 10 Mar. 1946 (FC). Levy county: on shell, Cedar keys, *M. A. Brannon 301,* 6 Jun. 1945 (FC PC); Way key, Cedar keys, *Drouet & Nielsen 11105, 11118, 11147a, 11155, 11158a, 11164, 11177, 11193,* 22 Jan. 1949 (FC, T). Monroe county: saline flats, Big Pine key, *E. P. Killip & J. F. Macbride,* Apr. 1951 (FC), *Killip 41765, 41810, 41858, 41950,* Jan.—Feb. 1952 (FC, US); plaster on stairway, Fort Jefferson, Garden key, Dry Tortugas, *W. R. Taylor 318,* 21 Jul. 1924 (FC, TA); corals and conch shells, Loggerhead and Garden keys, Dry Tortugas, *Taylor 505, 1061,* Jun. 1924, 1925 (TA); beachrock, west side of Loggerhead key, *R. N. Ginsburg 501,* Apr. 1952 (FC); Key Largo and North Key Largo, *Ginsburg LS1—3,* 1952 (FC); Largo sound, *L. B. Isham 26,* Nov. 1952 (FC); in brackish pond, Key West, *M. A. Howe 1736,* 14 Nov. 1902 (D); Plantation key, *T. A. & A. Stephenson PKIIA,* Jan.—Mar. 1947 (FC, UC); Saddlebunch keys, *J. A. Steyermark 63207b, 63208a,* 11 Mar. 1946 (FC); Soldier key, *Stephenson SKIIIC, SKIIIGa,* Jan.—Mar. 1947 (FC, UC); Teatable key, *Isham 24,* 1952 (FC); brackish pond, Key Vaca near Marathon, *Ginsburg NM1, NM2,* Apr. 1952 (FC). Okaloosa county: in Choctawhatchee bay at east end of Santa Rosa island, *Drouet, Nielsen, Madsen, Crowson, & Pates 10637,* 9 Jan. 1949 (FC, T); in Santa Rosa sound, Fort Walton, *Drouet, Nielsen, Madsen, Crowson, & Pates 10655a,* 9 Jan. 1949 (FC, T). Palm Beach county: in Lake Worth, West Palm Beach, *Drouet & Louderback 10193, 10196a, 10197,* 23 Dec. 1948 (FC); in Lake Worth, Palm Beach, *Drouet & Louderback 10219,* 24 Dec. 1948 (FC). St. Johns county: bay between Fort Marion and bridge in St. Augustine, *Madsen, Pates, & S. Parker 2035, 2036a,* 2 Jan. 1950 (FC, T); beach, northeast side of Anastasia island, *Madsen, Pates, & M. E. Thomas 2043,* 2 Jan. 1950 (FC, T). Taylor county: shore of Steinhatchee river, Steinhatchee, *Drouet & Nielson 11231,* 23 Jan. 1949 (FC, T). Volusia county: in shells, Mosquito inlet, New Smyrna beach, *Humm 9,* 28 Feb. 1946 (FC); Marineland, Daytona Beach, *T. A. & A. Stephenson MIB,* 20—29 Mar. 1947 (FC). Wakulla county: jetty, St. Marks lighthouse, *Nielsen & Crowson,* May 1949 (FC, T); Live Oak point, *Nielsen, R. W. Hanks, & H. Kurz,* 16 Jan. 1953 (FC, T), *G. Grice,* 10 Oct. 1953 (FC, T).

ALABAMA: on pilings in Mobile Bay, Point Clear, Baldwin county, *Drouet & Louderback 10162,* 20 Dec. 1948 (FC). MISSISSIPPI: shore of Mississippi sound between Bay St. Louis and Waveland, *Drouet 9809, 9813, 9821a,* 8 Dec. 1948 (FC); shore west of the lighthouse, Biloxi, *Drouet 9951,* 13 Dec. 1948 (FC); in Biloxi bay at Biloxi Oyster Laboratory, Biloxi, *Drouet & A. J. Bajkov 10051, 10067a, 10068, 10074a, 10081, 10083,* Dec. 1948 (FC); at the municipal docks, Ocean Springs, *Drouet & R. L. Caylor 9898,* 12 Dec. 1948 (FC); Pascagoula Beach, *Nielsen 1390,* 17 Jun. 1949 (FC, T). LOUISIANA: Calcasieu parish: in a cove in the southeast part of Lake Prien, *Drouet 8777,* 30 Oct. 1948 (FC); on the beach, north shore of Lake Charles, Lake Charles, *Drouet 8766,* 28 Oct. 1948 (FC). Cameron parish: in a pool beside Calcasieu river southwest of the school, Cameron, *Drouet 8807,* 2 Nov. 1948 (FC); Broussard's beach on the Gulf of Mexico east of Cameron, *Drouet 8874,* 3 Nov. 1948 (FC). Jefferson parish: in the bayou at Collins Camp west of Chenier Caminada, *Drouet & P. Viosca Jr. 9460, 9462, 9463,* 26 Nov. 1948 (FC); in Marais Cochon and the Gulf of Mexico southwest of Chenier Caminada, *Drouet & Viosca 9480, 9484,* 27 Nov. 1948 (FC). Lafourche parish: in the salt marshes between Leeville and Chenier Caminada, *Drouet & Viosca 9450,* 26 Nov. 1948 (FC). St. Tammany parish: on concrete-work, shore of Lake Pontchartrain, Mandeville, *Drouet & L. H. Flint 9543,* 2 Dec. 1948 (FC). Vermilion parish: in Vermilion river in the west part of Abbeville, *Drouet & R. P.*

Ehrhardt 9097, 13 Nov. 1948 (FC). TEXAS: Cameron county: on jetty, Brazos Santiago, *R. Runyon 3949a,* 7 Jun. 1945 (FC); in Laguna Madre at Point Isabel, *Runyon 3924, 3964a, 3965, 3965d,f,* 30 Jun. 1944, 17 Jul. 1945 (FC). Hidalgo county: La Sal del Rey, *E. S. Deevey,* 27 Nov. 1941 (FC). Willacy county: La Sal Viejo, *Deevey,* 29 Jan. 1941 (Type of *Pleurocapsa Deeveyi* Dr., FC). BRITISH COLUMBIA: Brandon island, Departure bay, near Nanaimo, *T. A. & A. Stephenson NB66,* Summer 1947 (FC). WASHINGTON: Neah Bay, *N. L. Gardner,* May 1917 (FC, L), *Gardner 3829a,* Jun. 1917 (Type of *Placoma violacea* Setch. & Gardn., UC; isotype, D).

CALIFORNIA: Alameda county: on logs in salt water, Alameda, *W. J. V. Osterhout & Gardner,* 26 Sept. 1903 (as *Gloeocapsa crepidinum* in Coll., Hold., & Setch., Phyc. Bor.-Amer. no. 1151, FC, L, TA), *Gardner 4119,* Oct. 1917 (Type of *Pleurocapsa gloeocapsoides* Setch. & Gardn., UC; isotypes, FC, L, TA); in jar of salt water, physiological laboratory, University of California, Berkeley, *Gardner 1702,* May 1906 (FC, UC); on shells, Bay Farm island, Alameda, *W. A. Setchell,* 6 Oct. 1898 (as *Hyella caespitosa* in Coll., Hold., Setch., Phyc. Bor.-Amer. no. LI, FC, L, TA, WU), *Setchell 2040—2043,* 3—6 Oct. 1898 (FC, UC), *Gardner 1195,* Sept. 1903 (FC, L, TA, UC), *Gardner 3683,* 2 Mar. 1917 (FU, UC). Humboldt county: on piles, Humboldt bay, Eureka, *Gardner 7843,* 1 Jun. 1934 (FC, UC). Los Angeles county: San Pedro, *Setchell,* Dec. 1905 (FC, UC); in old salt water cultures, La Verne, *G. J. Hollenberg 2179CD,* 14 Dec. 1937 (Type of *Microcystis ovalis* Hollenb. and *M. splendens* Hollenb. in herb. G. J. Hollenberg; isotype, FC); Point Firmin, San Pedro, *Gardner 1906,* 17 Jan. 1908 (FC, UC); on municipal breakwater, Cabrillo beach, San Pedro, *Hollenberg 1562a,* 3 Oct. 1936 (Type of *Entophysalis marginalis* Hollenb. in herb. G. J. Hollenberg: isotypes, FC, D), *Hollenberg 2379B,* 22 Oct. 1938 (Type of *Myxohyella seriata* Hollenb. in herb. G. J. Hollenberg; isotype, FC); rocky shore at Long point, Santa Catalina island, *E. Y. Dawson,* 30 Nov. 1948 (FC). Marin county: on shells, Bolinas, *Setchell & M. B. Nichols 6400a,* 16—19 Nov. 1906 (FC, UC). Monterey county: on *Littorina planaxis,* Carmel, *Gardner,* Dec. 1915 (as *Hyella Littorinae* Setch. & Gardn. in Coll., Hold., & Setch., Phyc. Bor.-Amer. no. 2255, FC, TA), *Gardner 3108,* Dec. 1916 (Type of *H. Littorinae* Setch. & Gardn., UC; isotype, FC, L, TA); on *Acinaea patina,* Carmel, *Gardner 4531,* Dec. 1919 (FC, UC); on rocks, Carmel bay, *Setchell & R. E. Gibbs 3219,* 12 Jan. 1899 (Type of *Pleurocapsa entophysaloides* Setch. & Gardn., UC; isotypes as *P. fuliginosa* in Coll., Hold., & Setch., Phyc. Bor.-Amer. no. 704, FC, L, TA, UC, WU), *C. P. B.,* 17 May 1897 (D, UC), *Gardner 4527,* Dec. 1919 (FC, UC); rocks near Pebble Beach, Carmel bay, *Gardner 6405,* Dec. 1927 (FC, UC); Mussel point, *T. A. & A. Stephenson PGM100,* Aug.—Nov. 1947 (FC); Point Labos, *Stephenson PGL13,* Aug.—Nov. 1947 (FC); dredged 0.5 mile east of Monterey, *Hollenberg 3018,* 11 Aug. 1939 (FC); Chatauqua beach, Monterey, *Setchell 1639,* 15 May 1897 (FC, UC). San Mateo county: on *Littorina* sp., Moss Beach, *Gardner 2101,* Dec. 1909 (FC, UC). San Francisco: on piles, U. S. Life Saving Station, North beach, *Setchell 1100,* Dec. 1895 (UC); Lands End, *Gardner 4392,* 25 Jan. 1919 (FC, UC), *Gardner 4480,* Nov. 1919 (FC, TA, UC); Presidio, *Gardner 4479,* Nov. 1919 (FC, L, UC), *Gardner 7033,* 1 Oct. 1932 (FC, UC); Fort Point, *Gardner 4894,* Feb. 1924 (FC, TA), *Gardner 6179, 6180,* Apr. 1926 (FC, UC). Santa Barbara county: mouth of Cañada Lobos, Santa Rosa island, *P. C. Silva 3937,* 26 Jan. 1949 (FC, UC); Willows Anchorage, Santa Cruz island, *Silva 6019,* 11 Mar. 1950 (FC, UC). Sonoma county: Fort Ross, *Setchell 714* (UC).

BAHAMA ISLANDS: on rocks, Atwood (Samana) cay, *M. A. Howe 5282,* 3 Dec. 1907 (Type of *Entophysalis violacea* Coll., NY; isotypes, D, UC); roof of cavern, Lignum Vitae cay, Berry islands, *Howe 3654,* 2 Feb. 1905 (FC, NY); rocks on ocean beach west of Lerner Laboratory, Bimini island, *H. J. Humm 4, 15,* Apr. 1948 (FC); border of a salt pond at eastern end of Mariguana, *Howe 5538,* 12 Dec. 1907 (Type of *Gloeocapsa bahamensis* Coll., NY; isotypes, FC, UC). BRITISH WEST INDIES: on pneumatophores of Avicennia, Cockburn Harbor, South Caicos, *Howe 5583,* Dec. 1907 (FH). PUERTO RICO: auf Kalksteinmauer beim Meere in Park, Santurce, *N. Wille 35,* 25 Dec 1914 (FC, UC); Sardinera, Mona island, *C. A. Kaye,* 1950 (D). GUADELOUPE: shore, Vieux Bourg, *A. Questel 255,* 22 Feb. 1944 (FC, TA). JAMAICA: on walls of caverns below Fort Clarence, Kingston, *Howe 4652, 4653,* 21 Dec. 1906 (D, NY); on coral, Drunkenman's cay near Kingston, *Howe 4734,* 29 Dec. 1906 (D, NY); littoral cavern, Montego Bay, *Howe 4898, 4899,* 8 Jan. 1907 (D, NY); Beach Rock, Middle cay, Morant cays, *V. J. Chapman 4, 4a,* 1939 (FC). NETHERLANDS WEST INDIES: Bonaire & Klein Bonaire, *P. Wagenaar Hummelinck,* 1930 (L). MEXICO: Baja California: Ojito near Punta Maria, *E. Y. Dawson & D. Fork 1595a,* 14 Apr. 1946 (FC); reef at southern tip of Melpomene cove (South Anchorage), Guadalupe island, *P. C. Silva,* 1 Feb. 1950 (FC, UC); Cape San Lucas, *Dawson 7255,* 1 Apr. 1949 (FC); Guadalupe island, *Dawson 8208,* 17 Dec. 1949 (FC). Chiapas: in small marine estuary, Barra de Cahuacan, *Dawson 3833,* 16 Jan. 1947 (FC). Colima: on rocks, Clarion island, Sulphur bay, Revilla Gigedo islands, *J. T. Howell 569a,* 24 Mar. 1932 (Type of *Polycystis clarionensis* Setch. & Gardn., CAS), *Howell 335* (FC, UC). Sonora: shore of island at entrance to harbor, Guaymas, *F. Drouet & D. Richards 3085, 3085a,b,* 2 Dec. 1939 (FC); shore of harbor near west end of the Paseo, Guaymas, *Drouet & Richards 3119,* 4 Dec. 1939 (FC); mangrove swamp at mouth of Rio de Sonora, Bahia Kino, *Drouet & Richards 2897a,* 9 Nov. 1939 (FC); rocks on south side of Isla Patos, *Dawson 651,* 17 Feb. 1946 (FC); rocks at Puerto Lobos, Bahia Tepoca,

Dawson 814, 19 Feb. 1946 (FC); tidal flats of Rio Mayo on north side of Yavaros, southwest of Huatabampo, *Drouet & Richards 3194, 3216,* 11 Dec. 1939 (FC). COSTA RICA: Braxillito bay, *J. T. Howell 778e,* 2 Jul. 1932 (FC, UC); on rocks and shells, Port Parker near Salinas bay, *W. R. Taylor 39-81, 39-88,* 24-25 Mar. 1939 (FC, MICH). PANAMA: rocky shore near the anchorage, Taboga island, Bahia de Panama, *Taylor 39-622,* 2 May 1939 (FC, MICH). VENEZUELA: Golfo de Cariaco, herb. W. F. R. Suringar, (D, L). COLOMBIA: on rocks in the bay near Cartagena, *F. A. Barkley 18B002c,* 10 Jul. 1948 (FC). ECUADOR: on southeast side of Narborough island, Galapagos islands, *J. T. Howell 576,* 1 Jun. 1932 (Type of *Dermocarpa sphaerica* var. *galapagensis* Setch. & Gardn., CAS; isotypes, FC, UC); on rock, Osborne islet in Gardiner bay, Hood island, Galapagos islands, *W. R. Taylor,* 31 Jan. 1934 (D, TA); on lava in brackish lagoon, Punta Albemarle, Isla Isabela, Galapagos islands, *Taylor 34-95,* 12 Jan. 1934 (FC, MICH). PERU: Lima, *A. Maldonado 11,* Dec. 1940 (FC); in salt lakes, Huacho, prov. Chancai, *Maldonado 80, 89, 94, 95a,* Feb. 1942 (FC); Las Salinas, Lago Chilca, dept. Lima, *Maldonado 62, 69,* Jan. 1942 (FC). URUGUAY: La Paloma, dept. Rocha, *W. G. Herter 95007,* Nov. 1934 (FC). HAWAIIAN ISLANDS: Kaneohe bay, Oahu, *M. S. Doty 8823, 10695,* 22 Apr. 1951, 9 May 1953 (FC); on breakwater, beach laboratory, University of Hawaii, Honolulu, *Doty 995,* 1952 (FC). CHRISTMAS ISLAND: lagoon bottom, Paris, *F. R. Fosberg 13260a,* 28 Aug. 1926 (FC). TUAMOTU ARCHIPELAGO: beach sand and rocks, Raroia atoll, *Doty & J. Newhouse 11351, 11146, 11832,* Jul.—Aug. 1952 (FC); cultures from beach rock and sand, Raroia, *Doty & Newhouse 11286, 11303, 11445, 11671, 11876, 11877,* 29 Sept. 1953 (FC). SOCIETY ISLANDS: on rocks near Huau, Tahiti, *W. A. Setchell & H. E. Parks 5112,* 5 Jun. 1922 (UC). WAKE ISLAND: *J. Randall 53-83, 53-92, 53-95,* 6 Jun. 1953 (FC). MARSHALL ISLANDS: lagoon reef and shore, Ine island, Arno atoll, *L. Horwitz 9074, 9569, 9571,* 15 Jul., 16 Aug. 1951 (FC); tidepools, Eniirikku island, Bikini atoll, *W. R. Taylor 46-53,* Mar. 1946 (FC, MICH); tidepools, Erik island, Bikini atoll, *R. F. Palumbo Y8a,* 1 Aug. 1949 (FC); pool back of beach rock, Giriinien island, Eniwetok atoll, *Taylor 46-339,* 28 May 1946 (FC, MICH); on rocks, Runit island, Eniwetok atoll, *Palumbo 52-16,* 22 Oct. 1952 (FC); on rocks, Kwajalein islet, Kwajalein atoll, *G. J. Hollenberg 48C1,* 2 Jul. 1948 (FC), *F. R. Fosberg 34422,* 15 Mar. 1952 (FC, US); coral boulder on lagoon beach, Sibylla islet, Pokak atoll, *Fosberg 34501,* 21 Jul. 1952 (FC, US); beach rock, Utirik islet, Utirik atoll, *Fosberg 33718,* 3 Dec. 1951 (FC, US); beach rock, Wotho and Enejelto islets, Wotho atoll, *Fosberg 34423, 34425a,* 18—19 Mar. 1952 (FC, US). PHOENIX ISLANDS: coral rock, Canton island, *O. Degener 21344b,* Apr. 1951 (FC).

NEW ZEALAND: vertical cliff face, Scott's Point, *V. J. Chapman,* 1949 (FC), *V. W. Lindauer 2830,* 26 Dec. 1942 (D); Bay of Islands, *V. W. Lindauer 26,* 20 Mar. 1938 (FC); shell-boring, Parua bay near Whangarei, *W. A. Setchell 6158, 6160,* 1904 (FC, UC); Oneroa, Bay of Islands, *L. M. Jones,* 18 Jan. 1935 (as *Entophysalis crustacea* in Tild., So. Pacific Pl., 2nd Ser., no. 374, FC); Tahapuke bay, Bay of Islands, *A. & V. K. Nash,* 15 Jan. 1935 (as *E. crustacea* in Tild., So. Pacific Pl., 2nd Ser., no. 347h, FC); rocks, Taharanui bay, North island, *Lindauer 1996, 1997,* 14 Dec. 1940 (FC); rocks below lighthouse, Half Moon bay, Stewart island, *I. B. Warnock,* 23 Feb. 1935 (as *E. crustacea* in Tild., So. Pacific Pl., 2nd Ser., no. 734, FC); Bounty islands, *Chapman* (FC); salt water, Bay of Islands, *Lindauer 675,* Jun. 1937 (UC); Long Beach, Bay of Islands, *Lindauer A84,* Feb. 1936 (UC). JAPAN: Tsuro, Cape Ashizuri, Kochi pref., *I. Umezaki 1473,* 19 Jun. 1954 (D). INDOCHINA: Bay Mien (Baichua), *Station Maritime de Cauda 147,* Jun. 1937 (L). INDONESIA: Dobo, Gisser, Kangean, Noesa-laut, Roma, Saleyer, Seget (New Guinea), Bima (Soembawa), Lamakwera (Solor), *A. Weber-van Bosse,* 1899—1900 (FC, L); Lamakwera, Solor, *Weber-van Bosse* (Types of *Entophysalis violacea* Weber-van Bosse and *Pleurocapsa magna* Weber-van Bosse, L); Noimini-bocht (Timor), Elat (Kei islands), *Weber-van Bosse,* 1899—1900 (L); Binnenmeer, Roesa Linguette island, *van der Sande,* 21 Jul. 1908 (FC, L); noorden van Kampong Gersik Poetik, Madoera, *L. G. M. Baas Becking,* Jul. 1936 (L); zuidkuist Ost-Java bij Kampong Nglijip, *P. Groenhart e,* Oct. 1936 (L). AUSTRALIA: marine on rocks at Miami, Queensland, *A. B. Cribb D,* 4 Jan. 1949 (FC); Long Reef, Collaroy, New South Wales, *C. M. Crosby & L. A. Doore,* 2 Jan. 1935 (with *Rivularia atra* Roth in Tild., So. Pacific Pl. 2nd ser., no. 325, FC); in Lithophyllum, Pt. Puer, Port Arthur, Tasmania, *Cribb 39.6,* 8 Mar. 1950 (FC); on shells, Pelican island, Southport, Tasmania, *Cribb 44.3,* 21 Mar. 1950 (FC); Mangrove Point, Spencers Gulf, South Australia, *H. B. S. Womersley A15046,* 31 Dec. 1950 (FC). PALESTINE: plage, Jaffa, *R. Joffé 152,* Jan. 1897 (PC).

2. ENTOPHYSALIS ENDOPHYTICA Drouet & Daily, Lloydia 11: 79. 1948. *Chlorogloea endophytica* Howe, Mem. Torr. Bot. Club 15: 13. 1914. *Myxohyella Howei* Geitler, Beih. z. Bot. Centralbl., II, 41: 247. 1925. —Type from Bay of Sechura, Peru (NY).

Chlorogloea lutea Setchell & Gardner, Univ. Calif. Publ. Bot. 6: 434. 1918. *Myxohyella lutea* Geitler, Beih. z. Bot. Centralbl., II, 41: 247. 1925. —Type from Carmel bay, California (UC).
FIG. 195.

110

Plantae endophyticae, microscopicae, olivaceae vel luteolae, cellulis solitariis sphaericis, diametro ad 2μ crassis; cellulis pulvini et filorum penetrantium praecipuius multiseriatorum sphaeroideis vel polyhedroideis, diametro $1—2\mu$ crassis; endosporangiis quaerendis; gelatino vaginale hyalino, homogeneo; protoplasmate luteolo, luteo, vel olivaceo. FIG. 175.

In the superficial layers of various large marine algae (*Iridophycus, Zanardinula* spp., etc.). The penetrating filaments and small cushions are the usual growth-form seen in *Entophysalis endophytica*. Endosporangia have not been observed in the specimens available.

Specimens examined:

CALIFORNIA: on *Iridaea minor*, Carmel bay, Monterey county, *N. L. Gardner 2981a*, May 1915 (Type of *Chloroglea lutea* Setch. & Gardn., UC; isotype, FC [Fig. 195]); on *Iridaea laminarioides*, Pyramid point, Monterey county, *W. A. Setchell 5464*, 16 Jun. 1901 (FC, UC). PERU: in *Leptocladia peruviana*, Bay of Sechura, *R. E. Coker 157*, Apr. 1907 (Type of *Chloroglea endophytica* Howe, NY); in *Chondrus canaliculata* and *Gymnogongrus furcellata*, Chincha islands, *Coker 192, 193*, Jun. 1907 (NY); on Gigartina, Chincha Norte island, *W. Vogt*, 1941 (FC); on *Lessonia nigrescens*, Pescadores islands near Ancon, *Coker 09150*, Feb. 1907 (NY); on *Gymnogongrus furcellatus*, Callao, *Coker 17*, Dec. 1906 (NY). HAWAIIAN ISLANDS: on *Chnoospora fastigiata*, Makua, Oahu, *M. Reed 197*, 14 Apr. 1905 (FC, UC).

3. ENTOPHYSALIS CONFERTA Drouet & Daily, Lloydia 11: 79. 1948. *Palmella conferta* Kützing, Phyc. Germ., p. 149. 1845. *Cagniardia conferta* Trevisan, Sagg. Monogr. Alg. Coccot., p. 50. 1848. *Pleurocapsa conferta* Setchell, Univ. Calif. Publ. Bot. 4: 229. 1912. *Chlorogloea conferta* Setchell & Gardner, in Gardner, Univ. Calif. Publ. in Bot. 6: 432. 1918. *Microcystis conferta* Geitler ex J. de Toni, Diagn. Algar. Nov., I. Myxophyc. 2: 105. 1937. —Type from Cuxhaven, Germany (L). FIG. 211.

Hydrococcus marinus Grunow, Verh. K. K. Zool.-Bot. Ges. Wien 11: 420. 1861. *Oncobyrsa marina* Rabenhorst, Fl. Eur. Algar. 2: 68. 1865. *Placoma marina* Bornet & Thuret, Mém. Soc. Nat. Sci. Nat. Cherbourg 28: 179. 1892. —Type from Leucadia, Greece (W).

Sphaenosiphon cuspidatus Reinsch, Contrib. Algol. & Fungol. 1: 15. 1874. *Dermocarpa cuspidata* Geitler, Rabenh. Krypt.-Fl. 14: 391. 1932. —Type from the Adriatic sea (K).

Sphaenosiphon smaragdinus Reinsch, Contrib. Algol. & Fungol. 1: 16. 1874. *Dermocarpa smaragdina* Tilden, Minn. Alg. 1: 54. 1910. —Type from central Africa (K).

Sphaenosiphon prasinus Reinsch, Contrib. Algol. & Fungol. 1: 17. 1874. *Dermocarpa prasina* Bornet in Bornet & Thuret, Notes Algol. 2: 73. 1880. —Type from the Adriatic sea (K).

Sphaenosiphon olivaceus Reinsch, Contrib. Algol. & Fungol. 1: 17. 1874. *Dermocarpa olivacea* Tilden, Minn. Alg. 1: 55. 1910. —Type from Marseilles, France (K).

Sphaenosiphon incrustans Reinsch, Contrib. Algol. & Fungol. 1: 18. 1874. *Dermocarpa prasina* var. *incrustans* Holmes & Batters ex Forti, Syll. Myxophyc., p. 132. 1907. —Type from St. Lawrence river, Quebec (K).

Sphaenosiphon roseus Reinsch, Contrib. Algol. & Fungol. 1: 18. 1874. *Dermocarpa rosea* Batters, Hist. Berwickshire Nat. Club 12: 141. 1889. *Chamaesiphon roseus* Hansgirg, Sitzungsber. K. Böhm. Ges. Wiss., Math.-Nat. Cl., 1892: 227. 1893. *Dermocarpa violacea* var. *rosea* Holmes & Batters ex Forti, Syll. Myxophyc, p. 130. 1907. —Type from central Africa (K).

Sphaenosiphon Leibleiniae Reinsch, Contrib. Algol. & Fungol. 1: 103. 1874. *Dermocarpa Leibleiniae* Bornet in Bornet & Thuret, Notes Algol. 2: 73. 1880. —Type from Atlantic North America (K).

Xenococcus Schousboei Thuret in Bornet & Thuret, Notes Algol. 2: 73, 75. 1880. *Coleonema arenifera* Schousboe pro synon. in Bornet & Thuret, ibid. p. 74. 1880. *Dermocarpa Schousboei* Bornet in Batters, Hist. Berwickshire Nat. Club 12: 11. 1889. —Type from Tangier (PC). FIG. 210.

Oncobyrsa adriatica Hauck, Meeresalg. Deutschl. & Oesterr., p. 515. 1885. —Type from Trieste (L). FIGS. 208, 209.

Chamaesiphon roseus var. *major* Hansgirg, Sitzungsber. K. Böhm. Ges. Wiss., Math.-Nat. Cl., 1892: 227. 1893. *Dermocarpa rosea* var. *major* Hansgirg ex Forti, Syll. Myxophyc., p. 130. 1907. —Type from St. Andrea, Trieste (W). FIG. 196.

Oncobyrsa adriatica var. *micrococca* Hansgirg, Sitzungsber. K. Böhm. Ges. Wiss., Math.-Nat. Cl., 1892: 228. 1893. —Type from Zengg, Croatia, Jugoslavia (W).

Palmella tuberculosa Hansgirg, Sitzungsber. K. Böhm. Ges. Wiss., Math.-Nat. Cl., 1892: 240. 1893. *Chlorogloea tuberculosa* Wille, Nyt Mag. Naturvid. 38(1): 5. 1900. —Type from Ragusa, Jugoslavia (W).

Pleurocapsa amethystea Rosenvinge, Meddel. om Grønl. 3: 967. 1893. *Chroococcopsis amethystea* Geitler, Beih. z. Bot. Centralbl. II., 41: 241. 1925. —Type from Fiskernaes, Greenland (C).

Radaisia Gomontiana Sauvageau, Journ. de Bot. 9: 374. 1895. —Type from Biarritz, France (PC).

Dermocarpa biscayensis Sauvageau, Journ. de Bot. 9: 403. 1895. —Type from Biarritz, France (PC).

Dermocarpa strangulata Sauvageau, Journ. de Bot. 9: 403. 1895. —Type from Biarritz, France (PC).

Pringsheimia scutata f. *Cladophorae* Tilden, Amer. Alg. 4: 382. 1900. *Xenococcus Cladophorae* Setchell & Gardner in Gardner, Univ. Calif. Publ. Bot. 6: 461. 1918. —Isotypes from Baird point, Vancouver island, British Columbia (D, FC).

Dermocarpa fucicola Saunders, Proc. Washington Acad. Sci. 3: 397. 1901. —Type from near Seattle, Washington (FH).

Hyella endophytica Børgesen, Bot. Faeroes 2: 525. 1903. *Myxohyella endophytica* Geitler, Beih. z. Bot. Centralbl., II, 41: 247. 1925. —Type from Thorshavn, Faeroes islands (in the collection of F. Børgesen).

Dermocarpa Farlowii Børgesen, Bot. Faeroes 2: 525. 1903. —Isotype from Suderø, Faeroes islands (UC).

Pleurocapsa amethystea var. *Schmidtii* Collins in Collins, Holden & Setchell, Phycotheca Boreali-Americana 35: 1704. 1911. —Type from Einarslon, Iceland (C).

Dermocarpa Vickersiae Collins, Rhodora 13: 184. 1911. —Type from Valentia, Barbados (NY). FIG. 214.

Dermocarpa solitaria Collins & Hervey, Proc. Amer. Acad. Arts & Sci. 53: 17. 1917. —Type from Bermuda (NY). FIG. 197.

Xenococcus Chaetomorphae Setchell & Gardner, Univ. Calif. Publ. Bot. 6: 436. 1918. —Type from Cypress point, Monterey county, California (UC). FIG. 213.

Dermocarpa hemisphaerica Setchell & Gardner, Univ. Calif. Publ. Bot. 6: 438. 1918. —Type from Moss Beach, San Mateo county, California (UC). FIG. 201 (paratype from Monterey county, California, FC).

Dermocarpa pacifica Setchell & Gardner, Univ. Calif. Publ. Bot. 6: 439. 1918. —Type from Cypress point, Monterey county, California (UC). FIG. 207.

Dermocarpa suffulta Setchell & Gardner, Univ. Calif. Pub. Bot. 6: 440. 1918. —Type from Moss Beach, San Mateo county, California (UC).

Dermocarpa sphaeroidea Setchell & Gardner, Univ. Calif. Publ. Bot. 6: 440. 1918. —Type from San Francisco, California (UC).

Hyella linearis Setchell & Gardner, Univ. Calif. Publ. Bot. 6: 442. 1918. *Myxohyella linearis* Geitler, Beih. z. Bot. Centralbl., II, 41: 247. 1925. —Type from Coos bay, Oregon (UC). FIG. 206.

Hyella socialis Setchell & Gardner, Univ. Calif. Publ. Bot. 6: 443. 1918. *Myxohyella socialis* Geitler, Beih. z. Bot. Centralbl., II, 41: 246. 1925. —Type from Carmel bay, California (UC).

Radaisia Laminariae Setchell & Gardner, Univ. Calif. Publ. Bot. 6: 444. 1918. *Nematoradaisia Laminariae* Geitler, Beih. z. Bot. Centralbl., II, 41: 242. 1925. —Type from San Francisco, California (UC). FIG. 204.

Radaisia clavata Setchell & Gardner, Univ. Calif. Publ. Bot. 6: 445. 1918. —Type from San Francisco, California (UC).

Radaisia subimmersa Setchell & Gardner, Univ. Calif. Publ. Bot. 6: 446. 1918. *Radaisiella subimmersa* Geitler, Beih. z. Bot. Centralbl., II, 41: 242. 1925. *Geitleriella subimmersa* J. de Toni, Noter. Nomencl. Algol. 8. 1930. —Type from Carmel bay, California (UC). FIG. 215.

Radaisia epiphytica Setchell & Gardner, Univ. Calif. Publ. Bot. 6: 447. 1918. —Type from Carmel Bay, California (UC). FIG. 205.

Dermocarpa protea Setchell & Gardner, Univ. Calif. Publ. Bot. 5: 456. 1918. —Type from Whidbey island, Washington (UC). FIG. 200.

Dermocarpa sphaerica Setchell & Gardner, Univ. Calif. Publ. Bot. 6: 457. 1918. —Type from San Pedro, California (UC).

Xenococcus acervatus Setchell & Gardner, Univ. Calif. Publ. Bot. 6: 459. 1918. —Type from Berkeley, California (UC). FIG. 212.

Xenococcus Gilkeyae Setchell & Gardner, Univ. Calif. Publ. Bot. 6: 462. 1918. —Type from Sitka, Alaska (UC).

Xenococcus pyriformis Setchell & Gardner, Univ. Calif. Publ. Bot. 6: 463. 1918. —Type from Coos bay, Oregon (UC).

Chlorogloea regularis Setchell & Gardner, Proc. Calif. Acad. Sci., ser. 4, 12: 698. 1924. —Type from Tortuga island, Baja California (CAS).

Dermocarpa Reinschii Setchell & Gardner, Proc. Calif. Acad. Sci., ser. 4, 12: 699. 1924. —Type from Tortuga island, Baja California (CAS).

Dermocarpa Marchantae Setchell & Gardner, Proc. Calif. Acad. Sci., ser. 4, 12: 700. 1924. —Type from Santa Rosalia, Baja California (CAS).

Xenococcus deformans Setchell & Gardner, Proc. Calif. Acad. Sci., ser 4, 12: 701. 1924. —Type from San Francisquito bay, Baja California (CAS).

Pleurocapsa violacea Weber-van Bosse, Vidensk. Medd. Dansk Naturh. Foren. Kjøbenhavn 81: 61. 1926. —Type from Aroi islands, Indonesia (C).

Radaisia pusilla Weber-van Bosse, Vidensk. Medd. Dansk Naturh. Foren. Kjøbenhavn 81: 62. 1926. —Type from Kai islands, Indonesia (C).

Chamaesiphon clavatus Setchell & Gardner, Proc. Calif. Acad. Sci., ser. 4, 19: 118. 1930. — Type from Guadalupe island, Baja California (CAS).

Dermocarpa simulans Setchell & Gardner, Proc. Calif. Acad. Sci., ser. 4, 22: 66. 1937. —Type from North Seymour island, Galapagos islands (CAS).

Xenococcus angulatus Setchell & Gardner, Proc. Calif. Acad. Sci., ser. 4, 22: 67. 1937. —Type from Santa Maria bay, Baja California (CAS).

Xenococcus endophyticus Setchell & Gardner, Proc. Calif. Acad. Sci., ser. 4, 22: 67. 1937. —Type from Albermarle island, Galapagos islands (CAS).

Xenococcus violaceus Anand, Journ. of Bot. 1937 (Suppl. 2): 38. 1937. —Type from Westgate, England (in the collection of F. E. Fritsch). FIG. 199.

Dermocarpa Enteromorphae Anand, Journ. of Bot. 1937 (Suppl. 2): 42. 1937. —Type from Westgate, England (in the collection of F. E. Fritsch). FIG. 198.

Microcystis minuta Kylin, K. Fysiograf. Sällsk. i Lund Förh. 7(12): 22. 1937. —Paratype and topotype from Kristineberg, Sweden (D).

Dermocarpa suecica Kylin, K. Fysiograf. Sällsk. i Lund Förh. 7(12): 23. 1937. —Paratype and topotype from Kristineberg, Sweden (D).

Xenococcus pulcher Hollenberg, Bull. Torr. Bot. Club 66: 492. 1939. —Type from San Pedro, California (in the herbarium of G. J. Hollenberg). FIG. 202.

Dermocarpa sublitoralis Lindstedt, Fl. Mar. Cyanophyc. Schwed. Westküste, p. 30. 1943. —Isotype from Kristineberg, Sweden (D). FIG. 203.

Original specimens have not been available to us for the following names; their original descriptions are here designated as the Types until the specimens are found:

Chamaesiphon hemisphaericus Lemmermann, Abh. Nat. Ver. Bremen 16: 353. 1900.

Dermocarpella hemisphaerica Lemmermann, Engler Bot. Jahrb. 38: 349. 1907. *Dermocarpa Lemmermannii* Geitler in Engler & Prantl, Natürl. Pflanzenfam., ed. 2, lb: 106. 1942.

Dermocarpa Leibleiniae var. *pelagica* Wille in Hansen, Ergebn. Plankton-Exped. Humboldt-Stift. 4Mf., Schizophyc., p. 50. 1904.

Xenococcus laysanensis Lemmermann, Engler Bot. Jahrb. 34: 618. 1905.

Dermocarpella incrassata Lemmermann, Engler Bot. Jahrb. 38: 350. 1907. *Dermocarpa incrassata* Geitler, Beih. z. Bot. Centralbl., II, 41: 248. 1925.

Dermocarpa minima Geitler, Rabenh. Krypt.-Fl. 14: 392. 1932.

Microcystis minuta var. *aeruginosa* Lindstedt, Fl. Mar. Cyanophyc. Schwed. Westküste, p. 17. 1943.

Myxohyella Cavanillesiana Gonzalez Guerrero, Anal. Jard. Bot. Madrid 6: 247. 1946.

Dermocarpa Cavanillesiana Gonzalez Guerrero, Anal. Jard. Bot. Madrid 6: 249. 1946.

Plantae epi- vel endophyticae, microscopicae vel macroscopicae, primo uni-cellulares aetate provecta in stratum vel pulvinum olivaceum, aerugineum, violaceum, vel rubrum increscentes, cellulis solitariis sphaericis vel ovoideis vel pyriformibus, diametro ad 20µ crassis; cellulis strati vel pulvini sphaeroideis, post divisiones praecipue sphaericis vel polyhedroideis, diametro (1—)3—6µ crassis; cellulis filorum penetrantium sphaericis, ovoideis, cylindraceis, vel polyhedroideis, diametro 3—8µ crassis; endosporangiis sphaericis, obovoideis, cylindraceis, pyriformibus, vel tubaeformibus, diametro ad 50µ crassis, solitariis vel disciformiter aggregatis; protoplasmate aerugineo, olivaceo, violaceo, vel roseo, homogeneo vel tenui-granuloso. FIGS. 196—215.

Epiphytic and endophytic on larger algae of marine and brackish waters, the smaller growth-forms developing on ephemeral hosts, the more massive growth-

113

forms on perennial hosts. Unicellular and few-celled growth-forms of *Entophysalis conferta* appear in a great variety of shapes on the larger annual algae of tide pools; many of the cells grow directly into endosporangia. Disc-like groups of new single-celled plants, often developing directly into endosporangia, grow from the released endospores. In some cases the endospores penetrate the outer layers of the cell walls of the host and grow there as endosporangia or as single—many-celled vegetative plants. On perennial algae, especially those subject to wave action, the penetrating filaments are often well developed. Large and bullose cushions are rarely found, since they develop on perennial algae in quiet waters.

Specimens examined:

NORWAY: paa *Fucus vesiculosus*, Biologisk Station, Dröbak, *N. Wille*, Apr. 1917 (S); in *Laminaria digitata* ad Dröbak, *H. G. Simmons*, Apr. 1898 (S). SWEDEN: in *Cladophora rupestri* epiphytica ad Väderö Storö in Bahusia, *G. Lagerheim*, Aug. 1882 (as *Dermocarpa violacea* in Wittr. & Nordst., Alg. Exs. no. 900, FC, L, NY, S, UC); on Membranipora on *Laminaria Cloustonii* and *L. saccharina*, Kristineberg, *A. Lindstedt*, 8 Jul. 1950 (paratypes and topotypes of *Microcystis minuta* Kylin and *Dermocarpa suecica* Kylin, D); culture from Kristineberg, *A. Lindstedt*, 1950 (isotype of *D. sublitoralis* Lindst., D [Fig. 203]). GERMANY: Helgoland, *F. Hauck* (L); auf *Callithamnion Rothii*, Cuxhaven, *F. T. Kützing* (Type of *Palmella conferta* Kütz., L [Fig. 211]). ADRIATIC SEA: in Sphacelaria in *Sargasso obtusato*, Adria, herb. P. Reinsch (K); in Laurencia in Rhodophyc. major. cresc., herb. Reinsch (Type of *Sphaenosiphon prasinus* Reinsch, K); in *Sargasso obtusato*, herb. Reinsch (Type of *Sphaenosiphon cuspidatus* Reinsch, K). TRIESTE: *F. Hauck*, Oct. 1879 (Type of *Oncobyrsa adriatica* Hauck, L [Figs. 208, 209]); on Lyngbya, *Hauck* (L); in *Catenella Opuntia* (Gooden. & Woodw.) Grev., *Hauck* (as *Dermocarpa prasina* in Wittr. & Nordst., Alg. Exs. no. 693, FC, L, NY, S); Servola, *Hauck*, Mar. 1883 (L); Salinen, Zaule, *Hauck* (L); on *Lyngbya luteo-fusca*, Umago, *Hauck*, 15 Jun. 1879 (D, FH); an *Enteromorpha fucicola* bei St. Andrea nächst Triest, *A. Hansgirg* (Type of *Chamaesiphon roseus* var. *major* Hansg., W; isotype, FC [Fig. 196]). JUGOSLAVIA: Istria: auf *Catenella Opuntia* bei Mostecnice, *V. Nabelek*, 26 Jul. 1914 (WU); Portalbona (Porto Rabaz), *Catenellae Opuntiae* insidens, *J. Baumgartner* (as *Dermocarpa prasina* in Mus. Vindob. Krypt. Exs. no. 3138, FC, L, NY, S, UC); prope Rovingo in *Catenella Opuntia*, *J. N. F. Wille* (as *Dermocarpa prasina* in Mus. Vindob. Krypt. Exs. no. 1518, FC, L, NY, S). Croatia: auf *Catenella Opuntia*, Insel Veglia bei der Stadt Veglia, *M. Lusina*, 28 Apr. 1915 (WU); an der Fluthgrenze bei Zengg, *Hansgirg*, 1891 (Type of *Oncobyrsa adriatica* var. *micrococca* Hansgirg, W; isotype, FC). Dalmatia: auf Halimeda, Spalato, *Hauck* (L); bei Ragusa, auf *Gelidium capillaceum*, Hansgirg, 1891 (Type of *Palmella tuberculosa* Hansg., W; isotype, FC). GREECE: auf *Cladophoram catenatam* e portu Leucadiae, *Mazziari 1912* (Type of *Hydrococcus marinus* Grun., W).

MEDITERRANEAN SEA: in *Hypnea musciformis*, herb. P. Reinsch (K). ITALY: Porto Mauritzio, *Strafforello*, Jun. 1874 (PC); Villa Rocco Romano, Posilippo, Napels, *J. T. Koster 422*, Apr. 1939 (L); Santa Lucia, Napels, *Koster 420*, May 1939 (L); Villa d'Ambra, Posilippo, Napels, *Koster 423*, Apr. 1939 (L); Palazza Donn'Anna, Posilippo, Napels, *Koster 416*, May 1939 (L). NETHERLANDS: Delfszijl, op *Rhodochorton Rothii*, *Koster 1660*, 5 Jun. 1950 (L), op *Elachista fucicola*, *M. Weber 201*, Jun. 1885 (L); op *Polysiphonia urceolata* en *Chaetomorpha aerea*, Hoek van Holland, *Koster 192*, Feb. 1938 (L); op *Rhodochorton Rothii*, Condorpe bij Ellewontsdijk, Westerschelde, *C. Brakman 65*, Feb. 1941 (L); on *Enteromorpha clathrata*, Koffiehoek, island of Tholen, *C. den Hartog & A. F. Mulder*, 27 Nov. 1951 (D, FC, L); op *Polysiphonia urceolata* etc., Oosterschelde, bij Kattendijke, Zuid-Beveland, *J. J. R. Walrecht*, 10 Aug. 1951 (FC, L); on *Polysiphonia fastigiata* (Roth) Grev., harbor, Zoological Station, den Helder, *Koster*, 2 Dec. 1936 (D, FH, L). FAEROES ISLANDS: on *Rhodochorton Rothii*, Hojirj, Strφmφ, *F. Bφrgesen 202*, 24 Jun. 1895 (K); Trangiscaafjorder, Suderφ, *Bφrgesen 546*, May 1896 (FH, UC); Krivig, Strφmφ, *Bφrgesen 757*, 31 May 1896 (C, D, DT); in *Polysiphonia fastigiata*, Tveraa, Suderφ, *Bφrgesen 536*, May 1896 (isotype of *Dermocarpa Farlowii* Bφrges., UC); in *Laminaria hyperborea* (Gunn.) Fosl., Thorshavn, Strφmφ, *H. Jonsson*, 20 Nov. 1897 (Type of *Hyella endophytica* Bφrges., slide in the collection of F. Bφrgesen; isotype, C). GREAT BRITAIN: Scotland: op *Ectocarpus* sp., haven van Leith, *J. T. Koster 129*, Aug. 1936 (L). England: on *Catenella Opuntia*, seaside rocks, Devonshire (in M. Wyatt, Algae Danmonienses no. 126, FC); on *Polysiphonia fastigiata* on *Fucus nodosus*, Devonshire (in Wyatt, Algae Danmonienses no. 177, FC); Hastings, *E. Batters*, Apr. 1883 (L), *Batters 64*, 1886 (PC); Berwick-on-Tweed, *Batters 92, 106* (PC), *129* (PC, UC), Feb. 1887, Jan. 1889; Sidmouth, *G. W. T.*, Feb. 1886 (NY); Yarmouth, *Batters 124*, Jul. 1888 (PC); Muirhead, *E. M. Holmes*, Jan. 1887 (PC); on *Callithamnion Rothii* on buttresses of iron wall, Teignmouth, *Holmes*, 1 Jan. 1880 (PC); in *Catenella Opuntia*, Corvallis, herb. P. Reinsch (K); on Enteromorpha, chalk cliffs at Westgate, *Anand*, Oct. 1934 (Types of *Xenococcus violaceus* Anand and *Dermocarpa Enteromorphae* Anand in the collection of F. E. Fritsch; isotypes,

114

D [Figs. 198, 199]). IRELAND: on *Rhodochorton floridulum*, Clare island, *A. D. Cotton*, 1910—11 (NY).
FRANCE: sur *Lyngbya confervoides*, Banyuls-sur-Mer (Pyrenées Orientales), *J. Feldmann 2659*, Sept. 1932 (L, NY, S, TA); sur Lyngbya, Biarritz, *E. Bornet*, 8 Jun. 1870 (FC, FH, NY, PC, UC), 6 Aug. 1870 (PC); Biarritz, *L. K. Rosenvinge 527b*, Mar. 1885 (PC); sur *Fucus platycarpus* Thur., Port Vieux, Guéthary au Phare de Biarritz, *Sauvageau*, 20 Feb.—25 Mar. 1894 (Type of *Radaisia Gomontiana* Sauvag., PC; isotype, FC); sur Sargassum rejeté au Port Vieux, Guéthary au Phare de Biarritz, *Sauvageau*, 20 Feb.—25 Mar. 1894 (Type of *Dermocarpa biscayensis* Sauvag. and *D. strangulata* Sauvag., PC); Cherbourg, sur *Catenella Opuntia*, *Bornet*, 11 Aug. 1874 (FC, FH, PC); Fouras (Charente Inférieure), *L. Marchand*, 15 Sept. 1879 (PC); Le Croisic, *Bornet*, Aug. 1873 (PC), Sept. 1873 (FH, L, NY, PC), Apr. 1877 (L), May 1877 (FC, L, NY, PC; as *Dermocarpa prasina* in Lloyd, Alg. de l'Ouest de la France no. 389, PC); in *Chorda filum*, Mare Mediteraneum, Marseille, herb. Reinsch (Type of *Sphaenosiphon olivaceus* Reinsch, K); St. Malo, le Grand Bey, *Bornet*, 4 Jul. 1872 (PC); ile Ste. Marguerite près Cannes, *G. Thuret*, 20 Apr. 1856 (PC).
SPAIN: sur *Cystoseira ericoides* à Rivadeo, *C. Sauvageau*, Oct. 1895 (PC); La Corogne, *Sauvageau*, Oct.—Nov. 1895 (PC). TANGIER: Tingi, *P. K. A. Schousboe*, Dec. 1825 (Type of *Xenococcus Schousboei* Thur., PC; isotypes, FH, L [Fig. 210], UC). SIERRA LEONE: on *Lyngbya confervoides* Ag., Freetown and Kent, *G. W. Lawson A622, A632*, 4 & 7 Jul. 1953 (FC). GOLD COAST: on *L. confervoides* Ag., Axim, *Lawson A472*, 17 Dec. 1952 (FC); Dixcove, *Lawson A292*, 29 Dec. 1951 (D); marine, Christiansborg, *Lawson A573*, 27 Mar. 1953 (FC). CENTRAL AFRICA: in *Leibleinia capillacea* Kütz. in *Laurencia obtusa* Ag. crescente, herb. P. Reinsch (Type of *Sphaenosiphon smaragdinus* Reinsch, K); in *Gelidio corneo*, herb. P. Reinsch (Type of *Sphaenosiphon roseus* Reinsch, K). MOZAMBIQUE: auf Rhizoclonium in Quilimaneflusz bei der Insel Pequene, *Stuhlmann 83*, Feb. 1889 (as *Xenococcus Kerneri* in Hauck & Richt., Phyk. Univ. no. 685, L). UNION OF SOUTH AFRICA: Cape of Good Hope: on Dictyota, Knysna, *A. Weber-van Bosse*, 1894 (L); on *Platysiphonia intermedia* (Grun.) Bфrges., Plettenberg bay, *Weber-van Bosse 18*, 1894 (FC, L); on *Lyngbya confervoides* Ag. on rocks, Muizenberg, *W. A. Setchell*, 19 May 1927 (FC, UC).
ICELAND: on *Rhodochorton Rothii*, Einarslon, *H. Jonsson 728*, 20 Jul. 1897 (Type of *Pleurocapsa amethystea* var. *Schmidtii* Coll., C). BERMUDA: on mangrove near Flats bridge, *F. S. Collins 7262*, May 1912 (NY); on Bostrychia, Gravelly bay, *A. B. Hervey* (UC); on *Spermothamnion gorgoneum* and *Ceramothamnion Codii*, Harrington sound, *Hervey*, Dec. 1914 (Type of *Dermocarpa solitaria* Coll. & Herv. in Coll., Hold., & Setch., Phyc. Bor.-Amer. no 2155, NY; isotypes, FC [Fig. 179], D, L, UC); on *Catenella Opuntia*, Hungry bay, *Hervey 27*, Feb. 1909 (UC), *Collins 7202*, May 1912 (UC), *Collins*, 2 May 1913 (as *Dermocarpa prasina* in Coll., Hold., & Setch., Phyc. Bor.-Amer. no. 2051, FC, L, NY); epiphyte, Red Hole, eastern St. Davids island, *W. R. Taylor 49-1888*, 27 May 1949 (FC, MICH); Shelly beach, *Collins*, Aug. 1910 (FH); on *Lyngbya confervoides* in ditch back of Shelly bay, *Collins*, 15 Aug. 1913 (as *Xenococcus Schousboei* in Coll., Hold., & Setch., Phyc. Bor.-Amer. no. 2052, FC, L, NY, TA); on *Enteromorpha intestinalis* in ditch in South Shore meadows, *Collins*, 1 Sept. 1913 (as *Chroococcus membraninus* in Coll., Hold., & Setch., Phyc. Bor.-Amer. no. 2151, FC, D, L). CANARY ISLANDS: Puerto Orotava, *C. Sauvageau*, Jan. 1905 (PC). FERNANDO DE NORONHA: on Rhizoclonium on rocks, *L. G. Williams*, Mar. 1945 (FC). FALKLAND ISLANDS: sur *Dumontia fastigiata* (in Hohenacker, Alg. Mar. no. 282, PC; in herb. Reinsch, K). GREENLAND: Fiskernaes, *L. K. Rosenvinge*, 31 May 1888 (Type of *Pleurocapsa amethystea* Rosenv., C). ATLANTIC NORTH AMERICA: in Leibleinia in Plocamio, herb. P. Reinsch (Type of *Sphaenosiphon Leibleiniae* Reinsch, K). SAINT PIERRE & MIQUELON: on Cladophora, Pointe Blanche, *P. Le Gallo 5*, 30 Oct. 1945 (FC, TA). NOVA SCOTIA: Sydney, *F. S. Collins*, Sept. 1899 (NY, UC); on Cladophora in tidepool, Prospect, *R. A. Lewin*, 1 Oct. 1952 (FC). MAINE: on *Polysiphonia fastigiata*, Pond Cove, *W. R. Taylor*, Feb. 1928 (TA); Inner Marsh island, Casco bay, *Collins*, Aug. 1903 (NY, UC); on *Rhodochorton Rothii*, Eagle island, *Collins 2113*, Jul. 1891 (NY), *Collins 2702*, Jul. 1893 (NY); Casco bay, *Collins*, Aug. 1903 (NY); Long island, Casco bay, *Collins*, Sept. 1902 (NY); Northport, *Collins 1012*, Aug. 1887 (NY); Cutler, *Collins*, Jul. 1902 (NY); Buck island, *Collins 2942*, Jul. 1894 (NY); Harpswell, *Collins 4805*, 13 Jul. 1903 (FH, NY), *Collins 5080*, Jul. 1904 (NY); Seal Harbor, *Collins*, Jul. 1899 (NY); on *Polysiphnoia fastigiata*, Seguin island, *M. A. Howe 1125*, Aug.—Sept. 1900 (FH, NY); Pemaquid Point, *H. B. Louderback & C. Graham*, 5 July 1949 (D, FC). NEW HAMPSHIRE: Appledore, Isles of Shoals, *W. G. Farlow*, Sept. 1900 (FH); Newcastle, *Collins 3727*, May 1900 (NY).
MASSACHUSETTS: on *Polysiphonia fastigiata*, Clifton, *Collins & W. A. Setchell*, 20 Apr. 1889 (FC, NY, TA, UC); on Cladophora in upper tide pool, rocky shore, Cohasset, *Collins*, 12 Oct. 1901 (as *Xenococcus Kerneri* in Coll., Hold., & Setch., Phyc. Bor.-Amer. no. 952, FC, L, UC); on *Polysiphonia fastigiata*, Little Nahant, *W. A. Setchell 319*, 18 Apr. 1891 (FC, UC; as *Dermocarpa prasina* in Coll., Hold., & Setch., Phyc. Bor.-Amer. no. 1, FC, L, NY, TA, UC); on *Rhodochorton Rothii*, Little Nahant, *Setchell 292*, Nov. 1890 (UC); on *R. Rothii*, Marblehead, *Collins*, 17 Jun. 1902 (FH), Aug. 1902 (NY, UC); on Cladophora, Magnolia, *Farlow*, Sept. 1903 (FC, FH); on *Polysiphonia fastigiata*, Nahant, *Farlow* (FH, PC), *Collins*, Apr. 1895 (as *Dermocarpa prasina* in Collins, N. Amer. Alg. no. 3, D, NY); on Porphyra, Nahant, *Setchell*, Jun. 1889 (UC); on *Rhizoclonium riparium*, Nahant, *Collins*, 17 Aug. 1889 (FC, FH, NY, UC); on *Rhodochorton*

115

Rothii, Nahant, *Collins 1518*, 30 Mar. 1890 (FC, UC), *L. N. Johnson 821*, 7 May 1892 (FC); on *Polysiphonia fastigiata*, Swampscott, *Collins 2032*, Apr. 1891 (NY, UC); on *Chaetomorpha aerea*, Woods Hole, *Setchell 815*, 16 Aug. 1894 (FC, UC); on *Rhodochorton Rothii*, Eel pond, Woods Hole, *E. T. Rose*, 16 Jul. 1936 (D); on *Lyngbya confervoides* Ag., steamboat wharf, Woods Hole, *R. N. Webster & F. Drouet 2150*, 7 Jul. 1938 (D). RHODE ISLAND: on *Lyngbya semiplena*, Newport, *Collins 4221*, Aug. 1901 (FH, NY, UC); on *Polysiphonia fastigiata*, Newport, *H. M. Richards*, Dec. 1890 (UC); on *Enteromorpha intestinalis*, Easton's point, Newport, *Mrs. W. C. Simmons*, Nov. 1896 (NY), Sept. 1898 (as *Dermocarpa violacea* in Coll., Hold., & Setch., Phyc. Bor. Amer. no. 556, FC, L, NY, TA, UC). CONNECTICUT: on *Rhodochorton Rothii*, Seaside park, Bridgeport, *I. Holden*, 4 Jul. 1892 (FH), *Holden 768*, 11 Dec. 1892 (FH, NY); Black Rock, Bridgeport, *Holden 667*, 28 Jul. 1892 (FH); on Cladophora, Mill river, Fairfield, *L. N. Johnson 194*, 12 Sept. 1891 (FC); culture from Long Island sound, New Haven, *R. A. Lewin C*, 29 Feb. 1948 (FC). LABRADOR: in *Plocamio coccineo*, herb. Reinsch (K). QUEBEC: in *Polysiphonia fastigiata*, St. Lawrence river, herb. Reinsch (Type of *Sphaenosiphon incrustans* Reinsch, K). NEW YORK: on *Cladophora expansa* in tide pools, Cold Spring Harbor, Long island, *Johnson 1055*, 14 Aug. 1894 (FC). NEW JERSEY: on Lyngbya, Atlantic City, *I. C. Martindale*, 12 Dec. 1886 (FH, UC); on Enteromorpha, Atlantic City, *J. N. Rose 19590*, Aug. 1925 (NY). NORTH CAROLINA: on Gelidium, Beaufort, *W. D. Hoyt* (NY); on *Lyngbya semiplena* on piling at west end of Shackleford Banks, Beaufort, *H. J. Humm 12*, 15 Sept. 1947 (FC); on *Calothrix prolifera* on wooden breakwater, Wrightsville Beach, *Humm*, 24 May 1946 (FC).

FLORIDA: on Ceramium washed up on the Atlantic ocean beach between Dania Beach and Hollywood Beach, Broward county, *F. Drouet & H. B. Louderback 10327*, 28 Dec. 1948 (FC); pilings 2 miles south of St. Johns river mouth, Jacksonville Beach, *Humm 3*, 19 Mar. 1948 (FC); on hydroid dredged 12 miles southeast of Alligator point, Franklin county, *Humm*, 27 Dec. 1952 (FC); Hendry creek about 10 miles south of Fort Myers, *P. C. Standley 73256*, Mar. 1940 (FC); on Bostrychia and Cladophora, Way key, Cedar keys, *Drouet & C. S. Nielson 11127, 11181*, 22 Jan. 1949 (FC, T); on Padina, moat, Garden key, Dry Tortugas, *W. R. Taylor 422b*, Jun. 1924 (NY, TA); on Chaetomorpha, Marineland, *Humm*, 19 Nov. 1950 (FC); on an ascidian on *Sargassum cymosum* dredged about 4 miles ESE of St. Marks Lighthouse, Wakulla county, *Humm*, 17 Aug. 1952 (FC). MISSISSIPPI: on Bostrychia on pilings, Gulf Coast Research Laboratory, Ocean Springs, Jackson county, *Drouet & R. L. Caylor 9904*, 12 Dec. 1948 (FC); on Bostrychia, Mississippi sound between Bay St. Louis and Waveland, *Drouet 9807, 9811*, 8 Dec. 1948 (FC); on Bostrychia in St. Louis bay, Bay St. Louis, *Drouet 9853*, 9 Dec. 1948 (FC). LOUISIANA: on *Lyngbya semiplena, L. confervoides*, and *Bostrychia* sp., Grand Isle, Jefferson parish, *L. H. Flint*, 2 Dec. 1945, 15 Aug. 1947, 13 Mar. 1953 (FC); on *Lyngbya semiplena* in Lake Pontchartrain at the south end of R. S. Maestri bridge, Orleans parish, *Drouet & P. Viosca Jr. 9324*, 24 Nov. 1948 (FC). TEXAS: on *L. confervoides* and *L. semiplena* on jetty, Brazos Santiago and South Padre island, Cameron county, *R. Runyon 3950, 3972*, 7 Jun., 18 Jul. 1945 (FC); on *Gracilaria foliifera* on jetty, Port Aransas, *H. Hildebrand*, 31 Dec. 1952 (FC). ALASKA: Sitka, *N. L. Gardner 3962*, Jul. 1917 (Type of *Xenococcus Gilkeyae* Setch. & Gardn., UC); on Fucus, Sitka, *D. Saunders* (UC). BRITISH COLUMBIA: on Cladophora in tide-pool, Baird point, strait of Juan de Fuca, *J. E. Tilden*, 3 Aug. 1898 (isotype of *Pringsheimia scutata* f. *Cladophorae* Tild. in Tild., Amer. Alg. no. 382, FC); on Cladophora, Nootka sound, Vancouver island, *W. A. Setchell & H. E. Parks*, 24 Jun. 1930 (FC, UC). WASHINGTON: on *Gigartina papillata*, Port Townsend, ex herb. Farlow (PC); on Cladophora, Port Townsend, *Miss Cooley*, Aug. 1891 (FH); on hairs of Phoca (the harbor seal), Puget sound, *J. W. Slipp*, Jul. 1942 (TA); on *Iridaea laminarioides*, Puget sound, herb. D. Saunders (FH); on *Fucus evanescens*, San Juan island, *W. C. Muenscher 1389*, Jun. 1915 (TA); on Fucus, Seattle, *D. Saunders* (UC); on *F. evanescens* f. *microcephalus* near Seattle, *D. Saunders*, Jun. 1899 (Type of *Dermocarpa fucicola* Saund. in Coll., Hold., & Setch., Phyc. Bor.-Amer. no 801, FH; isotypes, FC, L, MICH, TA); Three Tree point, 15 miles south of Seattle, *Gardner 7982*, Jul. 1936 (FC, UC); on Gelidium, Channel rocks, West Seattle, *Gardner 534* (FC, UC); on Spongomorpha, west coast of Whidbey island, *Gardner 467*, Jul. 1901 (Type of *Dermocarpa protea* Setch. & Gardn., UC; isotypes, FC [Fig. 200], FH, TA); on *Iridaea laminarioides*, Whidbey island, *Setchell & Gardner 92a, 291* (UC); Whidbey island, *Setchell & Gardner 301, 670* (UC). OREGON: on *Rhodochorton Rothii*, North Bend, Coos county, *Gardner 2756*, May 1914 (Type of *Xenococcus pyriformis* Setch. & Gardn., UC; isotypes, FC, TA); on Prionitis, Sunset Beach near the mouth of Coos bay, *Gardner 2769*, May 1914 (Type of *Hyella linearis* Setch. & Gardn., UC; isotype, FC [Fig. 206]).

CALIFORNIA: on Gigartina, *W. G. Farlow* (FH); herb. D. Saunders (FH); *N. L. Gardner 7315* (FC, UC). Alameda county: on Enteromorpha in pools in salt marsh near Berkeley, *Gardner 1580*, 15 Nov. 1905 (Type of *Xenococcus acervatus* Setch. & Gardn., UC; isotype, FC [Fig. 212], L, and as *Pleurocapsa amethystea* var. *Schmidtii* in Coll., Hold., & Setch., Phyc. Bor.-Amer. no. 1704, FC, L). Los Angeles county: on Dichothrix, Cabrillo Beach, San Pedro, *G. J. Hollenberg 1574*, 3 Oct. 1936 (FC, UC); on Chaetomorpha on piling, Cabrillo Beach, San Pedro, *Hollenberg 2145*, 30 Oct. 1937 (Type of *Xenococcus pulcher* Hollenb. in herb. G. J. Hollenberg;

isotypes, FC [Fig. 202], D); on Gigartina, Point Vincente, *W. G. Farlow* (FH); San Pedro, *Gardner 1906*, Jan. 1908 (Type of *Dermocarpa sphaerica* Setch. & Gardn., UC; isotype, TA); on Gelidium, Chaetomorpha, and Calothrix on breakwaters, San Pedro, *Hollenberg 486, 1576, 1579*, 2 Oct. 1934, 3 Oct. 1936 (D); on Gelidium, San Pedro, *W. A. Setchell*, Dec. 1895 (FC, UC); on *Hesperophycus Harveyanus*, Catalina harbor, Santa Catalina island, *P. C. Silva 4649*, 14 Feb. 1949 (FC, UC); Santa Catalina island, *E. Y. Dawson 5490*, 30 Oct. 1948 (FC). Marin county: on *Lyngbya confervoides*, Duxbury reef, Bolinas bay, *Setchell 1038*, Nov. 1895 (FC, UC); on Fucus, Bolinas, *Gardner 4599*, May 1920 (UC). Monterey county: on *Rhodochorton Rothii*, *Setchell 5489*, June 1901 (UC); on *Gelidium Coulteri*, Monterey, *Mrs. A. E. Bush*, Dec. 1885 (FH); on *Endocladia muricata* and *Iridaea laminariodes*, Pyramid point, Monterey, *Setchell 5464, 5456*, Jun. 1901 (UC); on Iridophycus, north end of beach, Carmel, *Gardner 7783, 7783a*, 18 Jul. 1935 (FC, UC); on *Calothrix crustacea*, Carmel bay, *Setchell & R. E. Gibbs*, 12 Jan. 1899 (as *Xenococcus Schousboei* in Coll., Hold., & Setch., Phyc. Bor.-Amer. no. 554, FC, L, TA, UC); on Iridaea, Carmel bay, *Gardner 2981*, May 1915 (Type of *Hyella socialis* Setch. & Gardn., UC; isotype, FC); on *Rhodochorton Rothii*, Carmel bay, *Gardner 2992*, May 1915 (FC [Fig. 201], UC; as *Dermocarpa hemisphaerica* in Coll., Hold., & Setch., Phyc. Bor.-Amer. no. 2253, FC, L, MICH, TA); on Rhodymenia, Carmel bay, *Gardner 3350a* (Type of *Radaisia subimmersa* Setch. & Gardn., UC; isotype, FC [Fig. 215]); on *Iridaea minor*, Carmel bay, *Gardner 3687*, May 1916 (Type of *Radaisia epiphytica* Setch. & Gardn., UC [Fig. 205]; isotype, TA); on Gelidium, Point Carmel, *Setchell*, 3 Jun. 1901 (as *Dermocarpa fuciola* in Coll., Hold., & Setch., Phyc. Bor.-Amer. no. 1251, FC, D, L, MICH, TA); on *Gastroclonium ovale*, Point Carmel, *Setchell 5448*, Jun. 1901 (UC); Carmel point, *I. F. Lewis*, 8 Jul. 1929 (FC); on *Chaetomorpha aerea* near Cypress point, *Gardner 3580*, Jan. 1917 (Types of *Xenococcus Chaetomorphae* Setch. & Gardn. and *Dermocarpa pacifica* Setch. & Gardn., UC; isotypes, FC [Fig. 207, 213], L, TA, UC); Cypress point, *Gardner 4528* (FC, L, TA, UC), *4543* (FC, L, TA, UC), *4656* (FC, UC), *4657* (FC, UC), *4694* (FC, UC), Dec. 1919, 1920; on *Rhodymenia californica*, Del Monte bay, *R. A. Lewin*, 19 Jul. 1948 (FC); on *Gelidium Coulteri* and *Gymnogongrus linearis*, Pacific Grove, *Setchell 5409* (FC, UC), May, Jun. 1901; on *Fucus furcatus*, Pacific Grove, *Gardner 4665*, Dec. 1920 (FC, UC); on *Bornetia secundiflora* near Pacific Grove, *J. M. Weeks 118a*, Mar. 1897 (UC); on *Chaetomorpha aerea*, mouth of Carmel river, Pacific Grove, *R. D. Wood*, 2 Jul. 1946 (FC); on *Gelidium Coulteri*, Pebble beach, Carmel bay, *Setchell 5482*, 17 Jun. 1901 (FC, UC); on *Chondrus canaliculatus*, Pebble beach, Carmel bay, *Gardner 7286*, May 1933 (FC, UC); on *Rhodochorton Rothii* in a cave, Pebble beach, Pacific Grove, *R. D. Wood*, 29 Jun. 1946 (FC). Orange county: on *Lyngbya aestuarii* in brackish water west of Newport Beach, *G. J. Hollenberg 1533, 1533c*, 26 Sept. 1934 (FC, D); on Gelidium, Laguna Beach, *Hollenberg 3094*, 20 Mar. 1940 (FC). San Diego county: on Dictyota, San Diego, *W. G. Farlow* (FH); on *Calothrix scopulorum*, La Jolla, *T. A. & A. Stephenson JBN41*, 1947 (FC); on Laurencia, La Jolla, *Hollenberg 2016*, 26 Dec. 1936 (D). San Francisco: on *Laminaria Sinclairii*, Fort point, *Setchell 2075*, 16 Nov. 1898 (FC, UC), *Gardner*, 25 Dec. 1905, Dec. 1918 (as *Radaisia Laminariae* in Coll., Hold., & Setch., Phyc. Bor.-Amer. no. 2254 a & b, FC [Fig. 204], TA), *Gardner 3158* (FC, L, TA), *3159* (D, FC), *3675* (Type of *Radaisia Laminariae* Setch. & Gardn., UC); on *Gymnogongrus linearis*, Fort point, *Setchell 6403*, 30 Nov. 1906 (FC, UC); on *Iridaea laminarioides*, Fort point, *Gardner 4482*, Nov. 1919 (FC, UC); on Ulva, Lands End, *Setchell & Gardner 1539*, 28 May 1905 (FC, UC); on various algae, Lands End, *Gardner 1538*, Aug. 1903 (FC, UC), *3706a* (Type of *Radaisia clavata* Setch. & Gardn., UC), *3710*, Apr. 1917 (Type of *Dermocarpa sphaeroidea* Setch. & Gardn., UC), *4447*, Apr. 1919 (FC, L, TA), *4780*, 25 May 1922 (FC, UC), *4857*, Oct. 1923 (FC, UC); on Rhodochorton in cave south of Point Lobos, *I. F. Lewis*, 22 Jul. 1929 (FC). San Luis Obispo county: on *Gymnogongrus linearis*, Bushnells beach, *R. S. Reed*, Jul. 1917 (UC); on *G. leptophyllus*, Cayucos, *P. C. Silva 2299*, 29 Nov. 1947 (FC, UC); on *Gelidium Coulteri*, Port San Luis, *Silva 2476*, 22 Feb. 1948 (FC, UC). San Mateo county: on *Rhodochorton Rothii*, Moss Beach, *Gardner 2183*, Apr. 1910 (Types of *Dermocarpa suffulta* Setch. & Gardn. and *D. hemisphaerica* Setch. & Gardn., UC; isotypes, FC, L, TA, UC), *Gardner 6258*, Apr. 1927 (FC, L, TA); Moss Beach, *Gardner 4308*, Jul. 1918 (FC, UC). Santa Barbara county: on *Gigartina Harveyana*, Point Arguello coast guard station, *P. C. Silva 2549*, 28 Dec. 1947 (FC, UC); on *Rhodochorton Rothii* in cave, Lady's harbor, Santa Cruz island, *Hollenberg 1311*, 19 Apr. 1936 (D); on *Corallina gracilis*, Forneys cove, Santa Cruz island, *Silva 5925*, 8 Mar. 1950 (FC, UC); on Corallina, Chaetomorpha, and Chondria, Santa Rosa island, *Silva 4069, 4079, 4142*, 27 Jan. 1949 (FC, UC). Sonoma county: on *Laurencia pinnatifida*, Fort Ross, *Setchell 1717*, Dec. 1897 (FC, UC).

BAHAMA ISLANDS: on Catenella, Little Harbor cay, Berry islands, *M. A. Howe 3592b*, 31 Jan. 1905 (D, NY). PUERTO RICO: on *Cladophoropsis membranacea*, Point Borinquen near Aguadilla, *Howe 2459b*, Jun. 1903 (NY); on *Lyngbya semiplena* on cement, Santurce beach, *H. L. Blomquist 12785*, 25 Feb. 1942 (FC); on *L. sordida*, Gaujatacca, *Blomquist 12923*, 21 Mar. 1942 (FC); on *Calothrix pilosa*, west side of Guanica bay, *Blomquist 13013*, 21 Mar. 1942 (FC); on Rhizoclonium on mudflat, Las Marias, *Blomquist 13099a*, 21 Mar. 1942 (FC). VIRGIN ISLANDS: Lt. Princess, St. Croix, *F. Børgesen 15x*, 1892 (UC). BARBADOS: on *Dictyopteris delicatula*,

Valentia, *A. Vickers,* 17 Jan. 1899 (FC, UC), Jun. 1899 (Type of *Dermocarpa Vickersiae* Coll., NY; isotypes in Coll., Hold., & Setch., Phyc. Bor.-Amer. no. 1602, FC [Fig. 214], FH, L, TA). JAMAICA: Northeast cay, Morant cays, *C. B. Lewis A1172,* Jun. 1950 (FC). MEXICO: Baja California: on *Gracilaria Cunninghamii,* Cabo Colnett, *E. Y. Dawson 5241,* 4 Sept. 1948 (FC); on *Lyngbya semiplena* on pilings, southeast side of Cerros island, *W. R. Taylor 39-2,* 14 Mar. 1939 (FC, MICH); Ensenada, *N. L. Gardner 4994f* (FC, UC); Guadalupe island, *H. L. Mason 63* (Type of *Chamaesiphon clavatus* Setch. & Gardn., CAS; isotype, UC); Punta Baja near Rosario, *Dawson 1196, 1199, 1200,* 9 Apr. 1946 (FC); San Bartolome bay, *J. T. Howell 759,* 14 Aug. 1932 (FC, UC); San Francisquito bay, *I. M. Johnston 13a* (Type of *Xenococcus deformans* Setch. & Gardn., CAS; isotype, UC); in *Callymenia angustata,* Santa Maria bay, *Howell 739a,* 12 Aug. 1932 (Type of *Xenococcus angulatus* Setch. & Gardn., CAS); Santa Rosalia, *Marchant 108* (Type of *Dermocarpa Marchantae* Setch. & Gardn., UC); on *Hesperophycus Harveyanus,* South island, Isla de Todos Santos, *P. C. Silva 4833,* 24 Feb. 1949 (FC, UC); Tortuga island, Gulf of California, *Johnston 34c* (Type of *Dermocarpa Reinschii* Setch. & Gardn., CAS), *Johnston 135a* (Type of *Chlorogloea regularis* Setch. & Gardn., CAS). Guerrero: on Sertularia, Acapulco, *Dawson 3853,* 3 Feb. 1947 (FC). Jalisco: on *Lyngbya confervoides,* Barra Navidad, *Dawson 3745,* 25 Dec. 1946 (FC). Oaxaca: on Chaetomorpha, Salina Cruz, *Dawson 3784,* 9 Jan. 1947 (FC). Sonora: shore of island at entrance of harbor, Guaymas, *F. Drouet & D. Richards 3352,* 21 Dec. 1939 (FC); on *Lyngbya confervoides,* Playa Miramar, Guaymas, *Drouet & Richards 3139,* 5 Dec. 1939 (FC); in tide-pool, Punta San Pedro, Guaymas, *Drouet & Richards 3382,* 22 Dec. 1939 (FC); on Chaetomorpha, Isla Patos, *Dawson 652,* 17 Feb. 1946 (FC).

GUATEMALA: on Chaetomorpha, Bay of Santo Tomás between Escobas and Santo Tomás, dept. Izabal, *J. A. Steyermark 39340,* 13 Apr. 1940 (FC). PANAMA: on *Lyngbya sordida* in tide-pools, Islas Secas near Puerto Nuevo, *W. R. Taylor 39-123,* 26 Mar. 1939 (FC). BRAZIL: Paraná: on *Lyngbya confervoides,* Ilha de Farol, Caiobá, *A. Brandão Joly 490,* 8 Feb. 1951 (FC); on Cladophora, Caiobá, *Brandão Joly 492, 528,* 9 Feb. 1951 (FC); on Gelidiella, Praia Medanha, *Brandão Joly 513,* 11 Feb. 1951 (FC); on *Lyngbya confervoides* and Pylaiella, Punta das Caieiras, *Brandão Joly 518, 519, 525,* 12 Feb. 1951 (FC). São Paulo: São Vicente near Santos, *F. Rawitscher 43,* 1941 (FC, TA); on Cladophora, Guarujá (Santos), *A. Padala 331,* 20 Oct. 1950 (FC); Praia de Parnapoan, S. Vicente, *Joly 150,* 7 Oct. 1953 (D); on *Lyngbya confervoides,* Peruíbe, *Brandão Joly 399, 459,* 29 Dec. 1950, 1 Jan. 1951 (FC). URUGUAY: on *L. confervoides,* Punta del Este, Maldonado, *D. Legrand 2,* Mar. 1945 (FC, TA). ECUADOR: Galapagos islands: among *Rhizoclonium robustum,* North Seymour island, *J. T. Howell 170b,* 11 Jun. 1932 (Type of *Dermocarpa simulans* Setch. & Gardn., CAS; isotypes, D, UC); on *Rhizoclonium riparium* in a tide-pool 5 miles northeast of Webb cove, Albemarle island, *Howell 429b,* 22 May 1922 (Type of *Xenococcus endophyticus* Setch. & Gardn., CAS; isotypes, D, UC); on Gelidium and mangrove, southeast side of Narborough island, *Howell 344, 587,* 1 & 2 June 1932 (FC, UC); tide-pool, Villemil, Albemarle island, *Howell 418,* 27 Apr. 1932 (FC, UC). PERU: on *Streblocladia spicata,* Pisco, *R. E. Coker 465,* Jul. 1908 (NY); Chincha Norte island, *W. Vogt,* 1941 (FC); Lima, *A. Maldonado 13,* Dec. 1930 (FC); on Rhizoclonium, Lago Chilca and Laguna Saraja, *Maldonado 65, 93,* Jan.—Feb. 1942 (FC). HAWAIIAN ISLANDS: Waikiki near Honolulu, *W. A. Setchell 5595,* Jul. 1900 (FC, UC). PHOENIX ISLANDS: on *Lyngbya confervoides,* southeast side of Canton island, *O. Degener 21343,* Apr. 1951 (FC). GUAM: on the reef at the mouth of Pago river, eastern coast, *A. E. Vatter Jr. 4,* 25 Jan. 1946 (FC). NEW ZEALAND: on Cladophora on floor of cave, Island, Half Moon bay, Stewart island, *I. B. Warnock 700A,* 19 Feb. 1935 (in Tild., So. Pacific Pl., 2nd ser. no. 700A, FC); New Zealand, on Rhizoclonium, *V. J. Chapman,* Dec. 1948 (FC); on Enteromorpha, Auckland, *Chapman,* Jun. 1946 (FC); Bay of Islands, *Chapman 9060b,* 1947 (FC); on Rhodochorton, Long Beach, Bay of Islands, *V. W. Lindauer 1725,* 26 Jan. 1940 (FC, UC).

JAPAN: on *Chordaria simplex,* Tokyo, *Yataba,* Feb. 1878 (FH, PC); on *Myelophycus caespitosus,* Japan, *K. Yendo,* Dec. 1898 (D, FC, UC); Shimoda, *Yendo,* Feb. 1907 (D, FC, UC); in *Gymnogongrus* sp., Aburataubo, and in *Carpopeltis* sp., Jyoga-shima, Miura peninsula, Misaki, *I. Umezaki 985, 987,* 29, 30 Mar. 1953 (D). PHILIPPINES: on *Lyngbya aestuarii* in fishponds, Manila, *E. D. Merrill 7458* (D, FC, UC), Aligaen near Manila, *G. T. Velasquez 144,* Dec. 1953 (D); on *Chaetomorpha brachygona* in fishponds at Dagat-Dagatan, Rizal, *Velasquez 624,* 19 Jan. 1941 (FC); on *Lyngbya aestuarii,* Parañaque, Rizal, *M. T. Cruz 19555,* Aug. 1933 (FC, UC). VIET NAM: on *Lyngbya confervoides* on jetty, Cauda, Bay of Nhatrang, *E. Y. Dawson 11300,* 10 Feb. 1953 (FC). INDONESIA: with *Bostrychia tenella,* Aroi Eilanden [Manumbai], *H. Jewell 234,* 1914—16 (Type of *Pleurocapsa violacea* Weber-van Bosse, C); on *Lyngbya confervoides,* Endek, Flores, *A. Weber-van Bosse 1124,* Dec. 1888 (L); Banda, Kai islands, on Cladophora, *T. Mortensen VIII,* 10 Jun. 1922 (Type of *Radaisia pusilla* Weber-van Bosse, C); drijvende op rif, Sanana, Soela Besi, *Weber-van Bosse,* Sept. 1899 (L); strand, Adja-Toening, Nieuw Guinea, *Weber-van Bosse,* 1899—1900 (L). AUSTRALIA: on *Calothrix pilosa,* Brampton island off Mackay, Queensland, *V. May 2701,* 16 Jun. 1948 (FC) in mari, Sydney, New South Wales, *A. Grunow 022,* Aug. 1884 (PC); on *Lyngbya confervoides* in rock pools, Wattamolla, New South Wales, *May 214,* 26 Aug. 1944 (D, FC); auf *Sphacelaria furcigera* auf *Sargassum Peronii,* Nordwest-Australien, *Gazellen-Expedition B312-*

118

(137)a, 26 Apr. 1875 (B); on *Calothrix crustacea*, Emu bay, Kangaroo island, South Australia, H. B. S. *Womersley O*, 10 Jan. 1946 (FC). ISRAEL: on *Lyngbya confervoides*, Tel Aviv, T. *Rayss 1*, Dec. 1943 (FC). MAURITIUS: on *L. confervoides*, Flacq, Baie de 4 Cocos, N. *Pike*, 23 Jan. 1870 (FC).

4. ENTOPHYSALIS RIVULARIS (Kütz.) Dr.

Plantae primum microscopicae aetate provecta in stratum aut pulvinum macroscopicum aerugineum, olivaceum, brunneum, violaceum, rubrum, vel nigrum increscentes, cellulis solitariis et ad basem strati aut pulvini sphaericis, ellipsoideis, obovoideis, vel cylindraceis, diametro 2—8μ crassis; cellulis superioribus praecipue sphaeroideis vel post divisiones polyhedroideis; cellulis filorum intus substratum penetrantium ovoideis, sphaericis, cylindraceis, vel polyhedroideis, diametro (1—)2—15μ crassis; endosporangiis sphaericis vel ovoideis, diametro ad 30μ crassis; gelatino vaginale primum hyalino demum lutescente vel raro rubrescente seu coerulescente, homogeneo vel lamelloso; protoplasmate aerugineo, olivaceo, luteolo, violaceo, vel roseo, homogeneo vel sparse grossi-granuloso. FIGS. 216—220, 222—234, 374, 375.

Strata and cushions are the most familiar growth-forms of *Entophysalis rivularis*. On limestone and shells, the basal penetrating filaments develop copiously; but they are absent or represented only by elongated and curved cells on non-calcareous rocks and on wood.

Key to forms of *Entophysalis rivularis*:

Solitary and basal cells ovoid or spherical ....f. RIVULARIS

Solitary and basal cells cylindrical ....f. PAPILLOSA

4a. ENTOPHYSALIS RIVULARIS f. RIVULARIS. *E. rivularis* Drouet, Amer. Midl. Nat. 30: 671. 1943. *Hydrococcus rivularis* Kützing, Linnaea 8: 380. 1833. *Oncobyrsa rivularis* Meneghini, Mem. R. Accad. Sci. Torino, ser. 2, 5 (Sci. Fis. & Math.): 47, 96. 1843. —Type from Schleusingen, Germany (L). FIG. 217.

Pleurococcus julianus Menegh., Giorn. Toscano Sci., Med., Fis. e Nat. 1: 186. 1840. *Protococcus julianus* Kützing, Sp. Algar., p. 198. 1849. *Chroococcus julianus* Nägeli, Gatt. Einzell. Alg., p. 46. 1849. *Gloeocapsa juliana* Kützing ex Rabenhorst, Fl. Eur. Algar. 2: 42. 1865. *Bichatia juliana* Kuntze, Rev. Gen. Pl. 2: 886. 1891. —Type from the Julian springs, Italy (FI). FIG. 225.

Pleurococcus glomeratus Meneghini, Mem. R. Accad. Sci. Torino, ser. 2, 5 (Sci. Fis. & Mat.): 40. 1843. *Chroococcus glomeratus* Forti, Syll. Myxophyc., p. 20. 1907. —Type from the Italian Tyrol (FI). FIG. 234.

Palmella duriuscula Kützing, Phyc. Gener., p. 172. 1843. *Cagniardia duriuscula* Trevisan, Sagg. Monogr. Alg. Coccot., p. 50. 1848. —Type from Schleusingen, Germany (L). FIG. 220.

Hydrococcus ulvaceus Kützing, Phyc. Gener., p. 177. 1843. *Oncobyrsa ulvacea* Rabenhorst, Fl. Eur. Alg. 2: 67. 1865. *O. rivularis* var. *ulvacea* Hansgirg, Sitzungsber. K. Böhm. Ges. Wiss., Math.-Nat. Cl., 1890(2): 93. 1891. —Type from Monfalcone, Italy (L). FIG. 223.

Protococcus carneus Kützing, Phyc. Germ., p. 146. 1845. —Type from Silesia (L).

Palmella Brebissonii Kützing, Tab. Phyc. 1: 24. 1846. —Type from Falaise, France (L).

Exococcus ovatus Nägeli, Die Neuern Algensyst., p. 170. 1847. —Type from Zürich, Switzerland (ZT).

Gloeocapsa dermochroa Nägeli in Kützing, Sp. Algar., p. 224. 1849; Nägeli, Gatt. Einzell. Alg., p. 51. 1849. *Bichatia dermochroa* Kuntze, Rev. Gen. Pl. 2: 886. 1891. *Pleurococcus dermochrous* Nägeli pro synon. ex Wille, Nyt Mag. Naturvid. 62: 186. 1925. —Type from Zürich, Switzerland (L).

Gloeocapsa rubicunda Nägeli, Die Stärkekörner (Pflanzenphys. Unters. 2), p. 282. 1858. —Type from St. Moritz, Switzerland (ZT).

Polycystis fuscolutea Hansgirg, Oesterr. Bot. Zeitschr. 38: 87. 1888. *Microcystis fuscolutea*

119

Forti, Syll. Myxophyc., p. 92, 1907. —Type from Modran, Bohemia (W).
Chroococcus fuscoater var. *fuscoviolaceus* Hansgirg, Oesterr. Bot. Zeitschr. 38: 115. 1888.
C. fuscoviolaceus Hansgirg, Sitzungsber. K. Böhm. Ges. Wiss., Math.-Nat. Cl., 1890(1): 20. 1890.
Chamaesiphon fuscoviolaceus Margalef, Collect. Bot. (Barcelona) 3(2): 221. 1952. —Type from
Tannwald, Bohemia (W). FIG. 226.
    Cyanoderma rivulare Hansgirg, Notarisia 4: 658. 1889. *Pleurocapsa rivularis* Hansgirg,
Sitzungsber. K. Böhm. Ges. Wiss., Math.-Nat. Cl., 1890(2): 91. 1891. *Xenococcus rivularis*
Geitler, Beih. z. Bot. Centralbl., II, 41: 245. 1925. —Type from Pocatek, Bohemia (W).
FIG. 231.
    Chroococcus fuscoviolaceus var. *cupreofuscus* Hansgirg, Sitzungsber. K. Böhm. Ges. Wiss., Math.-
Nat. Cl., 1890(1): 20. 1890. *Pleurocapsa cuprea* Hansgirg, Prodr. Algenfl. Böhmen 2: 128.
1892. —Type from Tannwald, Bohemia (W). FIG. 216.
    Pleurocapsa minor Hansgirg, Sitzungsber. K. Böhm. Ges. Wiss., Math.-Nat. Cl., 1890(2): 89.
1891. *Scopulonema minus* Geitler in Engler & Prantl, Natürl. Pflanzenfam., ed. 2, 1b: 93. 1942.
—Type from Kuchelbad, Bohemia (W). FIG. 224.
    Pleurocapsa concharum Hansgirg, Sitzungsber. K. Böhm. Ges. Wiss., Math.-Nat. Cl., 1890(2):
90. 1891. *Scopulonema concharum* Geitler in Engler & Prantl, Natürl. Pflanzenfam., ed. 2, 1b:
93. 1942. —Type from near Roztok, Bohemia (W).
    Aphanocapsa anodontae Hansgirg, Sitzungsber. K. Böhm. Ges. Wiss., Math.-Nat. Cl., 1890(2):
99. 1891. *Microcystis anodontae* Elenkin, Monogr. Algar. Cyanophyc., Pars Spec. 1: 138. 1938.
—Type from near Roztok, Bohemia (W).
    Chamaesiphon fuscus var. *auratus* Hansgirg, Sitzungsber. K. Böhm. Ges. Wiss., Math.-Nat. Cl.,
1891(1): 538. 1891. —Type from Martinscica near Fiume, Jugoslavia (W).
    Aphanocapsa anodontae var. *major* Hansgirg, Sitzungsber. K. Böhm. Ges. Wiss., Math.-Nat. Cl.,
1891(1): 358. 1891. —Type from near Cilli, Slovenia, Jugoslavia (W). FIG. 219.
    Entophysalis Cornuana Sauvageau, Bull. Soc. Bot. Fr., ser. 2, 14: cxvii. 1892. *Radaisia
Cornuana* Sauvageau, Journ. de Bot. 9: 376. 1895. —Type from near Biskra, Algeria (PC).
FIG. 218.
    Dermocarpa Flahaultii Sauvageau, Bull. Soc. Bot. Fr., ser. 2, 14: cxix. 1892. —Type from
near Biskra, Algeria (PC). FIG. 218.
    Hyella fontana Huber & Jadin, Journ. de Bot. 6: 279. 1892. *H. fonticola* Huber & Jadin ex
Forti, Syll. Myxophyc., p. 127. 1907. —Type from near Montpellier, France (PC).
    Chroococcus (Rhodococcus) Hansgirgii Schmidle, Allgem. Bot. Zeitschr. 6: 79. 1900; Hedwigia
39: 189. 1900. —Type from Poona, India (ZT). FIG. 229.
    Coelosphaerium holopediforme Schmidle, Beih. z. Bot. Centralbl. 10: 180. 1901. —Type from
Schwenningen, Württemberg (ZT). FIG. 227.
    Placoma africana Wille, Oesterr. Bot. Zeitschr. 53: 90. 1903. —Type from the Kachembe
river, South Africa (O).
    Hyellococcus niger Schmidle, Allgem. Bot. Zeitschr. 1905: 64. 1906. —Type from Mammern,
Switzerland (ZT). FIG. 230.
    Guyotia singularis Schmidle, Allgem. Bot. Zeitschr. 1905: 64. 1906. —Type from the Sinai
peninsula, Egypt (ZT). FIG. 222.
    Palmophyllum foliaceum G. S. West, Journ. Linn. Soc. Bot. 38: 145. 1907. —Type from
Niamkolo, Tanganyika (BM).
    Pleurocapsa polonica Raciborski, Phyc. Polon. 1: 11. 1910. *Siphononema polonicum* Geitler,
Beih. z. Bot. Centralbl., II, 41: 251. 1925; Geitler, Arch. f. Protistenk. 51: 332. 1925. *Scopulonema
polonicum* Geitler in Engler & Prantl, Natürl. Pflanzenfam., ed. 2, 1b: 93. 1942. —Type from
Morskiem Okiem, Galicia, Poland (WU).
    Entophysalis samoensis Wille, Hedwigia 53: 144. 1913; in Rechinger, Bot. & Zool. Ergebn.
Samoa- & Salomoninseln, Süsswasseralg., p. 7. 1914. —Type from Savaii, Samoa (O).
    Aphanocapsa constructrix Bremekamp, Teysmannia 25: 73. 1914. —Type from Idjen
mountain in eastern Java, Indonesia (in the Mineralogisch-Geologisch Instituut, University of
Utrecht).
    Chroococcopsis gigantea Geitler, Arch. f. Protistenk. 51: 342. 1925; Beih. z. Bot. Centralbl.,
II, 41: 241: 1925; in Pascher, Süsswasserfl. 12: 125. 1925. —Paratype named by the author,
from Ranau lake, Sumatra (in the collection of L. Geitler).
    Chamaesiphon polymorphus Geitler, Arch. f. Protistenk. 51: 327. 1925. —Paratype named by
the author from Tjibodas, Java (in the collection of L. Geitler).
    Chroococcus minutissimus Gardner, Mem. New York Bot. Gard. 7: 8. 1927. —Type from
near Arecibo, Puerto Rico (NY). FIG. 232.
    Entophysalis chlorophora Gardner, Mem. New York Bot. Gard. 7: 30. 1927. —Type from
Coamo Springs, Puerto Rico (NY). FIG. 233.
    Radaisia Willei Gardner, Mem. New York Bot. Gard. 7: 32. 1927. —Type from near
Humacao, Puerto Rico (NY).
    Radaisia confluens Gardner, Mem. New York Bot. Gard. 7: 32. 1927. —Type from Maricao,
Puerto Rico (NY). FIG. 228.

Chamaesiphon ferrugineus Fritsch, New Phytol. 28: 178. 1929. —Type from Devonshire, England (in the collection of F. E. Fritsch).

Pseudoncobyrsa fluminensis Fritsch, New Phytol. 28: 181. 1929. —Type from Devonshire, England (in the collection of F. E. Fritsch). The young, largely unicellular, hormogonia of *Amphithrix janthina* (Mont.) Born. & Flah. make up a large part of this specimen; they account for at least part of the author's description.

Chroococcopsis fluminensis Fritsch, New Phytol. 28: 183. 1929. —Type from Devonshire, England (in the collection of F. E. Fritsch).

Chlorogloea purpurea var. *minutissima* Geitler, Arch. f. Hydrobiol., Suppl. XII, 4: 623. 1933. —Type from Ranu Pakis, east Java (in the collection of L. Geitler).

Chlorogloea major Geitler, Arch. f. Hydrobiol., Suppl. XII, 4: 623. 1933. *C. minor* Geitler, ibid., Suppl. XIV., 6: 386. 1935. —Type from Dijeng plateau, Java (in the collection of L. Geitler).

Myxosarcina spectabilis Geitler, Arch. f. Hydrobiol., Suppl. XII, 4: 624. 1933. —Type from Bukit Kili, Sumatra (in the collection of L. Geitler).

Chamaesiphon fallax Geitler, Arch. f. Hydrobiol., Suppl. XII, 4: 625. 1933. —Type from Pasergede, Java (in the collection of L. Geitler).

Radaisia gigas Geitler, Arch. f. Hydrobiol., Suppl. XII, 4: 626. 1932. —Type from Ranau lake, Sumatra (in the collection of L. Geitler).

Cyanodermatium gelatinosum Geitler, Arch. f. Hydrobiol., Suppl. XII, 4: 627. 1933. —Type from Ranau lake, Sumatra (in the collection of L. Geitler).

Cyanodermatium violaceum Geitler, Arch. f. Hydrobiol., Suppl. XII, 4: 627. 1933. —Type from Ranau lake, Sumatra (in the collection of L. Geitler).

Dermocarpa Gardneriana Drouet, Field Mus. Bot. Ser. 20: 128. 1942. —Type from San Francisco, California (FC).

Dermocarpa Setchellii Drouet, Field Mus. Bot. Ser. 20: 129. 1942. —Type from Harlem hot springs, San Bernardino county, California (FC).

Original specimens have not been available to us for the following names; their original descriptions are here designated as the Types until the specimens are found:

Microcystis pulchra Flotow, Nova Acta Acad. Caes. Leop.-Car. Nat. Cur. 20(2): 478. 1842. *Protococcus pulcher* Kützing, Phyc. Germ., p. 147. 1845. *Pleurococcus pulcher* Trevisan, Sagg. Monogr. Alg. Coccot., p. 32, 1848.

Palmella dura Wood, Smiths. Contrib. Knowl. 241: 80. 1872.

Chroococcus rubrapunctis Wolle, Bull. Torr. Bot. Club 6: 181. 1877.

Sphaerogonium subglobosum Rostafinski, Rozpr. Akad. Umiej. Krakow, Wydz. Mat.-Przyr., 10: 291. 1883. *Chamaesiphon subglobosus* Lemmermann, Krypt.-Fl. Mark Brandenb. 3: 98. 1907.

Sphaerogonium polonicum Rostafinski, Rozpr. Akad. Umiej. Krakow, Wydz. Mat.-Przyr., 10: 299. 1883. *Chamaesiphon polonicus* Hansgirg, Oesterr. Bot. Zeitschr. 37: 100. 1887.

Askenasya polymorpha Möbius, Ber. Deutsch. Bot. Ges. 5: lxii. 1887. —See Möbius, ibid. 6: 358. 1888.

Hyella jurana Chodat, Bull. Herb. Boissier 6: 446. 1898. *H. fontana* var. *rubra* Nadson, Scripta Bot. Hort. Univ. Imp. Petropol. 18: 27. 1900.

Chroococcus Goetzei Schmidle, Engler Bot. Jahrb. 30: 242. 1901.

Coelosphaerium Goetzei Schmidle, Engler Bot. Jahrb. 30: 243. 1901.

Pleurocapsa salevensis Chodat, Étude sur de Polymorph. des Alg., p. 133. 1909.

Oncobyrsa sarcinoides Elenkin, Not. Syst. Inst. Crypt. Horti Bot. Petropol. 2: 11. 1923. *Chroococcus sarcinoides* Wislouch *pro synon.* in Elenkin, loc. cit. 1923; Wislouch, Acta Soc. Bot. Polon. 2: 110. 1924. *Chlorogloea sarcinoides* Troitzkaja, Dnevn. Vsesoiuzn. Sezda Bot., p. 162. 1928.

Oncobyrsa sarcinoides var. *sparsa* Elenkin, Not. Syst. Inst. Crypt. Hort. Bot. Petropol. 2: 12. 1923.

Oncobyrsa sarcinoides var. *fulvo-cubica* Elenkin, Not. Syst. Inst. Crypt. Hort. Bot. Petropol. 2: 12. 1923.

Oncobyrsa sarcinoides var. *irregulariter-consociata* Elenkin, Not. Syst. Inst. Crypt. Hort. Bot. Petropol. 2: 12. 1923.

Oncobyrsa sarcinoides var. *irregulariter-consociata* f. *pallida* Elenkin, Not. Syst. Inst. Crypt. Hort. Bot. Petropol. 2: 12. 1923.

Oncobyrsa sarcinoides var. *irregulariter-consociata* f. *fusca* Elenkin, Not. Syst. Inst. Crypt. Hort. Bot. Petropol. 2: 12. 1923.

Chroococcus lithophilus Ercegovic, Acta Bot. Inst. Bot. Univ. Zagreb. 1: 75. 1925. *Gloeocapsa lithophila* Hollerbach in Elenkin, Monogr. Algar. Cyanophyc., Pars Spec. 1: 207. 1938.

Chroococcus lithophilus f. *achromaticus* Ercegovic, Acta Bot. Inst. Bot. Univ. Zagreb. 1: 76. 1925.

Chroococcus lithophilus f. *coloratus* Ercegovic, Acta Bot. Inst. Bot. Univ. Zagreb. 1: 76. 1925.

Chroococcus schizodermaticus f. *pallidus* Ercegovic, Acta Bot. Inst. Bot. Univ. Zagreb. 1: 77. 1925.

Lithocapsa fasciculata Ercegovic, Acta Bot. Inst. Bot. Univ. Zagreb. 1: 82. 1925.

Lithococcus ramosus Ercegovic, Acta Bot. Inst. Bot. Univ. Zagreb. 1: 83. 1925.
Pseudocapsa dubia Ercegovic, Acta Bot. Inst. Bot. Univ. Zagreb. 1: 95. 1925.
Chlorogloea microcystoides Geitler, Arch. f. Protistenk. 51: 357. 1925.
Aphanocapsa endolithica Ercegovic, Acta Bot. Inst. Bot. Univ. Zagreb. 1: 81. 1925.
Aphanocapsa endolithica var. *violacescens* Ercegovic, Acta Bot. Inst. Bot. Univ. Zagreb. 1: 82. 1925.
Aphanocapsa distans Zalessky, Rév. Gén. de Bot. 38: 35. 1926.
Hyella fontana var. *maxima* Geitler, Arch. f. Protistenk. 62: 100. 1928.
Aphanocapsa endolithica var. *rivulorum* Geitler, Arch. f. Protistenk. 60: 441. 1928.
Pleurocapsa cuprea f. *pirinica* Petkoff, Bull. Soc. Bot. Bulgarie 3: 37. 1929.
Chlorogloea microcystoides f. *pallida* Skuja, Acta Horti Bot. Univ. Latv. 4: 14. 1929.
Radaisia violacea Frémy, Arch. de Bot. Caen 3(Mém. 2): 56. 1930.
Pleurocapsa aurantiaca Geitler, Rabenh. Krypt.-Fl. 14: 354. 1931. *Scopulonema aurantiacum*
Geitler in Engler & Prantl, Natürl. Pflanzenfam., ed. 2, 1b: 93. 1942.
Myxosarcina spectabilis f. *regularis* Geitler ex J. de Toni, Diagn. Algar. Nov., I, Myxophyc. 9: 849. 1946.

Cellulae solitariae et ad basem strati aut pulvini sphaericae vel ovoideae. FIGS. 216—220, 222—234.

On rocks, shells, woodwork, backs of turtles, etc. in fresh water, mostly in clear streams, springs, fountains, and watering-tanks. The upper cells of the stratum or cushion, especially where the gelatinous matrix has hydrolyzed, often simulate those of *Anacystis montana, Coccochloris elabens,* certain small growth-forms of *C. stagnina,* and *Palmogloea protuberans* (Sm. & Sow.) Kütz. The unicellular primordia of *Amphithrix janthina* (Mont.) Born. & Flah. have also been mistakenly identified with this form.

Specimens examined:

NORWAY: in rivulo frigido prope Giövik in praefectione tromsoensi lapidibus affixum, *G. Lagerheim,* Sept. 1893 (as *Sphaerogonium fuscum* in Wittr., Nordst., & Lagerh., Alg. Exs. no. 1611, FC, L, MIN, NY). POLAND: Galicia: na granitach wýscielajacych brzegi Czarnego stawu nad Morskiem Okiem, *M. Raciborski,* 5 Oct. 1909 (Type of *Pleurocapsa polonica* Racib. in Racib., Phyc. Polon. no. 11, WU; isotype, FH). Silesia: in einem Flasche Brunnenwasser, *v. Flotow,* 29 Feb. 1844 (Type of *Protococcus carneus* Kütz., L). GERMANY: Saxony: an Pflanzenstengeln im Salzsee bei Halle, *P. Richter* (UC). Thuringia: an Brünnenufern, Schleusingen, *F. T. Kützing* (Type of *Palmella duriuscula* Kütz., L [Fig. 220]; isotype, UC); in rivulo prope Hirschbach, Schleusingen, *Kützing,* 1830 (Type of *Hydrococcus rivularis* Kütz., L [Fig. 217]; isotype, UC). Württemberg: Neckar [-quelle bei Schwenningen], *G. Schmidle,* Aug. 1900 (Type of *Coelosphaerium holopediforme* Schmidle in the Schmidle collection, ZT [Fig. 227]). CZECHOSLOVAKIA: Moravia: Oslava, *F. Nováček,* 29 Aug. 1933 (FC). Bohemia: Chotzen, *A. Hansgirg,* Jul. 1887 (FC, W); Desná u Tannwaldu, *Hansgirg,* Jul. 1885 (FC, W); Dux, *Hansgirg,* Aug. 1883 (FC, W); Elbfall im Riesengebirge, *Hansgirg,* Sept. 1883 (FC, W); Hlubocepy, *Hansgirg,* Jun. 1885 (FC, W); in lapidibus calcareis viaductus magni ad Hlubocep, *Hansgirg* (as *Gloeocapsa rupicola* in Kerner, Fl. Exs. Austro-Hung. no. 3199, Z); Johannisbad, *Hansgirg,* Jul. 1885 (FC, W); Karlovary, Studeñ, *Hansgirg,* Aug. 1883 (FC, W); ad rupes thermarum Carolinarum, *Welwitsch* (DT); Koda prope Karlstein, Sedlec prope Lodenic, *Hansgirg,* May 1890 (FC, W); in offenen Quellen an Steinen oberhalb Kuchelbad nächst Prag, *Hansgirg,* 3 May 1884 (Type of *Pleurocapsa minor* Hansg., W; isotype, FC [Fig. 224]); an von lauwarmen Wasser bespritzten Steinen und Eisenplatten in der Moldau bei der Zuckerraffinerie bei Modran nächst Prag, *Hansgirg,* Nov. 1887 (Type of *Polycystis fusculotea* Hansg., W; isotype FC); Mummelfall bei Harrachsdorf, *Hansgirg,* Jul. 1886 (FC, W); Pantzschefall, *Hansgirg,* Sept. 1883 (FC, W); ad lapides in rivulo prope Pocatek, *Hansgirg,* Aug. 1888 (Type of *Cyanoderma rivulare* Hansg., W; isotype, FC [Fig. 231]); Praha, na stenach verej. vodovodu i v pissoirech, *Hansgirg,* Apr. 1883 (FC, W); Praha, stoky, *Hansgirg,* Feb. 1886 (FC, W); St. Paka, *Hansgirg,* Nov. 1883 (FC, W); Skt. Wenzens Brunnen, Prag, *Hansgirg,* Feb. 1884 (FC, W); Seegrund bei Eichwald, *Hansgirg,* Jul. 1883 (FC, W); Spindelmühle im Riesengebirge, *Hansgirg,* Nov. 1883 (FC, W); skály proti Srbsku u Karlsteina, *Hansgirg,* Jul. 1884 (FC, W); Stríbro, *Hansgirg,* Aug. 1883 (FC, W); Tábor, *Hansgirg,* Aug. 1883 (FC, W); in Bergbachen an Steinen, Tannwald, *Hansgirg,* Jul. 1885 (Types of *Chroococcus fuscoater* var. *fuscoviolaceus* Hansg. and *C. fuscoviolaceus* var. *cupreofuscus* Hansg., W; isotypes, FC [Figs. 216, 226]); an alten Schalen von Anodonta, Tuchomeric—Ounetic nächst Roztok, *Hansgirg,* Jun. 1890 (Types of *Pleurocapsa concharum* Hansg. and *Aphanocapsa anodontae* Hansg., W; isotypes, FC); Vöslau, *Hansgirg,* May 1887 (FC, W); Zel. Brod, *Hansgirg,* Jul. 1885 (FC, W).

AUSTRIA: Styria: in rivulo Gaiswinkelbach prope Grundlsee, *K. & L. Rechinger*, Sept. (as *Chamaesiphon polonicus* in Mus. Vindob. Krypt. Exs. no. 1760, FC, L, S, WU), *L. & K. Rechinger 31*, Sept. 1909 (FC, W); ad ostium lacus Leopoldsteiner See, *K. de Keissler*, Jul. (as *C. polonicus* in Mus. Vindob. Krypt. Exs. no. 1760b, FC, L, NY, S, WU). Tyrol: Hall, Innsbruck, und Jenbach, *A. Hansgirg*, 1891 (FC, W). JUGOSLAVIA: Slovenia: Podnart, Kaltenbrunn (prope Laibach), and Steinbrück, *Hansgirg*, Aug. 1890 (FC, W); an Schalen von Süsswasserschnecken bei Tremersfeld nächst Cilli, *Hansgirg*, Aug. 1890 (Type of *Aphanocapsa anodontae* var. *major* Hansg., W; isotype, FC [Fig. 219]). Istria: Pisino, *Hansgirg*, Aug. 1889 (FC, W); in aqua dulci, Parenzo, *Hansgirg*, Apr. 1889 (FC, W). Croatia: in aqua dulce, Fiume, *Hansgirg*, 1891 (FC, W); auf in schnell fliessendem Wasser untergetauchten Steinen in Bächen bei Martinscica nächst Fiume, *Hansgirg*, 1891 (Type of *Chamaesiphon fuscus* var. *auratus* Hansg., W; isotype, FC). Dalmatia: ad cataractam fluminis Kerkae ad Scardonam, *Hansgirg*, Aug. 1888 (FC, W); Clissa prope Spalato, *Hansgirg*, Aug. 1889 (FC, W); Cannosa—Valdinoce prope Ragusa, Castelnuovo, Metkovic, Ombla prope Ragusa, Ragusa, & Topla—Gruda prope Castelnuovo, *Hansgirg*, 1891 (FC, W). ALBANIA: Karbounara presso Lushnjë (prov. Berat), corassa cornea di tartaruga palustre, presso la sorgente di H_2S, *G. de Toni 47*, 30 May 1941 (D, DT, FC); Fushes-Dukati (prov. Valona), m. Shendelliut, velo d'acqua nella sorgente, in cavitá naturale, *G. de Toni & L. Boldori 81*, 6 Jul. 1941 (D, DT, FC). SWITZERLAND: St. Moritz, an einem hölzernen Trog, *C. von Nägeli*, Jul. 1849 (Type of *Gloeocapsa rubicunda* Näg., with *Leptothrix muralis*, ZT); Zürich, Pfiffens in einer Brunnen, *Nägeli*, Sept. 1845 (Type of *Exococcus ovatus* Näg., ZT); Zürich, [Küsnacht], *Nägeli 107*, Jul. 1847 (Type of *Gloeocapsa dermochroa* Näg. in herb. Kützing, L; isotypes, FI, M, UC); auf Steinen am Bodenseeufer bei Mammern, *G. Schmidle*, Mar. 1905 (Type of *Hyellococcus niger* Schmidle, ZT [Fig. 230]). ITALY: acqua santa near Rome (L); S. Margarita prope Ala, Auer, and Brixen, *A. Hansgirg*, 1891 (FC, W); fontana della piazza del Borghetti (Tirolo italiano), *D. Clementi* (Type of *Pleurococcus glomeratus* Menegh., FI [Fig. 234]); in fluvio Timavo prope Monfalcone, *F. T. Kützing, Thomasini, & Biasoletto*, 23 Apr. 1835 (Type of *Hydrococcus ulvaceus* Kütz., L [Fig. 223]); fons prope Monfalcone, *Hansgirg*, Aug. 1889 (FC, W); nel bagno freddo thermarum Julianarum, *J. Corinaldi 33* (Type of *Pleurococcus julianus* Menegh., FI [Fig. 225]; isotypes, L, S, W).

NETHERLANDS: in kanaal achter de dijk ten Linden van Roptazijl, *J. T. Koster 750*, Aug. 1942 (L). ENGLAND: on stones in rather slow-flowing Badgeworthy Water, Devonshire, *F. E. Fritsch*, 3 Sept. 1918 (Type of *Chamaesiphon ferrugineus* Fritsch in the collection of F. E. Fritsch; isotype, D); on smooth rock-surfaces in the midst of the rushing torrent in the East Lyn, north Devonshire, *Fritsch*, 1 Sept. 1920 (Type of *Chroococcopsis fluminensis* Fritsch and *Pseudoncobyrsa fluminensis* Fritsch in the collection of F. E. Fritsch; isotype, D). FRANCE: carrières, Falaise, *A. de Brébisson 526* (Type of *Palmella Brebissonii* Kütz., L; isotype, UC); environs de Paris, *M. Cornu*, 1878 (D, FH); sources du Lez près Montpellier, *F. Jadin*, Jan. 1892 (Type of *Hyella fontana* Hub. & Jad., PC), Jul. 1894 (FC); source de Font Méjane près Montarnaud (Hérault), *J. Huber & F. Jadin*, May 1893 (PC); Montpellier, *C. Flahault 150* (PC); dans une source fraiche se jetent dans le Lac (Lauvitel) près le Bourge d'Oisans, *C. Sauvageau*, Aug. 1900 (PC). TUNISIA: Tozeur, *M. Serpette TL59*, 26 Mar. 1948 (D). ALGERIA: source d'Ain Oumach près Biskra, *Sauvageau*, Apr. 1892 (Types of *Dermocarpa Flahaultii* Sauvag. and *Entophysalis Cornuana* Sauvag., PC; isotypes D [Fig. 218]). TANGANYIKA: with *Cladophora inconspicua* G. S. West on stones, shore, Niamkolo, *W. A. Cunnington* (Type of *Palmophyllum foliaceum* G. S. West, in slide collection, BM). SOUTH AFRICA: im Flussbette der Kachembe, *J. Menyhardt 335* (Type of *Placoma africana* Wille, in the slide collection of N. Wille, O).

NEW BRUNSWICK: boulder in stream, Grand Falls, *H. Habeeb 10159*, 20 Jul. 1947 (FC, HA); in Little river, Grand Falls, *Habeeb 10424, 10562*, 17 Jul., 16 Aug. 1948 (FC, HA); ledges in canyon, Grand Falls, *Habeeb 10379, 10707*, 12 Jul., 3 Sept. 1948 (FC, HA); shore of lake 8 miles south of Grand Falls, *Habeeb 11664*, 22 Jun. 1951 (FC, HA); in Rapide de Femme brook, fish hatchery, Grand Falls, *Habeeb 10447*, 21 Jul. 1948 (FC, HA); in Falls brook, Grand Falls, *Habeeb 10453, 10500, 10667, 13353, 13372, 13373, 13375, 13405*, Jun.—Aug. 1948 (FC, HA). MASSACHUSETTS: on stones, shore of Spot pond, Medford [parasitized material], *F. S. Collins*, 14 Aug. 1892 (as *Pleurocapsa fluviatilis* in Coll., Hold., & Setch., Phyc. Bor.-Amer. no. 555, FC, D, L, WU); in stone watering-trough, corner of Washington and Cedar streets, Woburn, *Collins 5381*, 9 Sept. 1905 (D, NY). RHODE ISLAND: on wet rocks, Mackerel cove, Conanicut island, *Collins*, 21 Apr. 1898 (FC, UC). CONNECTICUT: in shells, Twin lakes, Salisbury, *W. A. Setchell & I. Holden*, 24 Aug. 1895 (as *Hyella fontana* in Coll., Hold., & Setch., Phyc. Bor.-Amer. no. 303, FC, L, WU). QUEBEC: on ledge in falls of Camp No. 3 river, 10.5 miles north of the hotel, road to Mt. Albert, Gaspé, *H. Habeeb 1967*, 29 Jun. 1951 (FC, HA). NEW YORK: Cattaraugus county: in Beeline creek near lake, Allegany state park, *J. Blum 185 (H153)*, 29 Dec. 1948 (FC). Erie county: on wet sandstone, McKinley monument, Buffalo, *I. Knobloch 80020*, 16 Sept. 1944 (FC); escarpment below Busti avenue, Buffalo, *Blum 171 (H167)*, 20 Jan. 1949 (FC); in south branch of Smoke creek at highway no. 20A near Orchard Park, *Blum 209 (H192)*, 18 Feb. 1949 (FC). Genesee county: in Tonawanda creek at Indian Falls, *Blum 330 (H273)*, 28 Jul. 1949 (FC). Ontario county: in Reed's creek, *D. Haskins 42, 47a, 62*, 1944 (FC). NEW JERSEY:

on a water-tank, Tricker's Water Gardens, Saddle River, Bergen county, *F. Drouet, H. Bold, & J. Rubinstein 3869*, 9 Jul. 1941 (FC); in the edge of Toms river at Toms River, Ocean county, *J. C. Bader 206*, 10 Sept. 1938 (FC). PENNSYLVANIA: Dauphin county: on submerged stems in spring, Derry Church, *F. Wolle*, 21 Oct. 1887 (FC, D). Lancaster county: *J. L. Blum 113A*, 18 Aug. 1948 (FC, PH); springs in Springs park, Lititz, *Blum 106, J. Wallace 107*, 9 Aug. 1948 (FC, PH), *R. Patrick 106*, Oct. 1948 (FC, PH); Lititz run and tributary east of Lititz, *Blum 95, 110B*, 5 & 6 Jul. 1948 (FC, PH); Mill creek 2.5 miles south of New Holland, *Blum 88*, 28 Jun. 1948 (FC, PH). Philadelphia: in basins of Wm. Leonidas springs and a spring at Strawberry Mansion bridge, Fairmount Park, *Drouet, Patrick, D. Richards, & C. Hodge 5523, 5526*, 5 Aug. 1944 (FC).

VIRGINIA: in stream entering Lake Matoaka, Williamsburg, *J. C. Strickland 1240, 1239*, 28 Jul. 1942 (FC, ST); on rocks in Moormans river below Sugar Hollow, Albemarle county, *Strickland 657*, 6 Oct. 1940 (FC, ST); on rock in Crab run west of McDowell, Highland county, *E. S. Luttrell & Strickland 936*, 20 Jul. 1941 (FC, ST); on rock in spring at Hawksbill Gap on Skyline drive, Madison county, *Luttrell 3400*, 25 Apr. 1942 (FC, ST). FLORIDA: Gainesville, *M. A. Brannon 40, 160*, 16 Feb. 1942, 11 Apr. 1943 (FC, PC); St. Marks river at Little Natural Bridge, *C. S. Nielsen, G. C. Madsen, & D. Crowson 559*, 30 Oct. 1948 (FC, T); Log Sulphur springs, Newport, Wakulla county, *Nielsen 251*, Aug. 1948 (FC, T); on stones in a large spring bathing pool north of Newport, *Drouet, Crowson, & R. Thornton 11400a*, 25 Jan. 1949 (FC, T). OHIO: in Lake Erie on the south end of South Bass island, Ottawa county, *J. Blum 300 (H244)*, 15 Jun. 1949 (FC); on cement wall, Lakeview fish hatchery, Cincinnati, *W. A. Daily 195*, 31 Oct. 1939 (DA, FC). KENTUCKY: concrete aquarium in greenhouse, University of Kentucky, Lexington, *B. B. McInteer 1045*, 19 Sept. 1939 (DA, FC). TENNESSEE: on concrete of pond, Mack's, 5 miles east of Knoxville, *H. Silva 854*, 18 Jun. 1949 (FC, TENN); concrete in lake outflow, Chilowee park, Knoxville, *Silva 831*, 18 Jun. 1949 (FC, TENN); concrete sides of fishpool, Ijam's place, Knoxville, *Silva 1510*, 20 Jul. 1949 (FC, TENN); rock in spring, Shadow farm near Decatur, Meigs county, *Silva 961*, 28 Jun. 1949 (FC, TENN); rock, Road Prong, Sevier county, *Silva 448*, 11 Jan. 1947 (FC, TENN); in drinking fountain, Newfound Gap, Sevier county, *Silva 60, 317, 1891*, 25 Aug. 1941, 3 Nov. 1946, 10 Sept. 1949 (FC, TENN); rocks in spray of falls, State Teachers College campus, Johnson City, Washington county, *Silva 462*, 14 Jan. 1947 (FC, TENN). MICHIGAN: Cheboygan county: in Black lake, *H. K. Phinney 12M41/7, 206, 208*, 24 Jul. 1941, 11 Aug. 1942 (FC, PHI); on pilings, Pigeon river near Indian River, *Phinney 518*, 11 Aug. 1943 (FC, PHI). Presque Isle county: on pipe of artesian spring, Cliff's place, Ocquioc Lake, *Phinney 509*, 29 Jul. 1943 (FC, PHI). INDIANA: Lowell, Lake county, *M. A. Brannon 313*, 10 Sept. 1945 (FC, PC); on dam and in mill race, Avoca state fish hatchery, Lawrence county, *F. K. & W. A. Daily 2280, 2296, 2299*, 1 Aug. 1950 (DA, FC); laboratory culture, Indiana University, Bloomington, *R. C. Starr*, 29 Jan. 1953 (FC); in Canyon creek near old quarry and in small falls of Echo canyon, McCormicks Creek state park, Owen county, *F. K. & W. A. Daily 1772, 1792*, 9 Aug. 1947 (DA, FC); in spring and fountain, Dunes state park, Porter county, *L. J. King*, 17 Sept. 1941 (FC); in a stream north of Whiting Boy Scout Camp, Dunes state park, *Drouet & H. B. Louderback 5634a*, 15 Jul. 1945 (FC); in the fountain at the courthouse, Richmond, *King*, 10 Nov. 1940 (EAR, FC); on rock, Thistlethwaites falls, Richmond, *King*, 3 Oct. 1940 (EAR, FC).

WISCONSIN: Madison: in fountain in B. B. Clarke park, *Louderback & Drouet 5498*, 23 Jul. 1944 (FC); dredged, Lake Mendota, *Forbes & A. B. Seymour*, Sept. 1885 (FC); in the source of Yahama river, Tenney park, *Louderback & Drouet 5519, 5519a,b*, 23 Jul. 1944 (FC). ILLINOIS: Cook county: in drinking fountains (Chicago & Michigan avenues, Jackson park, Grant park, Lincoln park), Chicago, *Drouet & Louderback 5228, 5236, Louderback 4*, Oct. 1943, 18 Jul. 1944 (D, FC), *K. C. Fan 10630, 10640, 10645, 10646*, 12 Sept. 1954 (D); laboratory culture, University of Chicago, *P. D. Voth*, May 1945 (FC); on tank in greenhouses, University of Chicago, *Voth, J. M. Beal, & Drouet 2375*, 5 Oct. 1938 (FC); rocks in quarry at Sag Bridge at 111th street, *King & J. O. Young 457*, 3 Aug. 1941 (FC); on a sink in the greenhouse, Northwestern University, Evanston, *Drouet, H. K. Phinney, F. A. Barkley, & D. Richards*, 17 Mar. 1945 (FC). Kane county: in a fountain in Pottawatomie park and in a shell in Fox river south of Main street, St. Charles, *Louderback & Drouet 5313, 5333*, 14-15 Apr. 1944 (FC). Livingston county: in shells in Vermilion river near Mill street, Pontiac, *Drouet & Louderback 5201, 5211a, 5248*, Sept.—Oct. 1943 (FC). MISSISSIPPI: outlet of the freshwater reservoir in the western part of Gulfport, *Drouet & R. L. Caylor 9944a, 9945*, 12 Dec. 1946 (FC). MINNESOTA: Mississippi river headwaters, Itasca state park, *Drouet, Fan & J. Rowley 11974, Fan 10160*, 21 June 1954; in Sucker creek west of Lake Itasca post office, Clearwater county, *Drouet, T. Morley & Fan 11901*, 17 June 1954 (D, MIN). IOWA: in a drinking fountain in the city park, McGregor, Clayton county, *Drouet & Louderback 5162*, 20 Aug. 1943 (D, FC). MISSOURI: on rocks in Gravois springs, Gravois Mills, Morgan county, *F. Drouet 1234, 1237*, 19 Aug. 1934 (D); on rocks in spring, Monegaw Springs, St. Clair county, *Drouet 730*, 4 Oct. 1930 (D); on pebbles in White river southeast of Mincy, Taney county, *J. A. Steyermark 40099*, 20 Jun. 1941 (FC). LOUISIANA: on turtle, Baton Rouge, *L. H. Flint*, 13 Oct. 1947 (FC).

NEBRASKA: Banner county: rock in streamlet north of Harrisburg, *W. Kiener 22142*, 22 May

1947 (FC, KI). Cherry county: pebbles in Minnechaduza creek, fish hatcheries, Valentine, *Kiener* 23825, 18 Jun. 1948 (FC, KI). Cheyenne county: rocks in Lodgepole creek east of Potter, *Kiener* 22613a, 22614, 22615, 22618a, 22621, 16 Aug. 1947 (FC, KI). Dakota county: back of Mississippi Map Turtle, Crystal lake, Dakota City, *Kiener* 21778, 21778b, 17 Feb. 1947 (FC, KI). Dawes county: rock in Chadron creek, Chadron state park, *Kiener* 20370, 20370a, 20371a, 20375a, 20413, 20415, 15 May 1946 (FC, KI). Dundy county: on pebbles in creek 8 miles north of Max, *Kiener* 10595, 29 Jul. 1941 (FC, KI); pebbles in outlet of fishponds, Rock Creek Hatchery, *Kiener* 19566, 3 Aug. 1945 (FC, KI); on rock and bark in water, Rock Creek Lake state park, 25064, 25115b, 25161, 21 May 1947, 21—22 Nov. 1949 (FC, KI); rocks in seepage near park north of Parks, *Kiener* 21886, 28 Mar. 1947 (FC, KI). Garden county: pebbles in Blue creek north of Oshkosh, *Kiener* 23106, 23108, 2 Apr. 1948 (FC, KI). Keith county: on rock in Otter creek northwest of Ogallala, *Kiener* 21056, 6 Jul. 1946 (FC, KI); pebbles, Lonergan creek springs, Lemoyne, *Kiener* 23389, 23408, 4 May 1948 (FC, KI). Lancaster county: watering-tank 9 miles northwest of Lincoln, *Kiener* 12937, 20 Aug. 1942 (FC, KI); on wood in runoff of tank 14 miles northwest of Lincoln, *Kiener* 12979b, 29 Aug. 1942 (FC, KI); bird fountain, Lincoln, *Kiener* 23915, 28 Jun. 1948 (FC, KI); on turtle, Lincoln, *Kiener* 24101, 10 Aug. 1948 (FC, KI). Morrill county: pebbles in Lawrence Fork creek, southwest of Redington, *Kiener* 22737a, 22738, 21 Aug. 1947 (FC, KI); drainage ditch, west edge of Bayard, *Kiener* 25136, 23 Nov. 1949 (FC, KI). Scotts Bluff county: pebbles in marsh pond northeast of Morrill, *Kiener* 22793, 24 Aug. 1947 (FC, KI); rocks in Akers Draw, Morrill, *Kiener* 22124b, 22126a, 25063, 25064, 25115b, 25161, 21 May 1947, 21—22 Nov. 1949 (FC, KI); rocks in seepage near Mitchell, *Kiener* 25161, 25164, 21 Nov. 1949 (FC, KI).

TEXAS: rocks from shallow stream, Austin, *F. A. Barkley T1004*, 5 Dec. 1942 (FC, TEX); New Braunfels, *L. H. Flint 3*, 18 Nov. 1954 (D); on tub and rocks at bath house, Hot Springs, Brewster county, *E. Whitehouse 24643, 24645*, 24 Dec. 1950 (FC). MONTANA: rocks in running water 0.5 mile east of Potter canyon slaughterhouse, Missoula county, *S. R. Ames & F. A. Barkley 18*, 26 Aug. 1940 (FC). COLORADO: wet cliff, Chasm gorge, Long's Peak, *W. Kiener 4187a*, 21 Sept. 1936 (FC, KI); culture from cold streamlet, Granite Basin, Longs Peak, *Kiener 1281*, 21 Mar. 1943 (FC, KI); in drinking fountains, Denver, *Drouet, D. Richards, & J. Rubinstein 4097*, 6 Sept. 1941 (FC), *Drouet & H. B. Louderback 5731*, 18 Aug. 1946 (FC), *Louderback 5*, 22 Aug. 1947 (FC). NEW MEXICO: on rocks in a pond in the lava-beds 6—8 miles east of Grant, Valencia county, *A. A. Lindsey 2E*, 26 Aug. 1944 (FC). UTAH: on rock in pothole at falls, Mt. Timpanogos, Utah county, *A. O. Garrett 73*, 16 Aug. 1925 (FC, UC); in fountain in front of main building, Utah Agricultural college, Logan, *G. Piranian*, 30 Sept. 1933 (FC). NEVADA: in a drinking-fountain in the park along the Truckee river, Reno, *Drouet & Richards 4112*, 9 Sept. 1941 (FC). WASHINGTON: in Columbia river, Richland, *R. Genoway*, 15 Dec. 1952 (FC). CALIFORNIA: Alameda county: in a brook, Lake Chabot, San Leandro, *W. J. V. Osterhout & N. L. Gardner 541*, 28 Jun. 1902 (FC, UC); water-trough, Oakland, *Gardner 20*, Oct. 1900 (UC). Napa county: on a cement runway, Seigler hot springs, *Gardner 7474*, 7 Aug. 1933 (FC, UC). San Bernardino county: Harlem hot springs, *W. A. Setchell 1560*, 19 Dec. 1896 (Type of *Dermocarpa Setchellii* Dr.; isotype, UC). San Francisco: culture from Lake Merced, *Gardner 6837*, 2 Oct. 1931 (Type of *D. Gardneriana* Dr., FC, isotype, UC); on shells, Mountain lake (parasitized material), *Osterhout & Gardner 545*, 24 June 1902 (FC, UC; as *Pleurocapsa* 'concharum' in Coll., & Setch., Phyc. Bor.-Amer. no. 1051, FC, L, UC, WU); culture from a dripping water-pipe, Golden Gate park, *Gardner 890a*, 21 Feb. 1933 (FC, UC). Santa Clara county: in rapids of Coyote creek, San José, *J. F. Macbride 7806a*, Sept. 1944 (FC). Siskiyou county: in seepage at the base of a cliff along Sacramento river, Dunsmuir, *Drouet & Richards 4203*, 15 Sept. 1941 (FC). Sonoma county: on dripping rocks in hot water, the Geysers, *Gardner & Bonar 7519, 7519a*, 8 Aug. 1933 (FC, UC). Tulare county: in Tule river, Porterville, *Drouet & M. J. Groesbeck 4443*, 4 Oct. 1941 (FC); in a water-fountain, Burnett park, east of Porterville, *Drouet & Groesbeck 4496a*, 5 Oct. 1941 (FC).

PUERTO RICO: on limestone, waterfall between Arecibo and Utuado, *N. Wille 1458b*, 4 Mar. 1915 (Type of *Chroococcus minutissimus* Gardn., NY [Fig. 232]); Morrillo de Cabo Rojo, *N. L. Britton 4375*, Feb. 1915 (UC); on pebbles in a stream of warm water by the hot spring, Coamo Springs, *Wille 381*, 12 Jan. 1915 (Type of *Entophysalis chlorophora* Gardn., NY [Fig. 233]; isotype, D, UC); auf Steine im Flusse bei Fajardo, *Wille 680a, 684*, 26 Jan. 1915 (FC, NY); on stones in a brook west of Humacao, *Wille 593*, Jan. 1915 (Type of *Radaisia Willei* Gardn., NY; isotypes, D, UC); in a fountain in the woods near Maricao, *Wille 1076*, Feb. 1915 (Type of *Radaisia confluens* Gardn., NY [Fig. 228]; isotypes, D, UC). MEXICO: from rocks, lower hot springs, Baños de Agua Caliente, Hildalgo, *R. Patrick 95*, 15 Jul. 1947 (FC, PH); spring, San José de Purua, Michoacan, *Patrick 427*, 4 Aug. 1947 (FC, PH). GUATEMALA: on pebbles in small stream, Quebrada Shusho above Chiquimula, *P. C. Standley 74337*, 14 Oct. 1940 (FC); on rocks in sulfur spring, shore of Laguna Petexbatúm south of Sayaxché, dept. Petén, *J. A. Steyermark 46232a,b*, 3 May 1942 (FC). HONDURAS: on metal water tanks, Comayagua, *Standley & J. Chacón P. 5632, 5781*, Mar. 1947 (FC); on side of hydrant, La Lima, dept. Cortés, *Standley &*

125

Chacón P. 7150, Apr. 1947 (FC). EL SALVADOR: in very warm water from sulfur spring, Ahuachapán, *Standley & E. Padilla V. 2451,* Jan. 1947 (FC); in shallow pool of stream, Metapán, dept. Santa Ana, *Standley & Padilla V. 3256,* 29 Jan.—1 Feb. 1947 (FC). VENEZUELA: dripping cave in lime bluff along Rio Zumbador near base of Piedra Blanca, northeast of Bergantín, Anzoátegui, *Steyermark 61284,* 1—2 Mar. 1945 (FC); PERU: La Encantada, Lago Chilca, *A. Maldonado 60,* Jan. 1942 (FC).

HAWAIIAN ISLANDS: Kaelepuhu stream, Kailua, Oahu, *M. S. Doty 12491,* 3 Aug. 1954 (D). SAMOA ISLANDS: [auf Felsen, Insel Savaii] *K. H. Rechinger 2834* (Type of *Entophysalis samoensis* Wille in the slide collection of N. Wille, O). NEW ZEALAND: in basin, Black Terraces, Taupo (East Taupo), *Setchell 6010a(101),* 13 May 1904 (FC, UC). INDONESIA: in oolithis in fonte Tjeding in monte Idjen in Java Orientali, *C. E. B. Bremekamp* (Type of *Aphanocapsa constructrix* Bremek. in the collections of the Mineralogisch-Geologish Instituut of the University of Utrecht); Tjibodas, Java, *Ruttner* (paratype of *Chamaesiphon polymorphus* Geitl., slides Tj21e and Tj211c ∝ in the collection of L. Geitler); Steine in Spritzwasser der Rinne, in der Bambu-Abflussrinne des Beckens, Gefasste Quelle am linken Seraju-Ufer, oberhalb der Dessa Patakbanteng, Dijeng-Plateau, Java, *Ruttner,* 3 Jun. 1929 (Type of *Chlorogloea major* Geitl., slide no. D6be in the collection of L. Geitler); Nostoc-Lager, Umgebung des T. Pasir, Wasserfall des Pagarede [Pasergede], Java, *Ruttner,* 9 Dec. 1928 (Type of *Chamaesiphon fallax* Geitl., slide no. SW1bβ in the collection of L. Geitler); Stein an dem Mittellauf des Quellbaches, starke Quelle am SE-Ufer, Ranu Pakis, Java, *Ruttner,* 29 Oct. 1928 (Type of *Chlorogloea purpurea* var. *minutissima* Geitl., slide no. KPQ1d in the collection of L. Geitler); Ranau-See, Süd-Sumatra, *Ruttner,* 1929 (paratype of *Chroococcopsis gigantea* Geitl., slide no. R4h∝ in the collection of L. Geitler), 27 Jan. 1929 (Type of *Cyanodermatium violaceum* Geitl., slide no. RA2 in the collection of L. Geitler), 28 Jan.—5 Feb. 1929 (Type of *Radaisia gigas* Geitl., slide no. R5aβ in the collection of L. Geitler), 1—5 Feb. 1929 (Type of *Cyanodermatium gelatinosum* Geitl., slide no. R4d in the collection of L. Geitler); an Steinen im unteren Becken, Bukit Kili, Sumatra, *Thienemann,* 2 Mar. 1929 (Type of *Myxosarcina spectabilis* Geitl., slide no. SkW1b in the collection of L. Geitler). INDIA: auf Schneckenschalen in einem See in der Stadt Poona, *A. Hansgirg* (Type of *Chroococcus Hansgirgii* Schmidle, ZT [Fig. 229]). EGYPT: Sinai[halbinsel, am Fusse des Dschebel Hamân, *A. Kneucker,* 16 Apr. 1904] (Type of *Guyotia singularis* Schmidle, ZT [Fig. 222]).

4b. ENTOPHYSALIS RIVULARIS f. PAPILLOSA (Kütz.) Drouet & Daily, STAT. NOV. *Palmella papillosa* Kützing, Phyc. Germ., p. 149. 1845. *Polycoccus papillosus* Roemer, Alg. Deutschl., p. 65. 1845. *Cagniardia papillosa* Trevisan, Sagg. Monogr. Alg. Coccot., p. 50. 1848. *Entophysalis papillosa* Drouet & Daily, Lloydia 11: 79. 1948.—Type from the Harz mountains, Germany (L). FIG. 375.

Sphaerogonium fuscum Rostafinski, Rozpr. Akad. Umiej. Krakow., Wydz. Math.-Przyr., 10: 295. 1883. *Chamaesiphon fuscus* Hansgirg, Oesterr. Bot. Zeitschr. 38: 117. 1888. —Topotypes from Kalatowkach, Poland (FH, WU).

Xenococcus britannicus Fritsch in West & Fritsch, Brit. Freshw. Alg., p. 467. 1927. *Oncobyrsa britannica* Fritsch *pro synon.,* loc. cit. 1927. *Chamaesiphonopsis regularis* Fritsch, New Phytol. 28: 173. 1929. *Chamaesiphon regularis* Geitler, Rabenh. Krypt.-Fl. Eur. 14: 440. 1932. —Type from Devonshire, England (in the collection of F. E. Fritsch).

Chamaesiphon pseudo-polymorphus Fritsch, New Phytol. 28: 176. 1929. —Type from Devonshire, England (in the collection of F. E. Fritsch).

The original specimen of the following has not been available to us; its original description is designated as the Type until the specimen can be found:

Chamaesiphon Geitleri H. Luther, Acta Bot. Fennica 55: 33. 1954.

Cellulae solitariae et ad basem strati aut pulvini cylindricae vel longi-ellipsoideae. FIGS. 374, 375.

On rocks, shells, woodwork, etc. in fresh water, chiefly in clear streams, springs, and fountains. *Entophysalis rivularis* f. *papillosa* may eventually prove to be only a growth-form of f. *rivularis* developing under special environmental conditions.

Specimens examined:

CRIMEA: in rivulo Ulu-usenj, *N. N. Woronichin,* 31 Aug. 1929 (FC, FH). POLAND: zbierane na kamykach granitowych w wywierzysku Dunajca Bialego w Kalatówkach, *M. Raciborski,* Oct. 1909 (topotype of *Sphaerogonium fuscum* Rostaf. as *Chamaesiphon fuscus* in Racib., Phyc. Polon. no. 10, FH,

126

WU). GERMANY: in rivulis montium ad saxa, Freiburg im Breisgau, *A. Braun 41*, 1848 (L, NY), 1849 (PC); in Bäche des Oberharzes, *Römer 148* (Type of *Palmella papillosa* Kütz., L [Fig. 375]; isotype, UC). CZECHOSLOVAKIA: Bohemia: am Wege von Deffernik zu Lackasee bei Eisenstein, *A. Hansgirg*, Aug. 1887 (FC, W); in aquaeductu ligneo fontis ad Deffernik prope Eisenstein, *Hansgirg*, 25 Aug. 1887 (as *Oncobyrsa rivularis* in Wittr. & Nordst., Alg. Exs. no. 999, FC, L, WU). ENGLAND: River Heddon, Devon, *F. E. Fritsch*, 5 Sept. 1920 (Type of *Chamaesiphon pseudopolymorphus* Fritsch in the collection of F. E. Fritsch [Fig. 374]; isotype, D); East Lyn, north Devon, *Fritsch*, 2 Sept. 1918 (Type of *Xenococcus britannicus* Fritsch in the collection of F. E. Fritsch; isotype, D). NEW BRUNSWICK: on rocks in a spring, Alma, Albert county, *H. Habeeb 11681*, 26 Jul. 1951 (FC, HA). NEW YORK: on rusty pipe in Hudson river, Warrensburg, *J. Bader 183*, 5 Aug. 1938 (FC). PUERTO RICO: on twigs in a stream near Maricao, *N. Wille 1260*, Feb. 1915 (NY).

5. ENTOPHYSALIS LEMANIAE (Ag.) Dr. & Daily.

Plantae primum microscopicae (saepius cellulae solitariae) deinde in pulvina macroscopica aeruginea, olivacea, brunnea, violacea, vel rubra increscentes, cellulis solitariis primo sphaericis aetate provecta ovoideis, cylindraceis, cylindraceo-ellipsoideis, vel lineari-tubaeformibus, diametro 1—6μ crassis, erectis vel varie curvatis cellulis pulvini juvenilis ellipticis vel cylindricis erectis, in pulvinis maturis sphaericis, ovoideis, vel polyhedroideis, diametro 2—8μ crassis; endosporangiis sphaericis, ovoideis, vel pyriformibus, diametro ad 25μ crassis; gelatino vaginale primum hyalino demum lutescente, homogeneo vel laminoso; protoplasmate aerugineo, olivaceo, violaceo, vel roseo, homogeneo. FIGS. 221, 235—246.

The commonly encountered growth-forms of *Entophysalis Lemaniae* are the solitary cells which develop in profusion on filamentous algae (*Cladophora* spp., *Oedogonium* spp., *Plectonema Wollei* Farl., etc.) and other plants which persist throughout a season in a habitat. On perennial vegetation, such as *Lemanea* spp., bryophytes, and certain Podostemaceae, the microscopic cushion-shaped masses are sometimes found; their rarity, however, may in large part be due to their destruction by fungi.

Key to forms of *Entophysalis Lemaniae*:

Solitary cells spherical to cylindrical .... f. LEMANIAE

Solitary cells linear-cylindrical and long-stipitate .... f. ELONGATA

5a. ENTOPHYSALIS LEMANIAE f. LEMANIAE. *E.* LEMANIAE (Agardh) Drouet & Daily, COMB. NOV. *Nostoc Lemaniae* Agardh, Syst. Algar., p. 20. 1824. *Nostocella Lemaniae* Gaillon, Apercu d'Hist. Nat., p. 28. 1833. *Oncobyrsa Lemaniae* Gomont, Bull. Soc Bot. Fr. 43: 377. 1896. *Hydrococcus Lemaniae* Bornet *pro synon.* in Gomont, loc. cit. 1896. —Type from Scania, Sweden (PC). FIG. 239.

Oncobyrsa Brebissonii Meneghini, Mem. R. Accad. Sci. Torino, ser. 2, 5(Sci. Fis. & Mat.): 96. 1843. *Nostoc fluviatile* Brébisson *pro synon.* in Meneghini, loc. cit. 1843. *Hydrococcus Brebissonii* Kützing, Tab. Phyc. 1: 23. 1846. *Oncobyrsa rivularis* var. *Brebissonii* Hansgirg, Sitzungsber. K. Böhm. Ges. Wiss., Math.-Nat. Cl., 1890(2): 93. 1891. *Entophysalis Brebissonii* Drouet & Daily, Lloydia 11: 79. 1948. —Type from Falaise, France (L). FIG. 245.

Hydrococcus Cesatii Rabenhorst, Alg. Sachs. 93 & 94: 922. 1860. *Oncobyrsa Cesatiana* Rabenhorst, Fl. Eur. Algar. 2: 68. 1865. —Type from Oropa, Piedmont, Italy (D). FIG. 221.

Chamaesiphon confervicola A. Braun in Rabenhorst, Fl. Eur. Algar. 2: 148. 1865. *Brachythrix confervicola* A. Braun *pro synon.* in Rabenhorst, loc. cit. 1865. —Type from Giessen, Germany (NY).

Chamaesiphon confervicola f. *elongatus* Rabenhorst, Fl. Eur. Algar. 2: 149. 1865. —Type from Giessen, Germany (NY).

Chamaesiphon incrustans Grunow in Rabenhorst, Fl. Eur. Algar. 2: 149. 1865. *Sphaerogonium incrustans* Rostafinski, Rozpr. Akad. Umiej. Krakow., Wydz. Mat.-przyr., 10: 282. 1883. —Type from Windisch-Garsten, Upper Austria (W).

Chamaesiphon Schiedermayeri Grunow in Rabenhorst, Fl. Eur. Algar. 2: 149. 1865. *C. confervicola* var. *Schiedermayeri* Borzi ex Hansgirg, Prodr. Algenfl. Böhmen 2: 124: 1892. —Type from Milchdorf, Upper Austria (W).

Chamaesiphon Schiedermayeri f. *subclavatus* Grunow in Rabenhorst, Fl. Eur. Algar. 2: 149. 1865. —Type from Vöcklabrück, Upper Austria (W).

Sphaenosiphon aquae-dulcis Reinsch, Contrib. Algol. & Fungol. 1: 15. 1874. *Dermocarpa aquae-dulcis* Geitler in Pascher, Süsswasserfl. 12: 142. 1925. —Type from Dalmatia (K).

Chamaesiphon curvatus Nordstedt, Alg. Aq. Dulc. & Charac. Sandvicens., p. 4. 1878. *Sphaerogonium curvatum* Rostafinski, Rozpr. Akad. Umiej. Krakow., Wydz. Math.-Przyr., 10: 292. 1883. *Chamaesiphon confervicola* var. *curvatus* Borzi ex Hansgirg, Prodr. Algenfl. Böhmen 2: 124. 1892. —Type from near Honolulu, Hawaiian islands (PC).

Chamaesiphon curvatus f. *elongatus* Nordstedt, Alg. Aq. Dulc. & Charac. Sandvicens., p. 4. 1878. *C. curvatus* var. *elongatus* Nordstedt ex Lemmermann, Krypt.-Fl. Mark Brandenb. 3: 100. 1907. —Type from near Honolulu, Hawaiian islands (PC).

Godlewskia aggregata Janczewski, Ann. Sci. Nat. VI. Bot. 16: 229. 1883. *Chamaesiphon aggregatus* Geitler in Pascher, Süsswasserfl. 12: 157. 1925. —Type from Krakow, Poland (PC).

Xenococcus Kerneri Hansgirg, Physiol. & Algol. Stud., p. 111. 1887. *Dermocarpa Kerneri* Hansgirg, Physiol. & Phycophyt. Unters., p. 234. 1893. *Pleurocapsa Kerneri* Drouet, Field Mus. Bot. Ser. 20: 125. 1942. —Type from Eisenbrod, Bohemia (FH).

Chamaesiphon incrustans f. *minor* Möbius, Hedwigia 27: 246. 1888. —Topotypes and paratypes from Coamo, Puerto Rico (D, MO, FC).

Pleurocapsa fluviatilis Lagerheim, Notarisia 3: 430. 1888. *Xenococcus fluviatilis* Geitler, Beih. z. Bot. Centralbl., II, 41: 245. 1925. —Type from Freiburg in Breisgau, Germany (DT). FIG. 240.

Oncobyrsa hispanica Lewin, Bih. t. Sv, Vet.-Akad. Handl., Afd. 3, 14(1): 4. 1888. —Type from Yunquera (Malaga), Spain (S).

Dermocarpa depressa W. & G. S. West, Journ. of Bot. 35: 301. 1897. —Type from Sassanga island, Loanda, Angola (BM). FIG. 244.

Chroococcus varius f. *samoensis* Wille, Hedwigia 53: 144. 1913. *C. varius* var. *samoensis* Wille ex J. de Toni, Diagn. Alg. Nov. I. Myxophyc. 3: 292. 1938. —Type from Upolu, Samoa (O).

Chroococcus turgidus var. *subviolaceus* Wille, Hedwigia 53: 144. 1913. —Type from Savaii, Samoa (O).

Xenococcus minimus Geitler, Ber. Deutsch. Bot. Ges. 40: 284. 1922. —Type from Vienna, Austria (W). FIG. 241.

Dermocarpa chamaesiphonoides Geitler, Ber. Deutsch. Bot. Ges. 40: 283. 1922. —Type from Vienna, Austria (W). FIG. 237.

Chamaesiphon cylindricus Boye Petersen, Bot. Icel. 2(2): 272. 1923. —Type (in the collection of J. Boye Petersen) from Husavik, Iceland. FIG. 238.

Pleurocapsa epiphytica Gardner, Mem. New York Bot. Gard. 7: 31. 1927. *Radaisia epiphytica* Gardner, New York Acad. Sci. Sci. Surv. Porto Rico 8(2): 264. 1932. —Type from near Mayaguez, Puerto Rico (NY).

Xenococcus Willei Gardner, Mem. New York Bot. Gard. 7: 33. 1927. —Type from near Coamo, Puerto Rico (NY). FIG. 236.

Chamaesiphon portoricensis Gardner, Mem. New York Bot. Gard. 7: 33. 1927. *C. minutus* var. *major* Geitler, Rabenh. Krypt.-Fl. Eur. 14: 429. 1932. —Type from Coamo Springs, Puerto Rico (NY). FIG. 243.

Xenococcus chroococcoides Fritsch, New Phytol. 28: 186. 1929. —Type from Devonshire, England (in the collection of F. E. Fritsch). FIG. 242.

Xenococcus minimus var. *Starmachii* Geitler, Rabenh. Krypt.-Fl. 14: 333. 1932. —Paratype named by the author from Danan Manindjau, Sumatra, Indonesia (in the collection of L. Geitler).

Dermocarpa clavata Geitler, Rabenh. Krypt.-Fl. 14: 406. 1932. *D. clavata* var. *aquae-dulcis* Geitler, ibid. 14: 1173. 1932. —Type from Ranau Bedali, Java, Indonesia (in the collection of L. Geitler). Both of these names appear to be based upon the same type.

Dermocarpa xenococcoides Geitler, Arch. f. Hydrobiol., Suppl. XII, 4: 625. 1933. —Type from Ranau Bedali, Java, Indonesia (in the collection of L. Geitler).

Chamaesiphon minutus var. *gracilior* Geitler, Arch. f. Hydrobiol., Suppl. XII, 4: 625. 1933. —Type from Singkarak lake, Sumatra, Indonesia (in the collection of L. Geitler).

Chamaesiphon subglobosus var. *major* Geitler, Arch. f. Hydrobiol., Suppl. XII, 4: 625. 1933. —Type from Tjibodas, Java, Indonesia (in the collection of L. Geitler).

Pleurocapsa subgelatinosa Geitler, Arch. f. Hydrobiol., Suppl. XII, 4: 626. 1933; ex Borge, Ark. f. Bot. 25A(17): 5. 1933. —Type from Kota Batu, Sumatra, Indonesia (in the collection of L. Geitler).

Chamaesiphon mollis Geitler, Arch. f. Hydrobiol., Suppl. XII, 4: 626. 1933. —Type from Tjibodas, Java, Indonesia (in the collection of L. Geitler).

Dermocarpa Hollenbergii Drouet, Field Mus. Bot. Ser. 20: 129. 1942. —Type from Mojave desert, California (D). FIG. 235.

Dermocarpa Solheimii Drouet, Field Mus. Bot. Ser. 20: 129. 1942. —Type from Medicine Bow national forest, Wyoming (FC).

Dermocarpa minuta Drouet, Field Mus. Bot. Ser. 20: 130. 1942. —Type from Tiburon, California (FC).

Original specimens have not been available to us for the following names; their original descriptions are here designated as the Types until the specimens can be found:

Linckia fragiformis Roth, Catal. Bot. 3: 345. 1806. —See Brébisson in Hedwigia 9: 65 (1870) and Hansgirg, Prodr. Algenfl. Böhmen 2: 130. 1892.

Chamaesiphon torulosus Borzi, N. Giorn. Bot. Ital. 14: 313. 1882.

Cyanocystis versicolor Borzi, N. Giorn. Bot. Ital. 14: 314. 1882. *Dermocarpa versicolor* Geitler in Pascher, Süsswasserfl. 12: 142. 1925.

Sphaerogonium minutum Rostafinski, Rozpr. Akad. Umiej. Krakow., Wydz. Mat.-Przyr., 10: 290. 1883. *Chamaesiphon minutus* Lemmermann, Krypt-Fl. Mark Brandenb. 3: 98. 1907.

Sphaerogonium amethystinum Rostafinski, Rozpr. Akad. Umiej. Krakow., Wydz. Mat.-Przyr., 10: 291. 1883. *Chamaesiphon amethystinus* Lemmermann, Krypt.-Fl. Mark Brandenb. 3: 99. 1907.

Chamaesiphon incrustans var. *laxus* Reinsch, Deutsche Polar Exped. 2: 334. 1890.

Xenococcus gracilis Lemmermann, Abh. Nat. Ver. Bremen 14: 510, footnote. 1898.

Chamaesiphon minimus Schmidle, Engler Bot. Jahrb. 30: 62. 1901. *C. africanus* var. *minimus* Lemmermann, Krypt.-Fl. Mark Brandenb. 3: 99. 1907.

Chamaesiphon africanus Schmidle, Engler Bot. Jahrb. 30: 62. 1901.

Chamaesiphon sphagnicola Maillefer, Bull. Herb. Boissier, ser. 2, 7: 44. 1906. *Dermocarpa sphagnicola* Geitler in Pascher, Süsswasserfl. 12: 143. 1925.

Chamaesiphon curvatus var. *Turneri* Forti, Syll. Myxophyc., p. 140. 1907.

Cyanocystis parva Conrad, Ann. Biol. Lac. 7: 130. 1914. *Dermocarpa parva* Geitler in Pascher, Süsswasserfl. 12: 142. 1925.

Chamaesiphon incrustans f. *asiaticus* Wille in Hedin, Southern Tibet 6(3, Bot.): 166. 1922.

Chamaesiphon incrustans f. *longissimus* Wille in Hedin, Southern Tibet 6(3, Bot.): 167. 1922.

Chamaesiphon oncobyrsoides Geitler, Arch. f. Protistenk. 51: 330. 1925.

Chamaesiphon incrustans var. *elongatus* Starmach, Spraw. Kom. Fizjog. Polsk. Akad. Umiej. 1926: 110. 1927.

Dermocarpa aquae-dulcis var. *tatrensis* Starmach, Spraw. Kom. Fizjog. Polsk. Akad. Umiej. 62: 10. 1928.

Chamaesiphon carpaticus Starmach, Acta Soc. Bot. Polon. 6: 34. 1929.

Chamaesiphon sideriphilus Starmach, Acta Soc. Bot. Polon. 6: 32. 1929.

Dermocarpa Plectonematis Frémy, Ann. de Bot. Caen 3(Mem. 2): 66. 1930.

Chamaesiphon sideriphilus var. *glabrus* B. Rao, Proc. Indian Acad. Sci. 6: 347. 1937.

Xenococcus Lyngbyae Jao, Sinensia 10: 180. 1939.

Chamaesiphon clavatus Jao, Sinensia 10: 181. 1939.

Chamaesiphon Duran-Mileri Gonzalez Guerrero, Anal. Jard. Bot. Madrid 6: 249. 1946.

Dermocarpa versicolor var. *subsalsa* Proschkina-Lavrenko, Akad. Nauk S. S. S. R. Otd. Sporovykh. Rost. Bot. Mater. 6: 70. 1950.

Chlorogloea sinensis Chu, Ohio Journ. Sci. 52(2). 101. 1952.

Cellulae solitariae primum sphaericae aetate provecta ovoideae, cylindricae, vel cylindraceo-ellipsoideae. FIGS. 221, 235—245.

On larger algae, mosses, and other plants in clear running fresh water of permanent streams, lakes, springs, and fountains. In old cultures containing this form, the endospores and the daughter cells produced by the solitary cells are often seen covering the walls of whatever larger plants are present. The solitary cells may be confused with those of *Clastidium setigerum, Stichosiphon sansibaricus,* and *Crenothrix polyspora* Cohn.

Specimens seen:

FINLAND: on *Fontinalis antipyretica* L. in stream 20 km. northwest of Helsingfors, *Collander & W. H. Welch 5660,* 8 Jul. 1938 (FC); in *Hygroamblystegio fluviatili* ad saxa submersa amnis Pusulanjoki, par. Pusula (Nylandia), *V. F. Brotherus,* 7 Jul. 1916 (FC). NORWAY: with *Schizothrix tinctoria* (Ag.) Gom. on *Fontinalis antipyretica,* Lillesand, *Schübeler,* 1846 (FC). SWED-

EN: in muscis aquaticis, Nacka, *G. Lagerheim*, Jun. 1883 (S); in rivulis Scaniae, *C. A. Agardh*, 1824 (Type of *Nostoc Lemaniae* Ag., PC [Fig. 239]); in *Cladophora fracta* in Lacu Valloxen Uplandiae, *V. Wittrock*, 27 Oct. 1878 (as *Chamaesiphon confervicola* in Wittr. & Nordst., Alg. Exs. no. 293, FC, L, S). POLAND: sur le Batrachospermum, Cracovie, *E. Janczewski*, 20 Apr. 1883 (Type of *Godlewskia aggregata* Jancz., PC), 22 Aug. 1885 (PC). ROMANIA: ad Cladophoram apud Baneasa prope Bucuresti, *Dna. O. Malinescu 1056*, 18 May 1896 (FC, W). GERMANY: Freiburg in Breisgau, *G. Lagerheim*, Jul. & Dec. 1887 (Type of *Pleurocapsa fluviatilis* Lagerh., DT [Fig. 240]); auf *Fontinalis antipyretica* L., Laubach (Hesse), *G. Roth*, Jul. 1908 (FC); auf *Cladophora gossypina* var. *intricata*, Giessen (Hesse), *A. Braun*, Dec. 1850 (Type of *Chamaesiphon confervicola* A. Br. and f. *elongatus* Rabenh., NY; isotypes, FH, PC, S); auf Chantransia, Leipzig, *P. Richter* (UC); auf *Cinclidotus aquaticus* in Gebirgsbächen im Schwarzwald (Württemberg), *P. Reinsch* (FC). CZECHOSLOVAKIA: Bohemia: Dobrá Voda u Budejovic, *A. Hansgirg*, Aug. 1883 (FC, W); in canalibus molarum ligneis ad Eisenbrod, *Hansgirg* (Type of *Xenococcus Kerneri* Hansg. on *Scytonema cincinnatum* in Kerner, Fl. Exs. Austro-Hungar. no. 1596, FH); auf Chantransia und Cladophora, Kaplitz, *Hansgirg*, Sept. 1885 (FC, W); Stríbro a Tábor, *Hansgirg*, Aug. 1883 (FC, W); Zel. Brod, *Hansgirg*, Jul. 1885 (FC, W). AUSTRIA: Upper Austria: auf *Calothrix Wrangelii* auf Steinen im Moosbache bei Milchdorf, *Schiedermayr* (Type of *Chamaesiphon Schiedermayeri* Grun., W); auf *Calothrix Brebissonii*, Vöcklabrück, *von Mörl* (Type of *Chamaesiphon Schiedermayeri* f. *subclavatus* Grun., W); Seebach am Gleinkersee bei Windisch-Garsten, *Schiedermayr*, Nov. 1865 (Type of *C. incrustans* Grun., W. isotypes in Rabenh. Alg. no. 1944, FH, PC). Lower Austria: auf *Cladophora putealis* in einem Brunnen zu Hörnstein, *A. Grunow*, Jul. 1864 (as *Chamaesiphon confervicola* in Rabenh. Alg. no. 1726, FH, PC); insidentes muscis sub aqua decavente canalis molendarii rivi Fischadagnitz ad Unter-Waltersdorf, *S. Stockmayer* (as *Oncobyrsa rivularis* in Mus. Vinbob. Krypt. Exs. no. 744, FC, L, WU); auf Cladophora in einem Kulturglas, Wien, *L. Geitler* (Type of *Dermocarpa chamaesiphonoides* Geitl. and *Xenococcus minimus* Geitl., W [Figs. 237, 241]).

JUGOSLAVIA: Slovenia: in *Lemanea fluviatili*, Kaltenbrunn prope Laibach, *A. Hansgirg*, Aug. 1889 (FC, W); in *Tolypothrice penicillata* in fonte parvo prope Kronau, *K. de Keissler* (as *Chamaesiphon minutus* in Mus. Vinbob. Krypt. Exs. no. 1949, FC, L, NY, S). Croatia: Mühlbach in Martinscica bei Fiume, *F. Hauck*, Sept. 1878 (L). Dalmatia: in Cinclidoto unacum *Chantransia violacea*, herb. *P. Reinsch* (Type of *Sphaenosiphon aquae-dulcis* Reinsch, K); Knin, *Hansgirg*, Aug. 1888 (FC, W); Ombla prope Ragusa et Topla—Gruda prope Castelnuovo, *Hansgirg*, 1891 (FC, W). TRIESTE: Zaule, *Hauck 1558*, 18 Mar. 1870 (FC, FH, PC). ITALY: on mosses, Treviso, *G. Venturi*, 1857 (FC); Roveredo, *A. Hansgirg*, 1891 (FC, W); ad *Fontinalem antipyreticam* in rivulis per prata excurrentibus all'Oropa, dition. Bugellens. (Piedmont), *Cesati*, Aug. 1859 (Type of *Hydrococcus Cesatii* Rabenh. in Rabenh. Alg. no. 922, D; isotypes, FC [Fig. 221], L WU). NETHERLANDS: in tropical aquarium, Hilversum, *Veldhuizen van de Wit*, 5 Mar. 1950 (L); on *Cladophora crispata*, waterworks filter, Aarden hout near Amsterdam, *G. J. van Huesden*, 30 Jul. 1940 (FC, L). BELGIUM: in *Fontinali squamosa* var. *tumida* submersa in rivo Amblève Arduennae, *F. Verdoorn*, Aug. 1927 (FC). GREAT BRITAIN: on *Cinclidotus fontinaloides* on stony bottom of Hellgill (Scotland), *G. Dixon*, 5 Mar. 1852 (FC); on *Cladophora glomerata* in River Coln at Fairford (Gloucester), *W. L. Tolstead 8472, 8504*, 24 Sept., 7 Oct. 1944 (FC); Baunton stream, Cirencester (Gloucester), *W. J. Joshua* (FC); on *Fontinalis squamosa*, Westmoreland, *P. Dreesin*, 1873 (FC); on *Cladophora glomerata*, East Lyn, north Devon, *F. E. Fritsch*, 2 Sept. 1918 (Type of *Xenococcus chroococcoides* Fritsch in the collection of F. E. Fritsch; isotype, D [Fig. 242]); on *Cinclidotus fontinaloides*, Plymouth, *Holmes*, 1867 (FC). FRANCE: sur les Lemanea et sur la mousse (as *Hydrococcus Brebissonii* in Desmazières, Pl. Cryptog. Fr., ser. 2, no. 1960, NY); sur *Trichostomum fontinaloides*, in rivulis ad saxa calcarea Jurassa, *J. B. Mougeot* (FC); with *Inactis tinctoria*, Vosges, ex herb. Lenormand (D); sur Cladophora, Berthenicourt-par-Moy (Aisne), *J. Mabille* 7, Aug. 1952 (FC); Vire (Calvados), *A. de Brébisson* (FC); Courteilles (Calvados), *Brébisson*, May 1839 (PC); Falaise (Calvados), *Brébisson* (Type of *Oncobyrsa Brebissonii* Menegh., L [Fig. 245]; isotypes, NY, UC); sur *Fontinalis antipyretica*, Mte. Rotondo (Corsica), *Solierol 5035* (FC); sur *Cinclidotus fontinaloides*, moulin de Kerauffray près Guingamp (Côtes du Nord), *E. Jeanpert*, 24 Jul. 1901 (FC); sur la *Lemanea incurvata* sur une écluse de la petite Vienne, Limoges (Haute Vienne), *Lamy de la Chapelle* (PC); sur *Cinclidotus aquaticus*, St. Guilhem (Hérault), *La Perraudière*, 11 Jun. 1857 (FC); Lozère, *Duby* (L); sur *Fontinalis Camusii* dans de lit de la Sèvre-Nantaise et de la Maine à Boussay et Aigrefeuille (Loire Inférieure), *E. Bureau & F. Camus* (FC).

SPAIN: Nacimiento del Rio Grande, nahe Yunquera in Serrania de Ronda (Malaga), *H. Nilsson 24*, 28 May 1883 (Type of *Oncobyrsa hispanica* M. Lewin, S). ANGOLA: Loanda, supra *Pithophorum radiantem* in aquariis aquae subdulcis insulae Cassanga prope Morro da Cruz, *F. Welwitsch 197*, Apr. 1854 (Type of *Dermocarpa depressa* W. & G. S. West, BM; isotypes, BIRM, D [Fig. 244], FC, UC). MADAGASCAR: auf *Rhizoclonium hieroglyphicum* in Süsswasser, Nosi-bé, *J. M. Hildebrandt 2*, Sept. 1879 (FC). ICELAND: on a lake-margin above Húsavik (North Amt), *J. Boye Petersen*, 26 Jul. 1914 (Type of *Chamaesiphon cylindricus* Boye Peters. in the collection of J. Boye Petersen; isotype, D [Fig. 238]). NEW BRUNSWICK: on *Tolypothrix penicillata* in brooklet 1 mile up Little river from Grand Falls, *H. Habeeb 10407*, 14 Jul. 1948 (FC, HA). NEW HAMP-

SHIRE: on *Fontinalis antipyretica* var. *gigantea, C. G. Pringle,* 22 Jun. 1880 (FC); in a brook, Glen Road, White mountains, *W. A. Setchell 364,* 9 Sept. 1891 (FC, UC); on mosses in stream on west slope of Moose mountain, Hanover, *L. H. Flint,* 12 Aug. 1945 (FC); on *Hypnum Lescurii,* Bowls and Pitchers, Shelburne, *W. G. Farlow,* Oct. 1899 (FC, FH); on *Hydrothyria venosa,* Ingalls brook, Shelburne, *E. Faxon,* 2 Sept. 1886 (D, FH). MASSACHUSETTS: on *Lemanea fucina,* Waverley, *W. H. Weston Jr.,* 1 Jun. 1923 (FC, TA); Gosnold pond, Cuttyhunk, *I. F. Lewis &* *R. H. Colley,* 5 Jul. 1915 (as *Chamaesiphon incrustans* in Coll., Hold., & Setch., Phyc. Bor.-Amer. no. 2101, FC, L). CONNECTICUT: on Stigonema, Stony brook, Montville, *W. A. Setchell 102,* 30 Jun. 1890 (FC, UC); Stony brook, Mohegan, *Setchell,* 9 Sept. 1892 (FC, UC); among Confervae, New Haven, *W. C. Sturgis,* 30 May 1889 (FC, UC). QUEBEC: submergé, Rivière Payne vers 71° 5″ long, O., Ungava, *J. Rousseau 1175,* 11 Aug. 1948 (FC); sur *Podostemon Ceratophyllum* dans les rapides près de l'Eglise, Saint Joseph, Rivière des Rapides, *F. Marie-Victorin & Rolland-Germain 33998,* 9 Sept. 1930 (FC). NEW YORK: on Oedogonium in aquarium in Barnard College, New York City, *H. C. Bold B71,* 1 May 1941 (FC); on *Cladophora crispata* in Green lake near Kirksville, *W. R. Maxon 6048,* 21 Oct. 1914 (FC, US). PENNSYLVANIA: on *C. glomerata* in Little Conestoga creek, Hempfield township, Lancaster county, *R. Patrick 63,* 30 Sept. 1948 (FC, PH); on *Podostemon Ceratophyllum* f. *chondroides,* Winona falls near Bushkill, Pike county, *N. C. Fassett & H. H. Calvert 19488,* 6 Sept. 1938 (FC).

VIRGINIA: James river, Richmond, *W. E. Wade,* 27 Jul. 1951 (DA); on *Fontinalis dalecarlica,* Dickey creek, Smyth county, *J. K. Small S14,* 17 Jun. 1892 (FC). WEST VIRGINIA: on *Audouinella violacea* on dripping rocks above Buckhannon river 5 miles below Buckhannon, Upshur county, *H. K. Phinney 530,* 30 Aug. 1938 (FC, PHI). NORTH CAROLINA: on *Fontinalis dalecarlica* in stream, Soco falls, Soco Gap, Jackson county, *W. H. Welch 2221,* 17 Jun. 1936 (FC); on *F. dalecarlica,* Montreat, Buncombe county, *P. C. Standley & H. Bollman 10123,* 3 Aug. 1913 (FC); near Round Bottom CCC camp, Swain county, *H. Silva 200,* Nov. 1941 (FC, TENN). SOUTH CAROLINA: on logs at 1st and 2nd dikes, Savannah river, Barnwell county, *J. H. Wallace 64,* 30 Jun. 1951 (DA). FLORIDA: Hamilton county: on *Lyngbya aestuarii,* West Lake, *J. H. Davis Jr.,* summer 1937 (D). Hernando county: on *Rhizoclonium fontanum, M. A. Brannon 574,* 23 Oct. 1948 (FC, PC); on moss, Weekiwachee Springs, *Standley 92737,* 6 Mar. 1946 (FC). Jackson county: on mosses in stream on highway no. 71 five miles south of Marianna, *C. S. Nielsen, G. C. Madsen, & D. Crowson 347,* 31 Aug. 1948 (FC, T). Leon county: *Nielsen 130,* Jun. 1948 (FC, T); Little Natural Bridge sink, *Nielsen 134, 136, 137,* Jun. 1948 (FC, T); St. Marks river at Little Natural Bridge, *Nielsen, Madsen, & Crowson 538,* 30 Oct. 1948 (FC, T). Marion county: Rainbow Springs near Dunnellon, *Brannon 375,* 20 Oct. 1946 (FC, PC), *Madsen, A. L. Pates, & M. N. Hood 1954,* 27 Nov. 1949 (FC, T). Orange county: on *Plectonema Wollei* in Rock spring at Kelly Park, *P. O. Schallert 2278,* 1 Jun. 1951 (FC). Seminole county: on *P. Wollei,* Rock Springs, *E. M. Davis,* May 1939 (FH). Taylor county: on Chantransia in a stream 1 mile northwest of Perry, *F. Drouet & Nielsen 10751,* 11 Jan. 1949 (FC, T). Wakulla county: on mosses in the outlet of a large sulfur spring bathing pool about one mile north of Newport, *Drouet, Crowson, & R. Thornton 11386,* 25 Jan. 1949 (FC, T). ONTARIO: St. Lawrence river, Kingston, *J. H. Wallace 22,* Jul. 1953 (D, PH). OHIO: on *Cladophora crispata* in Plum creek at the golf course, Oberlin, *P. Smith 2,* 29 Oct. 1938 (FC); on Chantransia on rock in the Scioto river, Chillicothe, *L. J. King 410,* 27 Jul. 1941 (FC, EAR). TENNESSEE: Sevier county: on moss in stream, Greenbrier cove, *H. Silva 418,* 28 Dec. 1946 (FC, TENN); on *Desmonema Wrangelii* in Buck Fork, *Silva 643,* 21 Apr. 1947 (FC, TENN); muddy stream at Sugarlands camp, *Silva 20,* 11—16 Aug. 1941 (FC, TENN); Ramsey Prong, *Silva 366, 425, 2353,* 3 Nov., 29 Dec. 1946, 9 Sept. 1950 (FC, TENN). MICHIGAN: on *Fontinalis antipyretica* var. *gracilis* in a brook near Bear Lake, Manistee county, *E. J. Hill 216,* 23 Aug. 1880 (FC); on Oedogonium, Moloney lake, Cheboygan county, *H. K. Phinney 79M4/5,* 15 Aug. 1941 (FC, PHI). INDIANA: on Pithophora in pool, Evansville College, Evansville, *R. K. Zuck,* May 1944 (FC). WISCONSIN: on Cladophora, Waunona (Monona) lake, Madison, *V. A. Latham,* 27 Apr. 1943 (FC); on Cladophora in the source of the Yahara river, Tenney park, Madison, *H. B. Louderback & F. Drouet 5516,* 23 Jul. 1944 (FC); on moss in Weber lake, Vilas county, *G. W. Prescott 3W104,* 21 Aug. 1938 (FC). ILLINOIS: on Cladophora in small stream, Biltmore subdivision near Barrington, *P. C. Standley 92845,* 19 May 1946 (FC); on *Pithophora oedogonia* in laboratory culture, Chicago Natural History Museum, Chicago, *F. A. Barkley 151117,* 24 Aug. 1945 (FC).

MINNESOTA: on *Lemanea torulosa,* Lester river, Duluth, *E. Butler,* Aug. 1902 (FC, FH); Mary lake and La Salle creek, Itasca state park, *Drouet 12172, K. C. Fan 10146,* 1954 (D). MISSOURI: on *Rhizoclonium hieroglyphicum* in Lake Hahatonka, Camden county, *Drouet 145,* 9 Aug. 1928 (D); on *R. hieroglyphicum,* Chouteau Springs, Cooper county, *Drouet & Louderback 5648,* 25 Aug. 1945 (FC); on Fontinalis in Fishing spring beside Meramec river 2 miles northwest of Steelville, Crawford county, *J. A. Steyermark 41394,* 16 Jun. 1941 (FC); Gravois spring, Gravois Mills, Morgan county, *Drouet 682, 882,* 7 Sept. 1930, 3 May 1931 (D); on *Rhizoclonium hieroglyphicum* in the creek below Roubidoux spring, Waynesville, Pulaski county, *Drouet 599,* 18 May 1930 (D). ARKANSAS: on *R. hieroglyphicum* in Saline river, Warren, Bradley county,

131

D. Demaree 24811, 24817, 23 Oct. 1943 (FC). LOUISIANA: in tank by Engineering laboratories, Louisiana State University, Baton Rouge, *L. H. Flint 5,* 18 Apr. 1945 (FC). NEBRASKA: on *R. hieroglyphicum* in culture from floodplain pond, Platte river south of Kearney, Buffalo county, *W. Kiener 16496d,* 6 Nov. 1945 (FC, KI); on *Cladophora glomerata* in Little Bordeaux creek near Chadron, Dawes county, *Kiener 20570,* 22 May 1946 (FC, KI); on Cladophora in floodplain creek, Platte river near Sutherland, Lincoln county, *Kiener 14875,* 1 Aug. 1943 (FC, KI); on Cladophora in culture from Pawnee slough 9 miles east of North Platte, *Kiener 16451a,* 16 Jul. 1944 (FC, KI); on *C. crispata* in drainage ditch, Northport, Morrill county, *W. Kiener 23056,* 4 Mar. 1948 (FC, KI); on *Basicladia Chelonum* on snapping turtle, 3 miles north of Henry, Sioux county, *Kiener 23524,* 11 May 1948 (FC, KI). TEXAS: swift stream at New Braunfels, *L. H. Flint 4,* 18 Nov. 1954 (D); Barton spring southwest of Austin, *E. M. Harper & F. A. Barkley 13229,* 23 Jul. 1943 (FC, TEX), *B. C. Tharp & Barkley 44alg11,* 5 Aug. 1944 (FC, TEX); on *Pithophora oedogonia* in Tucker lake, Brownsville, *R. Runyon 3865,* 18 Nov. 1944 (FC); on *Plectonema Wollei* in the head of the San Marcos river, Hays county, *C. M. Rowell & Barkley 16T469,* 2 Aug. 1946 (FC, TEX); on Vaucheria in running water, San Marcos, *G. L. Fisher 49233,* 24 Apr. 1949 (FC); on Pithophora in stream at sulphur spring, Palmetto state park near Gonzales, *B. F. Plummer & Barkley VII,* 22 May 1943 (FC, TEX); on *Cladophora glomerata,* Tunis Springs, about 20 miles east of Fort Stockton, Pecos county, *E. Whitehouse 25153,* 10 May 1951 (FC). MONTANA: on *Cladophora crispata* between Bowman lake and Mount Chapman, northwestern part of Glacier national park, *C. E. Fix,* Aug. 1949 (FC).

WYOMING: on Scytonema and mosses in creek near Emerald pool, Upper geyser basin, Yellowstone national park, *W. A. Setchell 1958a, 6138,* 28 Aug. 1898, 29 Jul. 1905 (FC, UC); on Vaucheria etc. in stream in an open meadow above University Camp, Medicine Bow national forest, *W. G. Solheim 53,* 27 Jun. 1933 (Type of *Dermocarpa Solheimii* Dr., FC), *Solheim 108,* 11 Jul. 1933 (FC). COLORADO: in shallow pool, Longs Peak, *W. Kiener 161,* 16 Sept. 1933 (FC, KI); on *Fontinalis antipyretica* in Cabin creek, Longs valley, *Kiener 4696,* 13 Oct. 1935 (FC, KI); in Alpine brook, Longs Peak valley, *Kiener 13126,* 20 Sept. 1941 (FC, KI). NEW MEXICO: on a liverwort in a small stream in the mountains near Santa Fe, *J. B. Routien,* Aug. 1945 (FC). IDAHO: on *Hygrohypum Bestii* in Adair creek, Custer district, Custer county, *F. A. Mac Fadden 18707,* 7 Aug. 1941 (FC); on hepatic with *Fontinalis antipyretica,* Little creek bridge, road from Deer Park to Idaho City, *Mac Fadden 19297,* 18 Oct. 1942 (FC). UTAH: on *Lemanea fucina* in small stream near head of small canyon above Brighton, Salt Lake county, *Prof. Vorhies,* Aug. 1910 (FC, UC); on Vaucheria, Twin creeks, Fish lake, Sevier county, *S. Wright,* 20 Aug. 1938 (FC). ARIZONA: on Chantransia, Fiftyfoot falls, Havasupai canyon, Coconino county, *E. U. Clover,* 1 Aug. 1940 (FC, MICH). NEVADA: on *Plectonema Wollei,* Rogers Spring, Lake Mead, Clark county, *R. Sumner 1614,* 12 Jul. 1952 (FC). ALASKA: on mosses in brook near Iliuliuk, Unalaska, *W. A. Setchell & A. A. Lawson 5036,* 9 Jun. 1899 (FC, UC). OREGON: on *Aneura pinnatifida,* Salem, *E. Hall,* 1871 (FC). CALIFORNIA: Alameda county: on Cladophora in laboratory cultures, University of California, Berkeley, *N. L. Gardner 3200, 3216,* Jan. 1915, 1916 (FC, UC); on Oedogonium in an aquarium in the courtyard of the Life Sciences building, Berkeley, *Gardner 8012,* 28 Nov. 1936 (FC, UC); on Cladophora in a fountain, campus, University of California, Berkeley, *Gardner,* Oct. 1905 (as *Chamaeisphon confervicola* in Coll., Hold., & Setch., Phyc. Bor.-Amer. no. 1705, FC, L, TA). Inyo county: on *Plectonema Wollei,* Furnace creek, Death valley, *S. B. Parish 10465,* 16 May 1915 (UC). Los Angeles county: Old El Encino hot springs, *B. C. Templeton 1,* Jul. 1944 (FC). Marin county: on moss in rapidly running stream, Mill Valley, *Gardner 1785,* 21 Jul. 1906 (FC, UC); laboratory culture from Tiburon (collected by H. E. Parks), *Gardner 6916,* 16 Oct. 1931 (Type of *Dermocarpa minuta* Dr., FC; isotype, UC). San Bernardino county: on Rhizoclonium in small pond at Old Woman Springs, Mojave desert, *G. J. Hollenberg 2084,* 2 May 1937 (Type of *Dermocarpa Hollenbergii* Dr., D [Fig. 235]). Shasta county: Pittsville bridge, Pitt river, *H. W. Shepherd 7975,* 15 Aug. 1936 (FC, UC). Tulare county: on *Desmonema Wrangelii* in Marble Rock creek, Sequoia national park, *R. Prettyman,* 30 Jul. 1938 (FC).

PUERTO RICO: Coamo, in den warmen Quellen von los Baños, *P. Sintenis A42,* 23 Dec. 1885 (topotype and paratype of *Chamaesiphon incrustans* f. *minor* Möb., MO, FC); Little Springs, Minillas, Mayaguez, *L. R. Almodovar 199,* 27 Jul. 1954 (D, T); in a ditch by the hot spring, Coamo Springs, *N. Wille 396,* Jan. 1915 (Type of *C. portoricensis* Gardn., NY [Fig. 243]; isotypes, FH, UC) on *Plectonema Wollei* in river 5 km. east of Coamo, *Wille 221e,* Jan. 1915 (Type of *Xenococcus Willei* Gardn., NY [Fig. 236]); in Tümpel, Jayuya, *Wille 1757,* 15 Mar. 1915 (FC, NY); in a pool about 4 km. north of Mayaguez, *Wille 1323b* (Type of *Pleurocapsa epiphytica* Gardn., NY), *1323x* (NY), Feb. 1915. JAMAICA: on Chantransia, Seaman's Valley, Portland, *W. R. Maxon & E. P. Killip 27,* 14 Feb. 1920 (FC); on *Pithophora oedogonia,* Salt river, St. Catherine, *C. B. Lewis A1236a,* 17 Jul. 1950 (FC). MEXICO: on *Plectonema Wollei,* Riito de Santa Rosa, Coahuila, *A. Schott 28,* Nov. 1852 (FC); on moss, waterfall, Barranca de Oblatos, Guadalajara, *W. Kiener 18130,* 21 Dec. 1944 (FC, KI); on turtle in a pool near the bridge, Unión, Hermosillo, Sonora, *F. Drouet & Richards 3020,* 24 Nov. 1939 (FC). GUATEMALA: on Cladophora in small stream, Quebrada Shusho above Chiquimula, *P. C. Standley 71933,* 22 Apr. 1939 (FC); on *C. crispata* in stream,

Rio Guacalate, dept. Escuintla, *Standley 60142,* 16 Dec. 1938 (FC); on Cladophora, *Marathrum foeniculaceum,* and *Tristicha hypnoides,* vicinity of Santa María de Jesús, dept. Quezaltenango, *Standley 87046,* 12 Feb. 1941 (FC), *J. A. Steyermark & A. E. Vatter 33240, 33363, 33368, 33576,* 31 Dec. 1939, 1—5 Jan. 1940 (FC); on *T. hypnoides* on rocks in river, Retalhuleu, *Standley 88817,* Feb.—Mar. 1941 (FC); on Marathrum in Rio Lima, Mazatenango, *W. C. Muenscher 12035,* 4 May 1937 (FC); on Oedogonium in small stream, Zacapa, *Standley 74268,* Oct. 1940 (FC). COSTA RICA: on *Marathrum Schiedeanum* in stream, San José, *Standley 41213,* Dec. 1925—Feb. 1926 (FC). PANAMA: on Chantransia in stream bed, Shannon trail, Barro Colorado island, Canal Zone, *G. W. Prescott CZ46,* Aug. 1939 (FC). VENEZUELA: on mosses in stream along Rio Karuai at base of Sororopán-tepuí west of La Laja, Bolivar, *Steyermark 60774,* 29 Nov. 1944 (FC). BRAZIL: on Oedogonium and Pithophora in jars of fish in the biology laboratory, Baptist College, Rio de Janeiro, *E. C. Jennings 21, 22,* 10 Aug. 1935 (D); sur *Lophogyne arculifera,* Rio Bengala entre le Alto et Novo Friburgo (état de Rio de Janeiro), *A. Glaziou 13147,* 10 Aug. 1881 (FC). PARAGUAY: on Chantransia, Colonia Risso, *G. A. Malme 92, 94B,* 14 & 18 Oct. 1893 (FC, S). ARGENTINA: in *Rhizoclonium hieroglyphicum* in aqua subsalsa, Laguna Colorada, prov. Jujuy, *R. E. Fries 26* (S). PATAGONIA (Chile?): Rio & Sierra Bagnales, *O. Borge 365, 367,* Mar. 1899 (S); Kark, *Borge 383,* Mar. 1899 (S). PERU: sobre *Pithophora oedogonia,* Salaverry, Trujillo, *N. Ibañez H.* "F", 18 May 1952 (FC); on *Rhynchostegium aquaticum* in irrigation ditch, Santa Eulalia, Chosica, *G. S. Bryan 9a,* 11—13 Mar. 1923 (FC); on *Rhizoclonium hieroglyphicum,* head of Mogollon quebrada in the Amotape mountains, prov. Paita, dept. Piura, *C. C. Sanborn,* Sept. 1943 (FC).

HAWAIIAN ISLANDS: Hawaii: on Pithophora, Punaluu, *J. E. Tilden,* 3 Jul. 1900 (FC), *M. Doty 8004,* 9 Sept. 1950 (FC). Kauai: on *Cladophora crispata* in a brook, Kumuela camp, *M. Reed,* 21 Jul. 1908 (FC, UC). Oahu: in *Cladophora longiarticulata* in piscinis in convalle Nuanu prope Honolulu, *S. Berggren,* 1875 (Types of *Chamaesiphon curvatus* Nordst. and f. *elongatus* Nordst., PC; isotypes on *Cladophora longiarticulata* in Wittr. & Nordst., Alg. Exs. no. 213, FC, MIN); on *Pithophora oedogonia* in a fishpond, Peninsula, Pearl City, *Tilden,* 6 Jun. 1900 (FC). CAROLINE ISLANDS: on *P. oedogonia* in Sapalap river, Sapalap, Metalanum dist., Ponape, *S. F. Glassman 2763,* 2 Aug. 1949 (FC). SAMOA ISLANDS: Savaii, Flussbette, *K. H. Rechinger 2969,* Jul. 1905 (Type of *Chroococcus turgidus* var. *subviolaceus* Wille in the slide collection of N. Wille, O); Upolu, *Rechinger 3165,* 25 Jun. 1905 (Type of *C. varius* f. *samoensis* Wille in the slide collection of N. Wille, O). NEW ZEALAND: *S. Berggren 145* (PC); Taupo lake, *Berggren 146,* 1875 (L, S); Roto-aira lake, *Berggren 267,* 1876 (S); in Cladophoris in Tokano river, *Berggren 319b,* Jan. 1875 (as *Sphaerogonium incrustans* and "*Xenococcus?* vel *Oncobrysa?*" in Wittr. & Nordst., Alg. Exs. no. 899, FC, L, NY, PC, S, UC); on *Cladophora glomerata* in cold stream, Wairakei geyser valley, Taupo county, *W. A. Setchell 5967(58),* 5 May 1904 (FC, UC); Whakarewarewa, Rotorua, *J. E. Tilden,* Nov. 1909 (as *Gloeocapsa gelatinosa cryptococcoides* in Tild., So. Pacific Alg. no. 2, L, N, UC). PHILIPPINES: on *Plectonema Wollei,* waterfall at Bisayaan, Puerto Galera, Mindoro, *G. T. Velasquez 1518,* 12 Apr. 1948 (FC, PUH); on *P. Wollei* in swimming pool, Pansol, Calamba, Laguna, *Velasquez 464,* 25 Feb. 1940 (FC). CHINA: small pond, Nine-Old-Men Cave, Omei, Szechwan, *H. Chu 1354,* 23 Aug. 1942 (FC). INDONESIA: Java: auf Moosen im fliessenden Wasser, Quelle bei Kandang Badak, Gebiet von Tjibodas, *Ruttner,* 12 Jul. 1929 (Type of *Chamaesiphon mollis* Geitl., slide no. TJ5b in the collection of L. Geitler; cotype, slide no. TJ5c in coll. L. Geitler); auf Oedogonium an triefender Wand, Wasserfälle von Tjibeureum, Gebiet von Tjibodas, *Ruttner,* 9 Jul. 1929 (Type of *Chamaesiphon subglobosus* var. *major* Geitl., slide no. Tj2Iα in the slide collection of L. Geitler); auf Lyngbya aus 5 m. Tiefe, Ranu Bedali, *Ruttner,* Nov. 1928 (Type of *Dermocarpa clavata* Geitl., slide no. KB2δ in the collection of L. Geitler; Type of *D. xenococcoides* Geitl., slide no. KB2δ in the collection of L. Geitler); Tengger, *Ruttner,* 25 Jun. 1929 (slide no. TG1d in the collection of L. Geitler). Sumatra: Dana Manindjau, *Ruttner,* 1929 (paratype of *Xenococcus minimus* var. *Starmachii* Geitl., slide no. FM2bα in the slide collection of L. Geitler); auf Cladophoraceae, von Strauchwerk tief beschattete Stelle am Ufer, Hauptzufluss des Ranau-Sees bei Kota Batu, *Ruttner* (Type of *Pleurocapsa subgelatinosa* Geitl., slide no. RE3aα in the collection of L. Geitler); auf Hydrella aus 7 m. Tiefe, Profil vor Panjingahan am Westufer des Singkarak-Sees, *Ruttner,* 22 Feb.—15 Mar. 1929 (Type of *Chamaesiphon minutus* var. *gracilior* Geitl., slide no. SK4d in the collection of L. Geitler); auf Cladophoraceae, Stromgebiet der Musi, Suban Ajer Panas, *Ruttner,* 7 May 1929 (slide no. M4gβ in the collection of L. Geitler). WESTERN TIBET: stream just below Phobrang, running into northwest end of Pang-gong Tso, *G. E. Hutchinson L46,* 8 Jul. 1932 (D, FH, L, NY, YU). INDIA: on Oedogonium in the Jumna, Allahabad, *P. Maheshwari 99, 138,* Apr. 1938 (FH).

5b. ENTOPHYSALIS LEMANIAE f. ELONGATA (Wille) Drouet & Daily, COMB. NOV. *Chamaesiphon gracilis* f. *elongatus* Wille, Bih. t. K. Sv. Vet.-Akad. Handl. 8(18): 28. 1884. *Entophysalis elongata* Drouet & Daily, Butler Univ. Bot. Stud. 10: 223. 1952. —Type from Nova Zembla (S). FIG. 246.

Original specimens have not been available to us for the following names; their original descriptions are here designated as the Types until the specimens can be found:

Chamaesiphon macer Geitler, Arch. f. Protistenk. 51: 331. 1925.
Chamaesiphon curvatus f. *polysporinus* Schirschov, Acta Inst. Bot. Acad. Sci. U. R. P. S. S., ser. 2, Pl. Crypt. 1: 81. 1933.
Chamaesiphon cylindrosporus Skuja, Symbolae Bot. Upsal. 9(3): 45. 1948.

Cellulae solitariae primum sphaericae aetate provecta lineari-cylindricae vel lineari-tubaeformes, pulvinis et endosporangiis quaerendis. FIG. 246.

On larger, chiefly perennial, algae and other plants in clear freshwater streams and lakes. The well developed solitary cells have the general aspect of plants of *Lyngbya versicolor* (Wartm.) Gom. and *L. Diguetii* Gom. *Crenothrix polyspora* Cohn may sometimes be confused with this form.

Specimens examined:
NOVA ZEMBLA: Norra Gäskap, *F. Kjellman 29b,* 1875 (Type of *Chamaesiphon gracilis* f. *elongatus* Wille, S [Fig. 246]). NORWAY: Olden in Nordfjord, *O. Nordstedt,* 17 Aug. 1878 (PC). SWEDEN: in *Fontinalis dalecarlica,* Kvarnbaacken, Högsjö, Angermanland, *A. Arvén,* 14 Jul. 1914 (FC). MASSACHUSETTS: with *Chantransia Hermannii* on mosses in brook, Sharon, *F. S. Collins,* May 1890 (in Collins, N. Amer. Alg. 153, D), Massapoag brook, Sharon, *W. A. Setchell 26,* 4 May 1890 (D). QUEBEC: sur les mousses au bord de la Rivière Payne, vers 73° 7' long. O., Ungava, *J. Rousseau 937,* 6 Aug. 1938 (FC). LOUISIANA: stream near Husser, Tangipahoa parish, *L. H. Flint,* 21 Mar. 1953 (FC).

FAMILY III. CLASTIDIACEAE

Drouet & Daily, Butler Univ. Bot. Stud. 10: 223. 1952.   —Type genus: *Clastidium* Kirchn.

Plantae microscopicae, solitariae, epiphyticae, primum sphaericae aetate provecta cylindricae, basim affixae, primum unicellulares demum interne in catenam cellularum sphaericarum, ovoidearum, vel cylindricarum uniseriatim (raro parce pauciseriatim) dividentes; vagina tenue, ad apicem clausa, ad basem incrassata, ad substratum adhaerente; reproductione a dissolutione vaginae demum catenae cellularum.

Plants of this family are elongate epiphytic unicells contained in thin gelatinous sheaths. The entire protoplast divides into a uniseriate (rarely few-seriate in part) chain of rounded cells which often remain united by their membranes after the hydrolyzation of the sheath. The cells, upon separation from each other, elongate and secrete new gelatinous sheaths.

Key to genera:

Plant terminating above in a hair-like extension of the sheath ....1. CLASTIDIUM
Plant smooth at the apex ....2. STICHOSIPHON

GENUS I. CLASTIDIUM

Kirchner, Jahresh. Ver. Vaterl. Naturk. Württemberg 36: 195. 1880.   —Type species: *C. setigerum* Kirchn.

Plantae microscopicae, solitariae, erectae, epiphyticae, primum sphaericae aetate provecta cylindricae, basim affixae, unicellulares demum interne in catenam

134

cellularum sphaericarum uniseriatim dividentes, vagina tenue, ad apicem clausa et in setam producta, ad basem incrassata et ad substratum adhaerente.

One species:

CLASTIDIUM SETIGERUM Kirchner, Jahresh. Ver. Vaterl. Naturk. Württemb. 36: 195. 1880. —Specimen presumably seen by the author from Malmagen in Herjedalia, Sweden (MIN), designated as the Type. Specimens from Bessvatn, Sweden (DA), photographed: FIGS. 251, 252.

Plantae microscopicae, aerugineae, solitariae, erectae, primum sphaericae aetate provecta cylindricae, diametro 2—5 μ crassae, ad 15(—38) μ altae, ad apicem acutae vel rotundae, basim affixae, primum unicellulares demum interne in catenam cellularum sphaericarum uniseriatam dividentes, vagina hyalina tenue, ad apicem clausa et in setam producta, ad substratum adhaerente; protoplasmate aerugineo, homogeneo. FIGS. 251, 252.

Epiphytic on larger algae, mosses, etc. in freshwater lakes, ponds, bogs, and streams. This species has been confused with solitary cells of *Entophysalis Lemaniae, Characium* spp., and *Characiopsis* spp., also with young filaments of *Amphithrix janthina* (Mont.) Born. & Flah., *Lyngbya versicolor* (Wartm.) Gom., and various bacteria.

Specimens examined:

SWEDEN: ad folia emortua Fontinalis in amne ad Malmagen in Herjedalia, *G. Lagerheim,* 28 Jul. 1897 (designated as Type of *Clastidium setigerum* Kirchn. in Wittr., Nordst., & Lagerh., Alg. Exs. no. 1536, MIN); on Dichothrix, Bessvatn, *K. Thomasson,* 18 Jul. 1951 (DA [Figs. 251, 252]). MISSISSIPPI: on Chantransia in swift water, stream between Cheraw and Sandy Hook, Marion county, *L. H. Flint,* 11 Oct. 1946 (FC). LOUISIANA: in a stream near Blond, St. Tammany parish, *Flint,* 7 Mar. 1953 (FC). ALASKA: Karluk lake, Kodiak, *D. Hilliard 11,* 1955 (D).

GENUS 2. STICHOSIPHON

Geitler, Rabenh. Krypt.-Fl. 14: 411. 1931. —Type species: *S. regularis* Geitl.

Plantae microscopicae, solitariae, erectae vel curvatae, epiphyticae, initio sphaericae aetate provecta cylindricae, basim affixae, primum unicellulares demum interne in catenam cellularum sphaericarum, ovoidearum, vel cylindricarum uniseriatam (vel raro parce pauci-seriatam) dividentes, vagina tenue, ad apicem clausa et rotunda, ad basem incrassata et ad substratum adhaerente.

One species:

STICHOSIPHON SANSIBARICUS Drouet & Daily, Butler Univ. Bot. Stud. 10: 223. 1952. *Chamaesiphon sansibaricus* Hieronymus in Engler, Pflanzenwelt Ostafr. C: 8. 1895. —Type from Zanzibar (BM). FIG. 253.

   Chamaesiphon filamentosus Ghose, Journ. Linn. Soc. Bot. 46: 337. 1924. *Stichosiphon filamentosus* Geitler, Rabenh. Krypt.-Fl. Eur. 14: 411. 1932. —Type from Lahore, Pakistan (in the collection of F. E. Fritsch). FIG. 254.

   Chamaesiphon Willei Gardner, Mem. New York Bot. Gard. 7: 34. 1927. —Type from Rio Piedras, Puerto Rico (NY).

   Stichosiphon regularis Geitler, Rabenh. Krypt.-Fl. 14: 411. 1932. —Type from Lake Ranau Bedali, east Java, Indonesia (in the collection of L. Geitler).

   Original specimens have not been available to us for the following names; their original descriptions are here designated as the Types until the specimens can be found:

   Stichosiphon indicus B. Rao, Proc. Indian Acad. Sci. 3: 167. 1936.

   Dermocarpa olivacea var. *gigantea* B. Rao, Proc. Indian Acad. Sci. 3: 167. 1936.

135

Plantae microscopicae, aerugineae, olivaceae, violaceae, vel roseae, solitariae, erectae vel varie curvatae, epiphyticae, initio sphaericae aetate provecta cylindricae, diametro 3—7μ crassae, ad 400μ altae, primum unicellulares demum interne in catenam cellularum sphaericarum, ovoidearum, vel cylindricarum uniseriatam (vel raro parce pauci-seriatam) dividentes, vagina hyalina tenue, ad apicem rotunda, primum clausa deinde aperta, ad basem incrassata saepe brevi-stipitata, ad substratum adhaerente; protoplasmate aerugineo, olivaceo, violaceo, vel roseo, homogeneo. FIGS. 253, 254.

Epiphytic on larger algae and other plants in freshwater lakes, ponds, and streams. Well developed plants of *Stichosiphon sansibaricus* are reminiscent in habit of young filaments of *Anabaena* and *Calothrix* spp. Smaller plants can be confused with large solitary cells of *Entophysalis Lemaniae*.

Specimens examined:

ZANZIBAR: auf Cladophora an einer undichten Stelle der Wasserleitung etwas nördlich der Stadt Sansibar, *Stuhlmann 21*, 31 May 1891 (Type of *Chamaesiphon sansibaricus* Hieron., BM [Fig. 253]). VIRGINIA: James river, Richmond, *W. E. Wade*, Jul. 1951 (DA). FLORIDA: on *Plectonema Wollei* and *Rhizoclonium hieroglyphicum*, Rock Springs, Seminole county, *E. M. Davis*, 1939 (FH); on Cladophora in the spring-pool, Wakulla Springs, Wakulla county, *F. Drouet, G. Madsen, & D. Crowson 11506*, 27 Jan. 1949 (FC, T); on *Plectonema Wollei* and *Rhizoclonium fontanum* in wet place at Rock Spring Park, Orange county, *P. O. Schallert 2300*, 1 Jun. 1951 (FC). TENNESSEE: on *R. hieroglyphicum*, Cumberland river at Clarksville, Montgomery county, *A. E. Clebsch 2023*, 1 Oct. 1949 (FC, TENN); on Pithophora in Blue Basin, Reelfoot lake, Lake county, *H. Silva 1156*, 2 Jul. 1949 (FC, TENN). ARKANSAS: on *Rhizoclonium hieroglyphicum* in Buffalo river, St. Joe, Searcy county, *D. Demaree 25307Aa*, 1 Oct. 1944 (FC). PUERTO RICO: on Oedogonium in a reservoir, Rio Piedras, *N. Wille 119b*, Dec. 1914 (Type of *Chamaesiphon Willei* Gardn., NY; isotypes, FH, UC). PHILIPPINES: Los Baños, Laguna, *E. Quisumbing 5434*, 13 Oct. 1929 (FC, UC); on Cladophora in little spring at base of Igkaras hill, Alimodian, Iloilo, *J. D. Soriano 1538*, 7 Dec. 1952 (FC, PUH). CHINA: on *Plectonema Wollei*, Wuhsien (Soochow), Kiangsu, *C. C. Wang 312*, 14 Oct. 1930 (FC, UC). INDONESIA: auf Lyngbya aus 3.6 m. Tiefe, Ranau Bedali-See, Ost-Java, *Ruttner*, 20—22 Nov. 1928 (Type of *Stichosiphon regularis* Geitl., slide no. KB2β in slide collection of L. Geitler). INDIA: on Oedogonium, McPherson lake, Allahabad, *P. Maheshwari 21*, Jul. 1939 (FH). PAKISTAN: on Pithophora in a stagnant pool in the Botanical Gardens, Lahore, *Ghose*, 13 Oct. 1918 (Type of *Chamaesiphon filamentosus* Ghose in the collection of F. E. Fritsch; isotype, D [Fig. 254]).

NOMINA EXCLUDENDA

The following names have been described on the bases of material not included in the families treated above. Most of these names were originally described as members of the Chroococcaceae, Chamaesiphonaceae, or Clastidiaceae; or were transferred into those families at one time or another; or have been suspected as being members of those families. Where the original specimens have not been available for study, we have designated the original description as the temporary Type, to serve until specimens seen by the author can be found.

The alga most frequently mistaken as one of the coccoid Myxophyceae is that classically referred to in the literature since 1849 as one or another species of Gloeocystis. In this study we could not obviate special efforts to understand the large numbers of specimens of this group from all parts of the world that have come to our attention. The genus and its single species are therefore treated here separately:

PALMOGLOEA Kützing, Phyc. Gener., p. 176. 1843. —Type species: *Ulva protuberans* Sm. & Sowerb., 1814.

Bromicolla Eichwald, Förh. Skandinav. Naturforsk., 3rd Möte, Stockholm (13—19 Juli 1842), p. 615. 1843. —Type species: *B. aleutica* Eichw., 1843.

Brachtia Trevisan, Sagg. Monogr. Alg. Coccot., p. 57. 1848. —Type species: *Palmella crassa* Nacc., 1828.

Hassallia Trevisan, Sagg. Monogr. Alg. Coccot., p. 67. 1848. —Type species: *Haematococcus Hookerianus* Berk. & Hass., 1845.

Urococcus Hassall ex Kützing, Sp. Algar., p. 206. 1849. *Haematococcus* Subgenus *Ouracoccus* Hassall, Hist. Brit. Freshw. Alg., p. 322. 1845. —Type species: *Haematococcus Allmanii* Hass., 1845.

Aphanocapsa Nägeli, Gatt. Einzell. Alg., p. 52. 1849. *Microcystis* Subgenus *Capsothece* Elenkin, Not. Syst. Inst. Crypt. Horti Bot. Petropol. 2: 67. 1923. —Type species: *Palmella parietina* Nägeli, 1849.

Dactylothece Lagerheim, Öfvers. K. Sv. Vet.-Akad. Förh. 40(2): 64. 1883. —Type species: *D. Braunii* Lagerh., 1883.

Rhodococcus Hansgirg in Wittrock & Nordstedt, Alg. Exs. 14: 697. 1884; in Wittrock & Nordstedt, Bot. Notiser 1884: 128. 1884; Oesterr. Bot. Zeitschr. 34: 314. 1884. *Chroococcus* Subgenus *Rhodococcus* Hansgirg, Prodr. Algenfl. Böhmen 2: 159. 1892. —Type species: *Rhodococcus caldariorum* Hansg., 1884.

Gloeothece Sectio Hyalothece Kirchner in Engler & Prantl, ed. 1, 1(1a): 55. 1898. *Gloeothece* Subgenus *Hyalothece* Kirchner ex Forti, Syll. Myxophyc., 60. 1907. —Type species: *Gloeocapsa palea* Kütz., 1843.

The following names are interpreted only on the bases of descriptions of their respective Type species:

Gloeocystis Nägeli, Gatt. Einzell. Alg., p. 65. 1849. —Type species: *G. vesiculosa* Näg., 1848.

Cyanostylon Geitler, Arch. f. Protistenk. 60: 441. 1928. —Type species: *C. microcystoides* Geitl., 1928.

Vanhoeffenia Wille, Deutsche Südpolar-Exped., 1901—03, 8: 422. 1928. —Type species: *V. antarctica* Wille, 1928.

Hormothece Jao, Sinensia 15: 77. 1944. —Type species: *H. rupestris* Jao, 1944.

Stilocapsa Ley, Bot. Bull. Acad. Sinica 1: 77. 1947. —Type species: *S. sinica* Ley, 1947.

Asterocapsa Chu, Ohio Journ. Sci. 52(2): 97. 1952. —Type species: *A. gloeotheceformis* Chu, 1952.

Plantae pauci- vel multicellulares, in strata vel pulvina increscentes, cellulis in divisione transversale primum hemisphaericis vel hemi-ellipsoideis, aetate provecta sphaericis, ovoideis, vel quasi-fusiformibus, per gelatinum vaginale irregulariter vel regulariter (nonnumquam uniseriatim) distributis; chloroplastidibus viridibus, irregulariter poculiformibus, parietalibus, solitariis, sine pyrenoideis; gelatino vaginale homogeneo vel lamelloso; reproductione a fragmentatione vel a zoosporis chlamydomonadoideis [?].

One species:

PALMOGLOEA PROTUBERANS (Sm. & Sow.) Kützing, Phyc. Gener., p. 176. 1843; Kützing, Linnaea 17: 85. 1843. *Ulva protuberans* Smith & Sowerby, Engl. Bot. 36: 2583. 1814. *Merrettia protuberans* S. F. Gray, Nat. Arr. Brit. Pl., p. 349. 1821. *Palmella protuberans* Greville, Fl. Edin., p. 323. 1824; Agardh, Syst. Algar., p. 14. 1824. *Coccochloris protuberans* Sprengel, Linn. Syst. Veget., ed. 16, 4(1): 373. 1827. —Type from Sussex, England: Uckfield, *W. Borrer* (BM).

Chaos bituminosus Bory, Dict. Class. d'Hist. Nat. 3: 16, 1823. *Palmella bituminosa* Meneghini, Mem. R. Accad. Sci. Torino, ser. 2, 5(Sci. Fis. & Mat.): 46, 56. 1843. *Protococcus bituminosus* Kützing, Tab. Phyc. 1: 4. 1846. *Pleurococcus bituminosus* Trevisan, Sagg. Monogr. Alg. Coccot., p. 32. 1848. *Gloeocapsa bituminosa* Kützing, Sp. Algar., p. 224. 1849. *Chroococcus bituminosus* Hansgirg, Oesterr. Bot. Zeitschr. 35: 116. 1885; Bot. Centralbl. 22: 25, 27. 1885. —Designated Type from France: sur les pierres calcaires de l'entrée des carrières de St.-Leu sous le camp de Cesar, *Bory de St.-Vincent*, Oct. 1815 (PC).

Palmella crassa Naccari, Algol. Adriat., p. 12. 1828; Fl. Veneta 6: 41. 1828. *Coccochloris crassa*

Meneghini in Zanardini, Bibl. Ital. 99: 197. 1840. *Brachtia crassa* Trevisan, Sagg. Monogr. Alg. Coccot., p. 57. 1848. *Palmophyllum crassum* Rabenhorst, Fl. Eur. Algar. 3: 49. 1868. —Paratype, here designated as the Type, from Istria: in Adriatico nella acqua del Golfo nominato Quarnero, *ex Naccari*, 1833 (PC). FIG. 289.

Palmella Grevillei Berkeley, Glean. Brit. Alg., p. 16. 1833. *Botrydina Grevillei* Meneghini, Mem. R. Accad. Sci. Torino, ser. 2, 5 (Sci. Fis. & Mat.): 47. 1843. *Botrydiopsis Grevillei* Trevisan, Nomencl. Algar. 1: 70. 1845. *Coccochloris Grevillei* Hassall, Hist. Brit. Fresh-w. Alg. 1: 318. 1845. *Pleococcus Grevillei* Trevisan, Sagg. Monogr. Alg. Coccot., p. 43. 1848. *Anacystis Grevillei* Kützing, Sp. Algar., p. 209. 1849. *Aphanocapsa Grevillei* Rabenhorst, Fl. Eur. Algar. 2: 50. 1865. *Palmogloea Grevillei* Crouan fr., Fl. Finistère, p. 110. 1867. *Microcystis Grevillei* Elenkin, Monogr. Algar. Cyanophyc., Pars Spec. 1: 126. 1938. *Gloeocystis Grevillei* Drouet & Daily, Lloydia 11: 79. 1948. —Type from Scotland: Kinnordy Moss, *R. K. Greville* (E); isotype, K. Both specimens are largely parasitized, but enough unparasitized material is present to indicate clearly the nature of the alga. FIG. 274.

Palmella furfuracea Berkeley, Glean. Brit. Alg., p. 18. 1833. *Anacystis furfuracea* Meneghini, Mem. R. Accad. Sci. Torino, ser. 2, 5 (Sci. Fis. & Mat.): 47, 94. 1843. *Botrydiopsis furfuracea* Trevisan, Nomencl. Algar. 1: 41. 1845. *Haematococcus furfuraceus* Hassall, Hist. Brit. Fresh-w. Alg. 1: 331. 1845. *Bichatia furfuracea* Trevisan, Sagg. Monogr. Alg. Coccot., p. 65. 1848. —Type from England: Milton, *herb. Berkeley* (K).

Palmella granosa Berkeley, Glean. Brit. Alg., p. 19. 1833. *Haematococcus granosus* Harvey, Man. Brit. Alg., p. 181. 1841. *Microcystis granosa* Meneghini, Mem. R. Accad. Sci. Torino, ser. 2, 5 (Sci. Fis. & Mat.): 47, 85. 1843. *Bichatia granosa* Trevisan, Nomencl. Algar. 1: 60. 1845. *Gloeocapsa granosa* Kützing, Tab. Phyc. 1: 25. 1846. *Microcystis granulosa* Meneghini ex Rabenhorst, Fl. Eur. Algar. 2: 61. 1865. *Gloeothece granosa* Rabenhorst, loc. cit. 1865. —Type from England: "Palmella granosa", *herb. Berkeley* (K). This specimen is chiefly composed of *Palmogloea protuberans* through which are distributed the cysts of Dinoflagellates.

Palmella grumosa Carmichael in Hooker, Engl. Fl. 5(1) [Brit. Fl. 2]: 397. 1833. *Sorospora grumosa* Hassall, Hist. Brit. Fresh-w. Alg. 1: 310. 1845. *Bichatia grumosa* Trevisan, Sagg. Monogr. Alg. Coccot., p. 60. 1848. —Type from Great Britain: "Palmella grumosa", *Carmichael* (K).

Palmella hyalina β muscicola Carmichael in Hooker, Engl. Fl., 5(1) [Brit. Fl. 2]: 397. 1833. *P. hyalina* var. *muscicola* Harvey ex Trevisan, Sagg. Monogr. Alg. Coccot., p. 56. 1848. *Coccochloris muscorum* Brébisson *pro synon.* ex Trevisan, loc. cit. 1848. —Type from Great Britain: *Carmichael* (K). FIG. 298.

Palmella rivularis Carmichael in Hooker, Engl. Fl. 5(1) [Brit. Fl. 2]: 397. 1833. *Coccochloris rivularis* Hassall, Hist. Brit. Fresh-w. Alg. 1: 317. 1845. *Bichatia rivularis* Trevisan, Sagg. Monogr. Alg. Coccot., p. 66. 1848. *Palmella hyalina δ rivularis* Kützing, Sp. Algar., p. 215. 1849. *Aphanocapsa rivularis* Rabenhorst, Fl. Eur. Algar. 2: 49. 1865. *Palmella hyalina* f. *rivularis* Crouan fr., Fl. Finistère, p. 109. 1867. *P. littoralis* Crouan fr. *pro synon.*, loc. cit. 1867. *Microcystis Grevillei* f. *rivularis* Elenkin, Monogr. Algar. Cyanophyc., Pars Spec., 1: 129. 1938. —Type from Great Britain: *Carmichael* (K). FIG. 294.

Palmella botryoides var. *uda* Kützing, Linnaea 8: 376. 1833. *Botrydina uda* Meneghini, Mem. R. Accad. Sci. Torino, ser. 2, 5 (Sci. Fis. & Mat.): 47. 1843. *Botrydiopsis uda* Trevisan, Nomencl. Algar. 1: 70. 1845. —Type from Thuringia: Hirschberg, *herb. Kützing*, 6 Jun. (L).

Palmella sordida Kützing, Linnaea 8: 377. 1833. *Coccochloris sordida* Meneghini, Mem. R. Accad. Sci. Torino, ser. 2, 5 (Sci. Fis. & Mat.): 46, 67. 1843. —Type from Germany: an Kellerwänden in Schleusingen, *F. T. Kützing* (L). FIG. 283.

Bromicolla aleutica Eichwald, Förh. Skandinav. Naturf., 3rd Möte, Stockholm, 13—19 Juli 1842, p. 615. 1843. —Type from the Aleutian islands: Ins. Unimah (LD). FIG. 290.

Gloeocapsa botryoides Kützing, Phyc. Gener., p. 173. 1843. *Bichatia botryoides* Trevisan, Nomencl. Algar. 1: 60. 1845. *Gloeocystis botryoides* Nägeli, Gatt. Einzell. Alg., p. 65. 1849. —Type from Germany: ad palos, Nordhausen, *herb. Kützing* (L). FIG. 255.

Gloeocapsa montana Kützing, Phyc. Gener., p. 173. 1843. *Bichatia montana* Trevisan, Nomencl. Algar. 1: 61. 1845. *B. Kuetzingiana* Trevisan, Sagg. Monogr. Alg. Coccot., p. 62. 1848. *Gloeocapsa caldarii* Suringar *pro synon.*, Obs. Phycol. p. 54. 1855. *Gloeocapsa caldariorum* Rabenhorst, Fl. Eur. Algar. 2: 37. 1865. *G. montana γ caldarii* Suringar *pro synon.* ex Rabenhorst, loc. cit. 1865. *G. montana* f. *caldarii* Suringar ex Kirchner, Krypt.-Fl. Schlesien 2(1): 257. 1878. *Bichatia caldariorum* Kuntze, Rev. Gen. Pl. 2: 886. 1891. *Gloeocapsa montana* var. *caldarii* Suringar ex Tilden, Amer. Alg. 2: 197. 1896. —Type from Germany: Nordhausen, *herb. Kützing* (L). FIG. 266.

Gloeocapsa palea Kützing, Phyc. Gener., p. 173. 1843. *Bichatia palea* Trevisan, Nomencl. Algar. 1: 61. 1845. *Coccochloris vernicosa* Brébisson *pro synon.* in Kützing, Sp. Algar., p. 217. 1849. *Gloeothece palea* Nägeli, Gatt. Einzell. Alg., p. 57. 1849. —Type from Bohemia: Carlsbad, *herb. Kützing* (L). FIG. 263.

Gloeocapsa salina Roemer, Alg. Deutschl., p. 63. 1845. *Bichatia salina* Trevisan, Sagg. Monogr. Alg. Coccot., p. 64. 1848. —Type from Germany: Salzboden bei Hildesheim (NY). FIG. 269.

Protococcus dissectus Kützing, Phyc. Germ., p. 144. 1845. *Pleurococcus dissectus* Trevisan, Sagg.
Monogr. Alg. Coccot., p. 34. 1848. *Protococcus dissectus* var. *articulatus* Suringar, Obs. Phycol., p.
56. 1855. *Protococcus viridis* var. *dissectus* Wolle, Fresh-w. Alg. U. S., p. 181. 1887. —Type
from Germany: Nordhausen, *F. T. Kützing*, 12 Oct. 1839 (L).

Palmella margaritacea Kützing, Phyc. Germ., p. 149. 1845. *Cagniardia margaritacea* Trevisan,
Sagg. Monogr. Alg. Coccot., p. 49. 1848. *Aphanothece margaritacea* Forti, Syll. Myxophyc., p.
79. 1907. —Type from Dalmatia: Lesina, *Botteri* (L). FIG. 286.

Gloeocapsa palea β *minor* Kützing, Phyc. Germ., p. 151. 1845. *G. confluens* β *palacea*
Kützing, Sp. Algar., p. 217. 1849. —Type from Germany: in Tannengesätz, Oldenburg, *Koch*,
Nov. 1844 (L). FIG. 262.

Haematococcus Allmanii Hassall, Hist. Brit. Freshw. Alg. 1: 322. 1845. *Hassallia Allmanii*
Trevisan, Sagg. Monogr. Alg. Coccot., p. 6. 1848. *Urococcus Allmanii* Hassall ex Kützing, Sp.
Algar., p. 206. 1849. —Type from England: "C. Allmanii", *herb. Hassall* (BM).

Haematococcus cryptophilus Hassall, Hist. Brit. Freshw. Alg. 1: 324. 1845. *Palmella cryptophila*
Carmichael *pro synon.* in Hassall, loc cit. 1845. *Hassallia cryptophila* Trevisan, Sagg. Monogr. Alg.
Coccot., p. 67. 1848. *Urococcus cryptophilus* Hassall ex Kützing, Sp. Algar., p. 206. 1849. —Type
from Great Britain: Appin, *Carmichael* (BM).

Haematococcus Hookerianus Berkeley & Hassall in Hassall, Hist. Brit. Freshw. Alg. 1: 325.
1845. *Hassallia Hookeriana* Trevisan, Sagg. Monogr. Alg. Coccot., p. 67. 1848. *Urococcus Hooker-*
ianus Berkeley & Hassall ex Kützing, Sp. Algar., p. 206. 1849. *Gloeocapsa Hookeri* Berkeley &
Hassall in Berkeley, Introd. Crypt. Bot., p. 117. 1857. —Paratype, here designated as the Type,
from England: Bristol, Jan. 1847, *in herb. Berkeley* (K). FIG. 264.

Palmella borealis Kützing, Tab. Phyc. 1: 6. 1846. *Cagniardia borealis* Trevisan, Sagg. Monogr.
Alg. Coccot., p. 49. 1848. —Type (with that of *Gloeocapsa rupicola* Kütz.) from Norway:
Sprengel (L). FIG. 279.

Palmella Botteriana Kützing, Tab. Phyc. 1: 10. 1846. *Cagniardia Botteriana* Trevisan, Sagg.
Monogr. Alg. Coccot., p. 49. 1848. —Type from Dalmatia: Lesina, *Botteri* (L). FIG. 282.

Palmella adriatica Kützing, Tab. Phyc. 1: 11. 1846. *Cagniardia adriatica* Trevisan, Sagg. Monogr.
Alg. Coccot., p. 49. 1848. *Aphanothece Le-Jolisii* Thuret *pro synon.* ex Chalon, Liste des Alg. Mar.,
p. 32. 1905. *Aphanocapsa Le-Jolisii* Frémy, Bull. Soc. Linn. Normandie, ser. 7, 7: 18. 1925.
—Type from Dalmatia: *ex herb. Meneghini* (L). FIG. 287.

Palmella laxa Kützing, Tab. Phyc. 1: 11. 1846. *Bichatia laxa* Trevisan, Sagg. Monogr. Alg.
Coccot., p. 66. 1848. *Aphanothece laxa* Rabenhorst, Krypt.-Fl. Sachsen 1: 76. 1863. —Type
from Netherlands: Goes, *van den Bosch 198*, Jan. 1846 (L). FIG. 291.

Gloeocapsa confluens Kütz., Tab. Phyc. 1: 14. 1846. *Bichatia confluens* Trevisan, Sagg.
Monogr. Alg. Coccot., p. 62. 1848. *Gloeothece confluens* Nägeli, Gatt. Einzell. Alg., p. 58. 1849.
Gloeocystis confluens Richter in Hauck & Richter, Phyk. Univ. 1: 81. 1886. *Dactylothece confluens*
Hansgirg, Sitzungsber. K. Böhm. Ges. Wiss., Math.-Nat. Cl., 1891(1): 322. 1891. —Type from
Germany: Nordhausen, *F. T. Kützing* (L) FIG. 259.

Gloeocapsa quaternata Kützing, Tab. Phyc. 1: 15. 1846. *Coccochloris quaternata* Brébisson
pro synon. in Kützing, loc. cit. 1846. *Bichatia quaternata* Trevisan, Sagg. Monogr. Alg. Coccot., p.
62. 1848. *Gloeocystis quaternata* Richter in Hauck & Richter, Phyk. Univ. 1: 81. 1886. *Gloeocapsa*
quaternaria Kützing ex Geitler, Rabenh. Krypt.-Fl. 14: 197. 1932. —Type from France: Falaise,
A. de Brébisson (L). FIG. 278.

Gloeocapsa cryptococca Kützing, Tab. Phyc. 1: 15. 1846. *Bichatia cryptococca* Trevisan, Sagg.
Monogr. Alg. Coccot., p. 62. 1848. —Type from Harz mountains, Germany: "G. cryptococca Kg."
(L). FIG. 258.

Gloeocapsa polydermatica Kützing, Tab. Phyc. 1: 15. 1846. *Bichatia polydermatica* Kuntze,
Rev. Gen. Pl. 2: 886. 1891. —Type from France: "Microcystis rupestris", Falaise, *Brébisson* (L).
FIG. 276.

Gloeocapsa muralis Kützing, Tab. Phyc. 1: 17. 1846. *Bichatia muralis* Kuntze, Rev. Gen.
Pl. 2: 886. 1891. —Type from Germany: an feuchten Mauern, Nordhausen, *Kützing* (L). FIG. 261.

Gloeocapsa squamulosa Brébisson in Kützing, Tab. Phyc. 1: 17. 1846. *Bichatia squamulosa*
Kuntze, Rev. Gen. Pl. 2: 886. 1891. —Type from France: Falaise, *A. de Brébisson* (L). FIG. 275.

Gloeocapsa microphthalma Kützing, Tab. Phyc. 1: 25. 1846. *Bichatia microphthalma* Kuntze,
Rev. Gen. Pl. 2: 886. 1891. —Type from Germany: ad cataracta pr. Nordhausen, *Kützing* (L).

Palmogloea vesiculosa Kützing, Tab. Phyc. 1: 25. 1846. —Type from Germany: in turfosis
humidis pr. Jever, *Koch*, Apr. 1846 (L).

Palmella (Gloeocapsa) chrysophthalma Montagne, Ann. Sci. Nat. III. Bot. 12: 286. 1849.
Gloeocystis chrysophthalma Montagne ex Farlow in Farlow, Anderson & Eaton, Alg. Exs. Am. Bor.
no. 230. 1889. *Bichatia chrysophthalma* Kuntze, Rev. Gen. Pl. 2: 886. 1891. —Type from
France: Etretat, *F. Baudry* (PC). FIG. 273.

Cryptococcus vernicosus Kützing, Sp. Algar., p. 146. 1849. —Type from France: Falaise,
Le Bailly (L).

Palmella parietina Nägeli in Kützing, Sp. Algar., p. 212. 1849. *Aphanocapsa parietina* Nägeli,

Gatt. Einzell. Alg., p. 52. 1849. *Microcystis parietina* Elenkin, Monogr. Algar. Cyanophyc., Pars Spec. 1: 135. 1938. —Type from Switzerland: Rheinfall, *Nägeli* (L). FIG. 280.

Gloeocapsa flavo-aurantia Kützing, Sp. Algar., p. 218. 1849. *G. montana* f. *flavo-aurantia* Kützing ex Rabenhorst, Fl. Eur. Algar. 2: 37. 1865. *G. montana* var. *flavo-aurantia* Kützing ex Hansgirg, Prodr. Algenfl. Böhmen 2: 152. 1892. —Type from France: au bas d'un mur humide, Chermont, *M. Tillette*, Apr. 1846 (in Desmazières, Pl. Crypt. Fr. 1956, NY).

Endospira bryophila Brébisson in Desmazières, Pl. Crypt. France, ser. 2, 1954. 1850. *Coccochloris muscorum* Brébisson pro synon., loc. cit. 1850. —Type from France: sur le mousse humide (in Desmaz., Pl. Crypt. Fr. 1954, FI).

Gloeocapsa palmelloides Rabenhorst, Alg. Sachs. 27—28: 262. 1853. *Gloeothece palmelloides* Rabenhorst, Krypt.-Fl. Sachsen 1: 75. 1863. —Type from Thuringia: am Falkenstein im Dintharzer Grund, *A. Roese*, 1852 (in Rabenh. Alg. 262, FC). FIG. 277.

Gloeocapsa insignis Thuret, Mém. Soc. Nat. Sci. Nat. Cherbourg 2: 388. 1854. —Type from France: Cherbourg, sur l'écorce des arbres de la route de Paris, *G. Thuret*, 19 Jan. 1853 (PC).

Gloeocapsa romana Montagne, Syll. Gen. Spec. Crypt., p. 470. 1856. *Bichatia romana* Kuntze, Rev. Gen. Pl. 2: 886. 1891. —Type from Italy: Roma, *E. Fiorini-Manzatti* (PC).

Gloeothece distans Stizenberger in Rabenhorst, Alg. Sachs. 97—98: 971. 1860. —Isotypes from Germany: Constanz, *E. Stizenberger*, Apr. 1860 (in Rabenh. Alg. 971, FC, TA). FIG. 272.

Gloeocystis riparia A. Braun in Rabenhorst, Alg. Eur. 172—173: 1729. 1864. —Isotypes from Germany: Sallentin, Insel Usedom, *A. Braun*, Aug. 1864 (in Rabenh. Alg. 1729, FC, NY). FIG. 271.

Palmella littorea Crouan fr. in Schramm & Mazé, Essai Class. Alg. Guadeloupe, p. 29. 1865; ibid., p. 67. 1866; in Mazé & Schramm, Essai Class. Alg. Guadeloupe, ed. 2, p. 12. 1870—77. —Type from La Guadeloupe: Lac Simpson, Saint-Martin (PC). FIG. 296.

Gloeocapsa Goeppertiana Hilse in Rabenhorst, Algen Eur. 221—222: 2220. 1870. *Gloeothece Goeppertiana* Forti, Syll. Myxophyc., p. 62. 1907. —Isotypes from Silesia (Poland): ad Vratislaviam, *Hilse*, May 1870 (FH, L, NY, UC).

Chroococcus granulosus Zeller, Journ. Asiatic Soc. Bengal 42(2): 176. 1873; Hedwigia 12: 169. 1873. —Type from Burma: auf blosser Erde, Thabyaegon, *S. Kurz* (in Rabenh. Alg. 2332, S).

Protococcus affinis Dickie, Journ. Linn. Soc. Bot. 14: 358. 1875. —Type from St. Paul's Rocks, Atlantic Ocean: on stalactitic guano in sheltered crevices, *H. N. Moseley*, 18 Aug. 1873 (BM).

Gloeothece involuta Reinsch, Journ. Linn. Soc. Bot. 15: 206. 1876. —Type from Kerguelen island: *Venus Expedition*, 1874—75 (K).

Dactylothece Braunii Lagerheim, Öfvers. K. Sv. Vet.-Akad. Förh. 40(2): 64. 1883. —Isotypes from Uppsala, Sweden, in Wittr. & Nordst., Alg. Exs. 531 (FC, UC). FIG. 256.

Rhodococcus caldariorum Hansgirg in Wittrock & Nordstedt, Alg. Exs. 14: 697. 1884; in Wittrock & Nordstedt, Bot. Notiser 1884: 128. 1884; Oesterr. Bot. Zeitschr. 34: 315. 1884. *Chroococcus caldariorum* Hansgirg, Bot. Centralbl. 22: 33. 1885. —Type from Bohemia: in parietibus caldariorum horti botanici Pragae, *A. Hansgirg*, 22 Dec. 1883 (in Wittr. & Nordst., Alg. Exs. 697, W; isotypes, FC, L, S, and in Kerner, Fl. Exs. Austro-Hung. 2399 & Nachtrag, S). FIG. 257.

Gloeocapsa salina Hansgirg, Oesterr. Bot. Zeitschr. 35: 115. 1885. —Type from Bohemia: Ouzic ad Kralup, *A. Hansgirg*, Oct. 1883 (W). FIG. 297.

Inoderma majus Hansgirg, Physiol. & Algol. Stud., p. 93. 1887; Oesterr. Bot. Zeitschr. 37: 122. 1887. —Type from Bohemia: Harrachsdorf, *A. Hansgirg*, Jul. 1886 (W). FIG. 288.

Gloeocystis rupestris var. *subaurantiaca* Hansgirg, Prodr. Algenfl. Böhmen 1(2): 136. 1888. —Type from Bohemia: Selc nächst Roztok, *A. Hansgirg*, Nov. 1886 (W). FIG. 295.

Gloeocystis vesiculosa var. *caldariorum* Hansgirg, Sitzungsber. K. Böhm. Ges. Wiss., Math.-Nat. Cl., 1891(1): 321. 1891. —Type from Bohemia: Tetschen, *A. Hansgirg*, Aug. 1890 (W). FIG. 293.

Spirotaenia closteridia var. *elongata* Hansgirg, Sitzungsber. K. Böhm. Ges. Wiss., Math.-Nat. Cl., 1891(1): 329. 1891. —Type from Bohemia: Edmundsklamm nächst Herrnskretschen, *A. Hansgirg*, Sept. 1890 (W).

Gloeocapsa nigra var. *minor* Hansgirg, Prodr. Algenfl. Böhmen 2: 152. 1892. —Type from Bohemia: Eichwald, *A. Hansgirg*, Jul. 1883 (W). FIG. 284.

Pleurococcus miniatus var. *virescens* Hansgirg, Prodr. Algenfl. Böhmen 2: 235. 1892. —Type from Bohemia: zwischen Debr und Josephsthal, *A. Hansgirg*, 1891 (W). FIG. 281.

Botryococcus Braunii var. *mucosus* Lagerheim in Hansgirg, Prodr. Algenfl. Böhmen 2: 242. 1892. —Type from Bohemia: Edmundsklamm nächst Herrnskretschen, *A. Hansgirg*, Sept. 1890 (W). FIG. 285.

Dactylococcus sabulosus Hansgirg, Sitzungsber. K. Böhm. Ges. Wiss., Math.-Nat. Cl., 1890(1): 12. 1890. *Keratococcus sabulosus* Pascher, Süsswasserfl. 5(Chlorophyc. 2): 217. 1915. —Type from Bohemia: zwischen Dittersbach und Hinter-Dittersbach, *A. Hansgirg*, Juli 1889 (W).

Urococcus pallidus Lagerheim in Wittrock, Nordstedt, & Lagerheim, Alg. Exs. 23: 1086. 1893. —Isotype from Ecuador: in rupibus humidis ad fluminem Pastasa prope Baños, prov. Tungurahua, *G. Lagerheim*, 2 Jan. 1892 (in Wittr., Nordst., & Lagerh., Alg. Exs. 1086, NY).

Dactylococcopsis montana W. & G. S. West., Journ. of Bot. 36: 337. 1898. —Type from Yorkshire, England: Cowgill Wold Moss, Widdale Fell, *herb. West*, Apr. 1898 (BIRM).

Chroococcus Simmeri Schmidle in Simmer, Allgem. Bot. Zeitschr. 1898: 158. 1899. *Gloeocapsa magma* var. *Simmeri* Novácek ex Geitler, Rabenh. Krypt.-Fl. 14: 1164. 1932. —Type from Carinthia, Austria: an einer alten, faulen Lärche im Gurskenthale, Kreuzeckgebiet, *H. Simmer,* 27 Aug. 1898 (Z).

Chroothece cryptarum Farlow in Collins, Holden, & Setchell, Phyc. Bor.-Amer. 16: 752. 1900. *Entophysalis cryptarum* Drouet, Field Mus. Bot. Ser. 20: 128. 1942. —Type from Bermuda: Aggar's cave, *W. G. Farlow,* 1881 (FH). FIG. 265.

Microcystis chroococcoidea W. & G. S. West, Brit. Antarct. Exped. 1907—09, 1: 296. 1911. *Chroococcus microcystoideus* W. & G. S. West, loc. cit. 1911. —Type from Antarctica: Green lake, Ross island (BIRM).

Dactylococcopsis antarctica Fritsch, Nat. Antarct. Exped. (1901—04), Nat. Hist. 6(Freshw. Alg.): 22. 1912. —Type from Antarctica: on *Nostoc commune* in damp spots, Granite Harbour, New Bay, *no. 47c,* 20 Jan. 1902 (BM).

Gloeocapsa sibogae Weber-van Bosse, Siboga Exped., Liste des Alg., 1: 6. 1913. —Type from Indonesia: Sabuda, *A. A. Weber-van Bosse,* 1899—1900 (L). FIG. 268.

Gloeothece Vibrio Carter in Compton, Journ. Linn. Soc. Bot. 46: 50. 1922. —Type from New Caledonia: river Dumbéa, *R. H. Compton 811* (BM).

Chroococcus constrictus Gardner, Mem. New York Bot. Gard. 7: 8. 1927. —Type from Puerto Rico: Rio Piedras, *N. Wille 1965,* 23 Mar. 1915 (NY). FIG. 267.

Gloeocapsa cartilaginea Gardner, Mem. New York Bot. Gard. 7: 9. 1927. —Type from Puerto Rico: Maricao, *N. Wille 1025,* 12 Feb. 1915 (NY). FIG. 260.

Gloeocapsa sphaerica Gardner, Mem. New York Bot. Gard. 7: 12. 1927. —Type from Puerto Rico: Arecibo—Utuado, *N. Wille 1482a,* 4 Mar. 1915 (NY). FIG. 270.

Gloeothece opalothecata Gardner, Mem. New York Bot. Gard. 7: 14. 1927. —Type from Puerto Rico: Hato Arriba near Arecibo, *N. Wille 1434,* Mar. 1915 (NY). FIG. 292.

Chroococcus schizodermaticus var. *incoloratus* Geitler, Arch. f. Hydrobiol., Suppl. 12, 4: 623. 1933. —Type from Java: Travertin-Hügel von Kuripan, *Ruttner* (slide BK2d in the collection of L. Geitler).

Cyanostylon cylindrocellularis Geitler, Arch. f. Hydrobiol., Suppl. 12, 4: 623. 1933. —Type from Java: Tjurug Dengdeng, Tjibeureum, Gebiet von Tjibodas, *Ruttner,* 9 Jul. 1929 (slide Tj2IIa ∝ in the collection of L. Geitler).

No original or authentic specimens have been available to us for the following names; the original descriptions are here designated as the Types until the specimens can be found:

Protococcus atrovirens Corda in Sturm, Deutschl. Fl., Abt. 2, 6: 43. 1833. *Microcystis atrovirens* Kützing, Linnaea 8: 374. 1833. *Protosphaeria atrovirens* Trevisan, Sagg. Monogr. Alg. Coccot., p. 30. 1848. *Gloeocapsa atrovirens* Richter in Hauck & Richter, Phyk. Univ. 1: 42. 1885. *Chroococcus atrovirens* Hansgirg, Bot. Centralbl. 22: 25. 1885.

Coccochloris Sczentzii Corda in Sturm, Deutschl. Fl., Abt. 2, 6: 49. 1833.

Gloeocapsa microcosmus Roemer, Alg. Deutschl., p. 62. 1845. *Bichatia microcosmus* Trevisan, Sagg. Monogr. Alg. Coccot., p. 63. 1848.

Haematococcus microsporus Hassall, Hist. Brit. Freshw. Alg. 1: 334. 1845. *Bichatia Hassallii* Trevisan, Sagg. Monogr. Alg. Coccot., p. 65. 1848.

Coccochloris variabilis Hassall, Hist. Brit. Freshw. Alg. 1: 441. 1845. *Bichatia variabilis* Trevisan, Sagg. Monogr. Alg. Coccot., p. 66. 1848.

Coccochloris obscura Hassall, Hist. Brit. Freshw. Alg. 1: 442. 1845. *Bichatia obscura* Trevisan, Sagg. Monogr. Alg. Coccot., p. 65. 1848.

Gloeocystis vesiculosa Nägeli, Gatt. Einzell. Alg., p. 65. 1849. *G. ampla* b. *vesiculosa* Kirchner, Krypt.-Fl. Schlesien 2(1): 112. 1878. *Gloeocapsa macrococca* Hansgirg, Prodr. Algenfl. Böhmen 1: 135. 1886.

Spirotaenia muscicola De Bary, Unters. Fam. Conjugat., p. 75. 1858.

Gloeocapsa ocellata Rabenhorst, Krypt.-Fl. Sachsen 1: 72. 1863. *Bichatia ocellata* Kuntze, Rev. Gen. Pl. 2: 886. 1891.

Cyanostylon microcystoides Geitler, Arch. f. Protistenk. 60: 442. 1928.

Vanhoeffenia antarctica Wille, Deutsche Südpolar-Exped. 1901—03, 8: 423. 1928.

Gloeocystis cylindrica Jao, Sinensia 11: 251. 1940.

Gloeocystis Bacillus Teiling, Bot. Notiser 1942: 64. 1942.

Chroococcus splendidus Jao, Sinensia 15: 75. 1944.

Hormothece rupestris Jao, Sinensia 15: 77. 1944.

Stilocapsa sinica Ley, Bot. Bull. Acad. Sinica 1: 77. 1947.

Asterocapsa gloeothaceformis Chu, Ohio Journ. Sci. 52(2): 97. 1952.

Entophysalis robusta Chu, Ohio Journ. Sci. 52(2): 100. 1952.

Cyanostylon sinensis Chu, Ohio Journ. Sci. 52(2): 100. 1952.

Plantae aerugineae, virides, nigrae, vel rubrae, crustaceae, pulvinatae, vel stratosae, gelatinosae, coriaceae, corneae, vel farinariae, cellulis sphaericis, ellipsoideis, vel fusiformibus, diametro 2—15µ crassis; chloroplastidibus viridibus, saepe cum haematochroma obscuratis; gelatino vaginale hyalino vel deinde coerulescente vel rubrescente vel fuscescente, homogeneo aut circa unamquidque cellulam (saepe concentrice vel excentrice) pluri-lamellato, haud raro in caudicibus ramosis increscente. FIGS. 255—298.

On the ground (often among mosses), on wood and rocks, often in seepage, often in shaded places, and sometimes in intertidal zones on marine shores, throughout the world. The cells are bright green in sunny habitats, and starch may be seen as granules in the cytoplasm when stained with iodine. In shaded habitats, the starch is not evident except in the cells at the surface. Strata are often well developed in the mouths of moist caves where little light penetrates; here the chloroplasts appear blue-green or are obscured by haematochrome in the cytoplasm. Here also the excentric development of the sheath material often results in the formation of branched stalks bearing the cells imbedded in their upper ends. Where more light is available, concentrically or excentrically lamellose sheaths are conspicuous about the individual cells or small groups of cells. The gelatinous material is chiefly pectic and therefore subject to hydrolysis, especially during its formation; completely homogeneous, and even sirupy, gelatinous matrices are often encountered in seepage and continuously wet habitats. Brown and "indicator" red and blue pigments develop in the matrices in dry sunny locations. Fungi grow in the sheaths, and some of them parasitize the cells. With certain fungi the cells continue to grow as lichens. *Palmogloea protuberans* is most easily confused with *Coccochloris stagnina, Entophysalis deusta,* and the palmelloid growth-forms of *Stichococcus subtilis* (Kütz.) Klerck. and *Schizogonium murale* Kütz.

Other names excluded from the Chroococcaceae, Chamaesiphonaceae, and Clastidiacease are the following:

Achnanthes quadrijuga Turpin, Mém. Mus. d'Hist. Nat. Paris 16: 310. 1828. *Gonidium quadrijugum* Meneghini, Linnaea 14: 214. 1840. *Agmenellum quadrijugum* Trevisan, Prosp. della Fl. Eugan., p. 58. 1842. *Sphaerastrum pictum* Trevisan, Nomencl. Algar. 1: 28. 1845. —The original description is designated as the temporary Type = CHLOROPHYCEAE.

CHROOCOCCACEAE Subfam. *CHROOCYSTEAE* Hansgirg, Notarisia 3: 588. 1888. —Type genus: *ALLOGONIUM* Kützing, Phyc. Gener., p. 245. 1843. —Type species: *A. confervaceum* Kützing, loc. cit. 1843. *A. confervaceum* ∝ *tergestinum* Kützing, Sp. Algar., p. 364. 1849. *A. tergestinum* Kützing, Tab. Phyc. 3: 10. 1853. —Type from Istria: in aqua subsalsa, Zaule, *F. T. Kützing* (L) = MICROSPORA STAGNORUM (Kütz.) Lagerh.

Allogonium confervaceum β *Kochianum* Kützing, Sp. Algar., p. 364. 1849. *A. Kochianum* Kützing, Tab. Phyc. 3: 10. 1853. —Type from Oldenburg, Germany: in Gräben, Jever, *Koch,* Apr. 1844 (L) = MICROSPORA STAGNORUM (Kütz.) Lagerh.

Allogonium halophilum Hansgirg, Physiol. & Algol. Stud., p. 110. 1887; Hansgirg, Hedwigia 26: 22. 1887. *Asterocytis halophila* Forti, Syll. Myxophyc., p. 691. 1907. —Type from Bohemia: in den Salzwassersümpfen bei Auzitz nächst Kralup, *A. Hansgirg,* Jul. 1886 (W) = ASTEROCYTIS ORNATA (Ag.) Hamel. FIG. 310.

Allogonium halophilum var. *stagnale* Hansgirg, Prodr. Algenfl. Böhmen 2: 132. 1892. *Asterocystis halophila* var. *stagnalis* Hansgirg ex Forti, Syll. Myxophyc., p. 691. 1907. —The original description is designated here as the temporary Type = ASTEROCYTIS ORNATA (Ag.) Hamel.

Allogonium ramosum var. *crassum* Hansgirg, Oesterr. Bot. Zeitschr. 39: 4. 1889. —Type from Jugoslavia: im Hafen von Lussin-piccolo, *A. Hansgirg,* Aug. 1888 (W) = ASTEROCYTIS ORNATA (Ag.) Hamel. FIG. 302.

Allogonium smaragdinum var. *palustre* Hansgirg, Notarisia 3: 399. 1888. *A. smaragdinum* b. *palustre* Hansgirg, Prodr. Algenfl. Böhmen 2: 132. 1892. *Asterocytis smaragdina* var. *palustris*

Hansgirg ex Forti, Syll. Myxophyc., p. 691. 1907. —Type from Bohemia: Gross Wossek, *A. Hansgirg*, Aug. 1887 (W). = ASTEROCYTIS ORNATA (Ag.) Hamel. FIG. 303.

Allogonium Wolleanum var. *calcicola* Hansgirg, Sitzungsber. K. Böhm. Ges. Wiss., Math.-Nat. Cl., 1890(1): 18. 1890. *Asterocytis Wolleana* var. *calcicola* Hansgirg ex Forti, Syll. Myxophyc., p. 689. 1907. —Type from Dalmatia: Clissa nächst Spalato, *A. Hansgirg*, Aug. 1889 (W) = ASTEROCYTIS ORNATA (Ag.) Hamel. FIG. 305.

Allogonium Wolleanum var. *simplex* Hansgirg, Prodr. Algenfl. Böhmen 2: 131. 1892. —Type from Bohemia: Selc, *Hansgirg*, Nov. 1886 (W) = ASTEROCYTIS ORNATA (Ag.) Hamel.

Alternantia Geitleri Schiller, Sitzungsber. Oesterr. Akad. Wiss., Math.-Nat. Kl., I, 163: 134. 1954. —The original description is here designated as the temporary Type = BACTERIA.

Alternantia quadriga Schiller, Sitzungsber. Oesterr. Akad. Wiss., Math.-Nat. Kl., I, 163: 136. 1954. —The original description is here designated as the temporary Type = BACTERIA.

Alternantia ramifera Schiller, Sitzungsber. Oesterr. Akad. Wiss., Math.-Nat. Kl., I, 163: 136. 1954. —The original description is here designated as the temporary Type = BACTERIA.

ALTERNANTIACEAE Schiller, Sitzungsber. Oesterr. Akad. Wiss., Math.-Nat. Kl., I, 163: 134. 1954. —Type genus: *ALTERNANTIA* Schiller, loc. cit. 1954. —Type species: *A. serra* Schiller, loc. cit. 1954. —The original description is here designated as the temporary Type = BACTERIA.

Alternantia tabularis Schiller, Sitzungsber. Oesterr. Akad. Wiss., Math.-Nat. Kl., I, 163: 136. 1954. —The original description is here designated as the temporary Type = BACTERIA.

Anacystis paludosa Rabenhorst, Fl. Eur. Algar. 2: 52. 1865. *Microcystis paludosa* Forti, Syll. Myxophyc., p. 92. 1907. —The original description is here designated as the temporary Type = ZOOCHLORELLA PARASITICA Brandt in OPHRYDIUM sp. or STENTOR sp.

Aphanocapsa albida Zeller, Journ. Asiatic Soc. Bengal 42(2): 176. 1873; Zeller, Hedwigia 12: 169. 1873. —Type from Burma: bei Akgab, prov. Pegu, *G. Zeller* (in Rabenh. Alg. Eur. 2490, NY [FIG. 346]; isotypes, FC, L, UC) = BACTERIA.

Aphanocapsa crassa Ghose, Journ. Burma Res. Soc. 17(3): 243, 239. 1927. —The original description is here designated as the temporary Type = SCHIZOCHLAMYS AURANTIA (Ag.) Dr. & Daily.

APHANOCAPSA Subgenus *AUTAPHANOCAPSA* Hansgirg, Notarisia 3: 589. 1888. —Type species: *Aphanocapsa membranacea* Rabenhorst, Fl. Eur. Algar. 2: 49. 1865. *Aphanothece membranacea* Bornet, Bull. Soc. Bot. France 38: 249. 1891. *Gloeothece membranacea* Bornet, Mém. Soc. Nat. Sci. Nat. Cherbourg 28: 1751. 1892. —Topotype here designated as the Type, from France: fond d'un ruisseau à Mentone, *G. Thuret* (in Roumeguère, Alg. d'Eau Douce de France 827, L) = BACTERIA.

Aphanocapsa paludosa Rabenhorst, Krypt.-Fl. Sachsen 1: 73. 1863. —The original description is here designated as the temporary Type = palmelloid CHLOROPHYCEAE.

Aphanocapsa rufescens Hansgirg, Prodr. Algenfl. Böhmen 2: 157. 1892. —Type from Bohemia: Gross-Wenedig, *A. Hansgirg* (W) = spores of NOSTOC MUSCORUM Ag. FIG. 347.

Aphanothece clathrata W. & G. S. West, Trans. Roy. Irish Acad. 33(A): 111. 1906. —Type from Ireland: Lough Neagh, *G. S. West*, May 1900 (BIRM) = BACTERIA. FIG. 367.

PELODICTYON Lauterborn, Allgem. Bot. Zeitschr. 19: 98. 1913; Lauterborn, Verh. Nat.-Med. Ver. Heidelberg, N. F., 13: 431. 1916. —Type species: *Aphanothece clathratiformis* Szafer, Bull. Int. Acad. Sci. Cracovie, Cl. Sci. Math. & Nat., ser. B, 1910: 162. 1911. *Pelodictyon clathratiforme* Szafer ex Lauterborn, Allgem. Bot. Zeitschr. 19: 98. 1913; Lauterborn, Verh. Nat.-Med. Ver. Heidelberg, N. F., 13: 431. 1916. *P. Lauterbornii* Geitler in Pascher, Süsswasserfl. 12: 458. 1925. —Type from Galicia, Polish Ukraine: Lubién wielki pod Lwowem, *W. Szafer*, 9 Apr. 1910 (in Raciborski, Phyc. Polon. 53, FH) = BACTERIA.

Aphanothece clathratiformis var. *major* Liebetanz, Bull. Int. Acad. Polon. Sci., Cl. Sci. Math. & Nat., ser. B, 1925: 109. 1925. —The original description is here designated as the temporary Type = BACTERIA.

Aphanothece Goetzei Schmidle, Engler Bot. Jahrb. 30: 242. 1901. —The original description is here designated as the temporary Type = the 1—2-celled hormogonia of PHORMIDUM sp.

Aphanothece gracilis Schiller, Sitzungsber. Oesterr. Akad. Wiss., Math.-Nat. Kl., I, 163: 125. 1954. —The original description is here designated as the temporary Type = BACTERIA.

Aphanothece heterospora Rabenhorst, Alg. Eur. 168—171: 1673. 1864; Rabenhorst, Hedwigia 1864: 144. 1864. —Type from Schleswig, Germany: in lacu prope Lutzhöst fl. Flensburgensis, *R. Haecker* (in Rabenh. Alg. 1673, L; isotypes, FC [FIG. 371], NY) = SCHIZOCHLAMYS AURANTIA (Ag.) Dr. & Daily.

SCHMIDLEA Lauterborn, Allgem. Bot. Zeitschr. 19: 98. 1913. —Type species: *Aphanothece luteola* Schmidle, Beih. z. Bot. Centralbl. 10: 179. 1901. *Schmidlea luteola* Lauterborn, Allgem. Bot. Zeitschr. 19: 98. 1913. —Type from Germany: Ludwigshafen, *Lauterborn* (ZT) = BACTERIA. FIG. 369.

Aphanothece nidulans var. *endophytica* W. & G. S. West, Journ. Linn. Soc. Bot. 40: 432. 1912. *A. saxicola* f. *endophytica* Elenkin, Monogr. Algar. Cyanophyc., Pars Spec. 1: 151. 1938. —Type from Scotland: Loch Katrine, *W. & G. S. West*, 16 Mar. 1909 (BIRM) = the organism called PHORMIDIUM MUCICOLA Naum. & Hub.

Aphanothece nidulans var. *thermalis* Hansgirg, Prodr. Algenfl. Böhmen 2: 137. 1892. *A. saxicola* f. *thermalis* Elenkin, Monogr. Algar. Cyanophyc., Pars Spec. 1: 151. 1938. —Type from Bohemia: an warmen Quellen in Karlsbad, *A. Hansgirg,* Aug. 1883 (W) = BACTERIA.

PEDIOCHLORIS Geitler in Pascher, Süsswasserfl. 12: 457. 1925. —Type species: *Aphanothece parallela* Szafer, Bull. Int. Acad. Sci. Cracovie, Cl. Sci. Math. & Nat., ser. B, 1910: 163. 1911. *Pediochloris parallela* Geitler in Pascher, Süsswasserfl. 12: 457. 1925. —Type from Polish Ukraine: Lubień wielki pod Lwowem, *W. Szafer,* 9 Apr. 1910 (in Raciborski, Phyc. Polon. 53, FH) = BACTERIA.

Aphanothece piscinalis Rabenhorst, Fl. Eur. Algar. 2: 66. 1865. *Coccochloris piscinalis* Richter in Hauck & Richter, Phyk. Univ. 5: 240. 1888. *Aphanothece stagnina* f. *piscinalis* Elenkin, Monogr. Algar. Cyanophyc., Pars Spec. 1: 145. 1938. —Type from the Netherlands: in piscina ad Rosendaal, *van den Bosch* (L) = NOSTOC CARNEUM (Lyngb.) Ag. FIG. 368.

CLATHROCHLORIS Geitler in Pascher, Süsswasserfl. 12: 457. 1925. —Type species: *Aphanothece sulphurica* Szafer, Bull. Int. Acad. Sci. Cracovie, Cl. Sci. Math. & Nat., ser. B, 1910: 162. 1911. *Clathrochloris sulphurica* Geitler in Pascher, Süsswasserfl. 12: 457. 1925. —Type from Polish Ukraine: Lubień wielki pod Lwowem, *W. Szafer,* 9 Apr. 1910 (in Raciborski, Phyc. Polon. 53, FH) = BACTERIA.

Aphanothece viennensis Schiller, Sitzungsber. Oesterr. Akad. Wiss., Math.-Nat. Kl., I, 163: 124. 1954. —The original description is here designated as the temporary Type = BACTERIA.

Arthrodesmus glaucescens Wittrock, Bih. K. Sv. Vet.-Akad. Handl. 1(1): 55. 1872. *Tetrapedia glaucescens* Boldt ex de Toni, Nuova Notarisia 3: 105. 1892. —The original description is here designated as the temporary Type = TETRAEDRON sp.

Arthrodesmus glaucescens var. *papilliferus* Gutwinski, Spraw. Kom. Fizyjogr. Akad. Umiej. Krakowie 25(2): (16). 1890. *Tetrapedia glaucescens* var. *papillifera* Gutwinski, ibid. 30(2): 162. 1895. —The original description is here designated as the temporary Type = CHLORO-PHYCEAE.

Asterocytis africana G. S. West, Journ. Linn. Soc. Bot. 38: 196. 1907. —Type from Tanganyika: Kituta bay, *W. A. Cunnington 79,* 26 Aug. 1904 (in the slide collection, BM) = A. ORNATA (Ag.) Hamel.

Asterocytis antarctica W. & G. S. West, Brit. Antarctic Exped. 1907—09, 1: 296. 1911. —The original description is here designated as the temporary Type. The original sketches and notes by the Wests in the British Museum (Natural History) are reproduced here: FIG. 311 = NOSTOC sp.

Bacularia thermalis Frémy, Explor. Parc Nat. Albert, Mission H. Damas, 19: 39. 1949. —The original description is here designated as the temporary Type = the 1—2-celled hormogonia of PHORMIDIUM TENUE (Menegh.) Gom.

Family *PALMELLEAE* Kützing, Phyc. Gener., p. 166. 1843. *PALMELLACEAE* Nägeli, Die Neuern Algensyst., p. 125. 1847. —Type genus: *PALMELLA* Lyngbye, Tent. Hydrophyt. Dan., p. 203. 1819. Synonym: *CLUZELLA* Bory, Dict. Class. Hist. Nat. 3: 14; 4: 234. 1823. —Type species: *Batrachospermum Myosurus* Ducluzeau, Essai s. l'Hist. Nat. Conf. Env. Montpellier, p. 76. 1805. *Palmella Myosurus* Lyngbye, Tent. Hydrophyt. Dan., p. 203. 1819. *Cluzella Myosurus* Bory, Dict. Class. Hist. Nat. 4: 234. 1823. *Coccochloris Myosurus* Sprengel, Linn. Syst. Vegetab., ed. 16, 4(1): 373. 1827. *Hydrurus Ducluzeli* Ag., Consp. Crit. Diatomac. 2: 27. 1830. *Hydrurus Myosurus* Trevisan, Nomencl. Algar. 1: 57. 1845. —It was not considered necessary here to present the complete synonymy of this species. Nägeli in Gatt. Einzell. Alg., p. 66 (1849), selected *Palmella mucosa* Kütz. as the Type for *Palmella* Lyngb.; his selection cannot be considered appropriate under the present International Rules of Nomenclature, since Lyngbye did not treat this species published twenty-four years after his own work appeared. The Type of *Batrachospermum Myosurus* Ducluz. is from France: dans les Cévennes, *Ducluzeau* (PC) = PALMELLA MYOSURUS (Ducluz.) Lyngb.

BICHATIA Turpin, Mém. Mus. d'Hist. Nat. Paris 16: 163. 1828; Turpin, ibid. 15: 376. 1827. GLOEOCAPSA Subgenus BICHATIA Forti, Syll. Myxophyc., p. 50. 1907. —Type species: *Bichatia vesiculinosa* Turpin, Dict. Sci. Nat., Végét. Acot., 65: pl. 10. 1816—29; Mém. Mus. d'Hist. Nat. Paris 15: 376. 1827; ibid. 16: 163. 1828. —The original description of Turpin is here designated as the temporary Type = the coccoid growth-form of STICHOCOCCUS SUBTILIS (Kütz.) Klerck.

BOANEMA Ercegovic, Acta Bot. Inst. Bot. Univ. Zagreb. 2: 83. 1927. —Type species: *B. scoparium* Ercegovic, ibid., p. 84. 1927. —The original description is here designated as the temporary Type = primordia of RHODOPHYCEAE. Geitler in Engler & Prantl, Natürl. Pflanzenfam., ed. 2, 1b: 98 (1942) expresses a similar opinion.

PHYTOCONIS Bory, Mém. s. 1. Genres Conferva & Byssus, p. 54. 1795. *COCCODEA* Palisot de Beauvois in Desvaux, Journ. de Bot. 1: 124. 1808. *OLIVIA* S. F. Gray, Nat. Arr. Brit. Pl., p. 349. 1821. *CHAOS* Bory, Dict. Class. d'Hist. Nat. 3: 13, 15. 1823. *SPHAERELLA* Sommerfelt, Mag. f. Naturvid. 4: 252. 1824. *CHLOROCOCCUM* Fries, Syst. Orb. Veget. 1(Appendix): 356. 1825. *GLOBULINA* Turpin, Mém. Mus. d'Hist. Nat. Paris 15: 25, 62. 1827. *PRIESTLEYA* Meyen, Linnaea 2: 301. 1827. *MONASELLA* Gaillon, Aperçu d'Hist. Nat., p. 29. 1833. *BOTRYDINA* Brébisson, Mém. Soc. Acad. Sci., Art. & Belles-Lettres Falaise, p. 3 (of reprint).

1839. *BOTRYDIOPSIS* Trevisan, Nomencl. Algar. 1: 70. 1845. *DESMOCOCCUS* Brand, Arch. f. Protistenk. 52: 344. 1925. *PLEUROCOCCUS* Untergattung *TETRACHOCOCCUS* Nägeli, Die Neuern Algensyst., p. 127. 1847. —Type species: *Byssus botryoides* Linnaeus, Sp. Pl., p. 1169. 1753. *Tremella botryoides* Schreber, Spicil. Fl. Lips., p. 141. 1771. *Byssus viridis* Lamarck, Fl. Franc., ed. 1, 1: (103). 1778. *Lepra botryoides* Weber in Wiggers, Primit. Fl. Holsat., p. 97. 1780. *Lichen botryoides* Hagen, Tent. Hist. Lichen., p. xl. 1782. *Phytoconis botryoides* Bory, Mém. s. 1. Genres Conferva & Byssus, p. 54. 1795. *Lepraria botryoides* Acharius, Lichenogr. Suec. Prodr., p. 10. 1798. *Monilia viridis* Sprengel, Fl. Halens., p. 386. 1806. *Globulina botryoides* Turpin, Dict. Sci. Nat. 65 (Végét. Acot.): pl. 4, 6, 7. 1816—29. *Nostoc botryoides* Agardh, Synops. Algar. Scand., p. 135. 1817. *Palmella botryoides* Lyngbye, Tent. Hydrophyt. Dan., p. 205. 1819. *Olivia botryoides* S. F. Gray, Nat. Arr. Brit. Pl., p. 349. 1821. *Chaos primordialis* Bory, Dict. Class. d'Hist. Nat. 3:16. 1823. *Sphaerella botryoides* Sommerfelt, Mag. f. Naturvid. 4: 252. 1824. *Priestleya botryoides* Meyen, Linnaea 2: 401. 1827. *Chlorococcum vulgare* Greville, Scott. Crypt. Fl. 5: 262. 1827. *Monasella uva* Gaillon, Aperçu d'Hist. Nat., p. 29. 1833. *Coccophysium botryoides* Link, Handb. d. Erkenn. Gewächs. 3: 342. 1833. *Chlorococcum botryoides* Fries, Fl. Scan., p. 330. 1835. *Anacystis botryoides* Meneghini, Consp. Algol. Eugan., p. 324. 1837. *Botrydina vulgaris* Brébisson, Mém. Soc. Acad. Sci. Art. & Belles-Lettres Falaise, p. 3 (of reprint). 1839. *Pleurococcus vulgaris* Meneghini, Mem. R. Accad. Sci. Torino, ser. 2, 5(Sci. Fis. & Mat.): 38. 1843. *Pleococcus viridis* Kützing, Phyc. Gener., p. 170. 1843. *Haematococcus vulgaris* Hassall, Hist. Brit. Freshw. Alg. 1: 333. 1845. *Lepraria viridis* Jenner, Fl. Tunbridge Wells, p. 150. 1845. *Botrydiopsis vulgaris* Trevisan, Nomencl. Algar. 1: 70. 1845. *Protococcus communis* and *P. communis β Pleurococcus* Kützing, Tab. Phyc. 1: 4. 1846. *Protosphaeria viridis* Trevisan, Sagg. Monogr. Alg. Coccot., p. 29. 1848. *Pleococcus botryoides* Trevisan, ibid., p. 42. 1848. *Cagniardia botryoides* Trevisan, ibid., p. 50. 1848. *Protococcus vulgaris* and *P. vulgaris β Pleurococcus* Kützing, Sp. Algar., p. 199. 1849. *P. vulgaris* var. *Pleurococcus* Kützing ex Mougeot & Nestler, Stirp. Crypt. Vogeso-Rhen., no. 1294. 1850. *Desmococcus vulgaris* Brand, Arch. f. Protistenk. 52: 344. 1925. —Silva & Starr have enlarged the generic synonymy here by designating, in Sv. Bot. Tidskr. 47(2): 245 (1953), types for *Sphaerella* Sommerf. and *Chlorococcum* Fr. from among the above synonyms of *Byssus botryoides* L. The Type specimen upon which all these names depend is that labeled "Byssus botryoides" in herb. Linnaeus in the Linnaean Society of London = PHYTOCONIS BOTRYOIDES (L.) Bory.

BYSSUS Linnaeus, Sp. Pl., p. 1168. 1753. —Type species: *B. flos-aquae* Linnaeus, loc. cit. 1753. *Phytoconis natans* Bory, Mém. s. 1. Genres Conferva & Byssus, p. 54. 1795. *Micraloa flos-aquae* Trevisan, Nomencl. Algar. 1: 76. 1845. —It is unnecessary to the purposes of this paper to present the complete synonymy of *B. flos-aquae* L. here. The Type specimen is that labeled "Byssus flos-aquae" in herb. Linnaeus in the Linnaean Society of London = OSCILLATORIA PROLIFICA (Grev.) Gom.

Byssus lactea Linnaeus, Sp. Pl., p. 1169. 1753. *Lichen lacteus* Schreber, Spicil. Fl. Lips., p. 139. 1771. *Lepra lactea* Weber in Wiggers, Primit. Fl. Holsat., p. 97. 1780. *Phytoconis lactea* Bory, Mém. s. 1. Genres Conferva & Byssus, p. 56. 1795. *Globulina lactea* Turpin, Dict. Sci. Nat. 65(Végét. Acot.): pl. 6, fig. 1. 1816—29. *Protococcus lacteus* Kützing, Linnaea 8: 369. 1833. —The Type, here labeled "Byssus lactea", is in the herbarium of Linnaeus in the Linnaean Society of London = a young LICHEN.

Callonema Itzigsohnii Reinsch, Contrib. Algol. & Fungol. 1: 41. 1874. *Hormospora pusilla* Itzigsohn *pro synon.* in Reinsch, loc cit. 1874. *Allogonium Itzigsohnii* Hansgirg, Hedwigia 26: 23. 1887; Physiol. & Algol. Stud., p. 110. 1887. *Asterocytis Itzigsohnii* Forti, Syll. Myxophyc., p. 690. 1907. —The original description is here designated as the temporary Type = ASTEROCYTIS ORNATA (Ag.) Hamel.

CALLONEMA Reinsch, Contrib. Algol. & Fungol. 1: 40. 1874. —Type species: *C. smaragdinum* Reinsch, ibid. 1: 41. 1874. *Allogonium smaragdinum* Hansgirg, Hedwigia 26: 23. 1887; Physiol. & Algol. Stud., p. 110. 1887. *Asterocytis smaragdina* Forti, Syll. Myxophyc., p. 691. 1907. —The original description is here designated as the temporary Type = ASTEROCYTIS ORNATA (Ag.) Hamel.

CATELLA Alvik, Bergens Mus. Årbok 1934 (Naturvid. Rekke 6): 37. 1934. —Type species: *C. rubra* Alvik, ibid. p. 38. 1934. —The original description is here designated as the temporary Type = NOSTOCACEAE.

Chaetococcus hyalinus Kützing, Osterprogr. Realsch. Nordhausen, p. 6. 1863; Hedwigia 2: 86. 1863. —Type from Cherbourg, France: *A. Le Jolis 1783* (L) = ACROCHAETE sp.

CHALARODORA Pascher, Jahrb. f. Wiss. Bot. 71(3): 460. 1929. —Type species: *C. azurea* Pascher, loc. cit. 1929. —The original description is here designated as the temporary Type = CHLOROPHYCEAE.

Chamaesiphon crenothrichoides Zopf, Morphol. Spaltpfl., p. 55. 1882; Bot. Centralbl. 10: 35. 1882. —The original description is here designated as the temporary Type = CRENOTHRIX sp.

Chamaesiphon gracilis Rabenhorst, Fl. Eur. Algar. 2: 149. 1865. —Type from Germany: Radeberg bei Dresdam, *Rabenhorst* (L) = filamentous BACTERIA.

Chamaesiphon gracilis f. *major* Magnus & Wille, Sitzungsber. Naturf. Fr. Berlin 1882: 101.

1882. —The original description is here designated as the temporary Type = AMPHITHRIX sp.
Chamaesiphon hyalinus Scherffel, Ber. Deutsch. Bot. Ges. 25: 232. 1907. —The original description is here designated as the temporary Type = filamentous BACTERIA.
Chamaesiphon marinus Wille & Rosenvinge, Alg. Novaia-Zemlia og Kara-Havet, p. 4. 1885. —The original description is here designated as the temporary Type = filamentous BACTERIA.
Chamaesiphon Rostafinskii var. *minor* Hansgirg, Prodr. Algenfl. Böhmen 2: 123. 1892. —Type from Bohemia: Sct. Prokop, *A. Hansgirg*, Apr. 1887 (W) = FERROBACTERIACEAE. FIG. 366.
Chaos sanguinarius Bory, Dict. Class. d'Hist. Nat. 3: 16. 1823. —Type from France: Lille, *D. Letestiboudois* (PC) = PORPHYRIDIUM CRUENTUM (Sm. & Sow.) Näg.
CHLOROBIUM Nadson, Bull. Jard. Imp. Bot. St. Petersbourg 6: 190. 1906. —Type species: *C. limicola* Nadson, loc. cit. 1906. —The original description is here designated as the temporary Type = BACTERIA.
Chlorococcum Montagnei Meneghini, Mem. R. Accad. Sci. Torino, ser. 2, 5(Sci. Fis. & Mat.): 30. 1843. *Protosphaeria Montagnei* Trevisan, Sagg. Monogr. Alg. Coccot., p. 30. 1848. *Protococcus Montagnei* Kützing, Sp. Algar., p. 200. 1849. —Type from Cuba: *M. R. de la Sagra*, 1836 (PC) = ASTEROCYTIS ORNATA (Ag.) Hamel.
Chlorococcum Orsinii f. *fimicola* Rabenhorst, Fl. Eur. Algar. 3: 60. 1868. —Type from France: Vache, *M. Roberge* (in Desmazières, Pl. Crypt. Fr., ed. 3, no. 531, UC) = FLAGELLATA.
Chlorogloea purpurea Geitler, Arch. f. Protistenk. 62: 98. 1928. —Paratype from eastern Java: Ranau Pakis, *Ruttner* (in the slide collection of L. Geitler) = PARASITIZED ALGAE.
CHROOCOCCIDIUM Geitler, Arch. f. Hydrobiol., Suppl. XII, 4: 624. 1933. —Type species: *C. gelatinosum* Geitler, loc cit. 1933. —Type from southern Sumatra: Ranau-See, *Ruttner* (no. R5b in the slide collection of L. Geitler) = palmelloid FLAGELLATA.
Chroococcus confervoides A. Braun, Sitzungsber. Ges. Naturf. Fr. Berlin 1875: 99. 1875. —The original description (although very brief) is here designated as the temporary Type = primordia of TRENTEPOHLIA sp.
Chroococcus lageniferus Hildebrandt ex A. Braun, Sitzungsber. Ges. Naturf. Fr. Berlin 1875: 99. 1875. —The original description is here designated as the temporary Type = primordia of TRENTEPOHLIA sp.
Chroococcus macrococcus var. *aquaticus* Hansgirg, Oesterr. Bot. Zeitschr. 36: 333. 1886. —Type from Bohemia: Auzitz nächst Kralup, *A. Hansgirg*, Jul. 1886 (W) = cysts of DINOFLAGELLATES.
Chroococcus macrococcus var. *salinarum* Hansgirg, Sitzungsber. K. Böhm. Ges. Wiss., Math.-Nat. Cl., 1892: 230. 1892. —Type from Istria: in Salinen bei Strogniano, *A. Hansgirg*, Jun. 1889 (W) = PERIDINIUM sp. FIG. 345.
Chroococcus macrococcus f. *stipitatus* Enwald, Medd. Soc. pro Fauna & Fl. Fenn. 30: 151. 1904. —The original description is here designated as the temporary Type = DINOFLAGELLATES.
Chroococcus minimus var. *turfosus* Steinecke, Schrift. Phys.-ökon. Ges. Königsberg 56: 25. 1916. —The original description is here designated as the temporary Type = BACTERIA.
Chroococcus minor f. *major* W. & G. S. West, Trans. Roy Microsc. Soc. 1894: 17. 1894. —Type from England: Cronkley Fell, North Yorks, *W. & G. S. West*, 1888 (BIRM) = CHLOROPHYCEAE.
Chroococcus monetarum Reinsch, Flora 67: 176. 1884. —The original description is here designated as the temporary Type = BACTERIA.
Chroococcus pyriformis Bennett, Journ. Roy. Microsc. London, ser. 2, 8: 3. 1888. *Aphanothece pyriformis* Hansgirg, Prodr. Algenfl. Böhmen 2: 138. 1892. —The original description is here designated as the temporary Type = ASTEROCYTIS ORNATA (Ag.) Hamel.
Chroococcus siderochlamys Skuja, Symbolae Bot. Upsal. 9(3): 38. 1948. —The original description is here designated as the temporary Type = BACTERIA.
Chroococcus subtilissimus Skuja, Hedwigia, Sonderabdr. Bd. 77: 21. 1937. —The original description is here designated as the temporary Type = BACTERIA.
Chroococcus vacuolatus Skuja, Acta Horti Bot. Univ. Latv. 11—12: 42. 1939. —The original description is here designated as the temporary Type = ERYTHROCONIS LITTORALIS Oerst.
Chroococcus Zopfii Hansgirg, Bot. Centralbl. 22: 24. 1885. —The original description is here designated as the temporary Type = parasitized SCYTONEMA or STIGONEMA sp.
CHROODACTYLON Hansgirg, Ber. Deutsch. Bot. Ges. 3: 14. 1885. *ALLOGONIUM* Section *CHROODACTYLON* Hansgirg, Physiol. & Algol. Stud., p. 110. 1887. *ALLOGONIUM* Subgenus *CHROODACTYLON* Hansgirg, Notarisia 3: 588. 1888. —Type species: *Chroodactylon Wolleanum* Hansgirg, Ber. Deutsch. Bot. Ges. 3: 14, 15. 1885. *Asterocytis Wolleana* Lagerheim in Wittrock & Nordstedt, Alg. Exs. 16: 769. 1886. *Allogonium Wolleanum* Hansgirg, Hedwigia 26: 23. 1887; Oesterr. Bot. Zeitschr. 37: 56. 1887; Physiol. & Algol. Stud., p. 96, 110, 164. 1887. —Type from Bohemia: Pürglitz, *A. Hansgirg*, Sept. 1884 (in Wittr. & Nordst., Alg. Exs. 769, W) = ASTEROCYTIS ORNATA (Ag.) Hamel. FIG. 307.
Chroolepus coeruleum Nägeli in Kützing, Sp. Algar., p. 425. 1849. *Allogonium coeruleum* Hansgirg, Hedwigia 26: 23. 1887; Physiol. & Algol. Stud., p. 110. 1887. —Type from Switzerland: with *Gloeocapsa scopulorum*, an Felsen, Rheinfall, *Nägeli*, Jul. 1847 (M) = FUNGI.

CHROOMONAS Hansgirg, Physiol. & Algol. Stud., p. 117. 1887. —Type species: *C. Nordstedtii* Hansgirg, Bot. Centralbl. 23: 230. 1885. —Type from Bohemia: Nusle ad Pragam, *A. Hansgirg* (in Wittr. & Nordst., Alg. Exs. 800½, FC) = FLAGELLATA.

CHROOSTIPES Pascher, Ber. Deutsch. Bot. Ges. 32: 351. 1914. —Type species: *C. linearis* Pascher, loc. cit. 1914. —The original description is here designated as the temporary Type = BACTERIA.

Chroothece mobilis Pascher & Petrová, Arch. f. Protistenk. 74: 490. 1931. —The original description is here designated as the temporary Type = ASTEROCYTIS ORNATA (Ag.) Hamel.

CHROOCOCCACEAE Subfam. *EUCHROOCOCCACEAE* Trib. *THECINEAE* Hansgirg, Notarisia 3: 589. 1888. —Type genus: *CHROOTHECE* Hansgirg, Oesterr. Bot. Zeitschr. 34: 352. 1884; in Wittrock & Nordstedt, Alg. Exs. 14: 696. 1884; in Wittrock & Nordstedt, Bot. Notiser 1884: 128. 1884. —Type species: *C. Richteriana* Hansgirg, loc. cit. 1884. —Type from Bohemia: Auzitz prope Kralup, *A. Hansgirg,* 6 Apr. 1884 (FC) = ASTEROCYTIS ORNATA (Ag.) Hamel. FIG. 306.

Chroothece Richteriana var. *aquatica* Hansgirg, Oesterr. Bot. Zeitschr. 36: 333. 1886. —Type from Bohemia: Auzitz nächst Kralup, *A. Hansgirg,* Jul. 1886 (W) = ASTEROCYTIS ORNATA (Ag.) Hamel. FIG. 309.

Chroothece Richteriana f. *marina* Hansgirg, Oesterr. Bot. Zeitschr. 39: 5. 1889. —The original description is here designated as the temporary Type = ASTEROCYTIS ORNATA (Ag.) Hamel.

Chroothece rupestris Hansgirg, Oesterr. Bot. Zeitschr. 36: 110. 1886. *C. monococca* var. *rupestris* Hansgirg, Sitzungsber. K. Böhm. Ges. Wiss., Math.-Nat. Cl., 1891(1): 354. 1891. *C. monococca* b. *rupestris* Hansgirg, Prodr. Algenfl. Böhmen 2: 135. 1892. *C. monococca* f. *rupestris* Hansgirg ex Forti, Syll. Myxophyc., p. 31. 1907. —Type from Bohemia: Chuchelbad, *A. Hansgirg,* Nov. 1885 (W) = ASTEROCYTIS ORNATA (Ag.) Hamel.

Chroothece Willei Gardner, Mem. New York Bot. Gard. 7: 3. 1927. —Type from Puerto Rico: Hato Arriba, Arecibo, *N. Wille 1410* (NY) = ASTEROCYTIS ORNATA (Ag.) Hamel.

Clastidium cylindricum Whelden in Polunin, Bot. Canadian East. Arctic 2: 26. 1947. —The original description has been designated as the Type = AMPHITHRIX JANTHINA (Mont.) Born. & Flah.

Clastidium setigerum var. *rivulare* Hansgirg, Sitzungsber. K. Böhm. Ges. Wiss., Math.-Nat. Cl., 1890(1): 18. 1890. *C. rivulare* Hansgirg, Prodr. Algenfl. Böhmen 2: 125. 1892. —Type from Istria: Strogniano, *A. Hansgirg,* Apr. 1889 (W) = AMPHITHRIX sp.

Clathrocystis holsatica var. *minor* Lemmermann, Abh. Naturw. Ver. Bremen 18: 151. 1905. *Microcystis holsatica* var. *minor* Lemmermann, Krypt.-Fl. Mark Brandenb. 3: 77. 1907. *M. holsatica* f. *minor* Hollerbach, Acta Inst. Bot. Acad. Sci. U. R. P. S. S., ser. 2, Pl. Crypt. 3: 179. 1936. *M. pulverea* f. *minor* Hollerbach in Elenkin, Monogr. Algar. Cyanophyc., Pars Spec. 1: 123, 124. 1938. —The original description is here designated as the temporary Type = BACTERIA.

CLOSTERIOCOCCUS Schmidle, Allgem. Bot. Zeitschr. 1905: 64. 1906. —Type species: *C. virnheimensis* Schmidle, loc. cit. 1906. —The original description is here designated as the temporary Type = CHLOROPHYCEAE.

Coccochloris globosa Meneghini, Mem. R. Accad. Sci. Torino, ser. 2, 5(Sci. Fis. & Mat.): 46, 69. 1843. *Palmella globosa* Lenormand *pro synon.* in Meneghini, loc. cit. 1843. *Brachtia globosa* Trevisan, Sagg. Monogr. Alg. Coccot., p. 58. 1848. —Type from France: Vire, *herb. Lenormand* (FI) = TRACHELOMONAS sp.

Coccochloris Orsiniana Meneghini, N. Giorn. de' Letterati 38: 67. 1839. *Palmella Orsiniana* Kützing, Tab. Phyc. 1: 12. 1846. —Type from Italy: "Coccochloris Orsiniana Mgh." (FI) = FERROBACTERIACEAE.

Coccodea viridis Palisot de Beauvois in Link, Sitzungsber. Ges. Naturf. Fr. Berlin 1839—59: 12. 1839. —The original description is here designated as the temporary Type = BOTRYOCOCCUS sp.

Coccophysium expallens Link, Handb. z. Erkenn. Gewächse 3: 341. 1833. *Protococcus expallens* Rabenhorst, Deutschl. Krypt.-Fl. 2(2): 11. 1847. —The original description is here designated as the Type = CHLOROPHYCEAE.

COELOMORON Buell, Bull. Torr. Bot. Club 65: 379. 1938. —Type species: *C. regularis* Buell, loc. cit. 1938. —Para-topotype from Minnesota here designated as the Type: in Lake of the Isles lagoon, Minneapolis, *M. F. Buell, R. L. Cain, & M. F. Moore,* 26 Jul. 1938 (FC) = LAMPROCYSTIS ROSBA (Kütz.) Dr. & Daily.

Coelosphaerium compositum Liebetanz, Bull. Int. Acad. Cracovie, Cl. Sci. Math. & Nat., ser. B, 1925: 109. 1925. —The original description is here designated as the temporary Type = LAMPROCYSTIS ROSEA (Kütz.) Dr. & Daily.

Coelosphaerium confertum W. &. G. S. West, Journ. of Bot. 34: 382. 1896. —Type from Central Africa: Mwangadan river south of Fuladoga, *J. W. Gregory,* 1893 (BIRM) = LAMPROCYSTIS ROSEA (Kütz.) Dr. & Daily.

Coelosphaerium Dicksonii Archer, Quart. Journ. Microsc. Sci., N. S. 19: 440. 1879. —The original description is here designated as the temporary Type = NOSTOC sp.

Coelosphaerium dubium f. *majus* Thomasson, Sv. Bot. Tidskr. 46: 233. 1952. —Isotype from Sweden: Bessvatn, K. *Thomasson*, 18 Jul. 1951 (DA) = FLAGELLATA.

Coelosphaerium Leloupii Kufferath, Bull. Inst. Roy. Sci. Nat. Belg. 26(2): 5. 1950. — Topotype from Belgium: Boirs-sur-Geer east of Tongeren, L. *Brongersma*, 26 Sept. 1951 (D, DA, FC) = LAMPROCYSTIS ROSEA (Kütz.) Dr. & Daily. FIG. 363.

Coelosphaerium minutissimum Lemmermann, Ber. Deutsch. Bot. Ges. 18: 98. 1900. —The original description is here designated as the temporary Type = LAMPROCYSTIS ROSEA (Kütz.) Dr. & Daily.

Coelosphaerium minutissimum var. *liliaceum* Kufferath, Bull. Soc. Roy. Bot. Belg. 74: 95. 1942. —The original description is here designated as the Type = BACTERIA.

LEMMERMANNIA Elenkin, Acta Inst. Bot. Acad. Sci. U. R. P. S. S., ser. 2, Pl. Crypt. 1: 25. 1933. *LEMMERMANNIELLA* Geitler, in Engler & Prantl, Natürl. Pflanzenfam., ed. 2, Ib: 62. 1942. —Type species: *Coelosphaerium pallidum* Lemmermann, Bot. Centralbl. 76: 154. 1898. *Lemmermannia pallida* Elenkin, Acta Inst. Bot. Acad. Sci. U. R. P. S. S., ser. 2, Pl. Crypt. 1: 25. 1933. *Lemmermanniella pallida* Geitler in Engler & Prantl, Natürl. Pflanzenfam., ed. 2, Ib: 62. 1942. —The original description is here designated as the temporary Type = BACTERIA.

COCCOTALES Tribus *HYDRUREAE* Subtribus *EUHYDRUREAE* Trevisan, Sagg. Monogr. Alg. Coccot., p. 16, 71. 1848. —Type genus: *HYDRURUS* Agardh, Syst. Algar., p. xviii. 1824. —Type species: *Conferva foetida* Villars, Hist. Pl. Dauphiné 3: 1010. 1789. *Ulva foetida* Vaucher, Hist. Conf. d'Eau Douce, p. 244. 1803. *Rivularia foetida* Lamarck & De Candolle, Fl. Franc., ed. 3, 2: 5. 1805. *Hydrurus Vaucheri* Agardh, Syst. Algar., p. 24. 1824. *Bangia foetida* Sprengel, Linn. Syst. Vegetab. 4(1): 361. 1827. *Girodella foetida* Gaillon, Aperçu d'Hist. Nat., p. 35. 1833. *Hydrurus foetidus* Trevisan, Sagg. Monogr. Alg. Coccot., p. 75. 1848. —The original description is here designated as the temporary Type = PALMELLA MYOSURUS (Ducluz.) Lyngb.

Conferva ornata Agardh, Syst. Algar., p. 104. 1824. *Asterocytis ornata* Hamel, Rév. Algol. 1: 451. 1924. *Goniotrichum ornatum* Bornet *pro synon.* in Hamel, loc. cit. 1924. —Type labeled "Conf. ornata ded. Agardh" (S) = ASTEROĊYTIS ORNATA (Ag.) Hamel. FIG. 304.

CRYPTELLA Pascher, Jahrb. f. Wiss. Bot. 71(3): 459. 1929. —Type species: *C. cyanophora* Pascher, loc. cit. 1929. —The original description is here designated as the temporary Type = FLAGELLATA.

Cryptococcus brunneus Kützing, Sp. Algar., p. 146. 1849. —Type from Italy: in fontibus di Acqua Santa, *Meneghini 2* (L) = BACTERIA.

Cryptococcus coccineus Kützing, Sp. Algar., p. 146. 1849. —Type from Italy: in fonte presso la Capella ai Bagni di Armajolo (L) = BACTERIA.

Cryptococcus guttulatus Robin, Hist. Nat. Végét. Parasit., Atlas p. 7, pl. 3, fig. 5, pl. 6, fig. 2. 1853. —The original description is here designated as the temporary Type = BACTERIA.

Cryptococcus lobatus Roemer, Alg. Deutschl., p. 72. 1845. —The original description is here designated as the temporary Type = BACTERIA.

Cryptococcus major Kützing in Meneghini, Conspect. Algol. Eugan., p. 323. 1837. —Type from the Euganean springs, Italy: ad superficiem aquae etc., *Meneghini* (L) = FLAGELLATA.

Fam. *CRYPTOCOCCEAE* Kützing, Phyc. Gener., p. 147. 1843. —Type genus: *CRYPTOCOC-CUS* Kützing, Algar. Aq. Dulc. Germ. Dec. 3: 28. 1833. —Type species: *C. mollis* Kützing, loc. cit. 1833. —Type from Germany: Schleusingen, 6 Mar. 1832 (L) = BACTERIA and FUNGI.

Cryptococcus natans Kützing, Sp. Algar., p. 146. 1849. —Type from Holland: in superf. aqua, *van den Bosch 259* (L) = BACTERIA.

Cryptococcus sanguineus Kützing, Sp. Algar., p. 891. 1849. —Type from Holland: in limo maritimo, *van den Bosch 471* (L) = LAMPROCYSTIS ROSEA (Kütz.) Dr. & Daily.

CYANARCUS Pascher, Ber. Deutsch. Bot. Ges. 32: 351. 1914. —Type species: *C. hamiformis* Pascher, loc. cit. 1914. —The original description is here designated as the temporary Type = BACTERIA.

Cyanococcus pyrenogerus Hansgirg, Beih. z Bot. Centralbl. 18(2): 521. 1905. The original description is here designated as the temporary Type = BANGIACEAE or "syncyanose" according to Geitler in Engler & Prantl. Pflanzenfam., ed. 2, lb: 224. 1942.

CYANODICTYON Pascher, Ber. Deutsch. Bot. Ges. 32: 351. 1914. —Type species: *C. endophyticum* Pascher, loc. cit. 1914. —The original description is here designated as the temporary Type = degenerate trichomes of NOSTOCACEAE.

CYANOPHORA Korschikov, Arch. Russes Protistol. 3(1—2): 71. 1924. —Type species: *C. paradoxa* Korschikov, loc. cit. 1924. —The original description is here designated as the temporary Type = FLAGELLATES.

CYANOPTYCHE Pascher, Jahrb. f. Wiss. Bot. 71(3): 459. 1929. —Type species: *C. Gloeocystis* Pascher, ibid. p. 460. 1929. —The original description is here designated as the temporary Type = CHLOROPHYCEAE.

CYANOTHECA Pascher, Ber. Deutsch. Bot. Ges. 32: 351. 1914. —Type species: *C. longipes* Pascher, loc. cit. 1914. —The original description is here designated as the temporary Type = FLAGELLATA or FUNGI.

Cylindrocystis Le-Normandii Meneghini, Mem. R. Accad. Sci. Torino, ser. 2, 5(Sci. Fis. & Mat.):
92. 1843. —Type from France: Falaise, *herb. Lenormand* (FI) = spores of ANABAENA
OSCILLARIOIDES Bory.
CYSTOCOCCUS Nägeli, Gatt. Einzell. Alg., p. 84. 1849. —Type species: *C. humicola*
Nägeli, ibid. p. 85. 1849. *Chlorococcum humicola* Rabenhorst, Fl. Eur. Algar. 3: 58. 1868.
Protococcus viridis var. *humicola* Wolle, Fresh-w. Alg. U. S., p. 182. 1887. —Type from
Switzerland: Zürich, Burgholzli, am Fusse einer Buche auf feuchter Erde, *Nägeli*, Feb. 1846 (ZT) =
STICHOCOCCUS SUBTILIS (Kütz.) Klerck.
Dactylococcopsis acicularis Lemmermann, Ber. Deutsch. Bot. Ges. 18: 309. 1900. —The
original description is here designated as the temporary Type = ANKISTRODESMUS FALCATUS (Corda)
Ralfs.
Dactylococcopsis acicularis var. *grandis* Frémy, Arch. de Bot. Caen 3(Mem. 2): 8. 1930.
—The original description is here designated as the temporary Type = ANKISTRODESMUS FALCATUS
(Corda) Ralfs.
Dactylococcopsis africana G. S. West, Journ. Linn. Soc. Bot. 38: 184. 1907. —Type from
Lake Victoria Nyanza: in plankton near Bukoba, *W. A. Cunnington 252*, 21 Apr. 1905 (in the
slide collection, BIRM) = ANKISTRODESMUS FALCATUS (Corda) Ralfs.
Dactylococcopsis arcuata Gardner, Mem. New York Bot. Gard. 7: 3. 1927. —Type from
Puerto Rico: Caguas, *N. Wille 439c* (NY) = PLECTONEMA NOSTOCORUM Born. FIG. 365.
Dactylococcopsis Echinii Rosenvinge, K. Danske Videnskab. Selsk., Biol. Medd. 11(7): 10.
1934. —The original description is here designated as the temporary Type = FUNGI or CHLORO-
PHYCEAE.
Dactylococcopsis fascicularis Lemmermann, Bot. Centralbl. 76: 153. 1898. —The original
description is here designated as the temporary Type = ANKISTRODESMUS FALCATUS (Corda) Ralfs.
Dactylococcopsis irregularis G. M. Smith, Ark. f. Bot. 17(13): 6. 1922. —The original
description is here designated as the temporary Type = ANKISTRODESMUS FALCATUS (Corda) Ralfs.
Dactylococcopsis libyca Marchesoni, Mem. Ist. Ital. Idrobiol. Milano 3: 445. 1947. —Type
from Libya: in lacubus salatis Murzuch, *R. Cort*, May 1933 (DA) = FUNGUS SPORES.
Dactylococcopsis mucicola Hustedt, Hedwigia 48: 140. 1908. —The original description is
here designated as the temporary Type = ANKISTRODESMUS FALCATUS (Corda) Ralfs.
Dactylococcopsis pectinatellophila West, Journ. & Proc. Asiatic Soc. Bengal, N. S., 7: 83.
1911. —The original description is here designated as the temporary Type = CHLOROPHYCEAE.
Dactylococcopsis rhaphidioides Hansgirg, Notarisia 3: 540. 1888. —Type from Bohemia:
Schanzgräben von Prag, *A. Hansgirg*, Nov. 1883 (W) = spores of STICHOCOCCUS SUBTILIS (Kütz.)
Klerck. FIG. 359.
Dactylococcopsis rhaphidioides f. *falciformis* Printz, K. Norske Vidensk. Selsk. Skr. 1920(1):
35. 1920. —The original description is here designated as the temporary Type = ANKISTRODESMUS
FALCATUS (Corda) Ralfs.
Dactylococcopsis rhaphidioides f. *mucicola* Frémy, Arch. de Bot. Caen 3(Mem. 2): 8. 1930.
—The original description is here designated as the temporary Type = ANKISTRODESMUS FALCATUS
(Corda) Ralfs.
Dactylococcopsis rhaphidioides f. *subtortuosa* Printz, K. Norske Vidensk. Selsk. Skr. 1920(1):
35. 1920. —The original description is here designated as the temporary Type = ANKISTRODESMUS
FALCATUS (Corda) Ralfs.
DACTYLOCOCCOPSIS Hansgirg, Notarisia 3: 590. 1888. —Type species: *D. rupestris*
Hansgirg, loc. cit. 1888. —Type from Bohemia: Karlstein, *A. Hansgirg*, Jul. 1884 (W) = spores
of STICHOCOCCUS SUBTILIS (Kütz.) Klerck.
Dactylococcopsis rupestris f. *sigmoides* Ley, Bot. Bull. Acad. Sinica 2: 237. 1948. —The
original description is here designated as the temporary Type = primordia of CHLOROPHYCEAE.
KERATOCOCCUS Pascher, Süsswasserfl. 5(Chlorophyc. 2): 216. 1915. —Type species:
Dactylococcus caudatus Hansgirg, Physiol. & Algol. Stud., p. 86. 1887. *Keratococcus caudatus*
Pascher, Süsswasserfl. 5(Chlorophyc. 2): 216. 1915. —Type from Bohemia: Horovice, *A. Hansgirg*,
Aug. 1884 (W) = spores of STICHOCOCCUS SUBTILIS (Kütz.) Klerck.
Dactylococcus caudatus var. *minor* Hansgirg, Prodr. Algenfl. Böhmen 1(2): 146. 1888.
—Type from Bohemia: Hohenfurth, *A. Hansgirg*, Aug. 1884 (W) = spores of STICHOCOCCUS
SUBTILIS (Kütz.) Klerck.
Dactylococcus rhaphidioides Hansgirg, Oesterr. Bot. Zeitschr. 33: 122. 1887; Physiol. & Algol.
Stud., p. 93. 1887. *Keratococcus rhaphidioides* Pascher, Süsswasserfl. 5(Chlorophyc. 2): 218. 1915.
—Type from Bohemia: Mummelfall, *A. Hansgirg*, Jul. 1886 (W) = spores of STICHOCOCCUS
SUBTILIS (Kütz.) Klerck.
Dactylothece macrococca Hansgirg, Sitzungsber. K. Böhm. Ges. Wiss., Math.-Nat. Cl., 1891(1):
323. 1891. —Type from Bohemia: Nieder- und Mittelgrund, *A. Hansgirg* (W) = palmelloid
STICHOCOCCUS SUBTILIS (Kütz.) Klerck.
Dermocarpa fossae Fjerdingstad, Dansk Bot. Ark. 14(3): 34. 1950. —The original de-
scription is here designated as the temporary Type = CHARACIOPSIS sp.

DISCERAEA A. & C. Morren, Rech. Rubéfac. des Eaux, p. 37. 1841. —Type species: *D. purpurea* A. & C. Morren, loc. cit. 1841. —The original description is here designated as the temporary Type = FLAGELLATES.

Endonema gracile Pascher, Jahrb. f. Wiss. Bot. 70: 347. 1929. *Pascherinema gracile* J. de Toni, Noter. Nomencl. Algol. 8. 1936. —The original description is here designated as the temporary Type = FUNGI.

ENDONEMATACEAE Pascher, Jahrb. f. Wiss. Bot. 70: 346. 1929. *PASCHERINEMATACEAE* Geitler in Engler & Prantl, Natürl. Pflanzenfam., ed. 2, 1b: 99. 1942. —Type genus: *ENDONEMA* Pascher, Jahrb. f. Wiss. Bot. 70: 346. 1929. *PASCHERINEMA* J. de Toni, Noter. Nomencl. Algol. 8. 1936. —Type species: *Endonema moniliforme* Pascher, Jahrb. f. Wiss. Bot. 70: 346. 1929. *Pascherinema moniliforme* J. de Toni, Noter. Nomencl. Algol. 8. 1936. —The original description is here designated as the temporary Type = FUNGI.

Entophysalis zonata Gardner, Univ. Calif. Publ. Bot. 13: 369. 1927. —Type from Fukien province, China: Kushan near Foochow, *H. H. Chung F617*, Sept. 1926 (FH) = CAPSOSIRA BREBISSONII Kütz. FIG. 373.

ERNSTIELLA Chodat, Bull. Soc. Bot. Génève, ser. 2, 3: 126. 1911. —Type species: *E. rufa* Chodat, loc. cit. 1911. —The original description is here designated as the temporary Type = AMPHITHRIX sp.

ERYTHROCONIS Oersted, Naturh. Tidsskr. Kjøbenh. 3: 55. 1841. —Type species: *E. littoralis* Oersted, loc. cit. 1841. *Merismopoedia littoralis* Rabenhorst, Fl. Eur. Algar. 2: 57. 1865. *Sarcina littoralis* Winter, Rabenh. Krypt.-Fl. 1(1): 50. 1884. *Lampropedia littoralis* de Toni & Trev. in Saccardo, Syll. Fung. 8: 1049. 1889. —Type from Denmark: Kallebodstrand (C) = ERYTHROCONIS LITTORALIS Oerst.

Eucapsis minuta Fritsch, Nat. Antarctic Exped. (1901—04), Nat. Hist. 6 (Freshw. Alg.): 25. 1912. —Type from Antarctica: Cape Adare, 9 Jan. 1902 (in the collection of F. E. Fritsch) = BACTERIA.

Glaucocystis cingulata Bohlin, Bih. K. Sv. Vet.-Akad. Handl. 23(Afd. 3, 7): 13. 1897. —The original description is here designated as the temporary Type = OOCYSTACEAE.

Glaucocystis duplex Prescott, Farlowia 1: 371. 1944. —The original description is here designated as the temporary Type = OOCYSTACEAE.

GLAUCOCYSTIS Itzigsohn in Rabenhorst, Alg. Eur. 194 & 195: 1935. 1866. *CYANOCYSTIS* Rabenhorst *pro synon.*, loc. cit. 1866. —Type species: *Glaucocystis nostochinearum* Itzigsohn in Rabenhorst, loc. cit. 1866. *Cyanocystis Itzigsohniana* Rabenhorst *pro synon.*, loc. cit. 1866. *Oocystis cyanea* Nägeli *pro synon.* in Rabenhorst, loc. cit. 1866. —Type from Germany: im Grunewald, Berlin, *O. Kuntze*, Jul. 1864 (PC) = OOCYSTACEAE.

Glaucocystis nostochinearum var. *gigas* Gutwinski, Bull. Acad. Sci. Cracovie, Cl. Sci. Math. & Nat., 1909: 543. 1909. —The original description is here designated as the temporary Type = OOCYSTACEAE.

Glaucocystis nostochinearum f. *immanis* Schmidle, Engler Bot. Jahrb. 23: 79. 1902. —Type from Lake Nyasa: Mbasifluss, *Fulleborn*, 27 Apr. 1899 (ZT) = OOCYSTACEAE.

Glaucocystis oocystiformis Prescott, Farlowia 1: 372. 1944. —The original description is here designated as the temporary Type = OOCYSTACEAE.

GLAUCOSPHAERA Korshikov, Arch. f. Protistenk. 70: 222. 1930. —Type species: *G. vacuolata* Korshikov, loc. cit. 1930. —The original description is here designated as the temporary Type = CHLOROPHYCEAE.

Gloeocapsa ampla Kützing, Phyc. Gener., p. 174. 1843. *Bichatia ampla* Trevisan, Nomencl. Algar. 1: 60. 1845. *Gloeocystis ampla* Rabenhorst, Krypt.-Fl. Sachsen 1: 128. 1863. *Pleurococcus amplus* Rabenhorst, Fl. Eur. Algar. 3: 29. 1868. —Type from Germany: Nordhausen, *in herb. Kützing* (L) = palmelloid FLAGELLATES. ..FIG. 351.

Gloeocapsa conspicua Reinsch, Algenfl. Mittl. Th. Franken, p. 33. 1867. —The original description is here designated as the temporary Type = CHLOROPHYCEAE.

Gloeocapsa dubia Wartmann in Rabenhorst, Alg. Eur. 109—110: 1092. 1861. *Bichatia dubia* Richter in Kuntze, Rev. Gen. Pl. 2: 886. 1891. *Gloeothece dubia* Geitler, Rabenh. Krypt.-Fl. 14: 216. 1932. —Type from Switzerland: St. Gallen, *Wartmann* (FC) = ASTEROCYTIS ORNATA (Ag.) Hamel. FIG. 308.

Gloeocapsa fenestralis Kützing, Phyc. Gener., p. 173. 1843. *Bichatia fenestralis* Trevisan, Sagg. Monogr. Alg. Coccot., p. 63. 1848. *Gloeocystis fenestralis* A. Braun in Rabenhorst, Algen Eur. 246—248: 2468. 1876. *Gloeocapsa montana* f. *fenestralis* Hollerbach in Elenkin, Monogr. Algar. Cyanophyc., Pars Spec., 1: 226. 1938. —Type from Germany: an Mistbeet-Fenstern, Nordhausen, *F. T. Kützing* (L) = STICHOCOCCUS BACILLARIS Näg.

GLOEOTHECE Sect. *CHROMOTHECE* Kirchner in Engler & Prantl, Natürl. Pflanzenfam., ed. 1, 1a: 55. 1898. *GLOEOTHECE* Subgen. *CHROMOTHECE* Kirchner ex Forti, Syll. Myxophyc., p. 65. 1907. —Type species: *Gloeocapsa monococca* Kützing, Phyc. Gener., p. 175. 1843. *Bichatia monococca* Trevisan, Nomencl. Algar. 1: 61. 1845. *Gloeocapsa monococca* β *cohaerens* Kützing, Tab. Phyc. 1: 18. 1846. *Palmogloea monococca* Kützing, Sp. Algar., p. 229. 1849. *P. monococca* ∝ *aeruginea* Kützing, loc. cit. 1849. *Gloeothece monococca* Rabenhorst, Fl. Eur. Algar. 2: 62. 1865. *Chroothece monococca* Hansgirg, Sitzungsber. K. Böhm. Ges. Wiss., Math.-Nat. Cl., 1891(1): 354. 1891. —Type from Istria: in den Cataracten des Boschetto, Triest, *herb. Kützing* (L) = ASTEROCYTIS ORNATA (Ag.) Hamel.

Gloeocapsa stillicidiorum Kützing, Phyc. Gener., p. 173. 1843. *Bichatia stillicidiorum* Trevisan, Nomencl. Algar. 1: 62. 1845. —Type from Germany: Steinmühle Hercyniae, *F. T. Kützing* (L) = primordia of CHLOROPHYCEAE (STIGEOCLONIUM?). FIG. 350.

Gloeocapsa uvaeformis Roemer, Alg. Deutschl., p. 63. 1845. *Bichatia uvaeformis* Trevisan, Sagg. Monogr. Alg. Coccot., p. 67. 1848. —The original description is here designated as the temporary Type = CHLOROPHYCEAE.

Gloeocapsa zostericola Farlow, Bull. Torr. Bot. Club 9: 68. 1882. —An isotype: Woods Hole, Massachusetts, *W. G. Farlow* (S) = DINOFLAGELLATA.

Gloeochaete bicornis Kirchner, Jahresh. Ver. Vaterl. Naturk. Württemb. 44: 165. 1888. —The original description is here designated as the temporary Type = CHLOROPHYCEAE.

GLOEOCHAETE Lagerheim, Öfvers. K. Sv. Vet.-Akad. Förh. 40(2): 39. 1883. —Type species: *G. Wittrockiana* Lagerheim, loc. cit. 1883. —The original description is here designated as the temporary Type = CHLOROPHYCEAE.

Gloeococcus agilis Grunow in Rabenhorst, Fl. Eur. Algar. 3: 36. 1868. —Type from Austria: in einem kleinen Bache bei Berndorf, *A. Grunow*, 1 Sept. 1856 (PC) = FLAGELLATES.

Gloeocystis gigas var. *rufescens* A. Braun ex Hansgirg, Prodr. Algenfl. Böhmen 1: 136. 1886. —Type from Bohemia: Pisek, *A. Hansgirg* (W) = palmelloid FLAGELLATES.

Gloeocystis marina Hansgirg, Oesterr. Bot. Zeitschr. 39: 42. 1889. — Type from Istria: Pola, *A. Hansgirg*, 1888 (W) = basal parts of MONOSTROMA sp. FIG. 343.

Gloeocystis scopulorum Hansgirg, Sitzungsber. K. Böhm. Ges. Wiss., Math.-Nat. Cl., 1892: 239. 1892. —Type from Croatia: Zengg, *A. Hansgirg*, 1891 (W) = primordia of CHLOROPHYCEAE. FIG. 342.

Gloeothece Baileyana Schmidle, Flora 82: 312. 1896. —Type from Queensland, Australia: Jenkins quarry, Bundamba, *W. J. Byram* (ZT) = palmelloid FLAGELLATES. FIG. 340.

Gloeothece cystifera var. *maxima* West, Trans. Roy. Microsc. Soc. 1892: 743. 1892. *G. rupestris* var. *maxima* West ex Forti, Syll. Myxophyc., p. 64. 1907. *G. rupestris* f. *maxima* Hollerbach in Elenkin, Monogr. Algar. Cyanophyc., Pars Spec. 1: 250. 1938. —Type from the English Lake District: Bowness (BIRM) = ANABAENA FLOS-AQUAE (Lyngb.) Bréb., spores.

Gloeothece linearis f. *angusta* West, Proc. Roy. Irish Acad. 31: 40. 1912. —Type from Ireland: Lough Gall, Achill island (BIRM) = BACTERIA.

Gloeothece magna Wolle, Bull. Torr. Bot. Club 6: 138. 1877. —Type from Pennsylvania: with *Micrasterias multifida* Wolle in a Desmideen-Aufsammlung (Rabenh., Alg. Eur. 2510, FC) = palmelloid CHLOROPHYCEAE.

Gloeothece minor Beck in Becker, Hernstein in Niederösterreich, II, Fl. des Gebietes, p. 266. 1886. —Type from Hernstein, Lower Austria: hint. Grillenbergthal, *Beck 365* (in the Cryptogamic Herbarium, Charles University, Prague) = BACTERIA. FIG. 341.

Gloeothece trichophila Rabenhorst, Hedwigia 1867: 49. 1867. —Type from England: London, an Chignon-Haaren, *Beigel* (with *Pleurococcus Beigelii* in Rabenh. Alg. 2053, NY) = BACTERIA.

Gloeothece violacea Rabenhorst, Fl. Eur. Algar. 2: 61. 1865. —The original description is here designated the temporary Type = BACTERIA.

GLOIODICTYON Agardh, Consp. Crit. Diatomac. 2: 25. 1830. —Type species: *G. Blyttii Agardh*, ibid. p. 26. 1830. *Schizonema reticulatum* Agardh *pro synon.*, loc. cit. 1830. —Type from Norway: Tofledaljan, Nowi Trondhjem, *Blytt* (LD) = sterile ZYGNEMA sp.

Gloionema globiferum Agardh, Consp. Crit. Diatomac. 2: 31. 1830. —Type from Traneberg (LD) = DIATOMS in the sheaths of OPHRYDIUM sp.

Gloionema Leibleinii Agardh, Consp. Crit. Diatomac. 2: 31. 1830. —Type: *Leiblein 17* (LD) = DIATOMS.

GLOIONEMA Agardh, Dispos. Algar. Suec., p. 45. 1882. —Type species: *G. paradoxum* Agardh, loc. cit. 1812. —Type from Sweden: Jäder (LD) = EGGS OF ANIMALS.

Gloionema vermiculare Agardh, Flora 10(2): 630. 1827. —Type from Bohemia: Carlsbad (LD) = DIATOMS.

Gomphosphaeria aurantiaca Bleisch in Rabenhorst, Alg. Eur. 181—182: 1810. 1865. *G. aponina* var. *aurantiaca* Bleisch ex Kirchner, Krypt.-Fl. Schlesien 2(1): 255. 1878. *G. aponina* f. *aurantiaca* Forti, Syll. Myxophyc., p. 98. 1907. —Type from Silesia (Poland): auf dem Galgenberge bei Strehlen, *Bleisch* (L) = BOTRYOCOCCUS BRAUNII Kütz.

LAMPROPEDIA Schröter, Pilze Schlesiens 1: 151. 1886. —Type species: *Gonium hyalinum* Ehrenberg, Abh. K. Akad. Wiss. Berlin 1831: 75. 1832. *Merismopoedia hyalina* Kützing, Phyc. Germ., p. 142. 1845. *Sarcina hyalina* Winter, Rabenh. Krypt.-Fl. 1(1): 51. 1884. *Lampropedia hyalina* Schröter, Pilze Schlesiens 1: 151. 1886. —The original description is here designated the temporary Type = BACTERIA.

Gonium punctatum Ehrenberg, Abh. K. Akad. Wiss. Berlin 1833: 250. 1834. *Gonidium punctatum* Meneghini, Linnaea 14: 214. 1840. —The original description is here designated as the temporary Type = FLAGELLATES.

Haematococcus arenarius Hassall, Hist. Brit. Freshw. Alg. 1: 330. 1845. *Bichatia arenaria* Trevisan, Sagg. Monogr. Alg. Coccot., p. 66. 1848. *Gloeocapsa arenaria* Rabenhorst, Fl. Eur. Algar. 2: 39. 1865. —Type from England: Tunbridge Wells common, Dec. 1844 (BM) = palmelloid STICHOCOCCUS SUBTILIS (Kütz.) Klerck.

Haematococcus atomarius Flotow, Nova Acta Acad. Caes. Leop.-Car. Nat. Cur. 20(2): 424. 1842. *H. pluvialis* var. *atomarius* Flotow *pro synon.*, loc. cit. 1842. —The original description is here designated as the temporary Type = PROTOCOCCUS GREVILLEI (Ag.) Crouan.

Haematococcus fuliginosus Meneghini, Consp. Algol. Eugan., p. 323. 1837. *Protococcus fuligineus* Lenormand in Kützing, Tab. Phyc. 1: 3. 1846. *Chroococcus fuligineus* Rabenhorst, Fl. Eur. Algar. 2: 34. 1865. —Kützing's notation of the original locality is obviously an editorial error, since his name is based upon the same specimen as is Meneghini's. The Type is from Italy: aquis thermalibus Euganeis (L) = BACTERIA. FIG. 370.

COCCOTALES Tribus *PROTOCOCCEAE* Subtribus *HAEMATOCOCCEAE* Trevisan, Sagg. Monogr. Alg. Coccot., p. 16, 37. 1848. —Type genus: *HAEMATOCOCCUS* Agardh, Icon. Alg. Eur., no. 22. 1828—35. Synon.: *GLOIOCOCCUS* Shuttleworth, Bibl. Univ. Génève 25: 405. 1840. —Type species: *Haematococcus Grevillei* Agardh, Icon. Alg. Eur., no. 23. 1828—35. *Microcystis Grevillei* Kützing, Linnaea 8: 372. 1833. *Gloiococcus Grevillei* Shuttleworth, Bibl. Univ. Génève 25: 405. 1840. *Protosphaeria Grevillei* Trevisan, Sagg. Monogr. Alg. Coccot., p. 28. 1848. *Protococcus Grevillei* Crouan fr., Fl. Finistère, p. 109. 1867. —The type species of *Haematococcus* Ag. was designated by Silva & Starr, Sv. Bot. Tidskr. 47(2): 246 (1953). The Type is from Scotland: "Palmella nivalis" det. Greville, island of Lismore, *Carmichael* (K) = PROTOCOCCUS GREVILLEI (Ag.) Crouan. FIG 372.

Haematococcus insignis Hassall, Hist. Brit. Freshw. Alg. 1: 324. 1845. *Urococcus insignis* Hassall ex Kützing, Sp. Alg., p. 207. 1849. *Hassallia insignis* Trevisan, Sagg. Monogr. Alg. Coccot., p. 67. 1848. —The original description is here designated as the temporary Type = encysted FLAGELLATES.

Haematococcus mucosus Morren, Rech. Rubéf. des Eaux, p. 104. 1841. *Microcystis Morrenii* Meneghini, Mem. R. Accad. Sci. Torino, ser. 2, 5(Sci. Fis. & Mat.): 87. 1843. *Bichatia mucosa* Trevisan, Nomencl. Algar. 1: 61. 1845. —The original description is here designated as the temporary Type = FLAGELLATES.

MICROCYSTACEAE Elenkin, Acta Inst. Bot. Acad. Sci. U. R. S. S., II, 1: 19. 1923. *MICROCYSTIDACEAE* Elenkin, Monogr. Algar. Cyanophyc., Pars Spec. 1: 93. 1938. —Type genus: *MICROCYSTIS* Kützing, Linnaea 8: 372. 1833. *MICROCYSTIS* Unterabtheilung *MICRO-CYSTIS* Kützing, Tab. Phyc. 1: 7. 1846. *MICROCYSTIS* Subgenus *EUMICROCYSTIS* Elenkin, Not. Syst. Inst. Crypt. Horti Bot. Petropol. 2: 67. 1923. —Type species: *Haematococcus Noltii* Agardh, Icon. Alg. Eur., no. 22. 1828—35. *Microcystis Noltii* Kützing, Linnaea 8: 372. 1833. —Geitler in Engler & Prantl, Natürl. Pflanzenfam. ed. 2, 1b: 46 (1942) does not irrevocably designate *M. aeruginosa* as the "Leitart" of *Microcystis* Kütz. Kützing transferred his *Micraloa aeruginosa* into *Microcystis* ten years after he had described both; in another six years he transferred it into his genus *Polycystis*. The type species designated here is historically the correct one: *Microcystis Noltii* Kütz. Isotypes from Schleswig collected by Nolte are in several herbaria (FC, L, UC) = EUGLENA sp.

Haematococcus Orsinii Meneghini, Mem. R. Accad. Sci. Torino, ser. 2, 5(Sci. Fis. & Mat.): 22. 1843. *Protococcus Orsinii* Kützing, Phyc. Germ., 146. 1845. *Protosphaeria Orsinii* Trevisan, Sagg. Monogr. Alg. Coccot., p. 29. 1848. *Chroococcus Orsinii* Rabenhorst, Alg. Eur. 127—128: 1269. 1862. *Chlorococcum Orsinii* Meneghini ex Rabenhorst, Fl. Eur. Algar. 3: 60. 1868. —Isotype from Italy: Acqua Santa, *ex herb. Meneghini* (L) = EUGLENA sp.

Haematococcus pluvialis Flotow, Nova Acta Caes. Leop.-Car. Nat. Cur. 20(2): 415. 1842. *Protococcus pluvialis* Kützing, Phyc. Germ., p. 146. 1845. *P. pluvialis* ∝ *leprosus* Kützing, loc. cit. 1845. *Protosphaeria pluvialis* Trevisan, Sagg. Monogr. Alg. Coccot., p. 28. 1848. *Chlamidococcus pluvialis* A. Braun, Betracht. über Erschein. Verjüng. i. d. Natur, p. 147. 1849—50. —Topotype from Bohemia: Hirschberg, *Flotow* (in Rabenh. Alg. 71, FC) = PROTOCOCCUS GREVILLEI (Ag.) Crouan.

Haematococcus vesiculosus Morren, Rech. Rubéf. des Eaux, p. 104. 1841. *Coccochloris vesiculosa* Meneghini, Mem. R. Accad. Sci. Torino, ser. 2, 5(Sci. Fis. & Mat.): 70. 1843. —The original description is here designated as the temporary Type = FLAGELLATES.

HETEROCYANOCOCCUS Kufferath, Ann. Crypt. Exot. 2: 51. 1929. —Type species: *H. Haumanii* Kufferath, loc. cit. 1929. —The original description is here designated as the temporary Type = NOSTOC sp.

ASTEROCYTIS Gobi, Trudy St. Petersb. Obschech. Estestvoisp. 10: 86. 1879. *GLAUCONEMA* Reinhardt, Algol. Izsladov., p. 240. 1885. *ALLOGONIUM* Sect. *ASTEROCYTIS* Hansgirg, Physiol. & Algol. Stud., p. 109. 1887. *ALLOGONIUM* Subgenus *ASTEROCYTIS* Hansgirg, Notarisia 3: 588. 1888. —Type species: *Hormospora ramosa* Thwaites in Harvey, Phycol. Brit. 4: pl. CCXIII. 1849. *Asterocytis ramosa* Gobi, Trudy St. Petersb. Obschech. Estestvoisp. 10: 86. 1879. *Glauconema ramosum* Reinhardt, Algol. Izsladov., p. 240. 1885. *Goniotrichum ramosum* Hauck, Meeresalg., p. 519. 1885. *Chroodactylon ramosum* Hansgirg, Ber. Deutsch. Bot. Ges. 3: 19. 1885. *Allogonium ramosum* Hansgirg, Physiol. & Algol. Stud., p. 110. 1887. —The original description is here designated as the temporary Type = ASTEROCYTIS ORNATA (Ag.) Hamel.

HYALOCOCCUS Schröter, Pilze Schlesiens 1: 152. 1886. —Type species: *H. pneumoniae* Schröter, loc. cit. 1886. —The original description is here designated as the temporary Type = BACTERIA.

Hydrococcus ericetorum Kützing, Phyc. Germ., p. 154. 1845. *Oncobyrsa ericetorum* Rabenhorst, Fl. Eur. Algar. 2: 69. 1865. —Type from the Harz mountains, Germany: am Wege nach Sondershausen, *herb. Kützing* (L) = young LICHEN.

Hydrococcus guadelupensis Crouan fr. in Schramm & Mazé, Essai Class. Alg. de la Guadeloupe, p. 29. 1865; idem, p. 67. 1866. *Oncobyrsa guadelupensis* Crouan fr. in Mazé & Schramm, Essai Class. Alg. Guadeloupe, ed. 2, p. 13. 1870—77. —Type from La Guadeloupe: Basse Terre, *Mazé & Schramm* (PC) = NOSTOC SPHAERICUM Vauch.

Hydrococcus lacustris Brügger, Jahresber. Naturf. Ges. Graubündens, N. F., 8: 250. 1863. —The original description is here designated as the temporary Type = CHLOROPHYCEAE.

Hyella infestans Howe, Mem. Torrey Bot. Club 15: 14. 1914. *Myxohyella infestans* Geitler, Rabenh. Krypt.-Fl. 14: 381. 1932. —Type from Peru: Bay of Sechura, *Coker 157pp* (NY) = ERYTHROCLADIA sp.

Hyella terrestris Chodat, Bull. Soc. Bot. Génève, ser. 2, 13: 113. 1921. —The original description is here designated as the temporary Type = FISCHERELLA AMBIGUA (Näg.) Gom.

Hyphelia aurantiaca Libert, Crypt. Arduennae 4: 379. 1837. *Chroococcus aurantiacus* de Toni, Malpighia 1: 326. 1887. —Type from Belgium in Libert, Pl. Crypt. Arduennae no. 379 (BR) = BOTRYDIUM GRANULATUM (L) Grev. FIG. 355.

Inoderma fontanum Kützing, Phyc. Germ., p. 150. 1845. *Cagniardia fontana* Trevisan, Sagg. Monogr. Alg. Coccot., p. 51. 1848. *Inoderma lamellosum* b. *fontanum* Rabenhorst, Fl. Eur. Algar. 3: 38. 1868. *I. lamellosum* var. *fontanum* Rabenhorst ex Hansgirg, Prodr. Algenfl. Böhmen 1: 140. 1886. —Type from Weissenfels, Germany, in Kützing, Alg. Aq. Dulc. Dec. 4: 39 (L) = DIATOMS.

INODERMA Kützing, Alg. Aq. Dulc. Dec. 4: 39. 1833. —Type species: *I. lamellosum* Kützing, loc. cit. 1833. *Cagniardia lamellosa* Trevisan, Sagg. Monogr. Alg. Coccot., p. 51. 1848. —Type from Tennstaedt, Germany, in Kützing, Alg. Aq. Dulc. Dec. 4: 40 (L) = DIATOMS.

EMBRYOSPHAERIA Trevisan, Sagg. Monogr. Alg. Coccot., p. 36. 1848. —Type species: *Lepra infusionum* Schrank, Neue Ann. der Bot. 1(3): 4. 1794. *Lepraria infusionum* Schrank, Denkschr. K. Akad. Wiss. München 1811: 10. 1811. *Protococcus Monas β aquaticus* Kützing, Linnaea 8: 367. 1833. *Chlorococcus infusionum* Meneghini, Mem. R. Accad. Sci. Torino, ser. 2, 5(Sci. Fis. & Mat.): 9, 27. 1843. *Protococcus minor γ infusionum* Kützing, Tab. Phyc. 1: 3, 1846. *P. Meneghinii* Kützing, Tab. Phyc. 1: 4. 1846. *Protococcus Monas* var. *aquaticus* Kützing ex Trevisan, Sagg. Monogr. Alg. Coccot., p. 26. 1848. *Embryosphaeria Meneghinii* Trevisan, ibid., p. 37. 1848. *Chlorococcum infusionum* Brébisson *pro synon.* in Trevisan, loc. cit. 1848. *Protococcus infusionum* Kirchner, Krypt.-Fl. Schlesiens 2(1): 103. 1878. *P. viridis* var. *infusionum* Wolle, Fresh-w. Alg. U. S., p. 182. 1887. *Chlorella infusionum* Beijerinck, Bot. Zeit. 48: 758. 1890. *Gloeocystis infusionum* W. & G. S. West, Journ. of Bot. 37: 223. 1899. —The type species for *Embryosphaeria* Trevis. was designated by Silva & Starr, Sv. Bot. Tidskr. 47(2): 246 (1953). The original description of *Lepra infusionum* Schrank is here designated as the temporary Type of this species = FLAGELLATES.

Lepraria kermesina Wrangel, K. Vetenk.-Acad. Handl. 1823: 52. 1823. *Sphaerella Wrangelii* Sommerfelt, Mag. f. Naturvid. 4: 252. 1824. *Protococcus kermesinus* Agardh, Nova Acta Acad. Caes. Leop.-Car. Nat. Cur. 12(2): 749. 1825. *Byssus kermesina* Wahlenberg, Fl. Suecica 2: 924. 1833. *Haematococcus kermesinus* Flotow, Nova Acta Acad. Caes. Leop.-Car. Nat. Cur. 20(2): 426. 1842. —Type from Sweden: "Lepr. Kermesina. Wr. Act. Holm." in herb. Agardh (LD) = primordia of CHLOROPHYCEAE (STIGEOCLONIUM?).

Lichen roseus Schreber, Spicil. f. Lips., p. 140. 1771. *Lepra rosea* Hoffman *pro synon., Enum.* Lich. 1: 5. 1784. *Byssus rosea* Retz, Fl. Scand. Prodr., ed. al. p. 308. 1795. *Lepraria rosea* Acharius, Lichenogr. Suec. Prodr., p. 9. 1798. *Tubercularia rosea* Persoon, Syn. Meth. Fung. 1: 114. 1810. *Palmella rosea* Lyngbye, Tent. Hydrophyt. Dan., p. 207. 1819. *Coccochloris rosea* Sprengel, Linn, Syst. Veget., ed. 16, 4(1): 373. 1827. *Protococcus roseus* Corda in Sturm, Deutschl. Fl., Abt. 2, 6: 37. 1833. *Microcystis rosea* Kützing, Linnaea 8: 373. 1833. *Coccophysium cobaltinum* Link, Handb. Erkenn. Gewächse 3: 342. 1833. *Illosporium roseum* Link, ibid. 3: 488. 1833. *Haematococcus roseus* Meneghini, Consp. Algol. Eugan., p. 323. 1837. *Protococcus cobaltinus* Rabenhorst, Deutschl. Krypt.-Fl. 2 (2): 11. 1847. —The original description is here designated as the temporary Type = FUNGI.

COCCOPHYSIUM Link, Abh. K. Akad. Wiss. Berlin 1824: 184. 1826. —Type species: *Lichen rubens* Hoffman, Enum. Lich. 1: 4. 1784. *Lepraria rubens* Acharius, Meth. Lich. 1: 6. 1803. *Monilia cinnabarina* Sprengel, Fl. Halens., p. 386. 1806. *Globulina rubens* Turpin, Dict. Sci. Nat. 65(Végét. Acot.): pl. 6, fig. 7. 1816—29; Turpin, Mém. Mus. d'Hist. Nat. Paris 14: 26, 63. 1827.

Conferva rubens Sommerfelt, Suppl. Fl. Lappon., p. 194. 1826. *Protococcus rubens* Kützing, Linnaea 8: 369. 1833. —The original description is here designated as the temporary Type = FUNGI. *Limnodictyon obscurum* Dickie, Journ. Linn. Soc. Bot. 18: 125. 1880. —Type from Brazil: Obydos, *J. W. H. Trail,* 6 Feb. 1874 (BM) = CHLOROPHYCEAE.

LIMNODICTYON Kützing in Roemer, Alg. Deutschl., p. 61. 1845. —Type species: *L. Roemerianum* Kützing in Roemer, loc. cit. 1845. *Palmogloea Roemeriana* Kützing, Phyc. Germ., p. 153. 1845. *Protococcus infusionum* var. *Roemerianus* Hansgirg, Prodr. Algenfl. Böhmen 1: 143. 1886. —Type from Osterode, Germany: *Römer 142 in herb. Kützing* (L) = EUGLENA sp. FIG. 344.

LITHOMYXA Howe, U. S. Geol. Surv. Prof. Paper 170: 63. 1932. —Type species: *L. calcigena* Howe, loc. cit. 1932. —Type from West Virginia: in Furnace creek, above Harpers Ferry, *D. White, C. B. Read & M. A. Howe,* 7 Dec. 1930 (NY) = FERROBACTERIACEAE.

Lithomyxa calcigena f. *ferrifera* Howe, U. S. Geol. Surv. Prof. Paper 170: 63. 1932. —Type from West Virginia: Furnace creek, above Harpers Ferry, *D. White, C. B. Read & M. A. Howe,* 7 Dec. 1930 (NY) = FERROBACTERIACEAE.

Marssoniella andicola Tutin, Trans. Linn. Soc. London, ser. 3, 1(2): 200. 1940. —Type from Lake Titicaca, Bolivia: Lagunilla Lagunilla (BM) = CHARACIUM sp.

Marssoniella carpetana Gonzalez Guerrero, Bol. R. Soc. Española Hist. Nat. 29: 251. 1929. —The original description is here designated as the temporary Type = CHLOROPHYCEAE.

Merismopoedia aeruginea b. *violacea* Rabenhorst, Fl. Eur. Algar. 2: 57. 1865. —Type from France: Falaise, *de Brébisson* (FC) = ERYTHROCONIS LITTORALIS Oerst.

Merismopoedia aurantiaca Maggiora, Giorn. R. Soc. Ital. d'Igiene 2: 335. 1889. *Pediococcus Maggiorae* Trevisan in Saccardo, Syll. Fung. 8: 1051. 1889. —The original description is here designated at the Type = BACTERIA.

Merismopoedia chondroidea Wittrock in Wittrock & Nordstedt, Alg. Exs. 4: 200. 1878. —Type from Sweden: ad Upsaliam in fonte Slottsdämmen, *V. Wittrock,* Jul. 1876 (in Wittr. & Nordst., Alg. Exs. 200, S) = ERYTHROCONIS LITTORALIS Oerst.

Merismopoedia elegans var. *mandalensis* Wille, Nyt Mag. Naturvid. 38(1): 5. 1900. —The original description is designated here as the temporary Type = ERYTHROCONIS LITTORALIS Oerst.

Merismopoedia Marssonii Lemmermann, Ber. Deutsch. Bot. Ges. 18: 31. 1900. —The original description is here designated as the temporary Type = ERYTHROCONIS LITTORALIS Oerst.

Merismopoedia Massartii Kufferath, Bull. Soc. Roy. Bot. Belg. 74: 95. 1942. —The original description is designated here as the temporary Type = ERYTHROCONIS LITTORALIS Oerst.

Merismopoedia ochracea Mettenheimer, Abh. Senckenb. Naturf. Ges. 2: 141. 1856. *Lampropedia ochracea* Trevisan in Saccardo, Syll. Fung. 8: 1049. 1889. —The original description is designated here as the temporary Type = ERYTHROCONIS LITTORALIS Oerst.

Merismopoedia punctata var. *vacuolata* Playfair, Proc. Linn. Soc. New South Wales 43: 500. 1918. —The original description is designated here as the temporary Type = ERYTHROCONIS LITTORALIS Oerst.

Merismopoedia rectangularis Schiller, Sitzungsber. Oesterr. Acad. Wiss., Math.-Nat. Kl., I, 163: 127. 1954. —The original description is here designated as the temporary Type = BACTERIA.

Merismopoedia Reitenbachii Caspary, Schrift. Phys.-Ökon. Ges. Königsberg 15: 104. 1874. *Sarcina Reitenbachii* Winter, Rabenh. Krypt.-Fl. 1(1): 50. 1884. *Lampropedia Reitenbachii* de Toni & Trevisan in Saccardo, Syll. Fung. 8: 1048. 1889. —Type from East Prussia: Presberg bei Goldap, *R. Caspary,* 22 Sept. 1874 (L) = ERYTHROCONIS LITTORALIS Oerst. FIG. 300.

Merismopoedia sinica Ley, Bot. Bull. Acad. Sinica 1: 271. 1947. —The original description is here designated as the temporary Type = ERYTHROCONIS LITTORALIS Oerst.

Merismopoedia Trolleri Bachmann, Zeitschr. f. Hydrol. 1: 350. 1920. —The original description is designated here as the temporary Type = ERYTHROCONIS LITTORALIS Oerst.

Merismopoedia violacea Kützing, Sp. Algar., p. 472. 1849. *Agmenellum violaceum* Brébisson *pro synon.* in Kützing, loc. cit. 1849. *Micraloa violacea* Brébisson *pro synon.* in Rabenhorst, Fl. Eur. Algar. 2: 57. 1865. *Lampropedia violacea* de Toni & Trevisan in Saccardo, Syll. Fung. 8: 1048. 1889. *Merismopoedia aeruginea* var. *violacea* Rabenhorst ex Forti, Syll. Myxophyc., p. 107. 1907. —Type from France: Falaise, *Brébisson 532* (L) = ERYTHROCONIS LITTORALIS Oerst. FIG. 299.

Micraloa Pini-turinorum Biasoletto, Di Alc. Alghe Microsc., p. 48. 1832. *Microhaloa Pinii* Kützing, Tab. Phyc. 1: 9. 1846. *Micraloa Pinii* Kützing ex Trevisan, Sagg. Monogr. Alg. Coccot., p. 38. 1848. —Type from Trieste: *Biasoletto* (NY) = FLAGELLATES.

MICRALOA Biasoletto, Di Alc. Alghe Microsc., p. 44, 46. 1832. *SPHAEROTHROMBIUM* Kützing *pro synon.,* Linnaea 8: 370. 1833. *CALIALOA* Trevisan, Sagg. Monogr. Alg. Coccot., p. 41. 1848. —Type species: *Micraloa protogenita* Biasoletto, Di Alc. Alghe Microsc., p. 44. 1832. *Microhaloa protogenita* Kützing, Phyc. Germ., p. 147. 1845. *Calialoa Meneghinii* Trevisan, Sagg. Monogr. Alg. Coccot., p. 41. 1848. *Microcystis protogenita* Rabenhorst, Fl. Eur. Algar. 2: 51. 1865. *Chlorococcum protogenitum* Rabenhorst, ibid. 3: 58. 1868. *Protococcus protogenitus* Hansgirg, Prodr. Algenfl. Böhmen 1: 144. 1886. *Chroococcus protogenitus* Hansgirg, ibid. 2: 166. 1892. *Protococcus protogenitus* var. *minor* Forti, Syll. Myxophyc., p. 94. 1907. —Type from Trieste: *Biasoletto* (W) = FLAGELLATA.

Micraloa rosea Kützing, Linnaea 8: 371. 1833. *Sphaerothrombium roseum* Kützing *pro synon.*, ibid. 8: 370. 1833. *Cryptococcus roseus* Kützing, Phyc. Gener., p. 149. 1843; Kützing, Linnaea 17: 82. 1843. —Type from Germany: ad Charas putridas prope Tennstaedt fl. Halens., *Kützing,* 1832 (L) = LAMPROCYSTIS ROSEA (Kütz.) Dr. and Daily, COMB. NOV.

Micraloa teres Flotow, Nova Acta Acad. Caes. Leop.-Car. Nat. Cur. 20(2): 456. 1842. —Topotype from Bohemia: Hirschberg, *Flotow* (in Rabenh. Alg. 71, FC) = PROTOCOCCUS GREVILLEI (Ag.) Crouan.

Microcystis argentea Schiller, Sitzungsber. Oesterr. Akad. Wiss., Math.-Nat. Kl., I, 163: 116. 1954. —The original description is here designated as the temporary Type = LAMPROCYSTIS ROSEA (Kütz.) Dr. & Daily.

Microcystis austriaca Kützing, Tab. Phyc. 1: 7. 1846. *Haematococcus austriacus* Trevisan, Sagg. Monogr. Alg. Coccot., p. 39. 1848. —Type from Austria: *Welwitsch* (Sonder) (L) = EUGLENA sp.

Microcystis Brebissonii Meneghini, Mem. R. Accad. Sci. Torino, ser. 2, 5(Sci. Fis. & Mat.): 85. 1843. *Bichatia Brebissonii* Trevisan, Nomencl. Algar. 1: 60. 1845. *Phytoconis miniata* Brébisson ex Trevisan, Sagg. Monogr. Alg. Coccot., p. 61. 1848. *Gloeocapsa paroliniana β Brebissonii* Kützing, Sp. Algar., p. 223. 1849. *Gloeocystis paroliniana b. grumosa* Brébisson in Rabenhorst, Fl. Eur. Algar. 3: 30. 1868. *Gloeocapsa paroliniana* var. *Brebissonii* Hansgirg, Prodr. Algenfl. Böhmen 2: 150. 1892. —Type from France: *Palmella miniata* Leibl. Falaises de la cote du Calvados, *Brébisson* (FI) = ASTEROCYTIS ORNATA (Ag.) Hamel. FIG. 319.

Microcystis chroococcoidea var. *minor* Nygaard, K. Danske Vidensk. Biol. Skr. 7(1): 178. 1949. —Type from Denmark: Sønderborg, 26 Aug. 1944 (in the slide collection of G. Nygaard) = BACTERIA (LAMPROCYSTIS sp.?).

Microcystis Donnellii Wolle, Bull. Torrey Bot. Club 6: 282. 1879. —Type from Maryland: Garrett county, *J. D. Smith,* Jul. 1878 (PENN) = a ciliate PROTOZOAN containing ZOOCHLORELLA sp.

Microcystis lobata Dickie, Journ. Linn. Soc. Bot. 18: 128. 1880. —Type from the Rio Tapajos, Brazil: *J. W. H. Trail 74* (BM) = ANABAENA CIRCINALIS (Kütz.) Rabenh.

Microcystis mellea Meneghini, Mem. R. Accad. Sci. Torino, ser. 2, 5(Sci. Fis. & Mat.): 83. 1843. *Coccochloris mellea* Brébisson *pro synon.* in Meneghini, loc. cit. 1843. *Palmella mellea* Kützing, Phyc. Germ., p. 149. 1845. *Bichatia mellea* Trevisan, Nomencl. Algar. 1: 61. 1845. *Gloeocapsa mellea* Kützing, Tab. Phyc. 1: 18. 1846. *Cylindrocystis mellea* Brébisson *pro synon.* in Kützing, loc. cit. 1846. *Palmogloea monococca β mellea* Kützing, Sp. Algar., p. 229. 1849. *Gloeothece mellea* Rabenhorst, Fl. Eur. Algar. 2: 62. 1865. *Chroothece monococca* var. *mellea* Kützing ex Hansgirg, Prodr. Algenfl. Böhmen 2: 135. 1892. *Gloeothece monococca* var. *mellea* Kützing ex Geitler in Pascher, Süsswasserfl. 12: 94. 1925. —Paratypes from France: *Coccochloris mellea* Bréb. Falaise (FC, L, S, UC) = ASTEROCYTIS ORNATA (Ag.) Hamel.

Microcystis merismopedioides Fritsch, Journ. Linn. Soc. Bot. 40: 332. 1912. —Type from the South Orkney islands: Point Martin, Scotia bay (in the collection of F. E. Fritsch) = BACTERIA.

Microcystis minor Kützing, Tab. Phyc. 1: 9. 1846. *Haematococcus minor* Trevisan, Sagg. Monogr. Alg. Coccot., p. 39. 1848. —Type from the Netherlands: Goes, *van den Bosch* (L) = EUGLENA sp. FIG. 318.

Microcystis minuta var. *violacea* Lindstedt, Fl. Mar. Cyanophyc. Schwed. Westküste, p. 17. 1943. —Type from Sweden: Kristineberg (in the collection of A. Lindstedt) = LAMPROCYSTIS ROSEA (Kütz.) Dr. & Daily. FIG. 316.

Microcystis olivacea Kützing, Phyc. Gener., p. 170. 1843. —Type from Germany: Jever, *Jürgens* (L) = EUGLENA sp. FIG. 320.

Microcystis parasitica Kützing, Phyc. Gener., p. 170. 1843. *Anacystis parasitica* Trevisan, Sagg. Monogr. Alg. Coccot., p. 35. 1848. *Microcystis pulverea* f. *parasitica* Elenkin, Monogr. Algar. Cyanophyc., Pars Spec. 1: 124. 1938. —Type from Germany (L) = NOSTOC HEDERULAE Menegh. FIG. 317.

Microcystis rosea Kufferath, Bull. Soc. Roy. Bot. Belg. 74: 94. 1942. —The original description is here designated as the Type = LAMPROCYSTIS ROSEA (Kütz.) Dr. & Daily.

Microcystis rufescens Martens ex Wille, Nyt Mag. Naturv. 62: 202. 1925. —Type from Arracan, Burma: no. 4819, *in herb. Agardh* (LD) = EUGLENA sp.

Microcystis stereophysalis Meneghini, Giorn. Bot. Ital. I, 1: 297. 1844. *Bichatia stereophysalis* Trevisan, Nomencl. Alg. 1: 62. 1845. *Tetraspora stereophysalis* Kützing, Tab. Phyc. 1: 20. 1846. —Type from Italy: Ferrara, *Felisi* (FI) = MONOSTROMA sp.

Microcystis umbrina Kützing, Linnaea 8: 373. 1833. *Protococcus umbrinus* Kützing, Phyc. Gener., p. 169. 1843. *Chlorococcum umbrinum* Rabenhorst, Fl. Eur. Algar. 3: 60. 1868. —Type from Germany: in cortice Fagi sylvaticae prope Leucopetram (in Kütz. Alg. Aq. Dulc. Dec. 10: 91, L) = PHYTOCONIS BOTRYOIDES (L.) Bory, somewhat parasitized.

MICRODISCUS Steinecke, Schrift. Phys.-ökon. Ges. Königsberg 56: 25. 1916. —Type species: *M. parasiticus* Steinecke, loc. cit. 1916. —The original description is here designated as the temporary Type = BACTERIA.

Microhaloa aurantiaca Kützing, Phyc. Germ., p. 147. 1845. *Micraloa aurantiaca* Kützing ex

Trevisan, Sagg. Monogr. Alg. Coccot., p. 38. 1848. —Type from Germany: Jever, *Koch* (L) =
FLAGELLATES... FIG. 314.

MICROHALOA Kützing, Phyc. Gener., p. 169. 1843. —Type species: *M. botryoides* Kützing,
loc. cit. 1843. *Micraloa botryoides* Kützing ex Trevisan, Sagg. Monogr. Alg. Coccot., 38. 1848.
Cystococcus botryoides Rabenhorst, Krypt.-Fl. Sachsen 1: 137. 1863. *Chlorococcum botryoides*
Rabenhorst, Fl. Eur. Algar. 3: 57. 1868. —Type from Germany: Nordhausen, *Kützing* (L) =
PHYTOCONIS BOTRYOIDES (L.) Bory. FIG. 313.

Microhaloa major Kützing in Desmaziéres, Pl. Crypt. Fr. 11: 534. 1858. —Type: Algier?,
ex herb. Montagne (L) = encysted FLAGELLATES. FIG. 312.

Microhaloa pallida Kützing, Tab. Phyc. 1: 5. 1846. *Micraloa pallida* Kützing ex Trevisan, Sagg.
Monogr. Alg. Coccot., p. 37. 1848. —Type from France: Arromanches, *herb. Lenormand* (L)¯ =
EUDORINA ELEGANS Ehrenb. FIG. 315.

MONTANOA Gonzalez Guerrero, Anal. Jard. Bot. Madrid 8: 267. 1948. —Type species:
M. castellana Gonzalez Guerrero, ibid. 8: 268. 1948. —The original description is here designated
as the temporary Type = CALOTHRIX sp.

Myxobactron hirundiforme G. S. West, Ann. So. Afr. Mus. 9: 63. 1911. *Dactylococcopsis
hirundiformis* Geitler, Rabenh. Krypt.-Fl. 14: 284. 1932. —Type from Angola: Mossamedes,
Pearson (in the slide collection, BIRM) = ANKISTRODESMUS FALCATUS (Corda) Ralfs.

Myxobactron palatinum Schmidle, Allgem. Bot. Zeitschr. 1905: 65. 1906. *Dactylococcopsis
rhaphidioides* f. *longior* Geitler, Rabenh. Krypt.-Fl. 14: 282. 1932. —The original description is
here designated as the temporary Type = DIATOMS.

MYXOBACTRON Schmidle, Hedwigia 43: 415. 1904. —Type species: *M. Usterianum*
Schmidle, loc. cit. 1904. —Type from Negros island, Philippines: Urwald bei S. Carlos, *A.
Usteri 18* (ZT) = DIATOMS.

Myxosarcina burmensis Skuja, Nova Acta R. Soc. Sci. Uppsal., ser. 4, 14(5): 21. 1949.
—Type from Rangoon, Burma, *L. P. Khanna 529* (in the collection of L. P. Khanna) = BACTERIA.

CHROOCOCCACEAE Subfam. *EUCHROOCOCCACEAE* Hansgirg, Prodr. Algenfl. Böhmen 2:
129. 1892. *CHROOCOCCACEAE* Subfam. *EUCHROOCOCCACEAE* Gruppe *THECINEAE* Hansgirg,
loc. cit. 1892. —Type genus: ONCOBYRSA Agardh, Flora 10(2): 629. 1827. —Type species:
O. fluviatilis Agardh, loc. cit.. 1827. *Cagniardia fluviatilis* Trevisan, Sagg. Monogr. Alg. Coccot., p.
51. 1848. —Type from Bohemia: Carlsbad, *C. A. Agardh* (LD) = FUNGI with spores. FIG. 348, 360.

PSEUDONCOBYRSA Geitler, Beih. z. Bot. Centralbl., II, 41: 237. 1925; in Pascher, Süsswasserfl.
12: 121. 1925. —Type species: *Oncobyrsa lacustris* Schröter & Kirchner, Veget. des Bodensees, p.
102. 1896. *Pseudoncobyrsa lacustris* Geitler, Beih. z. Bot. Centralbl., II, 41: 237. 1925; in
Pascher, Süsswasserfl. 12: 121. 1925. —The original description is here designated as the temporary
Type = ASTEROCYTIS ORNATA (Ag.) Hamel.

Oncobyrsa sparsa Rabenhorst, Fl. Eur. Algar. 2: 69. 1865. *Coccochloris sparsa* Brébisson *pro
synon.* in Rabenhorst, loc. cit. 1865. —The original description is here designated as the temporary
Type = palmelloid STICHOCOCCUS SUBTILIS (Kütz.) Klerck.

ONKONEMA Geitler, Arch. f. Hydrobiol., Suppl. 12, 4: 627. 1933. —Type species: *O.
compactum* Geitler, loc cit. 1933. —Type from southern Sumatra: Ranau-See, 5 Feb. 1929 (slide no.
RW2a in the collection of L. Geitler) = primordia of HAPALOSIPHON LAMINOSUS (Kütz.) Hansg.

Oscillatoria clavata Corda, Almanach de Carlsbad (C. J. de Carro, ed.) 6: 203. 1836. —The
original description is here designated as the temporary Type = CLADOTHRIX sp.

Oscillatoria limnetica var. *acicularis* f. *brevis* Nygaard, K. Danske Vidensk. Selsk. Biol. Skr.
7(1): 191. 1949. *Dactylococcopsis inaequalis* Nygaard *pro synon.*, ibid. p. 192. 1949. —Type
from Denmark: Molledammen i Sønderborg, 26 Aug. 1944 (in the collection of G. Nygaard) =
BACTERIA.

PAGEROGALA Wood, Smiths, Contrib. Knowl. 241: 81. 1872. —Type species: *P. stellio*
Wood, ibid. p. 82. 1872. —The original description is here designated as the temporary Type =
ANIMAL MATERIAL.

TRYPOTHALLUS Hooker f. & Harvey in Hooker, Bot. Antarct. Voy. Erebus & Terror 1839—43,
1(2): 500. 1847. —Type species: *Palmella anastomosans* Hooker f. & Harvey, London Journ. Bot. 4:
298. 1845. *Trypothallus anastomosans* Hooker f. & Harvey, Bot. Antarct. Voy. Erebus & Terror
1839—43, 1(2): 500. 1847. *Palmodictyon Hookeri* Kützing, Sp. Algar., p. 234. 1849. *Gloiodictyon anastomosans* de Toni, Syll. Algar. 2: 653. 1889. —Type from Kerguelen island: moist caves
near the sea, Christmas Harbour, *no. 700*, May 1840 (K) = primordia of PRASIOLA sp.

Palmella aurantia Agardh, Syst. Algar., p. 14. 1824. *Coccochloris aurantia* Wallroth, Fl. Crypt.
Germ. 2: 6. 1833. *Cagniardia aurantia* Trevisan, Sagg. Monogr. Alg. Coccot., p. 49. 1848.
Aphanocapsa aurantia Rabenhorst, Fl. Eur. Algar. 2: 51. 1865. *Chroococcus aurantiacus* Wille, Nyt
Mag. Naturvid. 56(1): 50. 1918. *Schizochlamys aurantia* Drouet & Daily, Lloydia 11: 79. 1948.
—Type from Sweden: Traneberg, 30 Jul. 1823, *herb. Agardh* (LD) = SCHIZOCHLAMYS AURANTIA
(Ag.) Dr. & Daily. FIG. 361.

Palmella cylindrica Lyngbye, Tent. Hydrophyt. Dan., p. 205. 1819. *Coccochloris cylindrica*
Brébisson in Meneghini, Mem. R. Accad. Sci. Torino, ser. 2, 5(Sci. Fis. & Mat.): 46, 68. 1843.

156

—The original description is here designated as the temporary Type = CHLOROPHYCEAE. *CYLINDROCYSTIS* Meneghini, Cenni sulla Organogr. & Fisiol. d. Alg., p. 5, 26. 1838. —Type species: *Palmella cylindrospora* Brébisson in Brébisson & Godey, Mém. Soc. Acad. Sci., Art. & Belles-Lettres Falaise 1835: 256. 1836. *Cylindrocystis Brebissonii* Meneghini, Cenni sulla Organogr. & Fisiol. d. Alg., p. 5. 1838; in Desmazières, Pl. Crypt. Fr., ed. 2, no. 914. 1838. *Palmogloea Brebissonii* Kützing, Tab. Phyc. 1: 19. 1846. *P. Meneghinii* Kützing loc. cit. 1846. *Penium Brebissonii* Ralfs, Brit. Desmid., p. 153. 1848. *Cylindrocystis cylindrospora* Drouet & Daily, Leafl. Acadian Biol. 1: 16. 1953. —Type from France: Falaise, *ex herb. Lenormand* (FI) = CYLINDROCYSTIS CYLINDROSPORA (Bréb.) Dr. & Daily.

Palmella depressa Berkeley, Glean. Brit. Alg., p. 19. 1833. *Coccochloris depressa* Menegh., Mem. R. Accad. Sci. Torino, ser. 2, 5 (Sci. Fis. & Mat.): 46, 68. 1843. *Aphanocapsa depressa* Rabenhorst, Fl. Eur. Algar. 2: 51. 1865. —Type from England: "Palmella depressa", *herb. Berkeley* (K) = palmelloid plants of STIGEOCLONIUM sp.

Palmella effusa Kützing, Linnaea 8: 375. 1833. —Type from Germany: im Bruchteiche prope Tennstaedt, with *Oscillatoria alba, F. T. Kützing* (FC) = BACTERIA.

Palmella flava Kützing, Tab. Phyc., 1: 9. 1846. *Cagniardia flava* Trevisan, Sagg. Monogr. Alg. Coccot., p. 49. 1848. *Aphanocapsa flava* Rabenhorst, Fl. Eur. Algar. 2: 50. 1865. —Type from France: Vire, *herb. Lenormand* (L) = EUDORINA ELEGANS Ehrenb. FIG. 349.

Palmella fusiformis Roemer, Alg. Deutschl., p. 64. 1845. *P. fusiformis* Rabenhorst, Deutschl. Krypt.-Fl. 2: 60. 1847. *Cagniardia fusiformis* Trevisan, Sagg. Monogr. Alg. Coccot., p. 50. 1848. —The original description is here designated as the temporary Type = CHLOROPHYCEAE.

Palmella gigantea Sommerfelt, Suppl. Fl. Lappon., p. 203. 1826. —Type from Norway: Palmella gigantea. E. Norveg. Sommerf. (C) = ZOOCHLORELLA PARASITICA Brandt in OPHRYDIUM sp.

Palmella globosa Agardh, Syst. Algar., p. 13. 1824. *Coccochloris globosa* Crouan fr., Fl. Finistère, p. 110. 1867. —Type from Sweden: "Palmella globosa" (LD) = ZOOCHLORELLA PARASITICA Brandt in OPHRYDIUM sp.

Palmella heterospora Rabenhorst, Krypt.-Fl. Sachsen 1: 129. 1863. —Isotypes from Germany in Rabenh. Alg. Sachs. 970 (FC, FI, L, TA) = STICHOCOCCUS BACILLARIS Näg.

Palmella hormospora Meneghini, Mem. R. Accad. Sci. Torino, ser. 2, 5 (Sci. Fis. & Mat.): 52. 1843. *Phytoconis hormospora* Trevisan, Prosp. Fl. Eugan., p. 57. 1843. *Cagniardia hormospora* Trevisan, Sagg. Monogr. Alg. Coccot., p. 48. 1848. —Type from Italy: "Palmella hormospora" (FI) = NOSTOC HUMIFUSUM Carm. FIG. 356.

HOMALOCOCCUS Kützing, Oster-Progr. Realsch. Nordhausen, p. 6. 1863; Hedwigia 2: 86. 1863. —Type species: *Palmella hyalina* Lyngbye, Tent. Hydrophyt. Dan., p. 204. 1819. *Coccochloris hyalina* Meneghini, Mem. R. Accad. Sci. Torino, ser. 2, 5 (Sci. Fis. & Mat.): 46, 66. 1843. *Cagniardia Kuetzingiana* Trevisan, Sagg. Monogr. Alg. Coccot., p. 50. 1848. *Chaos hyalinus* Bory ex Trevisan, ibid., p. 56. 1848. *Homalococcus Hassallii* Kützing, Oster-Progr. Realsch. Nordhausen, p. 6. 1863; Hedwigia 2: 86. 1863. *Aphanothece hyalina* A. Braun ex Hansgirg, Prodr. Algenfl. Böhmen 2: 159. 1892. —Type from Faeroes islands: in fossis rivuli ad Thorshavn, 26 Jun. 1817 (C) = TETRASPORA GELATINOSA (Vauch.) Desv.

Palmella miniata Leiblein, Flora 13(1): 338. 1830. *Merrettia miniata* Trevisan, Sagg. Monogr. Alg. Coccot., p. 46. 1848. *Protococcus viridis* var. *miniatus* Wolle, Fresh-w. Alg. U. S., p. 181. 1887. —Type from Germany: aus einem Sümpfe, Würzburg, *Leiblein*, 12 Mar. 1827 (L) = encysted FLAGELLATES.

PALMELLA Untergattung *TETRATOCE* Nägeli, Die Neuern Algensyst., p. 131. 1847. —Type species: *Palmella aequalis* Nägeli in Kützing, Sp. Algar., p. 212. 1849. *P. miniata* var. *aequalis* Nägeli, Gatt. Einzell. Alg., p. 67. 1849. *Palmella miniata* b. *aequalis* Nägeli ex Rabenhorst, Fl. Eur. Algar. 3: 34. 1868. *Aphanocapsa aequalis* Hansgirg, Prodr. Algenfl. Böhmen 2: 158. 1892. —Type from Switzerland: feuchte Felsen, Zürich, *Nägeli* 307 (L) = CHLOROPHYCEAE (TETRASPORA sp?).

Palmella minuta Agardh, Flora 10(2): 630. 1827. *Coccochloris minuta* Wallroth, Fl. Crypt. Germ. 2: 5. 1833. *Anacystis minuta* Meneghini, Mem. R. Accad. Sci. Torino, ser. 2, 5 (Sci. Fis. & Mat.): 47, 94. 1843. *Cagniardia minuta* Trevisan, Sagg. Monogr. Alg. Coccot., p. 50. 1848. *Aphanothece minuta* Migula, Krypt-Fl. Deutschl., II, Alg. la: 35. 1905. —Type from Bohemia: Carlsbad, *herb. Agardh* (LD) = STICHOCOCCUS SUBTILIS (Kütz.) Klerck.

Palmella mirifica Rabenhorst, Algen Sachs. 55—56: 541. 1856. —Type from Germany: Dresden, *H. Richter*, 1856 (Rabenh. Alg. Sachs. 541, L) = BACTERIA.

PALMELLA Untergattung *DITOCE* Nägeli, Die Neuern Algensyst., p. 131. 1847. —Type species: *Palmella mucosa* Kützing, Phyc. Gener., p. 172. 1843. *Merrettia mucosa* Trevisan, Sagg. Monogr. Alg. Coccot., p. 46. 1848. —Type from Germany: in saxis rivulorum (Zorge), Nordhausen, *Kützing,* 1 Mar. 1841 (L) = young TETRASPORA GELATINOSA (Vauch.) Desv.

Palmella parvula Kützing, Phyc. Gener., p. 171. 1843. *Cagniardia parvula* Trevisan, Sagg. Monogr. Alg. Coccot., p. 50. 1848. —Type from Germany: Nordhausen, *Kützing* (L) = PALMODICTYON VIRIDE Kütz. FIG. 362.

Palmella pelludissima Bory in Mougeot & Roumeguère in Louis, Départ. des Vosges, Alg., p. 35 (of reprint). 1887. —Type from France: in lignis aqua immersis rivulorum circa Retournemer, Vosges, May 1822 (L) = FLAGELLATA.

Palmella pila Suhr, Flora 23(1): 297. 1840. *Coccochloris pila* Kützing, Phyc. Germ., p. 150. 1845. —Type from Germany: an der Eider, *Suhr* (L) = ZOOCHLORELLA PARASITICA Brandt in OPHRYDIUM sp.

Palmella rubescens Brébisson in Brébisson & Godey, Mém. Soc. Acad. Sci., Art. & Belles-Lettres 1835: 256. 1836. *Microcystis rubescens* Meneghini, Mem. R. Accad. Sci. Torino, ser. 2, 5(Sci. Fis. & Mat.): 47, 86. 1843. *Coccochloris rubescens* Brébisson *pro synon.* in Meneghini, loc. cit. 1843. *Bichatia rubescens* Trevisan, Nomencl. Algar. 1: 62. 1845. *Palmogloea rubescens* Kützing, Tab. Phyc. 1: 19. 1846. *Cagniardia Brebissonii* Trevisan, Sagg. Monogr. Alg. Coccot., p. 49. 1848. *Mesotaenium rubescens* Drouet & Daily in Drouet, Proc. Minnesota Acad. Sci. 22: 136. 1956. —Type from France: Falaise, *A. de Brébisson* (L) = MESOTAENIUM RUBESCENS (Bréb.) Dr. & Daily.

Palmella terminalis Agardh, Icon. Algar Eur. 14. 1828—35. *Coccochloris terminalis* Brébisson in Meneghini, Mem. R. Accad. Sci. Torino, ser. 2, 5(Sci. Fis. & Mat.): 46, 67. 1843. *C. thermalis* Rabenhorst, Deutschl. Krypt.-Fl. 2(2): 60. 1847. —Type labeled "Sept. 1825" in herb. Agardh (LD) = TETRASPORA GELATINOSA (Vauch.) Desv.

Palmella uvaeformis Kützing, Alg. Aq. Dulc. Dec. 11: 102. 1834. *Cagniardia uvaeformis* Trevisan, Sagg. Monogr. Alg. Coccot., p. 50. 1848. *Palmella partibilis* Suhr *pro synon.* ex Kützing, Sp. Algar., p. 211. 1849. —Type from Germany: Eilenburg (L) = primordia of CHLOROPHYCEAE.

Palmogloea aureo-nucleata Brébisson in Rabenhorst, Alg. Eur. 218—220: 2190. 1870. —Type from France: in paludibus turfosis, Falaise, *Brébisson* (in Rabenh. Alg. 2190, L) = ASTEROCYTIS ORNATA (Ag.) Hamel.

Palmogloea crassa Kützing, Tab. Phyc. 1: 19. 1846. —Type from Germany: Oberharz, *herb. Kützing* (L) = MESOTAENIUM RUBESCENS (Bréb.) Dr. & Daily.

Palmogloea dimorpha Kützing, Sp. Algar., p. 229. 1849. *Coccochloris macrocarpa* Brébisson *pro synon.* in Kützing, loc. cit. 1849. —Type from France: mousses humides, Falaise (L) = MESOTAENIUM RUBESCENS (Bréb.) Dr. & Daily.

TRICHOCYSTIS Kützing, Tab. Phyc. 1: 20. 1846. —Type species: *Palmogloea gigantea* Kützing, Bot. Zeitung 5: 221. 1847. *Trichocystis gigantea* Kützing, Tab. Phyc. 1: 20. 1846. —Type from Germany: Tolkwade bei Schleswig, *Suhr*, Apr. 1840 (L) = ZOOCHLORELLA PARASITICA Brandt in OPHRYDIUM sp. FIG. 364.

Palmogloea lurida Flotow in Kützing, Tab. Phyc. 1: 20. 1846. —Type from Sudetenland, Bohemia: unweit der Gibraltar-felsen, *Flotow*, 8 Jun. 1844 (L) = MESOTAENIUM RUBESCENS (Bréb.) Dr. & Daily.

Palmogloea macrococca Kützing, Phyc. Germ., p. 153. 1845. *Mesotaenium Braunii* de Bary, Unters. Fam. Conjugat., p. 74. 1858. *M. macrococcum* Kützing ex Roy (& Bisset), Ann. Scott. Nat. Hist. 1894: 253. 1894. —Type from Germany: Oberharz, *herb. Kützing* (L) = MESOTAENIUM RUBESCENS (Bréb.) Dr. & Daily.

Palmogloea micrococca Kützing, Tab. Phyc. 1: 20. 1846. *Mesotaenium macrococcum* var. *micrococcum* Kirchner ex W. & G. S. West, Trans. Yorkshire Nat. Union 5: 41. 1900. —Type from France: Vire, *Lenormand* (L) = palmelloid BOTRYDIUM GRANULATUM (L.) Grev.

Palmogloea rupestris Kützing, Phyc. Germ., p. 153. 1845. *Polycoccus marginatus* Kützing in Roemer, Alg. Deutschl., p. 65. 1845. —Type from Germany: *Römer 105* (L) = MESOTAENIUM RUBESCENS (Bréb.) Dr. & Daily.

PARACAPSA Naumann, Ark. f. Bot. 18(21): 6. 1924. —Type species: *P. siderophila* Naumann, loc. cit. 1924. *Pseudoncobyrsa siderophila* Geitler, Beih. z. Bot. Centralbl., II, 41: 237. 1925; in Pascher, Süsswasserfl. 12: 122. 1925. —Topotype, here designated as the Type, from Sweden: Förthultsjön, Aneboda, *G. Israelsson*, 2 Jul. 1928 (S) = CAPSOSIRA BREBISSONII Kütz.

PAULINELLA Lauterborn, Zeitschr. f. Wiss. Zool. 59: 537, 543. 1895. —Type species: *P. chromatophora* Lauterborn, loc. cit. 1895. —The original description is here designated as the temporary Type = RHIZOPODA.

PELIAINA Pascher, Jahrb. f. Wiss. Bot. 71(3): 458. 1929. —Type species: *P. cyanea* Pascher, ibid., p. 459. 1929. —The original description is here designated as the temporary Type = FLAGELLATES.

PILGERIA Schmidle, Hedwigia 40: 53. 1901. —Type species: *P. brasiliensis* Schmidle, ibid., p. 54. 1901. —The original description is here designated as the temporary Type = CHLOROPHYCEAE.

Placoma corticalis Shousboe ex Bornet, Mém. Soc. Sci. Nat. & Math. Cherbourg 28: 241. 1892. —Published as a synonym of RALFSIA VERRUCOSA Aresch.

Placoma fusca Schousboe ex Bornet, Mém. Soc. Nat. Sci. Nat. & Math. Cherbourg 28: 241. 1892. —Published as a synonym of RALFSIA VERRUCOSA Aresch.

Placoma lapidea Schousboe ex Bornet, Mém. Soc. Nat. Sci. Nat. & Math. Cherbourg 28: 347. 1892. —Published as a synonym of HILDBRANDTIA PROTOTYPUS Nardo.

Placoma membranacea Schousboe ex Bornet, Mém. Soc. Nat. Sci. Nat. & Math. Cherbourg 28: 231. 1892. —Published as a synonym of ZANARDINIA COLLARIS (Ag.) Crouan fr.

158

Pleurocapsa parenchymatica Geitler, Arch f. Hydrobiol., Suppl. XII, 4: 626. 1933. *Scopulonema parenchymaticum* Geitler in Engler & Prantl, Natürl. Pflanzenfam., ed. 2, 1b: 93. 1942. —Type from central Java: Seraja unterhalb Patagbanteng, *Ruttner*, 5 Jun. 1929 (slide no. D6aα in the collection of L. Geitler) = young LICHEN.

Pleurocapsa Usteriana Schmidle, Hedwigia 43: 414. 1904. —Type from Java: in einem kleinen Wasserfall, Dioputal, *A. Usteri 43*, 9 May 1903 (ZT) = young TETRASPORACEAE.

Pleurococcus angulosus var. *irregularis* Hansgirg, Prodr. Algenfl. Böhmen 1(2): 237. 1888. —Type from Bohemia: Plana nächst Tabor, *A. Hansgirg*, 4 May 1886 (W) = FLAGELLATES.

Pleurococcus angulosus f. *tectorum* Trevisan ex Kirchner, Krypt.-Fl. Schlesien 2(1): 115. 1878. —The original description is here designated as the temporary Type = STICHOCOCCUS SUBTILIS (Kütz.) Klerck.

Pleurococcus antarcticus W. & G. S. Wset, Brit. Antarct. Exped., 1907—09, 1: 276. 1911. *P. antarcticus* f. *simplex* Fritsch, Nat. Antarct. Exped. (1901—04), Nat. Hist. 6(Freshw. Alg.): 14. 1912. *Protococcus antarcticus* G. S. West, Journ. of Bot. 54: 2. 1916. —The original description is here designated as the temporary Type = CHLOROPHYCEAE.

Pleurococcus antarcticus f. *filamentosus* Fritsch, Nat. Antarct. Exped. (1901—04), Nat. Hist. 6(Freshw. Alg.): 14. 1912. —The original description is here designated as the temporary Type = CHLOROPHYCEAE.

Pleurococcus antarcticus f. *minor* Fritsch, Nat. Antarct. Exped. (1901—04), Nat. Hist. 6(Freshw. Alg.): 14. 1912. —The original description is here designated as the temporary Type = CHLOROPHYCEAE.

Pleurococcus antarcticus f. *robustus* W. & G. S. West, Brit. Antarct. Exped., 1907—09, 1: 276. 1911. —The original description is here designated as the temporary Type = CHLOROPHYCEAE.

Pleurococcus antarcticus f. *stellatus* Fritsch, Nat. Antarct. Exped. (1901—04), Nat. Hist. 6(Freshw. Alg.): 14. 1912. —The original description is here designated as the temporary Type = CHLOROPHYCEAE.

Pleurococcus antarcticus f. *typicus* Fritsch, Nat. Antarct. Exped. (1901—04), Nat. Hist. 6(Freshw. Alg.): 13. 1912. —The original description is here designated as the temporary Type = CHLOROPHYCEAE.

Pleurococcus aquaticus Snow, Bull. U. S. Fish Comm. 1902: 383. 1903. —The original description is here designated as the temporary Type = CHLOROPHYCEAE.

Pleurococcus Beigelii Kuchenmeister & Rabenhorst, Hedwigia, 1867: 49. 1867. *Hyalococcus Beigelii* Schröter, Pilze Schlesiens 1: 152. 1886. —Type from England: an Chignon-Haaren, London, *Dr. Beigel* (in Rabenh. Alg. Eur. 2053, NY) = BACTERIA.

CYANODERMA Weber-van Bosse, Natuurk. Verh. Holl. Maatsch. Wetensch. Haarlem, ser. 3, 5: 18. 1887. *CYANODERMA* Subgenus *EUCYANODERMA* Hansgirg, Notarisia 3: 588. 1888. —Type species: *Pleurococcus Bradypii* Kuhn in Welcker, Abh. Naturf. Ges. Halle 9: 66. 1866. *Cyanoderma Bradypodis* Weber-van Bosse, Natuurk. Verh. Holl. Maatsch. Wetensch. Haarlem, ser. 3, 5: 18. 1887. *Pleurocapsa Bradypodis* Hansgirg, Sitzungsber. K. Böhm. Ges. Wiss., Math.-Nat. Cl., 1890(2): 92. 1890. —The original description is here designated as the temporary Type = RHODOPHYCEAE.

Pleurococcus calcarius Boye Petersen, K. Dansk. Vidensk. Selsk. Skr., 7(Naturv.-Math. Afd.), 12: 320. 1916. —The original description is here designated as the temporary Type = CHLOROPHYCEAE.

Pleurococcus Choloepii Kuhn in Welcker, Abh. Naturf. Ges. Halle 9: 66. 1866. *Cyanoderma Choloepodis* Weber-van Bosse, Natuurk. Verh. Holl. Maatsch. Wetensch. Haarlem, ser. 3, 5: 18. 1887. *Pleurocapsa Choloepodis* Hansgirg, Sitzungsber. K. Böhm. Ges. Wiss., Math.-Nat. Cl., 1890(2): 91. 1890. —The original description is here designated as the temporary Type = RHODOPHYCEAE.

Pleurococcus cinnamomeus Meneghini, Mem. R. Accad. Sci. Torino, ser. 2, 5(Sci. Fis. & Mat.): 42. 1843. *Protococcus cinnamomeus* Kützing, Tab. Phyc. 1: 5. 1846. *Chroococcus cinnamomeus* Rabenhorst, Krypt.-Fl. Sachsen 1: 70. 1863. —Type from Italy: in olis seminarii (Padova), Jul. 1839, *herb. Meneghini* (FI) = PROTOSIPHON CINNAMOMEUS (Menegh.) Dr. & Daily, COMB. NOV.

Pleurococcus conglomeratus Artari, Bull. Soc. Imp. Nat. Moscou, N. S., 6: 244. 1893. —The original description is here designated as the temporary Type = CHLOROPHYCEAE.

Pleurococcus crenulatus Hansgirg, Oesterr. Bot. Zeitschr. 36: 110. 1886. —Type from Bohemia: Veseli, *A. Hansgirg*, Aug. 1884(W) = STICHOCOCCUS SUBTILIS (Kütz.) Klerck. FIG. 335.

Pleurococcus ellipticus Meneghini, Consp. Algol. Eugan., p. 338. 1837. *Protococcus ellipticus* Kützing, Sp. Algar., p. 198. 1849. —Type from the Euganean springs, Italy: "Pleurococcus ellipticus", *Meneghini* (FI) = CYLINDROSPERMUM MUSCICOLA Kütz., spores.

Pleurococcus frigidus W. & G. S. West, Brit. Antarct. Exped., 1907—09, 1: 276. 1911. *Protococcus frigidus* G. S. West, Journ. of Bot. 54: 2. 1916. —The original description is here designated as the temporary Type = CHLOROPHYCEAE.

Pleurococcus Koettlitzii Fritsch, Nat. Antarct. Exped. (1901—04), Nat. Hist. 6(Freshw. Alg.):

15. 1912. *Chlorella Koettlitzii* Wille, Deutsche Südpolar-Exped., 1901—03, 8: 401. 1928.
—The original description is here designated as the temporary Type = CHLOROPHYCEAE.
Pleurococcus marinus Collins, Rhodora 9: 197. 1907. —Type from Maine: in a marsh
pool, Stovers point, Harpswell, *F. S. Collins*, 16 Jul. 1906 (in Coll., Hold. & Setch., Phyc. Bor.-
Amer. 1316, NY) = CHLOROPHYCEAE.
Pleurococcus marinus Hansgirg in Foslie, Mar. Alg. Norway 1: 158. 1890. —The original
description is here designated as the temporary Type = CHLOROPHYCEAE.
Pleurococcus marinus f. *major* Hansgirg in Foslie, Mar. Alg. Norway 1: 158. 1890. —The
original description is here designated as the temporary Type = CHLOROPHYCEAE.
Pleurococcus miniatus var. *fuscescens* Hansgirg, Physiol. & Algol. Stud., p. 87. 1887. —Type
from Bohemia: Hortus Kinsky, Smichov, *A. Hansgirg,* Oct. 1883 (W) = FLAGELLATES. FIG. 336.
Pleurococcus miniatus var. *roseolus* Hansgirg, Prodr. Algenfl. Böhmen 1(2): 134. 1888. —Type
from Bohemia: Libsic, *A. Hansgirg,* Aug. 1886 (W) = FLAGELLATES. FIG. 337.
Pleurococcus monetarum Reinsch, Flora 67: 176. 1884. —The original description is here
designated as the temporary Type = CHLOROPHYCEAE.
Pleurococcus nimbatus de Wildeman, Notarisia 1893: 7. 1893. *Tetracoccus Wildemanii*
Schmidle, Flora 78: 45. 1894. —The original description is here designated as the temporary
Type = CHLOROPHYCEAE.
Pleurococcus pachydermus Lagerheim in Wittrock & Nordstedt, Alg. Exs. 9: 447. 1882.
Protococcus pachydermus G. S. West, Journ. of Bot. 54: 2. 1916. —Type from Sweden: ad
Bellevue prope Holmiam, *G. Lagerheim,* Jun. 1881 (in Wittr. & Nordst., Alg. Exs. 447, L) =
PHYTOCONIS BOTRYOIDES (L.) Bory, parasitized in part. FIG. 338.
Pleurococcus pachydermus f. *stipitatus* W. & G. S. West, Brit. Antarct. Exped., 1907—09, 1: 275.
1911. —The original description is here designated as the temporary Type = CHLOROPHYCEAE.
Pleurococcus pulcher Kirchner, Jahresb. Ver. Vaterl. Naturk. Württemberg 36: 170. 1880.
Protococcus viridis β *pulcher* Hansgirg in Wittrock & Nordstedt, Alg. exs. 15: 721. 1886. *P. viridis*
var. *pulcher* Hansgirg, Prodr. Algenfl. Böhmen 1: 141. 1886. —The original description is here
designated as the temporary Type = SCHIZOGONIUM MURALE Kütz.
Pleurococcus regularis Artari, Bull. Soc. Imp. Nat. Moscou, N. S., 6: 245. 1893. —The
original description is here designated as the temporary Type = CHLOROPHYCEAE.
Pleurococcus rufescens var. *sanguineus* W. & G. S. West, Journ. of Bot. 36: 336. 1898. —The
original description is here designated as the temporary Type = CHLOROPHYCEAE.
Pleurococcus seriatus Wood, Smiths. Contrib. Knowl. 241: 78. 1872. —The original description
is here designated as the temporary Type = PROTOSIPHON CINNAMOMEUS (Menegh.) Dr. & Daily.
Pleurococcus simplex Artari, Bull. Soc. Imp. Nat. Moscou, N. S., 6: 243. 1893. —The
original description is here designated as the temporary Type = CHLOROPHYCEAE.
Pleurococcus superbus Cienkowski, Bot. Zeitung 23: 21. 1865. —The original description
is here designated as the temporary Type = encysted FLAGELLATES.
Pleurococcus tectorum f. *antarcticus* Wille, Deutsche Südpolar-Exped., 1901-03, 8: 398. 1928.
—The original description is here designated as the temporary Type = CHLOROPHYCEAE.
Pleurococcus vestitus Reinsch, Algenfl. Mittl. Th. Franken, p. 56. 1867. *Protococcus viridis* var.
vestitus Wolle, Fresh-w. Alg. U. S., p. 183. 1887. —The original description is here designated
as the temporary Type = CHLOROPHYCEAE.
Pleurococcus vestitus f. *chlorophyllaceus* Lagerheim, Öfvers. K. Sv. Vet.-Akad. Förh. 1882(2): 79.
1882. —The original description is here designated as the temporary Type = CHLOROPHYCEAE.
Pleurococcus vulgaris f. *cohaerens* Wittrock in Lagerheim Öfvers. K. Sv. Vet.-Akad. Förh. 40(2):
59. 1883. *P. vulgaris* var. *cohaerens* Wittrock ex Hansgirg, Prodr. Algenfl. Böhmen 1: 133. 1886.
—The original description is here designated as the temporary Type = CHLOROPHYCEAE.
Pleurococcus vulgaris f. *glomeratus* Hansgirg, Sitzungsber. K. Böhm. Ges. Wiss., Math.-Nat. Cl.,
1891(1): 320. 1891. —Type from Dalmatia: Cannosa nächst Ragusa, *A. Hansgirg,* 1891 (W)
= primordia of STIGEOCLONIUM sp. FIG. 339.
Pleurococcus vulgaris f. *minor* Rabenhorst ex Kirchner, Krypt.-Fl. Schlesien 2(1): 115. 1878.
—The original description is here designated as the temporary Type = STICHOCOCCUS SUBTILIS
(Kütz.) Klerck.
Polycoccus lucidus Roemer, Alg. Deutschl., p. 65. 1845. —The original description is here
designated as the temporary Type = CHLOROPHYCEAE.
POLYCOCCUS Kützing, Phyc. Germ., p. 148. 1845. —Type species: *Polycoccus punctiformis*
Kützing, Natuurk. Verh. Holl. Maatsch. Wetensch. Haarlem, p. 67. 1841. *Microcystis punctiformis*
Kirchner, Krypt.-Fl. Schlesien 2(1): 256. 1878. —Type from Germany: Nordhausen, *F. T.*
Kützing (ZT) = NOSTOC sp., parasitized in part.
Polycystis marginata var. *minor* Hansgirg, Prodr. Algenfl. Böhmen 2: 145. 1892. *Microcystis*
marginata var. *minor* Hansgirg ex Forti, Syll. Myxophyc., p. 91. 1907. —Type from Bohemia:
Ouzic nächst Kralup, *A. Hansgirg,* Oct. 1883 (W) = LAMPROCYSTIS ROSEA (Kütz.) Dr. & Daily.
Polycystis violacea Itzigsohn in Rabenhorst, Alg. Sachs. 31 & 32: 306. 1853. *P. purpurascens*
A. Braun ex Rabenhorst, Krypt.-Fl. Sachsen 1: 74. 1863. *Aphanothece purpurascens* A. Braun *pro*

synon. in Rabenhorst, loc. cit. 1863. *Microhaloa jodes* Itzigsohn *pro synon.* in Rabenhorst, loc. cit. 1863. *Polycystis ichthyoblabe* b. *purpurascens* A. Braun ex Rabenhorst Fl. Eur. Algar. 2: 53. 1865. *Microcystis ichthyoblabe* var. *purpurascens* A. Braun ex Forti, Syll. Myxophyc., p. 89. 1907. —Type from Germany: im Rehwinkel, *Itzigsohn,* Aug. 1852 (in Rabenh. Algen 306, B) = BACTERIA. FIG. 352.

Porphyridium aerugineum Geitler, Oesterr. Bot. Zeitschr. 72: 84. 1923. —The original description is here designated as the temporary Type = P. CRUENTUM (Sm. & Sow.) Näg.

Porphyridium magnificum Wood, Proc. Amer. Philos. Soc. 11: 144. 1869. —Type from Texas (UC) = P. CRUENTUM (Sm. & Sow.) Näg.

RHODOPLAX Schmidle & Wellheim, Bull. Herb. Boiss., ser. 2, 1: 1012. 1901. —Type species: *Porphyridium Schinzii* Schmidle, Beih. z. Bot. Centralbl. 10: 180. 1901. *Rhodoplax Schinzii* Schmidle & Wellheim, Bull. Herb. Boiss., ser. 2, 1: 1012. 1901. —Type from the Rheinfall at Schaffhausen, Switzerland (slides in the collection of G. Schmidle, ZT) = primordia of CHANTRANSIA (BATRACHOSPERMUM?) sp.

Porphyridium sordidum Geitler, Arch. f. Protistenk. 76: 595. 1932. —The original description is here designated as the temporary Type = P. CRUENTUM (Sm. & Sow.) Näg.

Porphyridium Wittrockii Richter in Wittrock & Nordstedt, Alg. Exs. 9: 440. 1882. *P. cruentum* β *Wittrockii* Hansgirg, Oesterr. Bot. Zeitschr. 35: 315. 1884. *Aphanocapsa Wittrockii* Hansgirg, Bot. Centralbl. 22: 34. 1885. *A. cruenta* var. *Wittrockii* Hansgirg, Sitzungsber. K. Böhm. Ges. Wiss., Math.-Nat. Cl., 1890(2): 138. 1891. *A. cruenta* b. *Wittrockii* Hansgirg, Prodr. Algenfl. Böhmen 2: 155. 1892. —Isotype from Germany: Anger prope Lipsiam, *P. Richter,* 29 Nov. 1881 (in Wittr. & Nordstedt, Alg. Exs. 440 (FC) = PORPHYRIDIUM CRUENTUM (Sm. & Sow.) Näg.

Protococcus angulosus Corda in Sturm, Deutschlands Fl., II, 6: 41. 1833. *Microcystis angulosa* Kützing, Linnaea 8: 374. 1833. *Pleurococcus angulosus* Meneghini, Mem. R. Accad. Sci. Torino, ser. 2, 5(Sci. Fis. & Mat.): 10, 37. 1843. *P. Meneghinii* Brébisson *pro synon.* in Meneghini, loc. cit. 1843. *P. angulosus* f. *palustris* Kützing ex Kirchner, Krypt.-Fl. Schlesien 2(1): 115. 1878. *Protococcus viridis* var. *angulosus* Wolle, Fresh-w. Alg. U. S., p. 181. 1887. *Pleurococcus angulosus* a. *palustris* Kirchner ex Hansgirg, Prodr. Algenfl. Böhmen 1: 134. 1886. —The original description is here designated as the temporary Type = CHLOROPHYCEAE.

Protococcus annulatus Pascher, Süsswasserfl. 5(Chlorophyc. 2): 226. 1915. —The original description is here designated as the temporary Type = CHLOROPHYCEAE.

Protococcus atlanticus Montagne, Ann. Sci. Nat. III. Bot. 6: 267. 1846. *Sphaerella atlantica* Lemmermann, Abh. Nat. Ver. Bremen 17: 345. 1903. —Type from the Atlantic ocean, off Portugal: inter promontoria Spichel et la Rocca, *Turrell & de Freycinet* (PC) = FLAGELLATES.

Protococcus aurantio-fuscus Kützing, Phyc. Germ., p. 146. 1845. *Chroococcus aurantio-fuscus* Rabenhorst, Fl. Eur. Algar. 2: 34. 1865. —Type from Bohemia: ad parietes in caldariis horti Pragensis, *ex herb. Meneghini* (L) = encysted FLAGELLATES.

Protococcus aureo-viridis Kützing, Phyc. Germ., p. 146. 1845. *Chroococcus aureo-viridis* Rabenhorst, Krypt.-Fl. Sachsen 1: 70. 1863. *Pleurococcus aureo-viridis* Rabenhorst, Fl. Eur. Algar. 3: 26. 1868. —Type from Germany: Nordhausen, *herb. Kützing* (L) = primordia of CHLORO-PHYCEAE.

Protococcus aureus Kützing, Tab. Phyc. 1: 3. 1846. *Protosphaeria aurea* Trevisan, Sagg. Monogr. Alg. Coccot., p. 29. 1848. *Chroococcus aureus* Rabenhorst, Krypt.-Fl. Sachsen 1: 70. 1863. *C. macrococcus* f. *aureus* Rabenhorst, Fl. Eur. Algar. 2: 33. 1865. *C. macrococcus* var. *aureus* Rabenhorst ex Hansgirg, Prodr. Algenfl. Böhmen 2: 160. 1892. —Type from Bohemia: Hirschberg, *v. Flotow 3,* 1841 (L) = FLAGELLATES.

STICHOCOCCUS Nägeli, Gatt. Einzell. Alg., p. 76. 1849. —Type species: *Protococcus bacillaris* Nägeli in Kützing, Sp. Algar., p. 198. 1849. *Stichococcus bacillaris* Nägeli, Gatt. Einzell. Alg., p. 77. 1849. *Ulothrix bacillaris* Forest, Handb. Alg., p. 72. 1954. —The original description is here designated as the temporary Type = STICHOCOCCUS BACILLARIS Näg.

PROTOSIPHON Klebs, Beding. d. Fortpfl. b. Einig. Alg. & Pilz., p. 221. 1896. —Type species: *Protococcus botryoides* Kützing, Tab. Phyc. 1: 2. 1846. *Protosiphon botryoides* Klebs, Beding. d. Fortpfl. b. Einig. Alg. & Pilz., p. 222. 1896. —Type from Germany: Nordhausen, *F. T. Kützing* (L) = P. CINNAMOMEUS (Menegh.) Dr. & Daily.

CYANIDIUM Geitler, Arch. f. Hydrobiol., Suppl. XII, 4: 624. 1933. *PLUTO* Copeland ex Drouet, Amer. Midl. Nat. 30: 672. 1943. —Type species: *Protococcus botryoides* f. *caldarius* Tilden, Bot. Gaz. 25: 94. 1898. *Pleurocapsa caldaria* Setchell in Collins, Holden & Setchell, Phyc. Bor.-Amer. 18: 851. 1901. *Chroococcopsis caldaria* Geitler, Beih. z. Bot. Centralbl., II, 41: 241. 1925. *Cyanidium caldarium* Geitler, Arch. f. Hydrobiol., Suppl. XII, 4: 624. 1933. *Pluto caldarius* Copeland ex Drouet, Amer. Midl. Nat. 30: 672. 1943. *Dermocarpa caldaria* Drouet, loc. cit. 1943. *Chlorella caldaria* M. B. Allen, Rapp. & Comm. VIII. Congr. Int. Bot. Paris 17: 42. 1954. —Type from Yellowstone national park, Wyoming: Frying Pan basin, *J. E. Tilden,* Jul. 1896 (in Tild., Amer. Alg. 283, MIN) = CHLORELLA sp. The alga collected by Miss Tilden is definitely a green alga, its starch staining blue with iodine. The alga described by Setchell, Geitler, Copeland, Allen, and Drouet appears to be another species of Chlorella with blue-green pigment (see Allen, *loc. cit.*).

Protococcus botryoides var. *nidulans* Hansgirg, Prodr. Algenfl. Böhmen 1(2): 274. 1888.
—Type from Bohemia: Nepomuk, *A. Hansgirg,* Aug. 1887 (W) = FLAGELLATES. FIG. 331.
Protococcus bruneus Kützing, Phyc. Germ., p. 146. 1845. *P. brunneus* Kützing in Roemer, Alg.
Deutschl., p. 67. 1845. *Chroococcus bruneus* Rabenhorst, Fl. Eur. Algar. 2: 34. 1865. —Type
from Germany: auf einer Lache am Hundsberge (Hartz), *Römer* (L) = TRACHELOMONAS sp.
FIG. 323.
Protococcus caldariorum Magnus ex Braun, Sitzungsber. Ges. Naturf. Fr. Berlin 1875: 99. 1875;
ibid. 1877: 249. 1877. —Type from Germany in Rabenh. Alg. Eur. 2465 (L) = primordia (per-
haps parasitized?) of TRENTEPOHLIA sp. FIG. 325.
Protococcus Chlamidomonas Kützing, Phyc. Gener., p. 167. 1843. *Protosphaeria Chlamidomonas*
Trevisan, Sagg. Monogr. Alg. Coccot., p. 30. 1848. —Type from Germany: Nordhausen, *Kützing,*
12 Sept. 1842 (L) = FLAGELLATA (CHLAMYDOMONAS sp.?). FIG. 327.
Protococcus chnaumaticus Kützing in Roemer, Alg. Deutschl., p. 67. 1845. *Chroococcus*
chnaumaticus Rabenhorst, Fl. Eur. Algar. 2: 34. 1865. —Type from Germany: auf Quarzfels des
Bruchberges (Harz), *Römer 149,* 9 Sept. 1844 (L) = CHLOROPHYCEAE (lichen gonidia). FIG. 326.
Protococcus Clementii Meneghini in Kützing, Tab. Phyc. 1: 3. 1846. *Protosphaeria Clementii*
Trevisan, Sagg. Monogr. Alg. Coccot., p. 29. 1848. *Haematacoccus Clementii* Meneghini *pro synon.*
ex Trevisan, loc. cit. 1848. *Gloeocystis Clementii* Rabenhorst, Fl. Eur. Algar. 3: 31. 1868. —Type
from the Italian Tyrol: Campagna d'Aris, *Meneghini* (L) = FLAGELLATA.
Protococcus coccoma Kützing, Phyc. Gener., p. 168. 1843. *Palmella coccoma* Kunze *pro synon.*
in Kützing, loc. cit. 1843. *Protococcus aurantiacus* Wallroth *pro synon.* in Kützing, loc. cit. 1843.
Protosphaeria coccoma Trevisan, Sagg. Monogr. Alg. Coccot., p. 29. 1848. *Chlorococcum coccoma*
Rabenhorst, Fl. Eur. Algar. 3: 59. 1868. *Haematococcus coccoma* Meneghini *pro synon.* ex
Rabenhorst, loc. cit. 1868. —Type from Germany: Leipzig, *Kunze* (L) = primordia of
CHLOROPHYCEAE.
Protococcus crassus Kützing, Tab. Phyc. 1: 4. 1846. *Chlorococcum crassum* Meneghini *pro*
synon. in Kützing, loc. cit. 1846. *Chroococcus crassus* Nägeli, Gatt. Einzell. Alg., p. 46. 1849.
—Type from Italy: Verona, *Meneghini* (L) = FLAGELLATES. FIG. 334.
Protococcus crustaceus Kützing, Phyc. Germ., p. 146. 1845. —Type from Germany: ad
corticem arborum, Nordhausen, *herb. Kützing* (L) = TRENTEPOHLIA sp., parasitized.
Protococcus dissectus var. *cuneatus* Suringar, Obs. Phycol., p. 56. 1855. —Type from the
Netherlands: Boomstammen, Gagewater bij Leiden, *W. F. R. Suringar DDD58,* May 1854 (L) =
CHLOROPHYCEAE.
Protococcus Felisii Meneghini in Kützing, Tab. Phyc. 1: 4. 1846. —Type from Italy: Ferrara,
Meneghini (L) = CHLAMYDOMONAS sp. FIG. 332.
Protococcus fuscatus G. S. West, Mém. Soc. Neuchatel. Sci. Nat. 5: 1023. 1914. —The
original description is here designated as the temporary Type = CHLOROPHYCEAE.
Protococcus gigas Kützing, Phyc. Germ., p. 145. 1845. *Pleurococcus gigas* Trevisan, Sagg.
Monogr. Alg. Coccot., p. 34. 1848. *Chlorococcum gigas* Grunow in Rabenhorst, Alg. Eur. 143—144:
1436. 1863. *Gloeocystis gigas* Lagerheim, Öfvers. K. Sv. Vet.-Akad. Förh. 40(2): 63. 1883.
Protococcus viridis var. *gigas* Wolle, Fresh-w. Alg. U. S., p. 183. 1887. —Type from Italy: Battaglia,
Meneghini (L) = encysted FLAGELLATES.
Protococcus globosus Nägeli in Kützing, Sp. Algar., p. 200. 1849. —Type from Switzerland:
Zürich, *Nägeli 369* (L) = PHYTOCONIS BOTRYOIDES (L.) Bory. FIG. 333.
Protococcus glomeratus Agardh, Flora 10(2): 629. 1827. *Micraloa olivacea* Kützing, Linnaea
8: 371. 1833. *Chlorococcum glomeratum* Rabenhorst, Fl. Eur. Algar. 3: 59. 1868. —Type from
Bohemia: Carlsbad, *Agardh* (NY) = juvenile NOSTOC sp.
Protococcus Goetzei Schmidle, Engler Bot. Jahrb. 30: 252. 1901. —The original description
is here designated as the temporary Type = CHLOROPHYCEAE.
Protococcus grumosus Richter, Hedwigia 23: 65. 1884. —Type from Germany: ad parietes
caldarii in Anger prope Lipsiam, *P. Richter,* Mar. 1882 (in Wittr. & Nordst., Alg. Exs. 1090, L) =
FLAGELLATA.
Protococcus immanis Montagne, Ann. Sci. Nat. IV. Bot. 6: 179. 1856. —Type from Bolivia:
Andina, *Weddel* (L) = encysted FLAGELLATES (DINOFLAGELLATES?).
Protococcus lilacinus Rabenhorst, Alg. Sachs. 9—10: 81. 1851. *Chroococcus lilacinus* Rabenhorst,
Krypt.-Fl. Sachsen 1: 70. 1863. —Isotypes from Germany: in dem steinernen Wassertrog auf dem
böhmischen Bahnhofe in Dresden, *L. Rabenhorst* (FC, L, NY) = BACTERIA. FIG. 326.
CHROOCOCCUS Subgenus *CHRYSOCOCCUS* Hansgirg, Notarisia 3: 589. 1888; Prodr. Algenfl.
Böhmen 2: 159. 1892. —Type species: *Protococcus macrococcus* Kützing, Phyc. Gener., p. 169.
1843. *Protosphaeria macrococca* Trevisan, Sagg. Monogr. Alg. Coccot., p. 29. 1848. *Chroococcus*
macrococcus Rabenhorst, Krypt.-Fl. Sachsen 1: 70. 1863. —Type from Germany: Cohnstein bei
Nordhausen, *herb. Kützing* (L) = encysted PERIDINIUM sp.
Protococcus marinus Kützing, Phyc. Gener., p. 169. 1843. *Microcystis marina* Kützing *pro synon.,*
loc. cit. 1843. —Isotypes from Trieste, *herb. Kützing* (FC, MO, UC) = encysted PERIDINIUM sp.
Protococcus marinus f. *Foslieanus* Hansgirg in Foslie, Mar. Alg. Norway 1: 159. 1890. —Type

from arctic Norway: Bugonaes, *M. Foslie*, 10 Aug. 1889 (L) = FLAGELLATES.
Protococcus marinus var. *Foslieanus* Hansgirg, Sitzungsber. K. Böhm. Ges. Wiss., Math.-Nat.
Cl., 1892: 239. 1893. —Type from Croatia: Fiume, *A. Hansgirg*, 1891 (W) = primordia of
CHLOROPHYCEAE.
Protococcus marinus var. *virens* Hansgirg, Sitzungsber. K. Böhm. Ges. Wiss., Math.-Nat. Cl.,
1892: 239. 1892. —Type from Istria: Pola, *A. Hansgirg*, Apr. 1889 (W) = primordia of
CHLOROPHYCEAE. FIG. 328.
PLEUROCOCCUS Untergattung *DICHOCOCCUS* Nägeli, Die Neuern Algensyst., p. 127.
1847. —Type species: *Protococcus miniatus* Kützing, Sp. Algar., p. 203. 1849. *Pleurococcus
miniatus* Nägeli, Gatt. Einzell. Alg., p. 64. 1849. —Type from Germany: Freiburg-im-
Breisgau, *A. Braun* (NY) = FLAGELLATES.
Protococcus minor Kützing, Phyc. Germ., p. 144. 1845. *P. minor* ∝ *leprosus* Kützing, Tab.
Phyc. 1: 3. 1846. *P. minor* var. *leprosus* Kützing, Sp. Algar., p. 198. 1849. *Chroococcus minor*
Nägeli, Gatt. Einzell. Alg., p. 46. 1849. *Pleurococcus minor* Rabenhorst, Krypt.-Fl. Sachsen 1: 127.
1863. *P. minor* β *minor* Kirchner ex Lagerheim, Öfvers. K. Sv. Vet.-Akad. Förh. 1882(2): 79.
1882. *P. vulgaris* var. *minor* Kirchner ex Hansgirg, Prodr. Algenfl. Böhmen 1: 133. 1886.
P. Kuetzingii G. S. West, Journ. of Bot. 42: 287. 1904. *Protococcus Kuetzingii* G. S. West,
Journ. of Bot. 54: 2. 1916. *Gloeocapsa minor* Hollerbach in Elenkin, Monogr. Algar. Cyanophyc.,
Pars Spec. 1: 238. 1938. —Type from Germany: Nordhausen, *herb. Kützing* (L) = STICHOCOCCUS
SUBTILIS (Kütz.) Klerck. FIG. 324.
Protococcus minor β *mucosus* Kützing, Tab. Phyc. 1: 3, Tab. 3, fig. 22 β. 1846. *P. minor* var.
mucosus Kützing, Sp. Algar., p. 198. 1849. *Chroococcus minor* f. *mucosus* Rabenhorst, Fl.
Eur. Algar. 2: 30. 1865. *C. minor* var. *mucosus* Kützing ex Hansgirg, Prodr. Algenfl. Böhmen 2:
165. 1892. —Type from Germany: Hanau, *F. T. Kützing 159* (L) = FLAGELLATES. FIG. 321.
Protococcus Monas β *oblongus* Flotow, Nova Acta Acad. Caes. Leop.-Car. Nat. Cur. 20(2): 478.
1842. —Topotype from Hirschberg, Bohemia, in Rabenh. Alg. Sachs. 71 (FC) = PROTOCOCCUS
GREVILLEI (Ag.) Crouan fr.
Protococcus moniliformis Flotow, Nova Acta Acad. Caes. Leop.-Car. Nat. Cur. 20(2): 459.
1842. —Topotype from Hirschberg, Bohemia, in Rabenh., Alg. Sachs. 71 (FC) = PROTOCOCCUS
GREVILLEI (Ag.) Crouan fr.
Protococcus mucosus Kützing, Phyc. Germ., p. 145. 1845. *Pleurococcus mucosus* Rabenhorst,
Krypt.-Fl. Sachsen. 1: 127. 1863. —Type from Germany: Nordhausen, *Kützing* (L) = BOTRYDIUM
GRANULATUM (L.) Grev.
Protococcus natans Agardh ex Wille, Nyt Mag. Naturvid. 56(1): 55. 1918. —Isotype from
Bohemia: Carlsbad, *ex herb. Agardh* (NY) = palmelloid basal parts of STIGEOCLONIUM TENUE
(Ag.) Kütz.
Protococcus nebulosus Kützing, Linnaea 8: 365. 1833. *Cryptococcus nebulosus* Kützing, Phyc.
Gener., p. 147. 1843. —Type from Germany: in aquae superficie natans, Nordhusae, *Kützing*
(L) = BACTERIA.
Protococcus palustris Kützing, Phyc. Gener., p. 168. 1843. *Pleurococcus palustris* Trevisan, Sagg.
Monogr. Alg. Coccot., p. 34. 1848. —Type from Germany: in stagnis prope Halle, *herb. Kützing*
(L) = primordia of CHLOROPHYCEAE.
Protococcus persicinus Diesing in Meneghini, Mem. R. Accad. Sci. Torino, ser. 2, 5(Sci. Fis. &
Mat.): 13. 1843. *Pleurococcus persicinus* Rabenhorst, Fl. Eur. Algar. 3: 28. 1868. —Type from
the Euganean springs, Italy: "Protococcus persicinus Dies." (FI) = BACTERIA.
Protococcus pluvialis β *aquaticus* Kützing, Phyc. Germ., p. 146. 1845. —Topotype from
Hirschberg, Bohemia, in Rabenh. Alg. Sachs. 71 (FC) = PROTOCOCCUS GREVILLEI (Ag.)
Crouan fr.
COHNIA Winter, Rabenh. Krypt.-Fl. 1(1): 48. 1884; non Kunth, 1850, non Reichenbach,
1852. *LAMPROCYSTIS* Schröter, Pilze Schlesiens 1: 151. 1886. —Type species: *Protococcus
roseo-persicinus* Kützing, Phyc. Germ., p. 146. 1845. *Pleurococcus roseo-persicinus* Rabenhorst, Fl.
Eur. Algar. 3: 28. 1868. *Clathrocystis roseo-persicina* Cohn, Hedwigia 12: 58. 1873. *Beggiatoa
roseo-persicina* Zopf, Morph. Spaltpfl., p. 30. 1882. *Cohnia roseo-persicina* Winter, Rabenh. Krypt.-
Fl. 1(1): 48. 1884. *Lamprocystis roseo-persicina* Schröter, Pilze Schlesiens 1: 801. 1889.
Lankesteron roseo-persicinus Ellis, Sulphur Bacteria, p. 135. 1932. —Type from Germany: in
Infusionen unter Algen, Nordhausen, *F. T. Kützing* (L) = LAMPROCYSTIS ROSEA (Kütz.) Dr. & Daily.
Protococcus roseus Meneghini, Mem. R. Accad. Sci. Torino, ser. 2, 5(Sci. Fis. & Mat.): 14.
1843. *Pleurococcus roseus* Rabenhorst, Fl. Eur. Algar. 2: 27. 1868. *Chroococcus roseus* Wille,
Nyt Mag. Naturvid. 62: 178. 1925. —Type from Italy: ad muros calce illitos, *Meneghini* (L) =
primordia of a LICHEN.
Protococcus rubiginosus Suringar, Obs. Phycol., p. 57. 1855. *Chroococcus rubiginosus* Raben-
horst, Fl. Eur. 2: 43, 1865. —Type from Holland: Sloot Oenkerk, *W. F. R. Suringar DDD55c*,
Jul. 1854 (L) = FLAGELLATA.
Protococcus sabulosus Meneghini in Kützing, Tab. Phyc. 1: 3. 1846. *Protosphaeria sabulosa*
Trevisan, Sagg. Monogr. Alg. Coccot., p. 29. 1848. *Haematococcus sabulosus* Trevisan *pro synon.*,

163

loc. cit. 1848. *Gloeocapsa sabulosa* Richter in Wittrock & Nordstedt, Alg. Exs. 14: 698. 1884. *Chroococcus sabulosus* Hansgirg, Prodr. Algenfl. Böhmen 2: 164. 1892. —Type from Italy: in exsiccatis, Baenta, *Meneghini* (L) = FLAGELLATES.

Protococcus salinus Dunal, Ann. Sci. Nat. II. Bot. 9: 1838. —The original description is here designated as the temporary Type = FLAGELLATES.

Protococcus stercorarius Berkeley in Hooker, Bot. Antarct. Voy. Erebus & Terror, 1839—43, 1(2): 501. 1847. *Pleurococcus stercorarius* Berkeley ex Hariot, Mission du Cap Horn, Alg., p. 15. 1889. —Type from Falkland islands: on cow dung, *herb.* Berkeley (K) = encysted FLAGELLATES. FIG. 322.

Protococcus tectorum Kützing, Tab. Phyc. 1: 4. 1846. *Pleurococcus tectorum* Trevisan, Sagg. Monogr. Alg. Coccot., p. 34. 1848. *P. angulosus* b. *tectorum* Kirchner ex Hansgirg, Prodr. Algenfl. Böhmen 1: 133. 1886. —Type from Germany: Strohdächer, Clausthal, *Römer* (L) = primordia of CHLOROPHYCEAE.

Protococcus variabilis Hansgirg, Physiol. & Algol. Stud., p. 87. 1887; Oesterr. Bot. Zeitschr. 37: 122. 1887. —Type from Bohemia: Kinsky'scher Garten, Smichow, *A. Hansgirg,* Sept. 1886 (W) FLAGELLATES. FIG. 322.

Protococcus viridis Agardh, Syst. Algar., p. 13. 1824. *Chlorococcum murorum* Greville, Scott. Crypt. Fl. 6: 325. 1828. *Lepraria murorum* Greville ex Hooker, Engl. Fl. 5(1) [Brit. Fl. 2]: 163. 1833. *Chlorococcum murale* Greville in Meneghini, Consp. Algol. Eugan., p. 324. 1837. *C. Agardhii* Meneghini, Giorn. Toscano Sci. Med., Fis. & Nat. 1: 186. 1840. *Haematococcus murorum* Hassall, Hist. Brit. Freshw. Alg. 1: 323. 1845. *Hassallia murorum* Trevisan, Sagg. Monogr. Alg. Coccot., p. 68. 1848. *Pleurococcus murorum* Meneghini ex Trevisan *pro synon.,* loc. cit. 1848. *Lepra murorum* Schaerer, Enum. Crit. Lich. Eur., p. 241. 1850. *Pleurococcus viridis* Rabenhorst, Krypt.-Fl. Sachsen 1: 127. 1863. —No specimens labeled "Protococcus viridis" were found in the herbarium of the Agardhs except for some collected by A. Braun in the 1840's. The original description is here designated as the temporary Type = STICHOCOCCUS SUBTILIS (Kütz.) Klerck.

Protococcus viridis var. *insignis* Hansgirg, Prodr. Algenfl. Böhmen 1(2): 141. 1888. —Type from Bohemia: Wrsowic nächst Prag, *A. Hansgirg,* May 1886 (W) = FLAGELLATES. FIG. 329.

Protococcus Wimmeri Hilse in Rabenhorst, Alg. Eur. 103—104: 1031. 1861. *Chlorococcum Wimmeri* Rabenhorst, Fl. Eur. Algar. 3: 60. 1868. *Protococcus viridis* var. *Wimmeri* Wolle, Fresh-w. Alg. U. S., p. 183. 1887. —Type from Silesia (Poland): in fodinis prope Peterwitz, *Hilse* (in Rabenh., Alg. Eur. 1031, L) = encysted FLAGELLATES.

PROTOSPHAERIA Turpin, Dict. Sci. Nat., vol. 65 (Végét. Acot.), p. 1. 1816—29. —Type species: *P. simplex* Turpin, loc. cit. 1816—29. —The original description is here designated as the temporary Type = CHLOROPHYCEAE.

Rhabdoderma Cavanillesianum Gonzalez Guerrero, Anal. Jard. Bot. Madrid 6: 244. 1946. —The original description is here designated as the temporary Type = NOSTOCACEAE.

Rhabdoderma lineare var. *spirale* Woloszynskia, Bull. Int. Acad. Sci. Cracovie, Cl. Sci. Math. & Nat., ser B, 1912: 692. 1913. *Rhabdoderma lineare* f. *spirale* Hollerbach in Elenkin, Monogr. Algar. Cyanophyc., Pars Spec. 1: 44. 1938. —The original description is here designated as the temporary Type = LYNGBYA CONTORTA Lemm.

Rhabdoderma minimum Lemmermann, Arch. f. Hydrobiol. & Planktonk. 4: 187. 1909. —The original description is here designated as the temporary Type = BACTERIA.

Rhabdoderma salinum Tutin, Trans. Linn. Soc. London, ser. 3, 1(2): 200. 1940. —Type from Bolivia: plankton, Lake Poopó, Titicaca region, *no. 193/2* (BM) = BACTERIA.

Rhodococcus intermedius Bargagli-Petrucci, N. Giorn. Bot. Ital., N. S., 22(4): 399. 1915. —The original description is here designated as the temporary Type = BACTERIA.

Rhodosphaerium diffluens Nadson, Bull. Jard. Imp. Bot. St. Petersbourg 8: 113. 1908. —The original description is here designated as the temporary Type = LAMPROCYSTIS ROSEA (Kütz.) Dr. & Daily.

RHODOSPORA Geitler, Oesterr. Bot. Zeitschr. 76: 25. 1927. —Type species: *R. sordida* Geitler, loc. cit. 1927. —The original description is here designated as the temporary Type = RHODOPHYCEAE.

RHODOSTICHUS Geitler & Pascher, Arch. f. Protistenk. 73: 305. 1931. —Type species: *R. expansus* Geitler & Pascher, loc. cit. 1931. —The original description is here designated as the temporary Type = BACTERIA.

Rosaria clandestina Skuja in Handel-Mazzetti, Symbolae Sinicae 1: 18. 1937. —Type from Szechwan, China: heisse Quelle bei Lemoka in Lolo-Lande ostlich von Ningyüen, *H. Handel-Mazzetti 1621,* 23 Apr. 1914 (W) = HAPALOSIPHON LAMINOSUS (Kütz.) Hansg.

ROSARIA Carter in Compton, Journ. Linn. Soc. Bot. 46: 54. 1922; non Carmichael, 1833. *NELLIECARTERIA* J. de Toni, Not. Nomencl. Algol. 8. 1936. —Type species: *Rosaria ramosa* Carter, Journ. Linn. Soc. Bot. 46: 54. 1922. *Nelliecarteria ramosa* J. de Toni, Not. Nomencl. Algol. 8. 1836. —Type from New Caledonia: *R. H. Compton 1181* (in the slide collection, BM) = FUNGI.

Sarcina renis Hepworth, Quart. Journ. Microsc. Soc. 5: 2. 1857. *Merismopoedia renis*

Rabenhorst, Fl. Eur. Algar. 2: 59. 1865. —The original description is here designated as the temporary Type = BACTERIA.

SARCINA Goodsir, Edinburgh Med. & Surg. Journ. 57: 434, 437. 1842. —Type species: *S. ventriculii* Goodsir, ibid. p. 433, 437. 1842. *Merismopoedia ventriculii* Robin, Hist. Nat. Végét. Parasites, p. 331. 1853. *M. Goodsirii* Husemann, De Animal. & Vegetab. in Corp. Humano Parasitant., p. 13. 1853. —Paratype, designated here as the Type, in Rabenh., Alg. Sachs. 600 (L) = SARCINA VENTRICULII Goods. FIG. 353.

Sarcina Welckeri Rossmann, Zeitschr. f. Wiss. Bot. 40: 643. 1857. *S. vesicae* Rossmann *pro synon.,* loc. cit. 1857. *S. urinae* Rossmann *pro synon.,* loc. cit. 1857. *Merismopoedia urinae* Rabenhorst, Fl. Eur. Algar. 2: 59. 1865. —The original description is here designated as the temporary Type = BACTERIA.

SARCODERMA Ehrenberg, Ann. der Phys. & Chem. 18: 504. 1830. —Type species: *S. sanguineum* Ehrenberg, loc. cit. 1830. —The original description is here designated as the temporary Type = EUGLENA sp.

SCHRAMMIA Dangeard, Le Botaniste, ser. 1, 1889: 161. 1889. —Type species: *S. barbata* Dangeard, ibid., p. 158. 1889. —The original description is here designated as the temporary Type = GLOEOCHAETE WITTROCKIANA Lagerh., according to Lagerheim in Nuova Notarisia 1: 227 (1890) = CHLOROPHYCEAE.

Siderocapsa major f. *chromophila* Naumann, K. Sv. Vet.-Akad. Handl. 62(4): 16. 1921. —The original description is here designated as the temporary Type = BACTERIA.

Siderocapsa major f. *chromophoba* Naumann, K. Sv. Vet.-Akad. Handl. 62(4): 16. 1921. —The original description is here designated as the temporary Type = BACTERIA.

SIDEROTHECE Naumann, K. Sv. Vet.-Akad. Handl. 62(4): 17. 1921. —Type species: *S. major* Naumann, loc. cit. 1921. —The original description is here designated as the temporary Type = BACTERIA.

Siderothece minor Naumann, K. Sv. Vet.-Akad. 62(4): 17. 1921. —The original description is here designated as the temporary Type = BACTERIA.

Sorochloris vinosa Kufferath, Bull. Inst. Roy. Sci. Nat. Belg. 26(2): 10. 1950. —Topotype from Belgium: with *Coelosphaerium Leloupii* Kuff., Boirs-sur Geer, *L. Brongersma,* 28 Nov. 1951 (D, L) = BACTERIA.

Sorospora virescens Hassall, Hist. Brit. Freshw. Alg. 1: 310. 1845. *Microhaloa virescens* Kützing, Sp. Algar., p. 207. 1849. *Cagniardia virescens* Trevisan, Sagg. Monogr. Alg. Coccot., p. 50. 1848. *Aphanocapsa virescens* Rabenhorst, Fl. Eur. Algar. 2: 48. 1865. —Type from England: Dorset, 1847 (in the slide collection, BM) = FLAGELLATES.

Sphaenosiphon minimus Reinsch, Contrib. Algol. & Fungol. 1: 16. 1874. —Type on *Bryopsis Balbisiana* from Italy: ad Genuam, *J. G. Agardh* (FC) = BACTERIA.

Sphaenosiphon sorediiformis Reinsch, Contrib. Algol. & Fungol. 1: 16. 1874. *Dermocarpa sorediiformis* Geitler, Rabenh. Krypt.-Fl. 14: 396. 1932. —Type from the Mediterranean sea: in *Giraudia simplex* (K) = CHLOROPHYCEAE (ENTOCLADIA sp?).

Sphaerogonium gracile Rostafinski, Rozpr. Akad. Umiej. Krakow., Wydz. Math.-Przyr., 10: 293. 1883. *Chamaesiphon Rostafinskii* Hansgirg, Physiol. & Algol. Stud., p. 164. 1887; Oesterr. Bot. Zeitschr. 37: 56. 1887. —Topotype from Polish Ukraine: Siklawika pod Giewontem, *M. Raciborski,* 3 Oct. 1909 (WU) = juvenile AMPHITHRIX JANTHINA (Mont.) Born. & Flah.

Staurogenia Tetrapedia Kirchner, Jahresh. Ver. Vaterl. Naturk. Württemberg 36: 168. 1880. *Tetrapedia Kirchneri* Lemmermann, Ber. Deutsch. Bot. Ges. 18: 24. 1900. —The original description is here designated as the temporary Type = CHLOROPHYCEAE.

Stichosiphon Hansgirgii Geitler, Rabenh., Krypt.-Fl. 14: 414. 1932. —Type from Bohemia: Johannisbad, *A. Hansgirg,* Jun. 1885 (W) = AMPHITHRIX JANTHINA (Mont.) Born. & Flah. FIG. 301.

Synechococcus bigranulatus Skuja, Fedde Repert. Sp. Nov. 31: 7. 1933. —The original description is here designated as the temporary Type = PHORMIDIUM LAMINOSUM (Ag.) Gom., hormogonia.

Synechococcus Bosshardii Skuja, Fedde Repert. Sp. Nov. 31: 7. 1933. —The original description is here designated as the temporary Type = PHORMIDIUM LAMINOSUM (Ag.) Gom., hormogonia.

Synechococcus brunneolus Rabenhorst, Krypt.-Fl. Sachsen 1: 75. 1863. —Isotypes with *Sirosiphon coralloides* from Saxony in Rabenh., Alg. Sachs. 224 (FC) = MESOTAENIUM RUBESCENS (Bréb.) Dr. & Daily.

Synechococcus elongatus Geitler, Oesterr. Bot. Zeitschr. 70: 166. 1921. *S. Geitleri* J. de Toni, Noter. Nomencl. Algol. 3. 1936. —Paratype from western Java: Travertin-Hügel von Kuripan, *Ruttner* (slide BK2b in the collection of L. Geitler) = PHORMIDIUM TENUE (Menegh.) Gom., hormogonia.

Synechococcus fuscus Zeller, Journ. Asiatic Soc. Bengal 42(2): 176. 1873; Hedwigia 12: 169. 1873. —Type from Burma: Yomah, *S. Kurz 3258* (L) = mature spores of CYLINDROSPERMUM LICHENIFORME (Bory) Kütz.

Synechococcus Gaarderi Ålvik, Bergens Mus. Årbok 1934 (Naturvid. Rekke 6): 40. 1934. —The original description is here designated as the temporary Type = BACTERIA.
Synechococcus intermedius Gardner, Mem. New York Bot. Gard. 7: 3. 1927. —Type from Puerto Rico: Caguas, *N. Wille 439b*, 1915 (NY) = hormogonia of SCYTONEMA HOFMANNII Ag. FIG. 354.
Synechococcus kerguelensis Wille, Deutsche Südpolar-Exped., 1901—03, 8: 417. 1928. —The original description is here designated as the temporary Type = BACTERIA.
Synechococcus roseo-persicinus Grunow in Rabenhorst, Fl. Eur. Algar. 3: 418. 1863. —Type from Slovakia: in aqua stagnante ad Olascinum Sepusii, *C. Kalchbrenner* (W) = BACTERIA.
Synechococcus salinus Frémy in J. de Toni, Diagn. Alg. Nov., I, Myxophyc. 8: 703. 1946. —The original description is here designated as the temporary Type = BACTERIA.
Synechococcus ulcericola Rabenhorst, Fl. Eur. Algar. 3: 418. 1868. —The original description is here designated as the temporary Type = spores of FUNGI.
COCCOBACTREACEAE Elenkin, Acta Inst. Bot. Acad. Sci. U. R. S. S., II, 1: 19. 1933. —Type genus: *SYNECHOCYSTIS* Sauvageau, Bull. Soc. Bot. France, ser. 2. 14: cxv. 1892.
SYNECHOCOCCUS Subgenus *SYNECHOCYSTIS* Elenkin, Not. Syst. Inst. Crypt. Horti Bot. Petropol. 2: 65. 1923. —Type species: *Synechocystis aquatilis* Sauvageau, Bull. Soc. Bot. France, ser. 2, 14: cxvi. 1892. —Type from Algeria: Hamman-Salahin près Biskra, *C. Sauvageau*, Apr. 1892 (PC) = palmelloid CHLOROPHYCEAE. FIG. 357.
Synechocystis salina Wislouch, Acta Soc. Bot. Polon. 2: 111. 1924. —The original description is here designated as the temporary Type = BACTERIA.
Tetrachloris merismopedioides Skuja, Symbolae Bot. Upsal. 9(3): 27. 1948. —The original description is here designated as the temporary Type = BACTERIA.
Tetrapedia aversa W. & G. S. West, Journ. of Bot. 35: 301. 1897. —Type from Angola: in stagnis prope Anbilla (Condo), Pungo Andongo, *F. Welwitsch 177*, Mar. 1857 (BM) = TETRAEDRON sp.
Tetrapedia crux-Michaeli Reinsch, Algenfl. Mittl. Th. Franken, p. 38. 1867. *T. crux-melitensis* Reinsch ex Lemmermann, Krypt.-Fl. Mark Brandenb. 3: 89. 1907. —The original description is here designated as the temporary Type = PEDIASTRUM sp.
LEMMERMANNIA Chodat, Mém. Herb. Boissier 17: 3. 1900. —Type species: *Tetrapedia emarginata* Schröder, Ber. Deutsch. Bot. Ges. 15: 490, 492. 1897. *Lemmermannia emarginata* Chodat, Mém. Herb. Boissier 17: 5. 1900. —The original description is here designated as the temporary Type = CHLOROPHYCEAE.
Tetrapedia foliacea Turner, K. Sv. Vet.-Akad. Handl. 25(I, 5)): 12. 1892. —The original description is here designated as the temporary Type = CHLOROPHYCEÁE (TETRAEDRON sp.?).
TETRAPEDIACEAE Elenkin, Acta Inst. Bot. Acad. Sci. U. R. S. S., II, 1: 19: 1933. —Type genus: *TETRAPEDIA* Reinsch, Algenfl. Mittl. Th. Franken, p. 37. 1867; Act. Soc. Senckenb. 6: p. 30 of reprint. 1867. —Type species: *T. gothica* Reinsch, loc. cit. 1867. —The original descriptions are here designated as the temporary Type = TETRAEDRON sp.
Tetrapedia morsa W. & G. S. West, Journ. Linn. Soc. Bot., ser. 2, 5: 85. 1895. —Type from Madagascar: near Lake Alastra, *R. Baron 3* (BIRM) = TETRAEDRON sp.
Tetrapedia Penzigiana de Toni, Hedwigia 1891: 195. 1891. —The original description is here designated as the temporary Type = PEDIASTRUM sp.
Tetrapedia Reinschiana Archer, Quart. Journ. Microsc. Sci., N. S., 12: 364. 1872. —The original description is here designated as the temporary Type = TETRAEDRON sp.
Tetrapedia setigera Archer, Quart. Journ. Microsc. Sci., N. S., 12: 365. 1872. *Polyedrium trigonum* var. *setigerum* Schröder, Forschungsber. Biol. Sta. Plön 6: 23. 1898. *Tetraedron trigonum* var. *setigerum* Lemmermann, Arch. f. Bot. 2(2): 110. 1904. *Treubaria setigera* G. M. Smith, Fresh-w. Alg. U. S., p. 499. 1933. —The original description is here designated as the temporary Type = TETRAEDRON sp.
Tetrapedia trigona W. & G. S. West, Journ. Linn. Soc. Bot. 30: 277. 1894. —Type from St. Vincent, British West Indies: with *Cosmarium pseudopyramidatum, W. R. Elliott 477A* (in the slide collection, BM) = TETRAEDRON sp.
Tetrapedia Wallichiana Turner, K. Sv. Vet.-Akad. Handl. 25(I, 5): 12. 1892. —The original description is here designated as the temporary Type = PEDIASTRUM sp.
Tetraspora didyma Liebmann, Naturh. Tidskr. 2: 473. 1838—39. —Type from Denmark: in amne ad Nyemolle, *Liebmann* (C) = TETRASPORA GELATINOSA (Vauch.) Desv.
Thelephora sanguinea Persoon, Syn. Meth. Fung., p. 575. 1801. *Palmella sanguinea* Ehrenberg, Ann. der Physik & Chem. 18: 510. 1830. —Type labeled: "Thelephora sanguinea", *Greville*, in herb. Persoon (L) = FECES OF INSECTS.
THIOPEDIA Winogradsky, Beitr. a. Morphol. & Physiol. der Bacteria 1: 85. 1888. —Type species: *T. rosea* Winogradsky, loc. cit. 1888. *Lampropedia rosea* de Toni & Trevisan in Saccardo, Syll. Fung. 8: 1049. 1889. —The original description is here designated as the temporary Type = ERYTHROCONIS LITTORALIS Oerst.
MERRETTIA S. F. Gray, Nat. Arr. Brit. Pl., p. 348. 1821. —Type species: *Tremella adnata*

Linnaeus, Sp. Plant., ed. 2, 2: 1626. 1763. *Palmella adnata* Lyngbye, Tent. Hydrophyt. Dan., p. 205. 1819. *Merrettia adnata* S. F. Gray, Nat. Arr. Brit. Pl., p. 348. 1821. *Microcystis adnata* Meneghini, Mem. R. Accad. Sci. Torino, ser. 2, 5 (Sci. Fis. & Mat.): 47. 1843. *Bichatia adnata* Trevisan, Nomencl. Algar. 1:59. 1845. *Gloeocystis adnata* Nägeli, Gatt. Einzell. Alg., p. 65. 1849. —The original description is here designated as the temporary Type = LICHEN?
PORPHYRIDIUM Nägeli, Gatt. Einzell. Alg., p. 71. 1849. *RHODOCOCCUS* Sectio *PORPHY-RIDIUM* Hansgirg, Oesterr. Bot. Zeitschr. 34: 315. 1884. *APHANOCAPSA* Subgenus *PORPHY-RIDIUM* Hansgirg, Prodr. Algenfl. Böhmen 2: 154. 1892. —Type species: *Tremella cruenta* Smith & Sowerby, Engl. Bot. 25: 1800. 1807. *Olivia cruenta* S. F. Gray, Nat. Arr. Brit. Pl., p. 350. 1821. *Palmella cruenta* Agardh, Syst. Algar., p. 15. 1824. *Coccochloris cruenta* Sprengel, Linn. Syst. Vegetab., ed. 16, 5(1): 373. 1827. *Phytoconis cruenta* Trevisan, Prosp. Fl. Eugan., p. 57. 1842. *Porphyridium cruentum* Nägeli, Gatt. Einzell. Alg., p. 71. 1849. *Aphanocapsa cruenta* Hansgirg, Ber. Deutsch. Bot. Ges. 3: 16. 1885; Bot. Centralbl. 22: 55. 1885. —Type from London, England (BM) = PORPHYRIDIUM CRUENTUM (Sm. & Sow.) Näg.
TRICHODICTYON Kützing, Phyc. Germ., p. 153. 1845. —Type species: *T. rupestre* Kützing, loc. cit. 1845. *Penium rupestre* Rabenhorst, Fl. Eur. Algar. 3: 120. 1868. *Gyges rupestris* Kuntze, Rev. Gen. Pl. 2: 896. 1891. *Cylindrocystis rupestris* Dalla Torre & Sarnthein, Fl. Tirol, Vorarlb. & Liechtenst. 2: 59. 1901. —Type from Germany: an feuchten Felsen, Clausthal, *Römer* (L) = MESOTAENIUM RUBESCENS (Bréb.) Dr. & Daily.
TUBIELLACEAE Elenkin, Sovietsk. Bot. 2(5): 56. 1934. —Type genus: *TUBIELLA* Hollerbach in Elenkin, loc. cit. 1934; Hollerbach, Acta Inst. Bot. Acad. Sci. U. R. P. S. S., II, 2: 34. 1935. —Type species: *T. Elenkinii* Hollerbach, loc. cit. 1934, 1935. —The original description is here designated as the temporary Type = young NOSTOC sp.
HYDROGASTRUM Desvaux, Obs. sur les Pl. des Env. d'Angers, p. (18). 1818. *DES-MAZIERELLA* Gaillon, Aperçu d'Hist. Nat., p. 33. 1833. —Type species: *Ulva granulata* Linnaeus, Sp. Plant., p. 1164. 1753. *Tremella globosa* Weis, Pl. Crypt. Fl. Gottingens., p. 28. 1770. *Linckia granulata* Weber in Wiggers, Primit. Fl. Holsat., p. 94. 1780. *Hydrogastrum granulatum* Desvaux, Obs. sur les Pl. des Env. d'Angers, p. (19). 1818. *Vaucheria radicata* Agardh, Syst. Algar., p. 173. 1824. *Coccochloris radicata* Sprengel, Linn. Syst. Vegetabil., ed. 16, 4(1): 372. 1827. *Botrydium granulatum* Greville, Alg. Brit., p. 196. 1830. *Desmazierella granulata* Gaillon, Aperçu d'Hist. Nat., p. 33. 1833. —Type in the Linnaean herbarium in the Linnaean Society of London = BOTRYDIUM GRANULATUM (L.) Grev.
SCYTHYMENIA Agardh, Syst. Algar., xx. 1824. —Type species: *Ulva rupestris* Smith & Sowerby, Engl. Bot. 31: 2194. 1810. *Merrettia coriacea* S. F. Gray, Nat. Arr. Brit. Pl., p. 348. 1821. *Scythymenia rupestris* Agardh, Syst. Algar., p. 30. 1824. —Type from England; E. B. 2194, *Dr. Smith* in herb. Sowerby (BM) = STIGONEMA MINUTUM (Ag.) Hass.
PROTOCOCCUS Agardh, Syst. Algar., p. xvii. 1824. —Type species: *Uredo nivalis* Bauer, Quart. Journ. Lit., Sci. & Arts 7: 225. 1819. *Globulina sanguinea* Turpin, Dict. Sci. Nat. 65 (Végét. Acot.): pl. 7, fig. 1. 1816—29; Mém. Mus. d'Hist. Nat. Paris 14: 26, 63. 1827. *Protococcus nivalis* Agardh, Syst. Algar., p. 13. 1824. *Sphaerella nivalis* Sommerfelt, Mag. f. Naturvid. 4: 252. 1824. *Palmella nivalis* Hooker, Append. Parry's Journ., 2nd Voy. 1821—23, p. 428. 1825. *Coccochloris nivalis* Sprengel, Linn. Syst. Vegetabil., ed. 16, 4(1): 373. 1827. *Coccophysium nivale* Link, Handb. Erkenn. Gewächse 3: 342. 1833. *Chlamydomonas nivalis* Wille, Nyt Mag. Naturvid. 41(1): 147. 1903. —Type from Baffin bay, Canada: "specimen ex itinere Rossii" in herb. Agardh (LD) = PROTOCOCCUS NIVALIS (Bauer) Ag.
Xenococcus Schousboei var. *pallidus* Hansgirg, Oesterr. Bot. Zeitschr. 39: 5. 1889. —Type from Dalmatia: Zara, *A. Hansgirg*, Aug. 1888 (W) = primordia of CHLOROPHYCEAE. FIG. 358.
ZACHARIASIA Lemmermann, Forschungsber. Biol. Sta. Plön 3: 60. 1895. —Type species: *Z. endophytica* Lemmermann, loc. cit. 1895. —The original description is here designated as the temporary Type = RHODOPHYCEAE, according to Geitler & Mattfeld in Engler & Prantl, Natürl. Pflanzenfam., ed. 2, 1b: 225 (1942).

ILLUSTRATIONS

The photomicrographs were made from herbarium material mounted in water or dilute solutions of detergents on mica slips or glass slides. At the magnification (430 X) and with the equipment used, 1 millimeter equals 2.0 microns. Mrs. Fay K. Daily processed the films and made the prints reproduced here.

The prints for Figs. 15, 90A, 107, 142, 144, 162, 175 and 311 were prepared from negatives taken by members of the staffs of the British Museum (Natural History), London, and the Eidgenossische Technische Hochschule, Zürich.

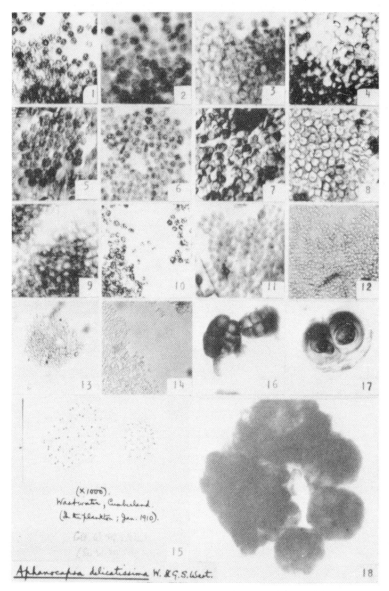

Figs. 1-18.—1. *Microcystis toxica* Stephens, Type, L. 2. *Polycystis scripta* Richter, Type, B.
3. *Polycystis ochracea* Brand, Type, B. 4. *Polycystis flos-aquae* Wittrock, Type, S. 5.
Microcystis protocystis Crow, Type in coll. of F. E. Fritsch. 6. *Polycystis prasina* Wittrock,
Type, S. 7. *Polycystis aeruginosa* var. *major* Wittrock, Isotype, FC. 8. *Chroococcus indicus* Zeller, Type, L. 9. *Micraloa aeruginosa* Kützing, Type, L. 10. *Polycystis (Clathrocystis) insignis* Beck, Isotype, FC. 11. *Clathrocystis robústa* Clark, Type, FC. 12. *Aphanothece clathrata* var. *rosea* Skuja, Paratype, D. 13. *Microcystis pallida* Migula, Isotype, FC.
14. *Aphanocapsa minima* Migula, Isotype, FC. 15. *Aphanocapsa delicatissima* W. & G. S.
West, West notes, BM. 16. *Gloeocapsa scopulorum* Nägeli, Type, L. 17. *Gloeocapsa alpina* Nägeli, Type, FC. 18. *Entophysalis atroviolacea* Novacek, Type, D.

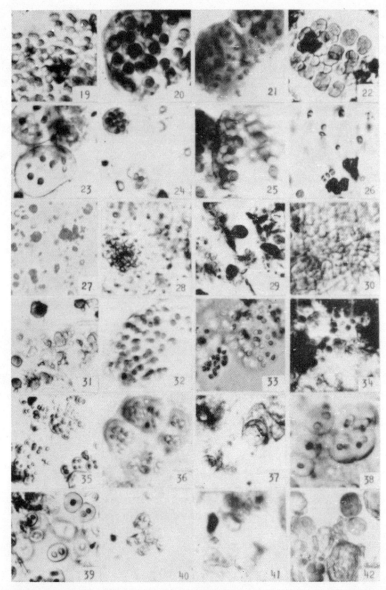

Figs. 19-42.—19. *Anacystis marginata* Meneghini, Type, L. 20. *Anacystis amplivesiculata* Gardner, Type, NY. 21. *Gloeocapsa multispherica* Gardner, Type, FH. 22. *Gloeocapsa saxicola* Wartmann, Type, FC. 23. *Gloeocapsa ianthina* Nägeli, Type, L. 24. *Endospora nigra* Gardner, Type, NY. 25. *Anacystis compacta* Gardner, Type, NY. 26. *Chroococcus muralis* Gardner, Type, NY. 27. *Protococcus pygmaeus* Kützing, Type, L. 28. *Aphanothece opalescens* Gardner, Type, NY. 29. *Endospora mellea* Gardner, Type, NY. 30. *Aphanocapsa biformis* A. Braun, Type, FC. 31. *Chroococcus cubicus* Gardner, Type, NY. 32. *Aphanocapsa intertexta* Gardner, Type, NY. 33. *Gloeocapsa geminata* Kützing, Type, L. 34. *Endospora bicoccus* Gardner, Type, NY. 35. *Gloeothece parvula* Gardner, Type, NY. 36. *Gloeocapsa tropica* Crouan, Type, PC. 37. *Gloeocapsa Shuttleworthiana* Kützing, Type, L. 38. *Haematococcus frustulosus* Harvey, Type, K. 39. *Gloeocapsa calcicola* Gardner, Type, NY. 40. *Pleurocapsa Mayorii* Setchell, Type, UC. 41. *Endospora olivacea* Gardner, Type, NY. 42. *Gloeocapsa violacea* Kützing, Type, L.

170

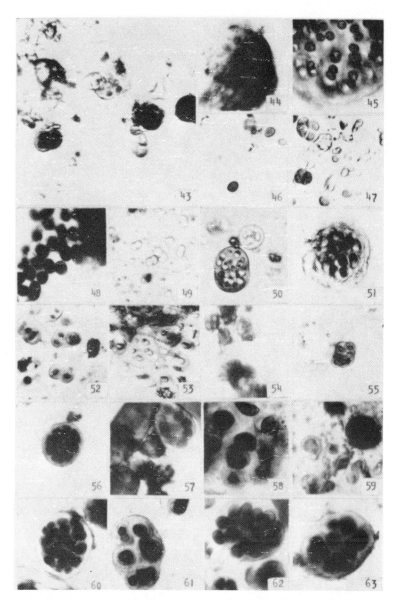

Figs. 43-63.—43. *Gloeocapsa compacta* var. *coeruleoatra* Novacek, Type, D. 44.
Anacystis gloeocapsoides Gardner, Type, NY. 45. *Gloeocapsa alpina* var. *mediterranea*
Hansgirg, Isotype, FC. 46. *Protococcus fusco-ater* Kützing, Type, L. 47. *Gloeocapsa
nigrescens* f. *vitrea* Novacek, Type, D. 48. *Monocapsa stegophila* Itzigsohn, Type, FC.
49. *Gloeocapsa compacta* Kützing, Type, L. 50. *Gloeocapsa versicolor* Nägeli, Type, L.
51. *Anacystis radiata* var. *major* Gardner, Type, NY. 52. *Gloeocapsa livida* var. *minor*
Gardner, Type, NY. 53. *Gloeothece endochromatica* Gardner, Type, NY. 54. *Gloeocapsa
chroococcoides* Novacek, Type, D. 55. *Aphanothece saxicola* var. *sphaerica* W. & G. S.
West, Type, BIRM. 56. *Anacystis radiata* var. *major* Gardner, Type, NY. 57. *Endospora
rubra* Gardner, Type, NY. 58. *Anacystis magnifica* Gardner, Type, NY. 59. *Gloeocapsa
Dvorakii* Novacek, Type, D. 60. *Anacystis radiata* Gardner, Type, NY. 61. *Gloeocapsa
sanguinolenta* Kützing, Type, L. 62. *Anacystis nidulans* Gardner, Type, NY. 63.
Anacystis pulchra Gardner, Type, NY.

171

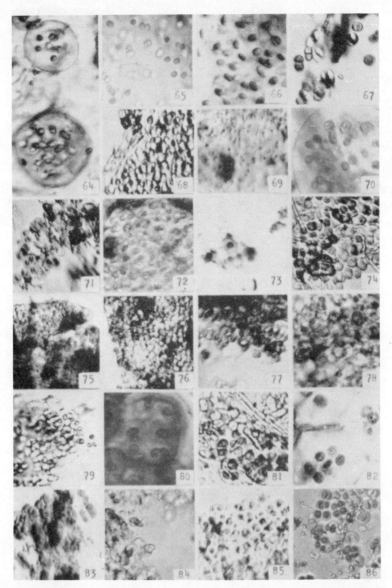

Figs. 64-86.—64. *Microcystis nigra* Meneghini, Type, FI. 65. *Palmella livida* Carmichael, Type, K. 66. *Placoma Willei* Gardner, Type, NY. 67. *Synechococcus ambiguus* Skuja, Isotype, FC. 68. *Gloeothece tepidariorum* var. *cavernarum* Hansgirg, Isotype, FC. 69. *Aphanocapsa nebulosa* A. Braun, Type, FC. 70. *Gloeocapsa haematodes* var. *violascens* Grunow, Type, W. 71. *Aphanocapsa thermalis* var. *minor* Hansgirg, Isotype, FC. 72. *Gloeothece rhodochlamys* Skuja, Type, FC. 73. *Anacystis minutissima* Gardner, Type, NY. 74. *Chroococcus varius* var. *luteolus* Hansgirg, Isotype, FC. 75. *Anacystis nigropurpurea* Gardner, Type, NY. 76. *Coelosphaerium limnicola* Lund, Type, in coll. of J. W. G. Lund. 77. *Coccochloris parietina* Meneghini, Type, FI. 78. *Chroococcus cohaerens* Nägeli, Type, FC. 79. *Coccochloris muscicola* Meneghini, Type, FI. 80. *Palmella sanguinea* Agardh, Type, LD. 81. *Gloeocapsa conglomerata* Kützing, Type, L. 82. *Chondrocystis Bracei* Howe, Type, NY. 83. *Pleurococcus communis* Meneghini, Type, FI. 84. *Haematococcus aeruginosus* Hassall, Type, K. 85. *Chroococcus varius* f. *atrovirens* A. Braun, Type, FC. 86. *Chroococcus violaceus* Rabenhorst, Type, L.

172

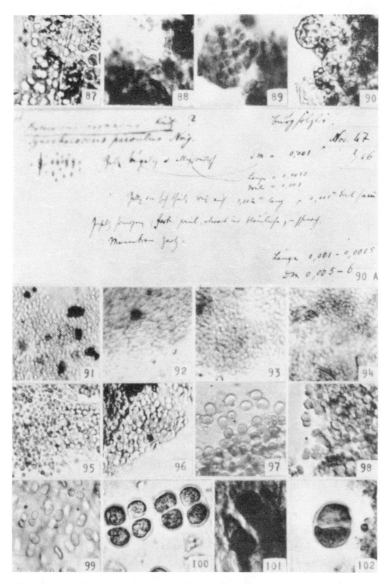

Figs. 87-102.—87. *Palmella brunnea* A. Braun, Type, L. 88. *Protococcus haematodes* Kützing, Type, L. 89. *Gloeocapsa pleurocapsoides* Novacek, Type, D. 90. *Gloeocapsa aeruginosa* Kützing, Type, L. 90A. *Synechococcus parvulus* Nägeli, from Nägeli notes in Zürich. 91. *Chroococcus parallelepipedon* Schmidle, Type, ZT. 92. *Aphanocapsa Farlowiana* Drouet & Daily, Type, D. 93. *Anacystis glauca* Wolle, Type, FC. 94. *Palmella sudetica* Rabenhorst, Type, L. 95. *Palmella hyalina* γ *seriata* Kützing, Type, L. 96. *Nostoc microscopicum* var. *linguiforme* Hansgirg, Isotype, FC. 97. *Aphanocapsa elachista* var. *irregularis* Boye Petersen, Type in coll. of J. Boye Petersen. 98. *Coccochloris stagnina* f. *gelatinosa* Hennings, Type, S. 99. *Synechocystis Willei* Gardner, Type, NY. 100. *Protococcus turgidus* Kützing & β *fuscescens* Kützing, Type, L. 101. *Chroococcus* (*Rhodococcus*) *insignis* Schmidle, Type, ZT. 102. *Chroococcus turgidus* var. *Hookeri* Lagerheim, Type, S.

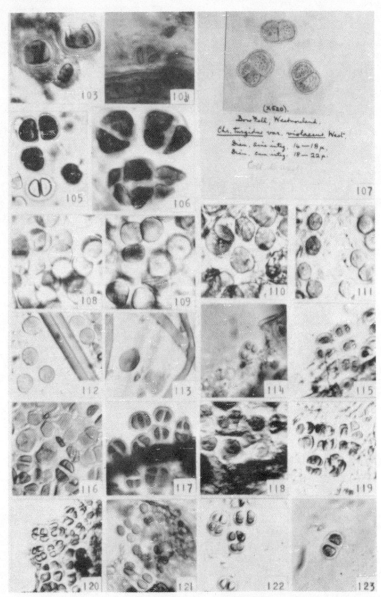

Figs. 103-123.—103. *Chroococcus schizodermaticus* var. *badio-purpureus* W. & G. S. West, Type, BM. 104. *Chroococcus turgidus* var. *submarinus* Hansgirg, Isotype, FC. 105. *Trochiscia dimidiata* Kützing, Type, L. 106. *Chroococcus giganteus* var. *occidentalis* Gardner, Type, NY. 107. *Chroococcus turgidus* var. *violaceus* West, in West notes, BM. 108. *Chroococcus smaragdinus* Hauck, Type, L. 109. *Aphanocapsa Howei* Collins, Type, NY. 110. *Palmogloea aeruginosa* Zanardini, Type, L. 111. *Aphanocapsa sesciacensis* Fremy, Type, NY. 112. *Chroococcus Raspaigellae* Hauck, Type, L. 113. *Aphanocapsa Feldmannii* Fremy, Type, NY. 114. *Chroococcus helveticus* Nägeli, Paratype, FC. 115. *Gloeocapsa ovalis* Gardner, Type, NY. 116. *Chroococcus virescens* Hantzsch, Type, L. 117. *Chroococcus subsphaericus* Gardner, Type, NY. 118. *Aphanocapsa salinarum* Hansgirg, Isotype, FC. 119. *Chroococcus helveticus* var. *aureo-fuscus* Hansgirg, Isotype, FC. 120. *Chroococcus turgidus* var. *glomeratus* Hansgirg, Isotype, FC. 121. *Chroococcus mediocris* Gardner, Type, NY. 122. *Chroococcus helveticus* var. *aurantio-fuscescens* Hansgirg, Isotype, FC. 123. *Pleurococcus membraninus* Meneghini, Type, FI.

174

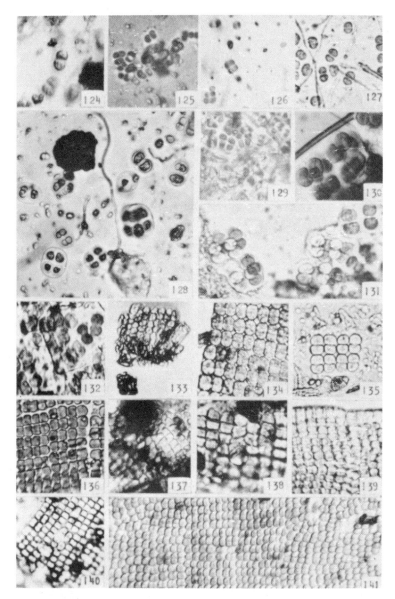

Figs. 124-141.—124. *Gloeocapsa calcarea* Tilden, Type, MIN. 125. *Pleurococcus cohaerens* Brébisson, Cotype, L. 126. *Chroococcus membraninus* var. *crassior* Hansgirg, Isotype, FC. 127. *Palmella testacea* A. Braun, Type, L. 128. *Chroococcus decorticans* A. Braun, Type, L. 129. *Protococcus pallidus* Nägeli, Type, L. 130. *Chroococcus Prescottii* Drouet & Daily, Type, D. 131, 132. *Chroococcus helveticus* f. *major* Lagerheim, Type, NY. 133. *Merismopoedia glauca* var. *fontinalis* Hansgirg, Isotype, FC. 134. *Merismopoedia elegans* A. Braun, Type, L. 135. *Merismopoedia major* Kützing, Type, L. 136. *Merismopoedia aeruginea* Brébisson, Type, L. 137. *Merismopoedia Willei* Gardner, Type, NY. 138. *Merismopoedia glauca* subsp. *amethystina* Lagerheim, Isotype, L. 139. *Merismopoedia convoluta* Brébisson, Type, L. 140. *Merismopoedia thermalis* Kützing, Type, L. 141. *Prasiola Gardneri* Collins, Type, NY.

175

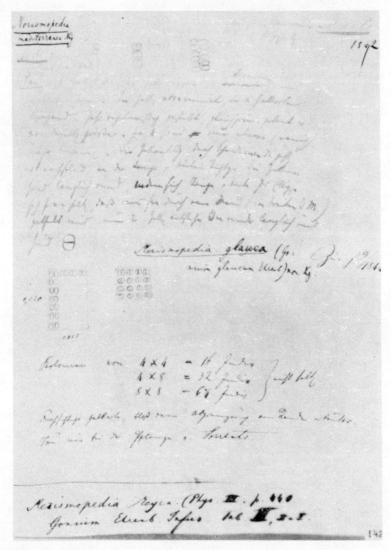

Fig. 142. *Merismopoedia mediterranea* Nägeli, from Nägeli notes, Zürich.

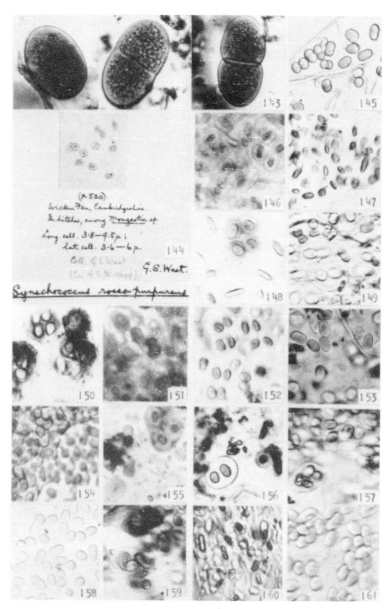

Figs. 143-161.—143. *Synechococcus major* f. *crassior* Lagerheim, Type, S. 144. *Synechococcus roseo-purpureus* G. S. West, from West notebooks, BM. 145. *Protococcus minutus* Kützing, Type, L. 146. *Gloeocapsa quaternata* var. *major* Gardner, Type, NY. 147. *Aphanothece Naegelii* Wartmann, Type, ZT. 148. *Palmella rupestris* Lyngbye, Type, C. 149. *Gloeothece decipiens* A. Braun, Type, L. 150. *Chroococcus montanus* Hansgirg, Isotype, FC. 151. *Aphanothece microscopica* var. *granulosa* Gardner, Type, NY. 152. *Microcystis paroliniana* Meneghini, Type, FI. 153. *Palmella heterococca* Kützing, Type, L. 154. *Coccochloris stagnina* Sprengel, Type, FC. 155. *Gloeothece interspersa* Gardner, Type, NY. 156. *Anacystis cylindracea* Gardner, Type, NY. 157. *Gloeocapsa tepidariorum* A. Braun, Type, L. 158. *Palmella Mooreana* Harvey, Type, K. 159. *Gloeocapsa Kuetzingiana* Nägeli, Type, L. 160. *Aphanothece conferta* var. *brevis* Gardner, Type, NY. 161. *Aphanothece* (*stagnina* var.) *prasina* A. Braun, Type, FC.

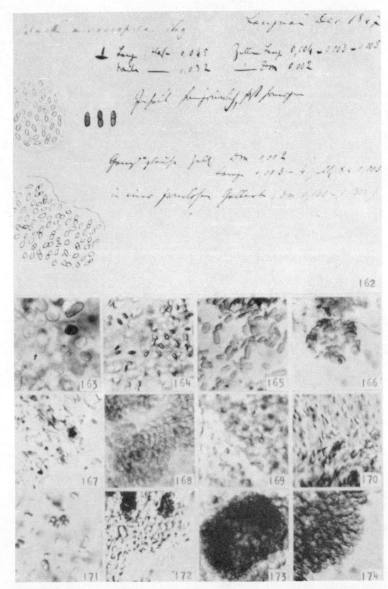

Figs. 162-174.—162. *Aphanothece microscopica* Nägeli, from Nägeli notes, Zürich. 163. *Gloeocapsa gelatinosa* Kützing, Type, L. 164. *Polycystis Packardii* Farlow, Type, FH. 165. *Aphanothece Lebrunii* Duvigneaud & Symoens, Type, BR. 166. *Aphanothece conglomerata* Rich, Type, D. 167. *Synechococcus curtus* Setchell, Type, UC. 168. *Micraloa elabens* Brébisson, Type, L. 169. *Aphanothece utabensis* Tilden, Type, MIN. 170. *Aphanothece saxicola* var. *aquatica* Wittrock, Type, S. 171. *Gloeothece prototypa* Gardner, Type, NY. 172. *Aphanothece subachroa* Hansgirg, Isotype, FC. 173. *Coelosphaerium Wichurae* Hilse, Type, FC. 174. *Hydroepicoccum genuense* de Notaris, Isotype, FC.

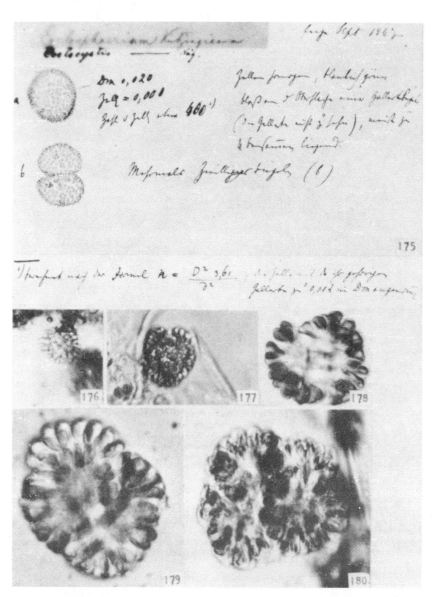

Figs. 175-180.—175. *Coelosphaerium Kuetzingianum* Nägeli, from Nägeli notes, Zürich. 176. *Coelosphaerium Collinsii* Drouet & Daily, Type, D. 177. *Gomphosphaeria fusca* Skuja, Topotype, D. 178. *Gomphosphaeria aponina* var. *cordiformis* Wolle, Type, FC. 179. *Gomphosphaeria aponina* Kützing, Type, L. 180. *Gomphosphaeria cordiformis* var. *olivacea* Hansgirg, Isotype, FC.

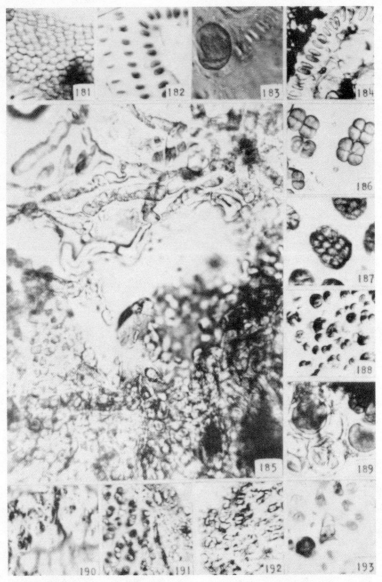

Figs. 181-193.—181. *Merismopoedia geminata* Lagerheim, Paratype, L, and *Microcrocis Dieteli* Richter, Type, L. 182. *Hormospora pellucida* Dickie, Type, BM. 183. *Nodularia ? fusca* W. R. Taylor, Type, TA. 184. *Cyanothrix primaria* Gardner, Type, NY. 185. *Hyella Balanii* Lehmann, Topotype, NY. 186. *Protococcus crepidinum* Thuret, Type, PC. 187. *Entophysalis granulosa* Kützing, Type, L. 188. *Pleurocapsa fluviatilis* var. *subsalsa* Hansgirg, Isotype, FC. 189. *Pleurocapsa fuliginosa* Hauck, Type, L. 190. *Palmella oceanica* Crouan, Type, PC. 191. *Coccochloris deusta* Meneghini, Type, Fl. 192. *Xenococcus concharum* Hansgirg, Isotype, FC. 193. *Palmella mediterranea* Kützing, Type, L.

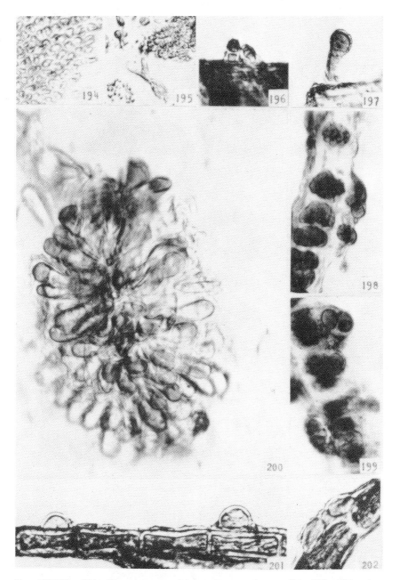

Figs. 194-202.—194. *Chroococcus atrochalybeus* Hansgirg, Isotype, FC. 195. *Chlorogloea lutea* Setchell & Gardner, Type, UC. 196. *Chamaesiphon roseus* var. *major* Hansgirg, Isotype, FC. 197. *Dermocarpa solitaria* Collins & Hervey, Type, NY. 198. *Dermocarpa Enteromorphae* Anand, Type, in coll. of F.E. Fritsch. 199. *Xenococcus violaceus* Anand, Type, in coll. of F.E. Fritsch. 200. *Dermocarpa protea* Setchell & Gardner, Type, UC. 201. *Dermocarpa hemisphaerica* Setchell & Gardner, Paratype, FC. 202. *Xenococcus pulcher* Hollenberg, Type, in herb. of G. J. Hollenberg.

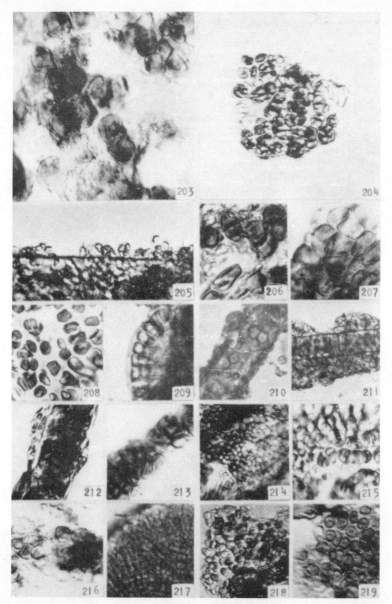

Figs. 203-219.—203. *Dermocarpa sublitoralis* Lindstedt, Isotype, D. 204. *Radaisia Laminariae* Setchell & Gardner, Type, UC. 205. *Pleurocapsa epiphytica* Gardner, Type, UC. 206. *Hyella linearis* Setchell & Gardner, Type, UC. 207. *Dermocarpa pacifica* Setchell & Gardner, Type, UC. 208, 209. *Oncobyrsa adriatica* Hauck, Type, L. 210. *Xenococcus Schousboei* Thuret, Type, PC. 211. *Palmella conferta* Kützing, Type, L. 212. *Xenococcus acervatus* Setchell & Gardner, Type, UC. 213. *Xenococcus Chaetomorphae* Setchell & Gardner, Type, UC. 214. *Dermocarpa Vickersiae* Collins, Type, NY. 215. *Radaisia subimmersa* Setchell & Gardner, Type, UC. 216. *Chroococcus fuscoviolaceus* var. *cupreofuscus* Hansgirg, Isotype, FC. 217. *Hydrococcus rivularis* Kützing, Type, L. 218. *Entophysalis Cornuana* Sauvageau & *Dermocarpa Flahaultii* Sauvageau, Types, PC. 219. *Aphanocapsa anodontae* var. *major* Hansgirg, Isotype, FC.

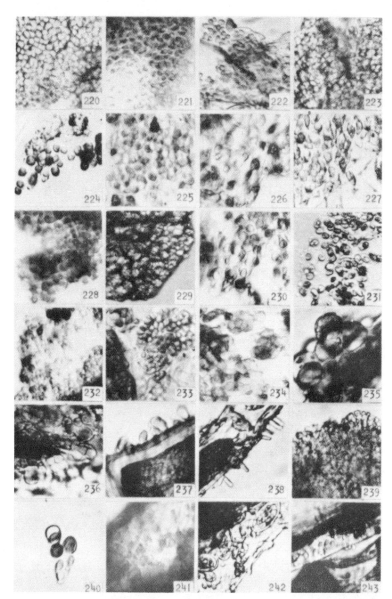

Figs. 220-243.—220. *Palmella duriuscula* Kützing, Type, L. 221. *Hydrococcus Cesatii* Rabenhorst, Type, D. 222. *Guyotia singularis* Schmidle, Type, ZT. 223. *Hydrococcus ulvaceus* Kützing, Type, L. 224. *Pleurocapsa minor* Hansgirg, Isotype, FC. 225. *Pleurococcus julianus* Meneghini, Type, FI. 226. *Chroococcus fuscoater* var. *fuscoviolaceus* Hansgirg, Isotype, FC. 227. *Coelosphaerium holopediforme* Schmidle, Type, ZT. 228. *Radaisia confluens* Gardner, Type, NY. 229. *Chroococcus (Rhodococcus) Hansgirgii* Schmidle, Type, ZT. 230. *Hyellococcus niger* Schmidle, Type, ZT. 231. *Cyanoderma rivulare* Hansgirg, Isotype, FC. 232. *Chroococcus minutissimus* Gardner, Type, NY. 233. *Entophysalis chlorophora* Gardner, Type, NY. 234. *Pleurococcus glomeratus* Meneghini, Type, FI. 235. *Dermocarpa Hollenbergii* Drouet, Type, D. 236. *Xenococcus Willei* Gardner, Type, NY. 237. *Dermocarpa chamaesiphonoides* Geitler, Isotype, FC. 238. *Chamaesiphon cylindricus* Boye Petersen, Type, in coll. of J. Boye Petersen. 239. *Nostoc Lemaniae* Agardh, Type, PC. 240. *Pleurocapsa fluviatilis* Lagerheim, Type, DT. 241. *Xenococcus minimus* Geitler, Isotype, W. 242. *Xenococcus chroococcoides* Fritsch, Type, in coll. of F. E. Fritsch. 243. *Chamaesiphon portoricensis* Gardner, Type, NY.

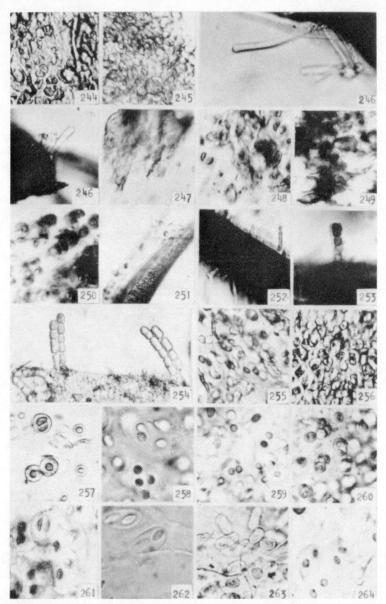

Figs. 244-264.—244. *Dermocarpa depressa* W. & G. S. West, Type, BM. 245.
Oncobyrsa Brebissonii Meneghini, Type, L. 246. *Chamaesiphon gracilis* f. *elongatus*
Wille, Type, S. 247. *Aphanocapsa concharum* Hansgirg, Isotype, FC. 248. *Aphanocapsa*
litoralis Hansgirg, Isotype, FC. 249. *Chroococcus calcicola* Anand, Type, in coll.
of F. E. Fritsch. 250. *Myrionema crustaceum* J. Agardh, Type, LD. 251, 252.
Clastidium setigerum Kirchner, from coll. of K. Thomasson, Sweden, DA. 253.
Chamaesiphon sansibaricus Hieronymus, Type, BM. 254. *Chamaesiphon filamentosus*
Ghose, Type, in coll. of F. E. Fritsch. 255. *Gloeocapsa botryoides* Kützing, Type, L.
256. *Dactylothece Braunii* Lagerheim, Isotype, FC. 257. *Rhodococcus caldariorum*
Hansgirg, Isotype, FC. 258. *Gloeocapsa cryptococca* Kützing, Type, L. 259. *Gloeocapsa*
confluens Kützing, Type, L. 260. *Gloeocapsa cartilaginea* Gardner, Type, NY. 261.
Gloeocapsa muralis Kützing, Type, L. 262. *Gloeocapsa palea β minor* Kützing, Type, L.
263. *Gloeocapsa palea* Kützing, Type, L. 264. *Haematococcus Hookerianus* Berkeley
& Hassall, Type, K.

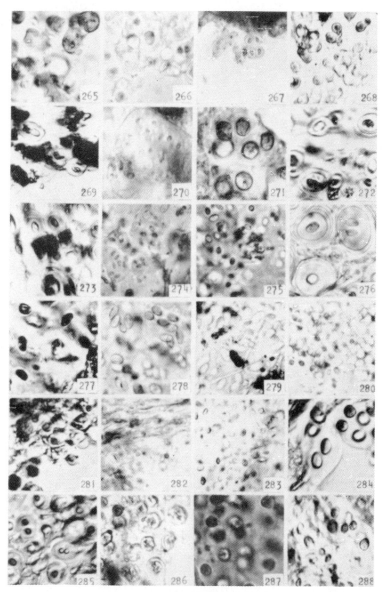

Figs. 265-288.—265. *Chroothece cryptarum* Farlow, Type, FH. 266. *Gloeocapsa montana* Kützing, Type, L. 267. *Chroococcus constrictus* Gardner, Type, NY. 268. *Gloeocapsa sibogae* Weber-van Bosse, Type, L. 269. *Gloeocapsa salina* Roemer, Type, NY. 270. *Gloeocapsa sphaerica* Gardner, Type, NY. 271. *Gloeocystis riparia* A. Braun, Isotype, FC. 272. *Gloeothece distans* Stizenberger, Isotype, FC. 273. *Palmella (Gloeocapsa) chrysophthalma* Montagne, Type, PC. 274. *Palmella Grevillei* Berkeley, Type, E. 275. *Gloeocapsa squamulosa* Brébisson, Type, L. 276. *Gloeocapsa polydermatica* Kützing, Type, L. 277. *Gloeocapsa palmelloides* Rabenhorst, Type, FC. 278. *Gloeocapsa quaternata* Kützing, Type, L. 279. *Palmella borealis* Kützing, Type, L. 280. *Palmella parietina* Nägeli, Type, L. 281. *Pleurococcus miniatus* var. *virescens* Hansgirg, Isotype, FC. 282. *Palmella Botteriana* Kützing, Type, L. 283. *Palmella sordida* Kützing, Type, L. 284. *Gloeocapsa nigra* var. *minor* Hansgirg, Isotype, FC. 285. *Botryococcus Braunii* var. *mucosus* Lagerheim, Isotype, FC. 286. *Palmella margaritacea* Kützing, Type, L. 287. *Palmella adriatica* Kützing, Type, L. 288. *Inoderma majus* Hansgirg, Isotype, FC.

185

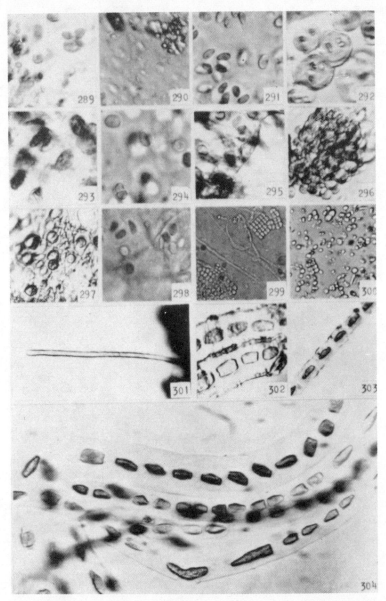

Figs. 289-304.—289. *Palmella crassa* Naccari, Type, PC. 290. *Bromicolla aleutica* Eichwald, Type, LD. 291. *Palmella laxa* Kützing, Type, L. 292. *Gloeothece opalothecata* Gardner, Type, NY. 293. *Gloeocystis vesiculosa* var. *caldariorum* Hansgirg, Isotype, FC. 294. *Palmella rivularis* Carmichael, Type, K. 295. *Gloeocystis rupestris* var. *subaurantiaca* Hansgirg, Isotype, FC. 296. *Palmella littorea* Crouan, Type, PC. 297. *Gloeocapsa salina* Hansgirg, Isotype, FC. 298. *Palmella hyalina β muscicola* Carmichael, Type, K. 299. *Merismopoedia violacea* Kützing, Type, L. 300. *Merismopoedia Reitenbachii* Caspary, Type, L. 301. *Stichosiphon Hansgirgii* Geitler, Isotype, FC. 302. *Allogonium ramosum* var. *crassum* Hansgirg, Isotype, FC. 303. *Allogonium smaragdinum* var. *palustre* Hansgirg, Isotype, FC. 304. *Conferva ornata* Agardh, Type, S.

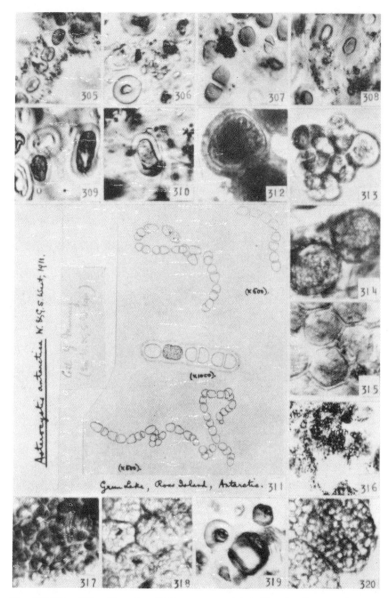

Figs. 305-320.—305. *Allogonium Wolleanum* var. *calcicola* Hansgirg, Isotype, FC. 306. *Chroothece Richteriana* Hansgirg, Type, FC. 307. *Chroodactylon Wolleanum* Hansgirg, Isotype, FC. 308. *Gloeocapsa dubia* Wartmann, Type, FC. 309. *Chroothece Richteriana* var. *aquatica* Hansgirg, Isotype, FC. 310. *Allogonium halophilum* Hansgirg, Isotype, FC. 311. *Asterocytis antarctica* W. & G. S. West, from West notes, BM. 312. *Microhaloa major* Kützing, Type, L. 313. *Microhaloa botryoides* Kützing, Type, L. 314. *Microhaloa aurantiaca* Kützing, Type, L. 315. *Microhaloa pallida* Kützing, Type, L. 316. *Microcystis minuta* var. *violacea* Lindstedt, Type, in coll. of A. Lindstedt. 317. *Microcystis parasitica* Kützing, Type, L. 318. *Microcystis minor* Kützing, Type, L. 319. *Microcystis Brebissonii* Meneghini, Type, FI. 320. *Microcystis olivacea* Kützing, Type, L.

187

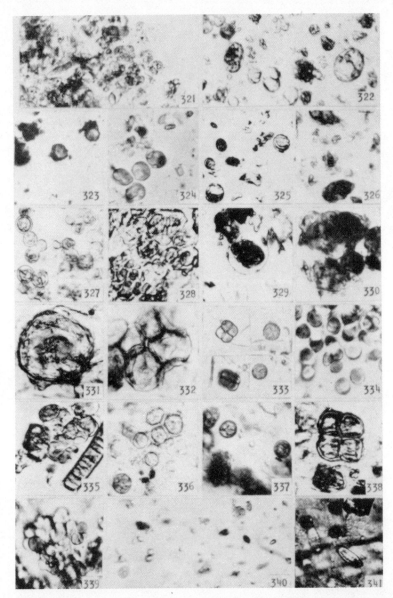

Figs. 321-341.—321. *Protococcus minor β mucosus* Kützing, Type, L. 322. *Protococcus variabilis* Hansgirg, Isotype, FC. 323. *Protococcus bruneus* Kützing, Type, L. 324. *Protococcus minor* Kützing, Type, L. 325. *Protococcus caldariorum* Magnus, Type, L. 326. *Protococcus lilacinus* Rabenhorst, Isotype, FC. 327. *Protococcus Chlamidomonas* Kützing, Type, L. 328. *Protococcus marinus* var. *virens* Hansgirg, Isotype, FC. 329. *Protococcus viridis* var. *insignis* Hansgirg, Isotype, FC. 330. *Protococcus chnaumaticus* Kützing, Type, L. 331. *Protococcus botryoides* var. *nidulans* Hansgirg, Isotype, FC. 332. *Protococcus Felisii* Meneghini, Type, L. 333. *Protococcus globosus* Nägeli, Type, L. 334. *Protococcus crassus* Kützing, Type, L. 335. *Pleurococcus crenulatus* Hansgirg, Isotype, FC. 336. *Pleurococcus miniatus* var. *fuscescens* Hansgirg, Isotype, FC. 337. *Pleurococcus miniatus* var. *roseolus* Hansgirg, Isotype, FC. 338. *Pleurococcus pachydermus* Lagerheim, Type, L. 339. *Pleurococcus vulgaris* f. *glomeratus* Hansgirg, Isotype, FC. 340. *Gloeothece Baileyana* Schmidle, Type, ZT. 341. *Gloeothece minor* Beck, Type, Charles University, Prague.

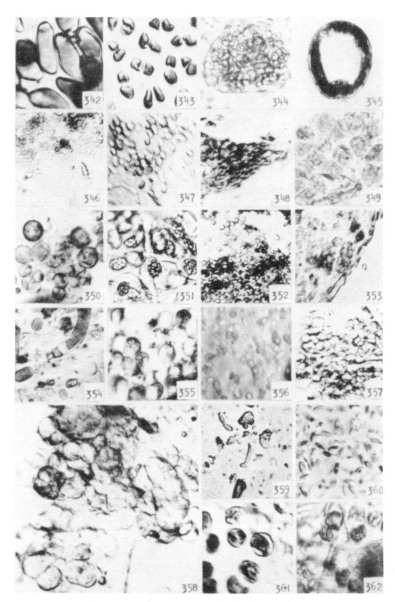

Figs. 342-362.—342. *Gloeocystis scopulorum* Hansgirg, Isotype, FC. 343. *Gloeocystis marina* Hansgirg, Isotype, FC. 344. *Limnodictyon Roemerianum* Kützing, Type, L. 345. *Chroococcus macrococcus* var. *salinarum* Hansgirg, Isotype, FC. 346. *Aphanocapsa albida* Zeller, Type, NY. 347. *Aphanocapsa rufescens* Hansgirg, Isotype, FC. 348. *Oncobrysa fluviatilis* Agardh, Type, LD. 349. *Palmella flava* Kützing, Type, L. 350. *Gloeocapsa stillicidiorum* Kützing, Type, L. 351. *Gloeocapsa ampla* Kützing, Type, L. 352. *Polycystis violacea* Itzigsohn, Isotype, FC. 353. *Sarcina ventriculi* Goodsir, Type, L. 354. *Synechococcus intermedius* Gardner, Type, NY. 355. *Hyphelia aurantiaca* Libert, Type, BR. 356. *Palmella hormospora* Meneghini, Type, FI. 357. *Synechocystis aquatilis* Sauvageau, Isotype, FC. 358. *Xenococcus Schousboei* var. *pallidus* Hansgirg, Isotype, FC. 359. *Dactylococcopsis rhaphidioides* Hansgirg, Isotype, FC. 360. *Oncobyrsa fluviatilis* Agardh, Type, LD. 361. *Palmella aurantia* Agardh, Type, LD. 362. *Palmella parvula* Kützing, Type, L.

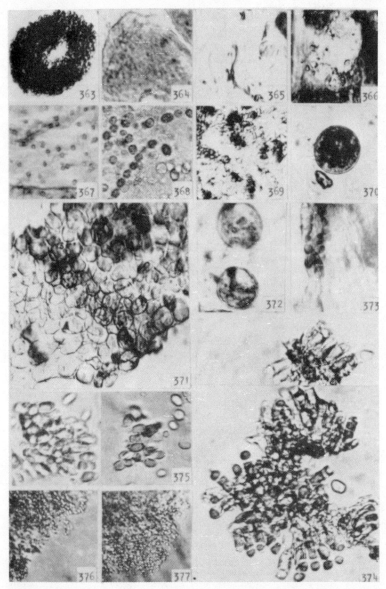

Figs. 363-377.—363. *Coelosphaerium Leloupii* Kufferath, Topotype, D. 364. *Palmogloea gigantea* Kützing, Type, L. 365. *Dactylococcopsis arcuata* Gardner, Type, NY. 366. *Chamaesiphon Rostafinskii* var. *minor* Hansgirg, Isotype, FC. 367. *Aphanothece clathrata* W. & G. S. West, Type, BIRM. 368. *Aphanothece piscinalis* Rabenhorst, Type, L. 369. *Aphanothece luteola* Schmidle, Type, ZT. 370. *Haematococcus fuliginosus* Meneghini, Type, L. 371. *Aphanothece heterospora* Rabenhorst, Isotype, NY. 372. *Haematococcus Grevillei* Agardh, Type, K. 373. *Entophysalis zonata* Gardner, Type, FH. 374. *Chamaesiphon pseudo-polymorphus* Fritsch, Type, in coll. of F. E. Fritsch. 375. *Palmella papillosa* Kützing, Type, L. 376. *Aphanocapsa marina* Hansgirg, Isotype, FC. 377. *Aphanocapsa Lewisii* Keefe, Type, FH.

ADDITIONS AND CORRECTIONS

Page 14, line 5 from bottom, after "1907." insert: *S. crassus* var. *maximus* Lemmermann ex Forti, Syll. Myxophyc., p. 27. 1907.

Page 15, line 6 from bottom, after "TA)." insert: WASHINGTON: floating in open water, Falls lake, Lower Grand Coulee, Grant county, *R. W. Castenholz 30*, 30 Apr. 1955 (D, DA, FC, W).

Page 32, line 28 from bottom, after "1938." insert: *A. clathrata* f. *longior* Elenkin, ibid., p. 149. 1938.

Page 44, line 26 from bottom, after "1929." insert: *Microcystis salina* Elenkin, Monogr. Algar. Cyanophyc., Pars Spec. 1: 122, 139. 1938.

Page 45, line 5 from bottom, after "1868." insert: *Protococcus frustulosus* de Toni, Syll. Algar. 2: 704. 1889.

Page 46, line 2 from top, after "1845." insert: *Gloeocapsa livida* Kützing, Tab. Phyc. 1: 17. 1846.

Page 53, line 9 from bottom, for "*A. nebulosa*" read: *Aphanocapsa nebulosa*

Page 56, line 10 from top, for "*G. nigrescens*" read: *Gloeocapsa nigrescens*

Page 59, line 25 from bottom, for "*Gleditschia*" read: *Gleditsia*

Page 71, line 6 from top, after "1926." insert: *Gloeocapsa turgida* f. *maxima* Hollerbach in Elenkin, Monogr. Algar. Cyanophyc., Pars Spec. 1: 212. 1938.

Page 71, line 16 from bottom, after "1932." insert: *Gloeocapsa turgida* f. *luteola* Hollerbach in Elenkin, Monogr. Algar. Cyanophyc., Pars Spec. 1: 212. 1938.

Page 83, line 20 from top, for "no. M4d" read: slide no. M4d.

Page 83, line 29 from bottom, for "*A. limnetica* f. *major*" read: *Anacystis limnetica* f. *major*

Page 86, line 5 from top, after "(FC)." insert: MICHIGAN: net collection from Fish lake, Orangeville, Barry county, *F. K. & W. A. Daily 2770*, 5 Sept. 1955 (D, DA).

Page 98, line 23 from top, after "1837." insert: *S. aponinum* Trevisan, Prosp. Fl. Eugan., p. 59. 1842.

Page 101, before line 26 from bottom, insert the new paragraph:
Phaeodermatium Hansgirg, Notarisia 4: 658. 1889. —Type species: P. *rivulare* Hansg.

Page 104, line 14 from bottom, after "1931." insert: *Ercegovicia litoralis* J. de Toni, Noter. Nomencl. Algol. 8. 1936.

Page 107, line 10 from top, after "shells," insert: Prospect, Halifax county

Page 120, before line 6 from top, insert the new paragraph:
Phaeodermatium rivulare Hansgirg, Notarisia 4: 658. 1889. —Type from near Cernosic, Bohemia (W).

Page 122, line 22 from bottom, before "Chotzen" insert: in einem Bächlein im oberen Theile des Solopisker-Thales nächst Cernosic, *A. Hansgirg,* Jun. 1885 (Type of *Phaeodermatium rivulare* Hansg., W; isotypes, D, FC);

Page 139, line 5 from top, after "1845." insert: *Coccochloris margaritacea* Meneghini *pro synon.* in Kützing, *loc. cit.* 1845.

Page 141, before line 20 from bottom, insert the following paragraph: *Haematococcus minutissimus* Hassall, Hist. Brit. Freshw. Alg. 1: 334. 1845. *Bichatia minutissima* Trevisan, Sagg. Monogr. Alg. Coccot., p. 65. 1848.

Page 145, line 22 from top, before "and *P. communis* β *Pleurococcus*" insert: *P. communis* α *leprarius,*

Page 145, line 31 from bottom, for "*B. flos-aquae*" read: *Byssus flos-aquae*

Page 146, before line 10 from top, insert the following paragraph: *Chlorella vulgaris* Beijerinck, Bot. Zeitung 48: 726, 758. 1890. *Pleurococcus Beyerinckii* Artari, Bull. Soc. Imp. Nat. Moscou N. S. 6: 246. 1893. —The Type, E. G. Pringsheim culture no. 211/11i (DA) = CHLORELLA VULGARIS Beij.

Page 150, before line 25 from bottom, insert the following new paragraph: *Glaucocystis nostochinearum* var. *minor* Hansgirg, Prodr. Algenfl. Böhmen 2: 140. 1892. —Type from Bohemia: Edmundsklamm, *A. Hansgirg,* Sept. 1890 (W) = spores of ANABAENA SP.

Page 153, line 14 from bottom, for "Spicil. f. Lips." read: Spicil. Fl. Lips.

Page 154, line 22 from top and line 32 from bottom, for "Trevisan" read: de Toni & Trevisan

Page 156, line 6 from top, after "1868." read: *Protococcus botryoides* Kirchner, Krypt. v. Schlesien 2(1): 103. 1878. *P. viridis* var. *botryoides* Wolle, Fresh-w. Alg. U. S., p. 182. 1887.

Page 156, lines 27 and 28 from top, delete: *CHROOCOCCACEAE* subfam. *EUCHROOCOC-CACEAE* Hansgirg, Prodr. Algenfl. Böhmen 2: 129. 1892.

Page 156, line 28 from bottom, after "1865." insert: *Anacystis sparsa* Brébisson *pro synon.* in Rabenhorst, *loc. cit.* 1865.

Page 157, lines 19 and 20 from top, delete: *P. fusiformis* Rabenhorst, Deutschl. Krypt.-Fl. 2: 60. 1847.

Page 157, line 26 from bottom, after "1892." insert: *Aphanocapsa hyalina* Hansgirg, *loc. cit.* 1892.

Page 161, line 3 from top, after "1907." insert: *M. ichthyoblabe* var. *violacea* Forti, *loc. cit.* 1907.

Page 161, line 7 from bottom, for "ex Drouet, Amer. Midl. Nat. 30: 672. 1943." read: Ann. New York Acad. Sci. 36: 72. 1936.

Page 163, line 8 from bottom, after "1868." insert: *P. miniatus* var. *roseus* de Toni ex Dalla Torre & Sarnthein, Fl. Tirol 2: 47. 1901.

INDEX

CHLOROGLOEACEAE

Chlorogloeaceae Geitl., 100
Chondrocystis Lemm., 102
C. Bracei Howe, 49 (fig. 82)
C. Schauinslandii Lemm., 104
CHROOCOCCACEAE Näg., 11
Chroococcaceae subfam. Chroocysteae Hansg., 142
Chroococcaceae I. Coccobactreae Elenk., 11
Chroococcaceae subfam. Euchroococcaceae Hansg., 11
Chroococcaceae subfam. Euchroococcaceae trib. Coccineae Hansg., 11
Chroococcaceae subfam. Euchroococcaceae trib. Phyllothecineae Hansg., 11
Chroococcaceae subfam. Euchroococcaceae Gruppe Thecineae Hansg., 156
Chroococcaceae subfam. Euchroococcaceae trib. Thecineae Hansg., 147
Chroococcaceae II. Gloeococceae-planimetreae Elenk., 11
Chroococcaceae III. Gloeococceae-stereometreae B. Heterogloeae Elenk., 11
Chroococcaceae III. Gloeococceae-stereometreae B. Heterogloeae 2. Tegumentocrassiores Elenk., 12
Chroococcaceae III. Gloeococceae-stereometreae B. Heterogloeae I. Tegumentotenuiores Elenk., 11
Chroococcaceae III. Gloeococceae-stereometreae A. Homoeogloeae 2. Excavatae Elenk., 11
Chroococcaceae III. Gloeococceae-stereometreae A. Homoeogloeae 2. Excavatae a) Distantes Elenk., 12
Chroococcidiopsis Geitl., 35
C. thermalis Geitl., 51
Chroococcidium Geitl., 146
C. gelatinosum Geitl., 146
Chroococcopsis Geitl., 101
C. amethystea (Rosenv.) Geitl., 112
C. caldaria (Tild.) Geitl., 161
C. fluminensis Fritsch, 121
C. gigantea Geitl., 120
Chroococcus Näg., 34
Chroococcus subgen. Chlorogloea (Wille) Elenk., 101
Chroococcus subgen. Chrysococcus Hansg., 162
Chroococcus subgen. Eucapsis (Clem. & Shantz) Elenk., 35
Chroococcus sect. Euchroococcus Hansg., 34
Chroococcus subgen. Euchroococcus Hansg., 34
Chroococcus subgen. Hydrococcus (Kütz.) Elenk., 101
Chroococcus subgen. Rhodococcus Hansg., 137
Chroococcus aeruginosus Gardn., 78 '
C. aeruginosus Rabenh., 70
C. alpinus Schmidle, 49
C. atrochalybeus Hansg., 103 (fig. 194)
C. atrovirens (Corda) Hansg., 141
C. aurantiacus Bern., 52
C. aurantiacus (Lib.) de Toni, 153
C. aurantio-fuscus (Kütz.) Rabenh., 161
C. aurantius (Ag.) Wille, 156
C. aureo-viridis (Kütz.) Rabenh., 161
C. aureus (Kütz.) Rabenh., 161
C. bataviae van Oye, 79

C. Bernardii van Oye, 71
C. bituminosus (Bory) Hansg., 137
C. bruneus (Kütz.) Rabenh., 162
C. calcicola Anand, 104 (fig. 249)
C. caldariorum Hansg., 140
C. chalybeus (Kütz.) Rabenh., 70
C. chnaumaticus (Kütz.) Rabenh., 162
C. cinnamomeus (Menegh.) Rabenh., 159
C. coerulescens (Bréb.) Forti, 32
C. cohaerens (Bréb.) Näg., 78 (fig. 78)
C. cohaerens Näg., 48
C. cohaerens var. antarcticus Wille, 52
C. confervoides A. Br., 146
C. consociatus Hariot, 18
C. constrictus Gardn., 141 (fig. 267)
C. crassus (Kütz.) Näg., 162
C. crepidinum (Thur.) Hansg., 103
C. cubicus Gardn., 50 (fig. 31)
C. cumulatus Bachm., 79
C. decolorans Mig., 49
C. decorticans A. Br., 78 (fig. 128)
C. Detonii Bern., 52
C. dimidiatus (Kütz.) Näg., 70
C. dispersus (Keissl.) Lemm., 83
C. dispersus var. minor G. M. Sm., 67
C. fuligineus (Lenorm.) Rabenh., 152
C. fuscescens (Kütz.) Richt., 70
C. fusco-ater (Kütz.) Rabenh., 46
C. fusco-ater var. fuscoviolaceus Hansg., 120 (fig. 226)
C. fuscoviolaceus Hansg., 120
C. fuscoviolaceus var. cupreofuscus Hansg., 120 (fig. 216)
C. giganteus West, 70
C. giganteus var. occidentalis Gardn., 71 (fig. 106)
C. glomeratus (Menegh.) Forti, 119
C. Goetzei Schmidle, 121
C. Gomontii Nyg., 83
C. granulosus Zell., 140
C. Hansgirgii Schmidle, 120 (fig. 229)
C. heangloios Gardn., 78
C. helveticus Näg., 78 (fig. 114)
C. helveticus var. aurantiofuscescens Hansg., 78 (fig. 122)
C. helveticus var. aureo-fuscus Hansg., 78 (fig. 119)
C. helveticus var. consociato-dispersus Elenk., 79
C. helveticus f. major Lagerh., 83 (fig. 131, 132)
C. Huberi J. de Toni, 52
C. indicus Bern., 71
C. indicus Zell., 36 (fig. 8)
C. indicus var. epiphyticus Ghose, 71
C. insignis Schmidle, 70 (fig. 101)
C. irregularis Hub., 79
C. julianus (Menegh.) Näg., 119
C. kerguelensis Wille, 29
C. lageniferus Hildebr., 146
C. lilacinus Rabenh., 162
C. limneticus Lemm., 83
C. limneticus var. carneus (Chod.) Lemm., 83
C. limneticus var. distans G. M. Sm., 83
C. limneticus var. elegans G. M. Sm., 83
C. limneticus var. fuscus Lemm., 83
C. limneticus var. multicellularis Chu, 52

CHROOCOCCUS

C. limneticus var. purpureus (Snow) Tiff. & Ahlstr., 83
C. limneticus var. subsalsus Lemm., 83
C. lithophilus Erceg., 121
C. lithophilus f. achromaticus Erceg., 121
C. lithophilus f. coloratus Erceg., 121
C. lithophilus var. rotae Gonz., 79
C. luteolus Woronich., 71
C. macrococcus (Kütz.) Rabenh., 162
C. macrococcus var. aquaticus Hansg., 146
C. macrococcus f. aureus (Kütz.) Rabenh., 161
C. macrococcus var. aureus (Kütz.) Rabenh., 161
C. macrococcus var. salinarum Hansg., 146 (fig. 345)
C. macrococcus f. stipitatus Enwald, 146
C. mediocris Gardn., 78 (fig. 121)
C. membraninus (Menegh.) Näg., 77
C. membraninus var. crassior Hansg., 78 (fig. 126)
C. membraninus var. salinus Erceg., 104
C. microcystoideus W. & G. S. West, 141
C. minimus (Keissl.) Lemm., 96
C. minimus var. crassus B. Rao, 18
C. minimus var. turfosus Steinecke, 146
C. minor (Kütz.) Näg., 163
C. minor var. dispersus Keissl., 83
C. minor f. glomeratus Frémy, 52
C. minor f. major W. & G. S. West, 146
C. minor f. minimus W. & G. S. West, 49
C. minor f. mucosus (Kütz.) Rabenh., 163
C. minor var. mucosus Kütz., 163
C. minor f. violaceus Wille, 52
C. minutissimus Gardn., 120 (fig. 232)
C. minutus (Kütz.) Näg., 16
C. minutus var. amethystaceus Wille, 79
C. minutus var. amethysteus Wille, 79
C. minutus var. carneus Chod., 83
C. minutus var. lacustris Chod., 83
C. minutus var. limneticus (Lemm.) Hansg., 83
C. minutus var. minimus Keissl., 96
C. minutus var. obliteratus (Richt.) Hansg., 78
C. minutus var. salinus Hansg., 78
C. minutus var. virescens (Hantzsch) Hansg., 78
C. mipitanensis (Wolosz.) Geitl., 71
C. monetarum Reinsch, 146
C. montanus Hansg., 17 (fig. 150)
C. montanus var. hyalinus B. Rao, 18
C. multicoloratus Wood, 71
C. muralis Gardn., 50 (fig. 26)
C. Mutisii Gonz., 71
C. niger Starm., 52
C. obliteratus Richt., 78
C. Orsinii (Menegh.) Rabenh., 152
C. pallidus Näg., 78
C. parallelepipedon Schmidle, 67 (fig. 91)
C. polyedriformis Schmidle, 78
C. Prescottii Dr. & Daily, 83 (fig. 130)
C. protogenitus (Bias.) Hansg., 154
C. purpureus Snow, 83
C. pyriformis Benn., 146
C. quaternarius Zalessk., 71
C. Raspaigellae Hauck, 76 (fig. 112)
C. Rechingeri Wille, 52
C. refractus Wood, 51

C. Rochei Vir., 18
C. roseus (Menegh.) Wille, 163
C. rubiginosus (Suring.) Rabenh., 163
C. rubrapunctis Wolle, 121
C. rufescens (Kütz.) Näg., 78
C. rufescens f. turicensis Näg., 78
C. rufescens var. turicensis Näg., 78
C. sabulosus (Menegh.) Hansg., 164
C. sarcinoides Hub., 52
C. sarcinoides Wisl., 121
C. Scherffelianus Kol, 79
C. schizodermaticus West, 49
C. schizodermaticus var. badio-purpureus W. & G. S. West, 70 (fig. 103)
C. schizodermaticus var. incoloratus Geitl., 141
C. schizodermaticus var. incoloratus f. paucistratosus Geitl., 78
C. schizodermaticus f. pallidus Erceg., 121
C. siderochlamys Skuja, 146
C. Simmeri Schmidle, 141
C. smaragdinus Hauck, 76 (fig. 108)
C. solitarius Eichl., 71
C. sonorensis Dr. & Daily, 76
C. spelaeus Erceg., 71
C. spelaeus var. aerugineus Erceg., 71
C. spelaeus var. violascens Erceg., 71
C. splendidus Jao, 141
C. subsphaericus Gardn., 78 (fig. 117)
C. subtilissimus Skuja, 146
C. tenax (Kirchn.) Hieron., 70
C. tenax var. boeticus Gonz., 71
C. thermalis (Menegh.) Näg., 70
C. thermophilus Wood, 70
C. turgidus (Kütz.) Näg., 70
C. turgidus a. chalybeus (Kütz.) Rabenh., 70
C. turgidus var. dimidiatus (Kütz.) Bréb., 70
C. turgidus var. fuscescens (Kütz.) Forti, 70
C. turgidus var. giganteus (West) Cedergr., 70
C. turgidus var. glomeratus Hansg., 78 (fig. 120)
C. turgidus var. Hookeri Lagerh., 70 (fig. 102)
C. turgidus var. japonicus Bern., 71
C. turgidus f. lamellosus Beck-Mannag., 71
C. turgidus var. maximus Nyg., 71
C. turgidus f. minor Wille, 71
C. turgidus var. mipitanensis Wolosz., 71
C. turgidus f. pallidus Beck-Mannag., 71
C. turgidus var. Pullei Bern., 71
C. turgidus var. rufescens Wartm., 78
C. turgidus var. solitarius Ghose, 71
C. turgidus var. submarinus Hansg., 70 (fig. 104)
C. turgidus δ subnudus Hansg., 70
C. turgidus f. subnudus Hansg., 70
C. turgidus var. subnudus Hansg., 70
C. turgidus var. subviolaceus Wille, 128
C. turgidus β tenax Kirchn., 70
C. turgidus var. tenax Kirchn., 70
C. turgidus γ thermalis (Menegh.) Rabenh., 70
C. turgidus f. thermalis (Menegh.) Rabenh., 70
C. turgidus var. thermalis (Menegh.) Rabenh., 70
C. turgidus var. uniformis Gardn., 78
C. turgidus var. violaceus West, 71 (fig. 107)
C. turgidus var. violaceus f. minor Wille, 71
C. turicensis (Näg.) Hansg., 79
C. vacuolatus Skuja, 146
C. varius A. Br., 48
C. varius f. atrovirens A. Br., 49 (fig. 85)

200

203

217

Reprinted from Transactions of the American Microscopical Society
Vol. LXXVI, No. 2, April, 1957

REVISION OF THE COCCOID MYXOPHYCEAE: ADDITIONS AND CORRECTIONS

Francis Drouet[1] and William A. Daily[2]

Since the revision of the coccoid Myxophyceae (Drouet & Daily, 1956) was published on June 30, 1956, we have examined the text carefully for typographical and other errors and for unintentional omissions. Dr. Joséphine Th. Koster pointed out numerous mistakes in our transcription of the Dutch language from handwritten herbarium labels, as well as certain other inaccuracies and omissions. Mrs. Fay K. Daily, Miss Rosalie Weikert, Dr. Kathleen M. Drew-Baker, Dr. Robert Ross, Mr. Harold B. Louderback, Dr. Joseph Rubinstein, Dr. Mary Belle Allen, Dr. G. T. Velasquez, M. & Mme. J. Feldmann, Dr. George W. Lawson, and Dr. Aaron J. Sharp called our attention to various items corrected herein. To all these friends and colleagues we are grateful. We wish to express our appreciation to Dr. G. W. Prescott for his careful work in editing this manuscript and to him and the Editorial Board of the American Microscopical Society for the opportunity to publish it in the *Transactions*.

CONSERVATION

Aphanothece Näg. vs. *Coccochloris* Spreng., page 13, line 19 from top. Although according to Lanjouw, *et al.*, International Code of Botanical Nomenclature, 1956, p. 199, *Aphanothece* has been conserved against *Coccochloris,* this action was taken on the recommendation of Geitler, Harms & Mattfeld in the treatment of this group by Geitler in Engler & Prantl, Natürl. Pflanzen., ed. 2, vol. 1b, p. 17, 1942, even though, as pointed out by Drouet in Doty, Lloydia, 13(1): 9 (footnote), 1950, the requirements of the Code (see Art. 14, Note 1) for a proposal for conservation were not in the least fulfilled. The present revision has enlarged considerably the conceptual limits of the group under consideration; therefore, as directed in Art. 14, Note 3 of the Code, a name must be chosen for the taxon in accordance with the principle of priority. *Coccochloris* Spreng. would appear to be the earliest name for the group as circumscribed here.

Gloeocapsa Kütz. vs. *Bichatia* Turp., page 34, line 22 from bottom. According to Lanjouw *et al.*, International Code of Botanical Nomenclature, 1956, p. 199, *Gloeocapsa* has been conserved against *Bichatia.* As pointed out by Drouet in Doty, Lloydia, 13(1): 9 (footnote), 1950, this action removes *Gloeocapsa* from consideration as a synonym of *Anacystis* Menegh. as treated in this revision. *Gloeocapsa* and its synonymy must be transferred to page 144, paragraph 3 from bottom, and inserted among the synonyms of *Bichatia.* It should be reiterated here that Geitler, Harms & Mattfeld in Geitler in Engler & Prantl, Natürl. Pflanzen., ed. 2, vol. 1b, p. 48, 1942, upon whose recommendation

---

[1] Cryptogamic Herbarium, Chicago Natural History Museum.
[2] Herbarium, Butler University.

219

Gloeocapsa Kütz. has been conserved against *Bichatia* Turp., did not in any sense meet the requirements for making such a proposal as set forth in Art. 14, Note 1 of the Code. Since *Gloeocapsa atrata* Kütz. has been retained specifically as the type species of *Gloeocapsa* Kütz. in the list of conserved names in the Code, it follows that the type specimen of *G. atrata* as indicated on pages 46 and 56 of this revision is nullified and that the type of *Bichatia vesiculinosa* Turp. must replace it.

ADDITIONS AND CORRECTIONS

P. 3, L. 17, 18 from top: for "accomodate" read accommodate

P. 4, L. 24 from top: after "polymorphic" insert (capable of being transformed into other species)

P. 10, L. 4 from top: before "TEX" insert TENN, University of Tennessee, Knoxville;

P. 10, L. 9, 10 from top: for "Eidgenossische", read Eidgenössische

P. 13, L. 12 from bottom: for "Natürl", read Natürl.

P. 17, L. 12 from bottom: for "Bard.", read Gard.

P. 18, L. 2 from top: for "Rico.", read Rico

P. 21, L. 1 at top: for "Krypt,", read Krypt.
P. 21, L. 21 from top: for "Slotte", read slooten

P. 21, L. 23 from top: for "culturtuin", read cultuurtuin

P. 21, L. 24, from top: for "*Leeutnaar*", read *Leentvaar*

P. 21, L. 25 from top: for "Oosthout", read Oostkant; for "Kaarwetering", read Kaaiwetering.

P. 21, L. 26 from top: for "Brenkleveen", read Breukelveen

P. 31, L. 28 from top: for "Haucho", read Huacho

P. 33, L. 25 from bottom: for "plashen", read plassen

P. 34, L. 27 from top: for "wand de grot", read wand van de grot

P. 45, L. 8 from bottom: for "Engl. Fl" read Engl. Fl.

P. 45, L. 3 from bottom: for "*Carmichael*", read Carmichael

P. 46, L. 27, 28 from bottom: delete this entire paragraph; transfer it, with the omission of all after "1850.", to page 144, line 14 from bottom, and insert immediately after "1828."

P. 46, L. 24 from bottom: for "Gener", read Gener.,

P. 51, L. 19 from bottom: insert after "1891.", The specific epithet was spelled "violacea" and "violaceus" by Kützing and subsequent authors.

P. 53, L. 20 from top: for "Schimp.", read Schimp.,

P. 53, L. 29 from top; for "aquaeducta", read aquaeductu

P. 53, L. 14 from bottom: for "Hauck & Richt.", read Hauck & Richt.,

P. 54, L. 26 from top: for "*Kütz.,*", read Kütz.,

P. 56, L. 10, 11 from top: delete "Type of *G. atrata* Kütz.," and ";isotypes"

P. 56, L. 33 from bottom: for "var", read var.

P. 59, L. 24, from bottom: for "abrk", read bark

P. 60, L. 15 from bottom: for "(FC TENN);", read (FC, TENN);

P. 62, L. 6 from bottom: for "Nov", read Nov.

P. 63, L. 34 from top: for "Gardn.", read Gardn.,

P. 63, L. 15 from bottom: for "Montoro", read Montoso

P. 64, L. 30 from top: for "truck", read trunk

P. 65, L. 12 from top: for "28,", read *28,*

P. 65, L. 27 from bottom: for "July.", read Jul.

P. 65, L. 24 from bottom: for "Ohinemuru", read Ohinemutu

P. 66, Last L.: for "Krummau", read Krumau

P. 67, L. 32 from top: for "*Alphanocapsa*" read *Aphanocapsa*

P. 68, L. 16 from top: for "*Krummau*" read Krumau

P. 68, L. 23 from top: for "Brenkelveen" read Breukelveen

P. 69, last L.: for "in" read im

P. 70, L. 20 from bottom: for "Ofvers.", read Öfvers.

P. 70, L. 15 from bottom: after "author,", insert from Silesia,

P. 71, L. 26 from top: for "Cl. Sci" read Cl. Sci.

P. 73, L. 8 from top: for "BRITISH TOGOLAND:", read GOLD COAST:

P. 76, L. 27 from top: for "(aut minoria)", read (aut minoria vel usque ad 20)"

P. 79, L. 30 from top: for "*crustacea*", read *deusta*

P. 80, L. 24 from top: for "NETHERLANDS:", read NORWAY:

P. 82, L. 30 from top: for "(FC)", read (FC, UC)
P. 86, L. 26 from bottom: for "1849.", read 1840.
P. 86, L. 12 from bottom: for "doubus", read duobus
P. 91, L. 26 from top: for "BRITISH TOGOLAND:", read GOLD COAST:
P. 92, L. 21 from top: for "goerdnet", read geordnet
P. 92, L. 23 from bottom: for "geworden . . .", read geworden, auch die Zellen etwas kleiner sind.
P. 93, L. 23 from top: for "2;", read 2:
P. 94, L. 21 from bottom: for "pseudovaculois," read pseudovacuolis
P. 94, L. 5 from bottom: for "FC", read D
P. 95, L. 17 from top: for "(N)", read (NY)
P. 95, L. 18 from bottom: for "221", read 2211
P. 96, L. 20 from top: for "1905", read 1935
P. 96: Transpose ¶ 4 (beginning "*Chroococcus minimus* var. *minimus* Keissler") and ¶ 5 beginning "*Gomphosphaeria lacustris* var. *compacta* Lemmermann").
P. 99, L. 5 from top: for "1959", read 1859
P. 99, L. 10 from top: after "Spalato" delete one of the two commas.
P. 103, L. 6 from top: for "Nomencl", read Nomencl.
P. 103, L. 24 from top: insert after "(FC).", See J. & G. Feldmann, Oesterr. Bot. Zeitschr. 100: 508. 1953.
P. 104, L. 4 from top: for "Myxophyc", read Myxophyc.
P. 106, L. 30 from top: for "Morsele", read Borsele
P. 106, L. 16 from bottom: for "1972", read 1872
P. 109, L. 16 from top: for "FU", read FC
P. 109, L. 22 from top: for "Hollenberg:", read Hollenberg;
P. 110, L. 33 from top: for "Labos", read Lobos
P. 111, L. 5 from top: for "FIG. 175.", read FIG. 195.
P. 111, L. 14 from bottom: for "Myxophyc,", read Myxophyc.,
P. 112, L. 31 from bottom: for "Pub.", read Publ.
P. 112, L. 12 from bottom: for "5: 456.", read 6: 456.
P. 113, after L. 13 from bottom: insert *Dermocarpella pedicellata* J. & G. Feldmann, Oesterr. Bot. Zeitschr. 100: 511. 1953., and *Xenococcus Laminariae* J. Feldmann, Bull. Soc. Bot. France 100: 292. 1953
P. 114, L. 19 from bottom: for "Ellewontsdijk", read Ellewoutsdijk

P. 114, L. 20 from bottom: for "Condorpe", read Coudorpe
P. 117, L. 20 from top: for "*fuciola*", read *fucicola*
P. 119, L. 22 from top: before "f. RIVULARIS", insert 4a.
P. 119, L. 23 from top: before "f. PAPILLOSA", insert 4b.
P. 119, L. 25 from bottom: for "Math.", read Mat.
P. 119, L. 15 from bottom: for "1948.", read 1848.
P. 122, L. 27 from bottom: for "Brünnenufern", read Brunnenufern
P. 123, L. 19 from top: for "einer", read einem
P. 123, L. 29 from top: for "Linden", read Zuiden
P. 125, L. 9 from top: delete and replace with, *Kiener 19566, 19824, 19824a, 19825, 19832, 3-5 Oct. 1945 (FC, KI); in spring seepage, state
P. 127, L. 20 from bottom: before "f. LEMANIAE", insert 5a.
P. 127, L. 19 from bottom: before "f. ELONGATA", insert 5b.
P. 127, L. 15 from bottom: for "Soc", read Soc.
P. 128, L. 27 from top: for "Sv,", read Sv.
P. 130, L. 20 from top: for "W.", read W;
P. 130, L. 29 from bottom: for "Aarden hout", read Aardenhout
P. 130, L. 30 from bottom: for "*Veldhuisen van de Wit*", read *Veldhuisen & de Wit*
P. 131, L. 10 from top: for "5‴", read 5′
P. 131, L. 11 from top: for "long,", read long.
P. 132, L. 4 from bottom: for "28", read *28*
P. 135, L. 7 & 11 from bottom: for "1932.", read 1931.
P. 142, L. 26 from bottom: for "Clastidiacease", read Clastidiaceae
P. 144, ¶ 3 from bottom: To the synonymy of *Bichatia* Turp. should be added *Gloeocapsa* Kütz. and its synonymy on page 34, paragraph 7 from bottom; everything after "1898." on page 34, line 19 from bottom should be deleted. Likewise, *G. atrata* Kütz. and its synonymy on page 46, lines 27 and 28 from bottom, with the exception of the part following "1850.", should be inserted here in line 14 from bottom, immediately after "1828."
P. 144, L. 2 from bottom: for "301", read 401
P. 146, L. 11 from bottom: before "*ALLOGONIUM*", insert *ASTEROCYTIS* Gobi ex Hansgirg *pro synon.*, Physiol. & Algol. Stud., p. 109. 1887. *ALLOGONIUM* sect. *ASTEROCYTIS*

222   FRANCIS DROUET AND WILLIAM A. DAILY

Hansgirg, *loc. cit.* 1887.
P. 146, L. 9 from bottom: after "1888.",
insert *ALLOGONIUM* subgenus *ASTER-OCYTIS* Hansgirg, *loc. cit.* 1888.
P. 151, L. 13 from bottom: after "*ANIMALS.*", insert See Willie in Festschr. z. P. Ascherson's 70. Geburtst., XXXVII, p. 439–450 (1904).
P. 152, before L. 5 from bottom: insert the paragraph: *HORMOSPORA* Brébisson, Mém. Soc. Acad. Sci. Art. & Belles Lettr. Falaise 1839: p. 2 of reprint. 1839.—Type species: *H. ovoidea* Brébisson, *loc. cit.* 1839 —The original description is here designated as the temporary Type = CHLOROPHYCEAE?
P. 152, L. 5 from bottom: delete "*ASTEROCYTIS* Gobi, Trudy St. Petersb. Obschech. Estestvoisp. 10: 86. 1879."
P. 152, L. 2–4 from bottom: delete "*ALLOGONIUM* sect. *ASTEROCYTIS* Hansgirg, Physiol. & Algol. Stud., p. 109. 1887. *ALOGONIUM* subgenus *ASTER-OCYTIS* Hansgirg, Notarisia 3: 588. 1888."
P. 153, L. 24 from top: for "(L)", read (L.)
P. 155, L. 14 from bottom: for "Type", read temporary Type
P. 156, L. 17 from bottom: for "Smiths',, read Smiths.
P. 158, L. 8 from bottom: for "Shousboe", read Schousboe
P. 161, L. 13, 14 from top: for "= primordia of CHANTRANSIA (BATRACHO-SPERMUM) sp.", read = PROTO-COCCUS? sp., as shown by Professor Otto Jaag in Ber. Schweiz. Bot. Ges. 41: 356 ff. (1932). Topotypic material of this has been studied in Mus. Vindob.

Krypt. Exs. Alg. no. 849 (FC) and in Migula, Crypt. Germ., Austr. & Helvet. Exs. 10, Alg. no. 42 (NY, ZT).
P. 161, L. 1–4 from bottom: delete all after "MIN)" and insert = CHLORELLA CALDARIA (Tild.) M. B. Allen.
P. 162, L. 2, 4 from bottom: for "PERIDINIUM sp.", read DINO-PHYCEAE.
P. 163, L. 3 from top: for "1893.". read 1892.
P. 165, L. 19 from bottom: for "Rabenh,", read Rabenh.
P. 167, after L. 19 from top: insert the new paragraph, *TROCHISCIA* Kützing, Linnaea 8: 592. 1833.—Type species: *T. solitaria* Kützing, *loc. cit.* 1833.—The specimen labeled thus in the Kützing herbarium from France: Falaise, *Lenormand* (L), is here designated as the Type = primordia of CHLOROPHYCEAE.
P. 168, L. 2 from top: for "Eidgenossische", read Eidgenössische
P. 169, L. 6 from bottom: for "dicus", read dicus
P. 185, L. 5 from bottom: for "FC,", read FC.
P. 191, L. 7 from bottom: for "P.", read *P.*
P. 192, L. 16 from bottom: for "read:", read insert:
P. 193, L. 16 from bottom: for "subgen", read subgen.
P. 193, L. 10 from bottom: for "FIG.", read (FIG.
P. 198, L. 28 from top: for "I. Tegumentotenuiores", read I. Tegumentotenuiores
P. 208, L. 2 from top: for "Willie.,", read Wille.
P. 209, L. 21 from bottom: for "Elenk., 152", read Elenk., 152

LITERATURE CITED

DROUET, F. AND DAILY, W. A. 1956. Revision of the coccoid Myxophyceae. Butler Univ. Bot. Stud., 12: 1–218.